IET ENERGY ENGINEERING SERIES 134

Hydrogen Passivation and Laser Doping for Silicon Solar Cells

Other volumes in this series:

Volume 1 **Power Circuit Breaker Theory and Design** C.H. Flurscheim (Editor)
Volume 4 **Industrial Microwave Heating** A.C. Metaxas and R.J. Meredith
Volume 7 **Insulators for High Voltages** J.S.T. Looms
Volume 8 **Variable Frequency AC Motor Drive Systems** D. Finney
Volume 10 **SF_6 Switchgear** H.M. Ryan and G.R. Jones
Volume 11 **Conduction and Induction Heating** E.J. Davies
Volume 13 **Statistical Techniques for High Voltage Engineering** W. Hauschild and W. Mosch
Volume 14 **Uninterruptible Power Supplies** J. Platts and J.D. St Aubyn (Editors)
Volume 15 **Digital Protection for Power Systems** A.T. Johns and S.K. Salman
Volume 16 **Electricity Economics and Planning** T.W. Berrie
Volume 18 **Vacuum Switchgear** A. Greenwood
Volume 19 **Electrical Safety: a guide to causes and prevention of hazards** J. Maxwell Adams
Volume 21 **Electricity Distribution Network Design, 2nd Edition** E. Lakervi and E.J. Holmes
Volume 22 **Artificial Intelligence Techniques in Power Systems** K. Warwick, A.O. Ekwue and R. Aggarwal (Editors)
Volume 24 **Power System Commissioning and Maintenance Practice** K. Harker
Volume 25 **Engineers' Handbook of Industrial Microwave Heating** R.J. Meredith
Volume 26 **Small Electric Motors** H. Moczala *et al.*
Volume 27 **AC-DC Power System Analysis** J. Arrillaga and B.C. Smith
Volume 29 **High Voltage Direct Current Transmission, 2nd Edition** J. Arrillaga
Volume 30 **Flexible AC Transmission Systems (FACTS)** Y-H. Song (Editor)
Volume 31 **Embedded generation** N. Jenkins *et al.*
Volume 32 **High Voltage Engineering and Testing, 2nd Edition** H.M. Ryan (Editor)
Volume 33 **Overvoltage Protection of Low-Voltage Systems, Revised Edition** P. Hasse
Volume 36 **Voltage Quality in Electrical Power Systems** J. Schlabbach *et al.*
Volume 37 **Electrical Steels for Rotating Machines** P. Beckley
Volume 38 **The Electric Car: Development and future of battery, hybrid and fuel-cell cars** M. Westbrook
Volume 39 **Power Systems Electromagnetic Transients Simulation** J. Arrillaga and N. Watson
Volume 40 **Advances in High Voltage Engineering** M. Haddad and D. Warne
Volume 41 **Electrical Operation of Electrostatic Precipitators** K. Parker
Volume 43 **Thermal Power Plant Simulation and Control** D. Flynn
Volume 44 **Economic Evaluation of Projects in the Electricity Supply Industry** H. Khatib
Volume 45 **Propulsion Systems for Hybrid Vehicles** J. Miller
Volume 46 **Distribution Switchgear** S. Stewart
Volume 47 **Protection of Electricity Distribution Networks, 2nd Edition** J. Gers and E. Holmes
Volume 48 **Wood Pole Overhead Lines** B. Wareing
Volume 49 **Electric Fuses, 3rd Edition** A. Wright and G. Newbery
Volume 50 **Wind Power Integration: Connection and system operational aspects** B. Fox *et al.*
Volume 51 **Short Circuit Currents** J. Schlabbach
Volume 52 **Nuclear Power** J. Wood
Volume 53 **Condition Assessment of High Voltage Insulation in Power System Equipment** R.E. James and Q. Su
Volume 55 **Local Energy: Distributed generation of heat and power** J. Wood
Volume 56 **Condition Monitoring of Rotating Electrical Machines** P. Tavner, L. Ran, J. Penman and H. Sedding
Volume 57 **The Control Techniques Drives and Controls Handbook, 2nd Edition** B. Drury
Volume 58 **Lightning Protection** V. Cooray (Editor)
Volume 59 **Ultracapacitor Applications** J.M. Miller
Volume 62 **Lightning Electromagnetics** V. Cooray
Volume 63 **Energy Storage for Power Systems, 2nd Edition** A. Ter-Gazarian
Volume 65 **Protection of Electricity Distribution Networks, 3rd Edition** J. Gers
Volume 66 **High Voltage Engineering Testing, 3rd Edition** H. Ryan (Editor)
Volume 67 **Multicore Simulation of Power System Transients** F.M. Uriate
Volume 68 **Distribution System Analysis and Automation** J. Gers
Volume 69 **The Lightening Flash, 2nd Edition** V. Cooray (Editor)
Volume 70 **Economic Evaluation of Projects in the Electricity Supply Industry, 3rd Edition** H. Khatib
Volume 72 **Control Circuits in Power Electronics: Practical issues in design and implementation** M. Castilla (Editor)
Volume 73 **Wide Area Monitoring, Protection and Control Systems: The enabler for Smarter Grids** A. Vaccaro and A. Zobaa (Editors)
Volume 74 **Power Electronic Converters and Systems: Frontiers and applications** A. M. Trzynadlowski (Editor)
Volume 75 **Power Distribution Automation** B. Das (Editor)
Volume 76 **Power System Stability: Modelling, analysis and control** A.A. Sallam and B. Om P. Malik
Volume 78 **Numerical Analysis of Power System Transients and Dynamics** A. Ametani (Editor)
Volume 79 **Vehicle-to-Grid: Linking electric vehicles to the smart grid** J. Lu and J. Hossain (Editors)
Volume 81 **Cyber-Physical-Social Systems and Constructs in Electric Power Engineering** S. Suryanarayanan, R. Roche and T.M. Hansen (Editors)
Volume 82 **Periodic Control of Power Electronic Converters** F. Blaabjerg, K.Zhou, D. Wang and Y. Yang

Volume 86 **Advances in Power System Modelling, Control and Stability Analysis** F. Milano (Editor)
Volume 87 **Cogeneration: Technologies, Optimisation and Implentation** C. A. Frangopoulos (Editor)
Volume 88 **Smarter Energy: from Smart Metering to the Smart Grid** H. Sun, N. Hatziargyriou, H. V. Poor, L. Carpanini and M. A. Sánchez Fornié (Editors)
Volume 89 **Hydrogen Production, Separation and Purification for Energy** A. Basile, F. Dalena, J. Tong and T.N.Veziroğlu (Editors)
Volume 90 **Clean Energy Microgrids** S. Obara and J. Morel (Editors)
Volume 91 **Fuzzy Logic Control in Energy Systems with Design Applications in Matlab/Simulink®** İ. H. Altaş
Volume 92 **Power Quality in Future Electrical Power Systems** A. F. Zobaa and S. H. E. A. Aleem (Editors)
Volume 93 **Cogeneration and District Energy Systems: Modelling, Analysis and Optimization** M. A. Rosen and S. Koohi-Fayegh
Volume 94 **Introduction to the Smart Grid: Concepts, technologies and evolution** S.K. Salman
Volume 95 **Communication, Control and Security Challenges for the Smart Grid** S.M. Muyeen and S. Rahman (Editors)
Volume 96 **Industrial Power Systems with Distributed and Embedded Generation** R Belu
Volume 97 **Synchronized Phasor Measurements for Smart Grids** M.J.B. Reddy and D.K. Mohanta (Editors)
Volume 98 **Large Scale Grid Integration of Renewable Energy Sources** A. Moreno-Munoz (Editor)
Volume 100 **Modeling and Dynamic Behaviour of Hydropower Plants** N. Kishor and J. Fraile-Ardanuy (Editors)
Volume 101 **Methane and Hydrogen for Energy Storage** R. Carriveau and D. S-K. Ting
Volume 104 **Power Transformer Condition Monitoring and Diagnosis** A. Abu-Siada (Editor)
Volume 106 **Surface Passivation of Industrial Crystalline Silicon Solar Cells** J. John (Editor)
Volume 107 **Bifacial Photovoltaics: Technology, applications and economics** J. Libal and R. Kopecek (Editors)
Volume 108 **Fault Diagnosis of Induction Motors** J. Faiz, V. Ghorbanian and G. Joksimović
Volume 110 **High Voltage Power Network Construction** K. Harker
Volume 111 **Energy Storage at Different Voltage Levels: Technology, integration, and market aspects** A.F. Zobaa, P.F. Ribeiro, S.H.A. Aleem and S.N. Afifi (Editors)
Volume 112 **Wireless Power Transfer: Theory, Technology and Application** N.Shinohara
Volume 114 **Lightning-Induced Effects in Electrical and Telecommunication Systems** Y. Baba and V. A. Rakov
Volume 115 **DC Distribution Systems and Microgrids** T. Dragičević, F.Blaabjerg and P. Wheeler
Volume 116 **Modelling and Simulation of HVDC Transmission** M. Han (Editor)
Volume 117 **Structural Control and Fault Detection of Wind Turbine Systems** H.R. Karimi
Volume 119 **Thermal Power Plant Control and Instrumentation: The control of boilers and HRSGs, 2nd Edition** D. Lindsley, J. Grist and D. Parker
Volume 120 **Fault Diagnosis for Robust Inverter Power Drives** A. Ginart (Editor)
Volume 121 Monitoring and Control using Synchrophasors in Power Systems with Renewables I. Kamwa and C. Lu (Editors)
Volume 123 **Power Systems Electromagnetic Transients Simulation, 2nd Edition** N. Watson and J. Arrillaga
Volume 124 **Power Market Transformation** B. Murray
Volume 125 **Wind Energy Modeling and Simulation Volume 1: Atmosphere and plant** P.Veers (Editor)
Volume 126 **Diagnosis and Fault Tolerance of Electrical Machines, Power Electronics and Drives** A.J. M. Cardoso
Volume 128 **Characterization of Wide Bandgap Power Semiconductor Devices** F. Wang, Z. Zhang and E.A. Jones
Volume 129 **Renewable Energy from the Oceans: From wave, tidal and gradient systems to offshore wind and solar** D. Coiro and T. Sant (Editors)
Volume 130 **Wind and Solar Based Energy Systems for Communities** R. Carriveau and D. S-K. Ting (Editors)
Volume 131 **Metaheuristic Optimization in Power Engineering** J. Radosavljević
Volume 132 **Power Line Communication Systems for Smart Grids** I.R.S Casella and A. Anpalagan
Volume 139 **Variability, Scalability and Stability of Microgrids** S. M. Muyeen, S. M. Islam and F. Blaabjerg (Editors)
Volume 145 **Condition Monitoring of Rotating Electrical Machines** P. Tavner, L. Ran, C. Crabtree
Volume 146 **Energy Storage for Power Systems, 3rd Edition** A.G. Ter-Gazarian
Volume 147 **Distribution Systems Analysis and Automation 2nd Edition** J. Gers
Volume 151 **SiC Power Module Design: Performance, robustness and reliability** Alberto Castellazzi and Andrea Irace (Editors)
Volume 152 **Power Electronic Devices: Applications, failure mechanisms and reliability** F Iannuzzo (Editor)
Volume 153 **Signal Processing for Fault Detection and Diagnosis in Electric Machines and Systems** Mohamed Benbouzid (Editor)
Volume 155 **Energy Generation and Efficiency Technologies for Green Residential Buildings** D. Ting and R. Carriveau (Editors)
Volume 156 **Lithium-ion Batteries Enabled by Silicon Anodes** Chunmei Ban and Kang Xu (Editors)
Volume 157 **Electrical Steels, 2 Volumes** A. Moses, K. Jenkins, Philip Anderson and H. Stanbury
Volume 158 **Advanced Dielectric Materials for Electrostatic Capacitors** Q Li (Editor)
Volume 159 **Transforming the Grid Towards Fully Renewable Energy** O. Probst, S. Castellanos and R. Palacios (Editors)
Volume 160 **Microgrids for Rural Areas: Research and case studies** R.K. Chauhan, K. Chauhan and S.N. Singh (Editors)
Volume 166 **Advanced Characterization of Thin Film Solar Cells** N Haegel and M Al-Jassim (Editors)
Volume 167 **Power Grids with Renewable Energy** Storage, integration and digitalization A. A. Sallam and B. OM P. Malik
Volume 170 **Reliability of Power Electronics Converters for Solar Photovoltaic Applications** Frede Blaabjerg, Ahteshamul Haque, Huai Wang, Zainul Abdin Jaffery and Yongheng Yang (Editors)
Volume 171 **Utility-scale Wind Turbines and Wind Farms** A. Vasel-Be-Hagh and D. S.-K. Ting
Volume 172 **Lighting interaction with Power Systems, 2 volumes** A. Piantini (Editor)
Volume 174 **Silicon Solar Cell Metallization and Module Technology** T. Dullweber (Editor)

Volume 193 **Overhead Electric Power Lines: Theory and practice** S. Chattopadhyay and A. Das
Volume 199 **Model Predictive Control for Microgrids: From power electronic converters to energy management** J. Hu, J. M. Guerrero and S. Islam
Volume 905 **Power system protection, 4 volumes**

Hydrogen Passivation and Laser Doping for Silicon Solar Cells

Edited by
Brett Hallam and Catherine Chan

The Institution of Engineering and Technology

Published by The Institution of Engineering and Technology, London, United Kingdom

The Institution of Engineering and Technology is registered as a Charity in England & Wales (no. 211014) and Scotland (no. SC038698).

First published 2021

The Institution of Engineering and Technology
Michael Faraday House
Six Hills Way, Stevenage
Herts, SG1 2AY, United Kingdom

www.theiet.org

British Library Cataloguing in Publication Data
A catalogue record for this product is available from the British Library

ISBN 978-1-78561-623-5 (hardback)
ISBN 978-1-78561-624-2 (PDF)

Typeset in India by Exeter Premedia Services Private Limited
Printed in the UK by CPI Group (UK) Ltd, Croydon

Contents

List of figures xv
List of tables xli
About the editors xliii

1 Industrial silicon solar cells **1**
Matthew Wright, Brett Hallam, Catherine Chan, and Martin Green

1.1 Climate change and the use of fossil fuels 1
1.2 Solar photovoltaics for future energy generation 4
1.3 The reducing cost of solar PV 6
1.4 Screen-printed silicon solar cells – the workhorse of the photovoltaics industry 9
1.5 Limitations to current commercial silicon solar cells 13
1.5.1 Performance-limiting defects and light-induced degradation 13
1.5.2 Selective emitter design 17
1.6 Overview of chapter structure 18
1.7 Acknowledgements 20
References 20

2 Hydrogen passivation mechanisms **25**
Phillip Hamer, Chang Sun, and Brett Hallam

2.1 The role of hydrogen passivation in high-efficiency silicon solar cells 25
2.2 Methods of introducing hydrogen in silicon solar cells 28
2.2.1 Annealing in a molecular hydrogen ambient 28
2.2.2 Exposure to a hydrogen containing plasma 29
2.2.2.1 Direct hydrogen plasma 29
2.2.2.2 Remote plasma hydrogenation 30
2.2.3 Dielectric layer deposition 30
2.2.4 Unintentional incorporation of hydrogen 32
2.3 Behaviour of interstitial hydrogen in silicon 32
2.3.1 Donor and acceptor levels of hydrogen in silicon 32
2.3.2 Neutralisation of dopant atoms 34
2.3.3 Hydrogen dimer formation 36
2.3.4 Hydrogen interactions with other defects 38
2.4 Hydrogen migration in silicon 40
2.4.1 Hydrogen diffusivity 40

2.4.2 Impact of charge state on hydrogen diffusivity 41
2.4.3 Uncertainties around hydrogen passivation 42
2.5 Detection of hydrogen in silicon 43
2.6 Fractional charge state concentrations of interstitial hydrogen in silicon in thermal equilibrium 46
2.7 Increased generation of H^0 using carrier injection 49
2.8 Unified modelling of hydrogen and defects 54
2.9 Production sequences for industrial silicon solar cells 55
2.9.1 Al-BSF solar cells 55
2.9.2 PERC solar cells 58
2.10 Summary 60
2.11 Acknowledgements 61
References 61

3 Hydrogen passivation of silicon surfaces **75**
Ruy S. Bonilla and Bram Hoex

3.1 Recombination and passivation at silicon interfaces 75
3.1.1 The surface as a defect 75
3.1.2 Characterisation of surface passivation in silicon 79
3.1.3 Dielectric coatings for surface passivation 82
3.1.4 State of the art in surface passivation of silicon 84
3.2 Hydrogen in silicon oxide thin films 86
3.2.1 Thermal silicon dioxide 86
3.2.2 Silicon oxides grown by PECVD, ALD, or chemical anodisation 88
3.3 Hydrogen in silicon nitride thin films 91
3.4 Hydrogen in amorphous silicon thin films 96
3.5 Hydrogen in aluminium oxide thin films 99
3.6 Hydrogen in tunnelling oxide passivating contact layers 101
3.7 Outlook for dielectric hydrogenation processes 102
References 102

4 Hydrogen passivation of bulk defects **115**
Brett Hallam, Matthew Wright, and Alison Ciesla

4.1 Hydrogenation of structural defects 116
4.1.1 Structural defects in cast silicon wafers 116
4.1.2 Passivation of grain boundaries and dislocation clusters 117
4.1.2.1 Characterisation of structural defects 117
4.1.2.2 Challenges of passivation 120
4.1.2.3 Impact of dielectric layer composition and thermal processing 122
4.1.3 Dehydrogenation during extended annealing or high-temperature processes 125
4.1.4 Development of cast-mono and high-performance multi-crystalline silicon material 126

4.1.5 Benefits of an additional hydrogenation processes after firing 128
4.2 Hydrogen passivation of impurity-related defects 130
4.2.1 Passivation of dopants 135
4.3 Hydrogenation of process-induced defects 137
4.3.1 Oxygen precipitates 137
4.3.2 Laser-induced defects 142
4.4 Hydrogenation of light- and carrier-induced defects 143
4.5 Complementary nature of gettering and hydrogenation for silicon solar cells 144
4.5.1 Impact of gettering and hydrogenation on lifetime 145
4.5.2 Gettering and hydrogenation in Al-BSF and PERC solar cells 147
4.5.3 Impact of evolving solar cell architectures and processes on gettering and hydrogen passivation 152
4.5.4 Gettering and hydrogenation in high-efficiency n-type cell technologies 155
4.6 Gettering and hydrogenation for heterojunction silicon solar cells 157
4.6.1 Impact of gettering and hydrogenation for SHJ solar cells 158
4.7 Summary 165
References 166

5 The boron-oxygen defect system **179**
Brett Hallam, Alison Ciesla, Moonyong Kim, and Axel Herguth

5.1 B-O-related degradation in p-type Cz silicon 182
5.1.1 Identification of the defect and its concentration 182
5.1.2 Correlation between boron doping, the interstitial oxygen concentration and both the NDD and minority carrier lifetime 184
5.1.3 Degradation characteristics 185
5.1.3.1 Two-stage degradation 185
5.1.3.2 Impact of the doping concentration and illumination intensity on the degradation rates 187
5.1.3.3 Inducing B-O degradation with carrier injection 188
5.1.4 Temporary recovery from B-O defects through dark annealing 188
5.1.5 Defect theories 189
5.1.6 Generalised model for the B-O defect system 190
5.2 Alternative silicon wafers to avoid B-O-related degradation 192
5.2.1 Methods to decrease the interstitial oxygen concentration 192
5.2.2 Casted multi-crystalline and quasi-mono-crystalline silicon 193
5.2.3 Decreasing the boron-doping concentration 194
5.2.4 Germanium and carbon co-doping 195
5.2.5 Summary of wafer alternatives 195

5.3 Multiple pathways of B-O LID mitigation during cell processing 196
5.3.1 Thermal processes to modify the B-O precursor concentration – A four-state model 196
5.3.1.1 High-temperature tube-furnace processes 197
5.3.1.2 Annealing at intermediate temperatures 198
5.3.1.3 Rapid thermal processing such as metallisation firing 198
5.3.1.4 Unknown nature and kinetics of precursor annihilation and reformation 200
5.3.2 Annealing with illumination or carrier injection for stable passivation of B-O defects – A three-state model 200
5.3.2.1 Determination of reaction rates for kinetic modelling 201
5.3.3.2 Basic kinetic modelling using ordinary differential equations 204
5.4 Fundamentals of illuminated annealing for B-O LID mitigation 209
5.4.1 The role of hydrogen 209
5.4.1.1 The importance of firing 210
5.4.1.2 The complicated impact of firing on B-O passivation 211
5.4.1.3 Impact of firing process parameters on passivation kinetics 212
5.4.1.4 Impact of film and dielectric stack composition 213
5.4.1.5 Modulating kinetics with prior annealing 215
5.4.1.6 Passivation with unintentionally introduced hydrogen 216
5.4.1.7 Summary of the role of hydrogen for B-O defect passivation 217
5.4.2 Role of hydrogen charge states 217
5.4.3 The critical role of defect formation 220
5.4.4 Development of rapid laser-based processes 222
5.4.5 Long-term stability of passivation 223
5.5 Eliminating B-O LID in commercial solar cells 224
5.5.1 Commercial B-O LID mitigation solutions for cells 225
5.5.2 B-O LID mitigation during module lamination 228
5.5.3 Self-recovery in the field 228
5.6 B-O LID in p-type silicon heterojunction solar cells 229
5.7 Summary 233
References 234

6 Negative impacts of hydrogen in silicon **249**
Ran Chen, Alison Ciesla, Michelle Vaqueiro-Contreras, Daniel Chen, Matthew Wright, Brett Hallam, and Tony Peaker

6.1 Hydrogen complexes with other species 250

6.1.1 Carbon-hydrogen (CH) complexes in silicon 250
6.1.2 Oxygen-hydrogen (OH) complexes in silicon 252
6.1.3 Carbon-oxygen-hydrogen (COH) complexes in silicon 252
6.1.4 Transition metal-hydrogen complexes in silicon 255
6.1.5 Vacancy-hydrogen complexes in silicon 255
6.2 Light- and elevated temperature-induced degradation 256
6.2.1 The impact of LeTID – cells, modules and systems 258
6.2.2 The key behaviours of LeTID 259
6.2.2.1 Dependence on firing on degradation extent 259
6.2.2.2 A universal defect in silicon 261
6.2.2.3 LeTID dependence on dielectrics 264
6.2.2.4 LeTID characterisation 264
6.2.3 The search for the root cause of LeTID 266
6.2.3.1 Metallic impurities 266
6.2.3.2 Hydrogen: a growing consensus 268
6.2.4 The role of hydrogen in LeTID 269
6.2.4.1 The impact of hydrogenated dielectric films 269
6.2.4.2 Direct correlation between hydrogen and LeTID 270
6.2.4.3 Analysis using DLTS 271
6.2.5 LeTID mitigation 272
6.2.6 Models for LeTID 275
6.2.7 LeTID in p-type silicon heterojunction solar cells 280
6.3 Negative effects due to hydrogen behaviour 282
6.3.1 Hydrogen-induced contact resistance 282
6.3.2 Neutralisation of charge and dopants 285
6.3.3 Formation of hydrogen-induced platelets 287
6.4 Summary 290
Acknowledgment 290
References 291

7 Laser doping for rapid diffusion in silicon 305
Michelle Vaqueiro-Contreras, Matthew Wright, and Catherine Chan

7.1 The origin of laser doping for silicon solar cell manufacturing 305
7.2 Dopant diffusion in crystalline silicon 306
7.2.1 Conventional doping methods 307
7.2.2 Laser doping 310
7.3 Approaches for laser doping 312
7.4 Impact of laser parameters on doping profiles 316
7.4.1 Characterisation of laser-doped regions 316
7.4.1.1 Sheet resistance 317
7.4.1.2 ECV/SIMS 317
7.4.1.3 SEM/EBIC 319
7.4.2 Type of laser 319
7.4.3 Diffusion profile in laser-doped regions 322

7.4.4 Impact of laser speed 324
7.4.5 Choice of wavelength 327
7.4.6 Beam optics 329
7.5 Chapter summary 329
References 330

8 Laser-doped selective emitter formation and the passivation of laser-induced defects 337
Arman Mahboubi Soufiani, Catherine Chan, Jingjia Ji, and Brett Hallam

8.1 The selective emitter concept 338
8.1.1 Trade-offs leading to the use of a selective emitter 339
8.1.2 Emitter recombination, surface recombination velocity and quantum efficiency 340
8.1.3 Contact resistivity and metal/Si interface recombination 342
8.1.4 Lateral emitter resistance 344
8.2 Techniques for selective emitter formation 345
8.2.1 Dielectric patterning using photolithography 346
8.2.2 Dielectric patterning through laser ablation 347
8.2.3 Dielectric patterning through inkjet or aerosol jet printing 348
8.2.4 Use of doped silicon inks 349
8.2.5 Ion implantation 349
8.2.6 Etch-back technique 350
8.3 Laser doping for selective emitter formation 350
8.3.1 Industrial LDSE PERC solar cells with aligned screen-printed contacts 351
8.3.2 LDSE solar cells with self-aligned plated contacts 354
8.4 Laser-induced defect generation 356
8.4.1 Characterisation of laser-induced defects 358
8.4.1.1 Electron beam microscopy-based characterisation 358
8.4.1.2 Combined QSSPC and photoluminescence-based characterisation 358
8.4.1.3 Spectral photoluminescence-based characterisation 362
8.4.1.4 DLTS-based characterisation 364
8.5 Techniques to minimise the impact of laser-induced defects 365
8.5.1 Reduction in laser-doping coverage and metal/Si interface area 365
8.5.2 Laser doping prior to dielectric deposition 366
8.5.2.1 Simultaneous laser doping and grooving 368
8.5.3 Localised passivated contact structures on laser-doped regions 369
8.5.4 Performing laser doping through dielectric layers 369
8.5.5 Deep-junction laser doping 371
8.5.6 Laser- and thermal-annealing for localised repair of laser damage 372

8.6 Hydrogen passivation of laser-induced defects 373
8.6.1 Performing laser doping prior to firing 374
8.6.2 Additional low-temperature belt furnace-annealing processes 375
8.6.3 Simultaneous laser-induced defect passivation and light-induced degradation mitigation 382
8.7 Conclusion 385
References 387

9 Applications of laser doping 395
Catherine Chan and Brett Hallam

9.1 Local back surface field formation for PERC-type cell structures 395
9.1.1 Laser doping as an alternative to the Al/Si alloying process for PERC 396
9.1.2 Aluminium oxide as a dopant source for p^{++} silicon formation 403
9.2 Semiconductor finger solar cells 407
9.2.1 Laser-doped semiconductor finger solar cells 409
9.2.2 Advanced laser-doped semiconductor finger cells 411
9.2.3 Future potential of the semiconductor finger technology 412
9.3 Interdigitated back-contact solar cells 412
9.3.1 Manufacturing challenges of IBC cells 414
9.3.2 Laser-doped IBC cells 416
9.4 Laser doping through doped surface layers of opposite polarity 418
9.4.1 Simplified interdigitated back-contact cells with laser-doped compensated contact 420
9.4.2 Isolation of shunts or high recombination areas such as edges 421
9.5 Laser doping for enhanced silicon gettering 424
9.6 Summary 428
References 430

10 Conclusion and future outlook 437
Matthew Wright, Brett Hallam, and Catherine Chan

10.1 Importance of hydrogenation and laser doping for silicon solar cells 438
10.1.1 Hydrogen passivation 440
10.1.2 Laser doping 443
10.2 Future outlook 445
10.2.1 Future issues for industrial PERC solar cells 445
10.2.1.1 Next-generation laser-doped selective emitters for PERC 445
10.2.1.2 Surface-related degradation 446
10.2.1.3 A shift to gallium-doped silicon wafers 447
10.2.2 N-type passivated contact solar cells 449

10.2.2.1 Selective emitter for TOPCon solar cells 450
10.2.2.2 Instability in n-type passivated contact solar cells 451
10.2.2.3 Post-cell treatments for SHJ solar cells 452
10.2.3 Silicon-based tandem solar cells 454
References 457

Index **461**

List of figures

Figure 1.1	Global average land-sea temperature anomaly, plotted over the time period from 1850 to 2019. The values are plotted relative to the average temperature during the time spanning from 1961 to 1990. The data has been replotted, original data from [3]	2
Figure 1.2	Historical global CO_2 emissions shown by region. Data is shown in the range from 1751 to 2017. The original graph was produced by Our World in Data from [4] and is reproduced according to the Creative Commons CC BY 4.0 license. Original data from [7] and [10]	3
Figure 1.3	World map indicating the direct normal irradiation (left). The area of Australia required to power the world using solar exclusively (right). Reproduced from [16]. © 2019 The World Bank, Source: Global Solar Atlas 2.0, Solar resource data: Solargis. Creative Commons CC BY 4.0	5
Figure 1.4	Learning curve for average PV module prices, shown as a function of cumulative PV module shipments. Reproduced with permission from [19]. Copyright 2020 VDMA Photovoltaic Equipment	7
Figure 1.5	Average quarterly selling price for multi-crystalline silicon photovoltaic modules. Reprinted from (25) Copyright (2019) with permission from Elsevier	9
Figure 1.6	Historical overview of the market share of solar cell technologies, spanning from 1980 to 2019. Data provided by Paula Mints [29]	10
Figure 1.7	Schematic comparison of the cell architecture for Al-BSF (top) and PERC solar cells (bottom)	13
Figure 1.8	Projected worldwide market share for different silicon solar cell technologies. Reproduced with permission from [19]. Copyright 2020 VDMA Photovoltaic Equipment	14
Figure 1.9	Progression of Hanwha Q-Cells production cell efficiency data for both Al-BSF (labelled 'BSF') and Q.ANTUM technology (labelled 'QNT', representing Hanwha Q-Cells' implementation of PERC). The labels on the x-axis represent December in the allotted calendar year. Graph kindly provided by Markus Fischer from Hanwha Q-Cells	14

Figure 1.10 1-sun calibrated iV_{OC} maps of a boron-doped p-type Cz wafer exhibiting oxygen precipitates (left) directly after deposition of SiN_x:H, and (right) following a subsequent high-temperature firing process for hydrogenation. Image reproduced with permission from [49] under Creative Commons CC-BY-NC-ND. © CSIRO 2018 Open Access 15
Figure 1.11 Effective minority carrier lifetime as a function of light soaking time for both Cz and FZ boron-doped p-type silicon wafers, illustrating the significant LID in p-type Cz silicon. Reproduced with permission from Tine Nærland [53] 16
Figure 1.12 Schematic diagram illustrating the selective emitter design in a PERC solar cell. Image reproduced with permission from [2] under Creative Commons CC BY. © 2020 The Authors. Progress in Photovoltaics: Research and Applications published by John Wiley & Sons Ltd. https://doi.org/10.1002/pip.3243 18
Figure 1.13 Projected market share for the different phosphorous emitter technologies for p-type silicon solar cells. This indicates the expected importance of laser doping for selective emitter formation. Reproduced with permission from [19]. Copyright 2020, VDMA Photovoltaic Equipment 19
Figure 2.1 Google scholar results per year of publication for search containing the terms 'hydrogen passivation', 'silicon' and 'solar cell' 26
Figure 2.2 Emitter saturation current density (J_{0e}) as a function of emitter sheet resistance for n^+pn^+ lifetime structures. Reprinted from [11] with the permission of AIP Publishing. © 2001 American Institute of Physics 27
Figure 2.3 (Left) Proposed mechanisms for splitting of hydrogen molecules by vacancies and surface damage adapted from [28]; (right) proposed alneal mechanism highlighting interactions between hydroxyl ions and Al layers 29
Figure 2.4 Schematic of PECVD equipment used for (left) direct and (right) remote plasma hydrogenation. Image reproduced with permission from PV-Manufacturing.org [31] 30
Figure 2.5 Energy levels of interstitial hydrogen according to work of both (left) Herring et al. [55] and (right) Bonde Nielsen et al. [56]. <E> denotes the energy level at which concentrations of H^+ and H^- are equal for each set of values 33
Figure 2.6 Minimum energy configurations of hydrogen charge states in silicon. (a) shows two silicon atoms with no hydrogen in the vicinity. (b) Shows H^+ at the BC site with stretching of the Si–Si bond. (c) Shows H^0, also at the BC site, also with stretching of the Si–Si bond. (d) Shows H^- at the tetrahedral site with negligible distortion of the silicon lattice. Figure adapted from [55] 35

Figure 2.7 Minimum energy configurations of hydrogen-dopant pairs in silicon. (a) Shows H bound to substitutional Boron while (b) shows H bound to a Si atom neighboring substitutional P. Figure adapted from [70] 36
Figure 2.8 Known stable H dimer formations in Si. (a) depicts isolated H_2 molecules at the tetrahedral sites. (a) shows with hydrogen bound to neighbouring Si atoms at the BC and AB sites 37
Figure 2.9 Hydrogen diffusivity data taken from the literature [54, 73, 97, 100, 104–119]. The solid line represents the extrapolation of the hydrogen diffusivity from the work of Van Wieringen et al. (V.W.W.) [104]. Adapted from [120] 40
Figure 2.10 APT datasets showing the enrichment of $^{2}H^{2+}$ and $^{29}Si^{2}H^{+}$ at three individual dislocations. (a) Viewing perpendicular to the dislocation. (b) Corresponding mass spectra of the matrix of the APT specimen. (c) Mass spectra of dislocation III, showing clear enrichment in ^{2}H at 4 Da, 31 Da and 32 Da. Figure adapted from [155]. Please see Tweddle et al. [155] for full color version and details of deconvolution 44
Figure 2.11 Room temperature Raman spectra of p-type FZ silicon after (a) exposure an 8 hour H_2 plasma treatment at 250 °C, (b) D_2 plasma, (c) a H_2:D_2 (50:50) plasma, and d) no plasma treatment. Excitation wavelength: 488 nm. Reprinted figure with permission from [81] Copyright (1998) by the American Physical Society. http://dx.doi.org/10.1103/PhysRevLett.81.421 45
Figure 2.12 (a) Spreading resistance and (b) SIMS profiles in boron-doped Czochralski-grown silicon with an initial uniform boron concentration of $5 \times 10^{18}/cm^3$. Reprinted figure with permission from [177] Copyright (1985) by the American Physical Society. http://dx.doi.org/10.1103/PhysRevB.31.5525 46
Figure 2.13 Fractional hydrogen charge state concentrations of interstitial hydrogen in silicon in thermal equilibrium at room temperature as a function of the Fermi level (E_{Fermi}) relative to the mid-gap energy level ($E_{mid\text{-}gap}$). Figure adapted from [120] 47
Figure 2.14 Fractional H^0 concentration in silicon in thermal equilibrium as a function of temperature for (top) p-type silicon as a function of acceptor dopant concentration (N_A), and (bottom) n-type silicon as a function of donor dopant concentration (N_D). It should be noted that the lower limit of the colour bar is truncated at 1×10^{-10}. Figure adapted from [120] 48
Figure 2.15 Fractional H^0 concentration in silicon with an excess minority carrier concentration of $\Delta n = 1 \times 10^{15}/cm^3$ as a function of temperature for (top) p-type silicon as a function of acceptor dopant concentration (N_A), and (bottom) n-type silicon as a function of donor dopant concentration (N_D). Note, a different scale has been used compared to that in Figure 2.14. Figure adapted from [120] 51

Figure 2.16	Fractional H^0 concentrations in 1 Ω·cm silicon material as a function of temperature and excess carrier concentration (Δn) for (top) p-type silicon and (bottom) n-type silicon using Sah-Shockley occupation statistics. Note, a different scale has been used compared to that in Figure 2.14, but the same scale as for Figure 2.15. Figure adapted from [120]	52
Figure 2.17	Fractional hydrogen (H) and boron-oxygen (B-O) charge state concentrations in 1 Ω.cm p-type silicon as a function of temperature for (a) thermal equilibrium and excess carrier densities of (b) $1 \times 10^{13}/cm^3$, (c) $1 \times 10^{15}/cm^3$ and (d) $1 \times 10^{17}/cm^3$. Figure based on modelling presented in [181]	54
Figure 2.18	Reaction products for H and B-O in 1 Ω.cm p-type silicon at temperature = 200 °C as a function of excess carrier density. Figure based on modelling presented in [181]	55
Figure 2.19	Process flow for industrial silicon Al-BSF solar cells	56
Figure 2.20	Process flow for PERC solar cells	59
Figure 3.1	(a) Schematic picture of a bare silicon surface with dangling bonds producing interface mid bandgap states. (b) Shockley's reconstruction of a one-dimensional lattice. Reprinted figure with permission from [2] Copyright (1939) by the American Physical Society. https://doi.org/10.1103/PhysRev.56.317	76
Figure 3.2	Calculated SRV for a range of capture cross sections and excess minority carrier concentration. SRV is calculated using the extended SRH statistics. N_A represents the acceptor doping concentration	78
Figure 3.3	Calculated SRV as a function of excess minority carrier concentration for different doping concentrations. SRV is calculated using the extended SRH statistics of (3.3)	79
Figure 3.4	Calculated effective SRV as a function of positive and negative dielectric charge concentration Q_f for *p*-type silicon with $S_{p0} = 0.1\ S_{p0}$, $\Delta n_d = 10^{15}/cm^3$, acceptor concentration $N_{acc} = 10^{16}/cm^3$, and uniform interface density D_{it}	80
Figure 3.5	Calculated effective lifetime as a function of wafer thickness and surface passivation quality from uniform interface density D_{it}, for *p*-type silicon, acceptor concentration $N_{acc} = 10^{16}/cm^3$. The dielectric charge concentration is set to $n_{i,eff} = 10^{11}\ q/cm^2$, with $S_{n0} = 10 \times S_{p0}$	82
Figure 3.6	Schematic diagram showing surface states in: (a) bare silicon; (b) chemically passivated silicon; and (c) chemically and field effect-passivated silicon	84
Figure 3.7	Timeline of the best passivation technologies available at the research level. It is evident here that SRV <5 cm/s are reasonably good, and those <1 cm/s are outstanding. Please refer to [30] for full details on the data from each author	85

Figure 3.8 (a) Density of interface defect states in wet oxidised <111>Si–SiO_2 as a function of hydrogenation conditions. Data adapted from Fussel [42, 44]. (b) Density of interface defect states in Si–SiO_2 for <111> and <100> surfaces, for wet and dry oxides, as a function of hydrogenation conditions. Data adapted from [7, 45, 46] 86

Figure 3.9 Summary of minimum SRV achieved as a result of reduced interface recombination from hydrogenation in thermal SiO_2 and the effect of FEP. Substrates are ~1 Ω·cm <100>. Data from [36, 38, 55, 59, 60] 89

Figure 3.10 Summary of effective SRV in PECVD and ALD synthesised SiO_2 with a strong hydrogenation effect. Data from [64–67] 90

Figure 3.11 Hydrogen effusion rate as a function of temperature. Dotted lines indicate maximal hydrogen effusion temperatures for each passivating film. Data sourced from [113, 114] 92

Figure 3.12 (a) Hydrogen content inside of SiN_x films obtained from FTIR measurements. Data reproduced from [117] with permission from Y. Wan. Copyright 2014 Y. Wan. (b) Total concentration of hydrogen of a SiN_x passivation layer as a function of refractive index at a wavelength of 633 nm. Reproduced with permission from [118]. © 2019 WILEY-VCH Verlag GmbH & Co. KGaA, Weinheim 93

Figure 3.13 Effective SRV as a function of refractive index at 632 nm for SiN_x PECVD passivation of *p*-type FZ silicon. Data from [90, 96, 103, 105, 106, 123, 124]. Reproduced from [102] under Creative Commons CC BY. © Author(s) 2013. http://aip.scitation.org/doi/10.1063/1.4795108 95

Figure 3.14 Measurements of SRV as a function of injection density calculated from PCD lifetime for top performing SiO_2–SiN_x double-layer films on Si. Data from [61, 62, 75, 137] 96

Figure 3.15 Influence of the post-deposition annealing temperature T_{ann} on the effective SRV, S_{eff}, and effective lifetime, τ_{eff}, of a symmetrically passivated c-Si wafer for a-Si:H films grown at various deposition temperatures T_{depo}. The annealing time was 30 minutes for all films while the deposition time t_{depo} was 12 minutes. Reprinted from [148], with the permission of AIP Publishing. © 2007 American Institute of Physics 97

Figure 3.16 Efficiency of industrial SHJ solar cells as a function of accelerated testing time. A clear difference in stability can be seen for solar cells that were annealed in the dark or under 1 sun illumination at 160 °C. Reprinted from [151] Copyright (2020), with permission from Elsevier 98

Figure 3.17 Minority carrier lifetime as (a) function of injection level and (b) surface potential for a symmetrically passivated *n*-type c-Si

	wafer. The surface potential was varied by the application of corona charging. Reprinted from [152] under Creative Commons CC BY. © 2016 The Authors. Phys. Status Solidi RRL published by WILEY-VCH Verlag GmbH & Co. KGaA, Weinheim. http://doi.wiley.com/10.1002/pssr.201600307	99
Figure 3.18	Hydrogen effusion measurements from an AlO_x film grown using ALD. It can be seen that hydrogen is released from the film for temperatures in excess of 400 °C mainly in the form of H_2O. Reprinted from [159], with the permission of AIP Publishing. © 2009 American Institute of Physics	100
Figure 3.19	Hydrogen, deuterium, and oxyen profiles of a *p*-type poly-Si/SiOx contact with a deuterated Al_2O_3 layer after a nitrogen anneal. Reprinted from [166], with the permission of AIP Publishing. © 2018 Author(s)	101
Figure 4.1	Calibrated iV_{OC} map of a diffused conventional multi-crystalline wafer after (left) SiN_x:H deposition and (right) subsequent firing, highlighting relatively large crystal grains and large dislocation clusters. Adapted from [32]	118
Figure 4.2	EBIC images of grain boundaries before and after hydrogen passivation for mc-Si wafers. Reproduced from [39] with the permission of John Wiley and Sons. © 2007 WILEY-VCH Verlag GmbH & Co. KGaA, Weinheim	121
Figure 4.3	Hydrogen desorption profiles obtained from TPD measurements for three different SiN_x:H layers. Reprinted from [50] Copyright (2009), with permission from Elsevier	123
Figure 4.4	PL images of mc-Si fired at different temperatures (unpublished) from the study of reference [54] presented at the 2017 IEEE Photovoltaic Specialists conference	125
Figure 4.5	Change in iV_{OC} (ΔiV_{OC}) induced by extended annealing at 700 °C after an initial firing process. Adapted from [57]	125
Figure 4.6	Calibrated PL images of cast-mono silicon wafers at various heights in the ingot highlighting the impact of gettering and hydrogenation. Reproduced from [20] under Creative Commons CC-BY-NC-ND. © 2019 The Authors. Published by Elsevier B.V. https://doi.org/10.1016/j.solmat.2019.110059	127
Figure 4.7	iV_{OC} maps of p-type HPM silicon (left) before and (right) after a belt furnace firing process for hydrogenation. Adapted from [42]	128
Figure 4.8	IQE (981 nm) from a LBIC measurement of (left) a finished screen-printed solar cell, and (right) after advanced hydrogenation. Adapted from [70]	129
Figure 4.9	PL images of a UMG Cz wafer with high impurity concentrations causing crystallographic defects (a) before and (b) after hydrogenation. © 2014 IEEE. Reprinted, with permission, from [86]	132

Figure 4.10 (left) iV_{OC} values of UMG Cz silicon wafer throughout an ingot directly after $SiON_x$:H deposition, after a CHP, and after a subsequent AHP, where x_n represents the fraction of solidified silicon. Image reproduced from [87]. (right) Reversibility of the bulk lifetime with different hydrogenation processes. Adapted from [87] 132

Figure 4.11 Lifetime curves of two n-type Cz wafers (W1 and W2) with low initial lifetimes and a high-lifetime n-type FZ wafer before and after a CHP at 400 °C, resulting in multi-millisecond lifetimes for the n-type Cz wafers. Reprinted from [13], Copyright (2015), with permission from Elsevier 133

Figure 4.12 (left) ECV measurement highlighting the passivation of boron dopants near the silicon surface after a shielded hydrogen passivation process. Reproduced from [102] with permission from John Wiley and Sons. © 2017 WILEY-VCH Verlag GmbH & Co. KGaA, Weinheim. (right) Change in sheet resistance with temperature induced by a hydrogen plasma for the formation of a selective emitter. © 2012 IEEE. Reprinted, with permission, from [103] 136

Figure 4.13 The recovered fraction N_R/N_0 for the dissociation of hydrogen-acceptor pairs with thermal annealing for 5 minutes. Reprinted with permission from [104]. Copyright (1988) by the American Physical Society. http://dx.doi.org/10.1103/PhysRevB.37.8313 136

Figure 4.14 Calibrated 1-sun implied V_{OC} mapping for p-type boron-doped silicon wafers exhibiting oxygen precipitates (a) directly after PECVD deposition of SiN_X:H, (b) after an AHP at 620 °C, (c) on a finished laser-doped selective emitter solar cell made with the same wafer. Adapted from [13] 139

Figure 4.15 Open circuit PL images of industrial n-PERT solar cells (left) before and (right) after the 20 s AHP treatment 141

Figure 4.16 IQE maps from spatially resolved LBIC measurements for mc-Si. Reprinted from [127], with the permission of AIP Publishing. © 2018 Author(s) 146

Figure 4.17 Calibrated implied V_{OC} mapping of pre-treated p-type mc-Si wafers (a) control, no gettering or hydrogenation, (b) G, pre-gettered only, (c) H, pre-hydrogenated only, (d) G + H both pre-gettered and pre-hydrogenated. All samples are measured with symmetrical SiN_X passivation. Reproduced from [128] under Creative Commons CC BY. © 2019 The Authors. Progress in Photovoltaics: Research and Applications published by John Wiley & Sons Ltd. https://doi.org/10.1002/pip.3230 147

Figure 4.18 Process flow for the fabrication of industrial screen-printed PERC solar cells and their relevance for defect engineering 148

Figure 4.19 Calibrated implied open-circuit voltage map of a p-type Cz wafer after an advanced hydrogen passivation process for the treatment of boron-oxygen defects and subsequent stability testing, where hydrogen was only injected into the top half of the wafer (as displayed in the figure), highlighting the effective use of the wafer as a storage for hydrogen to enable multi-stage hydrogenation processes. Reproduced with permission from [6], with the permission of John Wiley and Sons. © 2017 WILEY-VCH Verlag GmbH & Co. KGaA, Weinheim 152

Figure 4.20 (left) Diffusion profiles for an emitter suitable for screen-print (SP) or a lightly-doped emitter (LDE), as well as for both diffusions after thermal oxidation. (right) Corresponding impact on interstitial iron concentrations. Reproduced with permission from [6], with the permission of John Wiley and Sons. © 2017 WILEY-VCH Verlag GmbH & Co. KGaA, Weinheim 154

Figure 4.21 Schematic workflow illustrating the defect engineering processes and SHJ cell and lifetime precursor fabrication steps. Image reproduced from [128] under Creative Commons CC BY. © 2019 The Authors. Progress in Photovoltaics: Research and Applications published by John Wiley & Sons Ltd. https://doi.org/10.1002/pip.3230 159

Figure 4.22 Impact of pre-fabrication gettering (G) and hydrogenation (H) on (a) effective minority carrier lifetime extracted at a minority carrier density of $1 \times 10^{15}/cm^3$) and (b) 1-sun iV_{OC} on SHJ lifetime precursor structures fabricated using commercial grade p-type Cz silicon. (right) Corresponding external quantum efficiency (EQE) measurements on SHJ cells. Images reproduced from [154] under Creative Commons CC-BY-NC. © 2018 The Authors. Solar RRL Published by WILEY-VCH Verlag GmbH & Co. KGaA, Weinheim. https://doi.org/10.1002/solr.201700221 159

Figure 4.23 Calibrated iV_{OC} maps (left) and ΔiV_{OC} maps (right) of UMG mc-Si wafers passivated by SiN_x:H (a) without any pre-treatment, (b) with hydrogenation, (c) gettering, and (d) both gettering and hydrogenation. The ΔiV_{OC} maps on the right represent the absolute difference between (e) hydrogenation and the control, (f) gettering and the control, (g) gettering + hydrogenation and gettering only and (h) gettering + hydrogenation and control. Image reproduced from [159] under Creative Commons CC-BY-NC-ND. © 2019 The Authors. Published by WILEY-VCH Verlag GmbH & Co. KGaA, Weinheim. https://doi.org/10.1002/pssa.201900319 163

Figure 5.1 Efficiency of p-type Cz silicon Al-BSF and PERC solar cells (a) directly after cell fabrication and after a subsequent 48 h light soak with the activation of B-O-related LID (from [5]), and (b) relative performance degradation. Data from Münzer et al. [8] Cho et al. [9], Centrotherm [10], Fertig et al. [11] and Hallam et al. [12] 180

Figure 5.2 (a) Degradation before and during recovery (BR and DR, respectively). Stability testing using either (b) 1-sun illumination, or (c) 0.5 V forward voltage at room temperature on regenerated (R) or not regenerated (NR) samples. Adapted from [13] 181
Figure 5.3 Inverse effective minority carrier lifetime as a function of excess carrier concentration before and after degradation, and the corresponding normalised defect density and fit for SRH recombination 183
Figure 5.4 Correlation between the normalised defect density and (a) the substitutional boron (B_s)-doping concentration. Data from Glunz *et al.* [36] and Bothe et al. [22]. (b) Corresponding dependence on the O_i concentration. Data from Bothe et al. [22, 32]. Dashed lines represent the fit to the data. Figures adapted from [3] 184
Figure 5.5 Contour plot of the predicted minority carrier lifetime (based on [32] adding in an enhancement factor of $m = 2$) due to B-O defects as a function of O_i concentration and resistivity for p-type silicon solely doped with boron, also highlighting expected oxygen concentrations in Cz-grown silicon using data from LONGi [21] 186
Figure 5.6 Minority carrier lifetime, τ_{eff}, and NDD extracted at $\Delta n = 9.1 \times 10^{14}/cm^3$ as a function of light soaking duration at room temperature highlighting the two-stage degradation due to B-O defects with a fast degradation and slow degradation phase 186
Figure 5.7 (a) Impact of light intensity and (b) the equilibrium hole concentration on the degradation rate for B-O defect. For (b), open symbols are based on [B] and closed symbols are based on p_0. Data from Hashigami et al. [38] , Schmidt et al. [3], Palmer et al. [45], Glunz et al. [46], Lim et al. [47] and Bothe et al. [22] 187
Figure 5.8 Reversible B-O defect reactions with subsequent degradation following each dark-annealed lifetime recovery 189
Figure 5.9 Generalised model for the B-O defect system in Cz silicon. Reprinted from [57], with the permission of AIP Publishing. © 2018 Author(s) 190
Figure 5.10 A simplified four-state diagram to represent the system of B-O defects. Adapted from [20] 197
Figure 5.11 Impact of the peak firing temperature (T_{peak}) on the FDD of B-O defects, representing the reduction in NDD occurring through samples. Open symbols at high temperatures correspond to conditions that also activate other carrier-induced defects. Reprinted from [117], Copyright (2017), with permission from Elsevier 199
Figure 5.12 Original three-state model to describe the B-O defect system. Adapted from [13] 200

Figure 5.13 Loss of open-circuit voltage with light soaking as a function of dark annealing time after complete regeneration of cells. © 2006 IEEE. Reprinted, with permission, from [13] 202

Figure 5.14 Revised three-state diagram to represent the simplified system of B-O defects. Adapted from [20] 202

Figure 5.15 (a) Example transition from State B into State C for illuminated annealing at different temperatures. (b) Corresponding Arrhenius plot used to extract activation energy and reaction rate. Reproduced from [133], with permission from John Wiley and Sons. Copyright © 2007 John Wiley & Sons, Ltd 203

Figure 5.16 Temperature-dependent reaction rates for the B-O defect system assuming a degradation rate and defect annihilation from Bothe *et al.* ($v_{AB} = 10^{13}$/s for [B] = 1.5×10^{16}/cm^3) [22], and regeneration rate (2.7 suns) and destabilisation rate from Wilking et al. [125]. Figure adapted from [127] 208

Figure 5.17 Effectiveness of B-O LID mitigation with illuminated annealing at 2.7 suns as a function of temperature using data from Wilking et al. [125] and corresponding kinetic model of the expected effectiveness as presented in [126]. Corrected value of $v_{BC} = 1.33 \times 10^{10}$/s; t_{SS} shows the time to achieve steady state ($N_C = 99\%$) as a function of temperature with the same reaction rates 208

Figure 5.18 The Voc and efficiency loss without hydrogen/using a hydrogen-lean LPCVD SiN_X compared to using PECVD SiN_X (with hydrogen). Data from [8, 15] 210

Figure 5.19 (a) Evolution of NDD during illuminated annealing for samples coated with SiN_X:H layers with and without firing. Adapted from [149]. (b) Relative reduction in NDD after illuminated annealing (FDD) for samples with SiN_X:H and SiO_2 with and without firing, highlighting the critical role of a hydrogenated dielectric layer for regeneration. Adapted from [117] 211

Figure 5.20 (a) Impact of belt speed and the peak temperature (T_{peak}) and (b) cooling rate on the regeneration rate. (c) Hydrogen effusion loss of SiN_x:H layer (H_{loss}) as a function of T_{peak}. (d) The regeneration rate as a function of H_{loss}. (b) Data from [115] and [154]. Inset in (a) zooms into lower-temperature range. (a) and (c) Reprinted from [154], with the permission of AIP Publishing. © 2013 AIP Publishing LLC. (d) Adapted from [154] 212

Figure 5.21 Impact of SiN_x:H layer thickness on the regeneration rate showing a direct correlation between the hydrogen concentration at a depth of 80 nm as measured by NRRA. Adapted from [155] 213

Figure 5.22 (a) Impact of pre-annealing (dark annealing) temperature on the passivation rate of B-O defects during a subsequent illuminated anneal, and (b) close link between the passivation time constant and known boron-hydrogen (BH) forming kinetics. Reprinted from [125], Copyright (2014), with permission from Elsevier 216
Figure 5.23 Donor (E_d) and acceptor (E_a) energy levels of B-O defects [58] and interstitial hydrogen [166], in relation to the conduction band (E_c), valence band or mid-gap (E_m) 218
Figure 5.24 Illumination intensity as a function of regeneration rate with high- and low-bulk-lifetime samples. Adapted from [128] under Creative Commons CC BY. https://doi.org/10.1063/1.4978266 219
Figure 5.25 Correlation between the excess carrier concentration and the regeneration rate in high- and low-lifetime samples. Figure adapted from [128] under Creative Commons CC BY. https://doi.org/10.1063/1.4978266 220
Figure 5.26 (a) Simulated percentage of regeneration achieved for different regeneration temperatures for different defect formation rates (v_{AB}) through either high-intensity illumination or increasing the background-doping concentration. © 2016 IEEE. Reprinted, with permission, from [126]. (b) Corresponding impact on the time to achieve defect passivation as a function of temperature 222
Figure 5.27 Experimental results and simulated percentage of regeneration achieved for different regeneration temperatures for different defect formation rates through either high-intensity illumination or increasing the background-doping concentration. © 2016 IEEE. Reprinted, with permission, from [126] 223
Figure 5.28 Stability of lifetime after B-O passivation with light soaking at 160 °C 224
Figure 5.29 Production process for a typical commercial screen-printed PERC technology, noting steps that contribute to B-O defect passivation 225
Figure 5.30 (a) Thermal profile in a Centrotherm c.FIRE REG 9.600 tool with a peak set temperature of 815 °C for various illumination intensities and belt speeds. Figure adapted from [145]. (b) Corresponding impact on open-circuit voltage loss due to LID 226
Figure 5.31 (a) Stabilised efficiency of p-type Cz silicon Al-BSF and PERC solar cells with and without treatment to eliminate B-O-related LID. (b) Relative power (ΔP) or efficiency (Δη) change after stability test with treatment, ΔP, data from Münzer et al. [8], Cho et al. [9], Centrotherm [10], Fertig et al. [11] and Hallam et al. [12] 228
Figure 5.32 (a) In situ open-circuit voltage measured during 48 h light soaking for a p-type SHJ cell that is G gettered only (closed square) and gettered plus AHP-treated (open square), (b) effective lifetime extracted at both open circuit (OC) and

	MPP as a function of light soaking. Reproduced from [200], with permission from John Wiley and Sons. © 2020 WILEY-VCH Verlag GmbH & Co. KGaA, Weinheim	232
Figure 6.1	Low-energy sites (black dots) for hydrogen near C_S on bond centre (BC) and antibonding (AB) sites. C is represented in grey, and Si in white. Reproduced from Santos et al. [8] under Creative Commons CC BY. https://doi.org/10.1002/pssa.201700309 © 2017 The Authors	251
Figure 6.2	Low-energy sites (shown in black) for hydrogen near interstitial oxygen (O_i) (shown in grey). Reproduced from Santos et al. [8] under Creative Commons CC BY. © 2017 The Authors. https://doi.org/10.1002/pssa.201700309	253
Figure 6.3	(left) Ground-state structures for COH in the negative (a), neutral (b) and positive (c) charge states. Si, C, O and H atoms are shown in white, grey, light grey and black, respectively. Reproduced from Santos et al. [8] under Creative Commons CC BY. © 2017 The Authors. https://doi.org/10.1002/pssa.201700309; (right) Energy band diagram showing COH charge state energy levels located at E_C−0.12 eV, E_c−0.14 eV, E_V+0.38 eV and E_V+0.36 eV	253
Figure 6.4	(left) DLTS spectra measured from samples of a Cz n-type silicon wafer as it was received, after 20 min anneal at 650 °C and after 35 min at 550 °C. (right) MCTS spectra of samples from another C-rich Cz n-type silicon wafer after annealed at 700 °C, 500 °C and 550 °C for 30, 90 and 60 mins, respectively. The inset shows the 1104/cm absorption band measured from sister samples of the same wafer used for MCTS measurements after the same heat treatments. Both DLTS and MCTS spectra wre recorded after 1HF:7HNO_3 wet etch of the samples and with measurement parameters: rate window = 50/s; U_b= −2.0 V; t_p= 10 ms. Reproduced from Vaqueiro et al. [20]	254
Figure 6.5	Calculated neutral configurations of single-vacancy-hydrogen complexes, VH_n, with $n = 1, 2, 3, 4$ including the symmetry, internuclear distances and degrees of bonding. Reprinted with permission from [34], Copyright (1994) by the American Physical Society. https://doi.org/10.1103/PhysRevB.49.17040	257
Figure 6.6	LeTID testing of mc-Si lifetime test structures (left) τ_{eff} ($\Delta n = 1 \times 10^{15}$ cm^{-3}) of samples fired at 650 °C and 900 °C degraded under 1-sun at 75 °C. Figure adapted from Bredemeier et al. [65] under Creative Commons CC BY. https://doi.org/10.1063/1.4944839 © Author(s) 2016. (right) τ_{eff} ($\Delta n = 9.1 \times 10^{14}$ cm^{-3}) of samples after PECVD , after firing, and degraded under 44.8 kW/m^2 at 140 °C under 44.8; © [2016] IEEE. Reprinted, with permission, from [49]	260

Figure 6.7 LeTID in various wafer types: (top) Cz and mc-Si τ_{eff} (normalised to the value directly after firing) fired at various temperatures and degraded using 175 °C dark anneal. Reprinted from [43] Copyright (2017), with permission from Elsevier. (bottom) FZ τ_{eff} (at $\Delta n = 5 \times 10^{15}$ cm^{-3}) with various thermal treatments after deposition degraded at 75 °C and 1-sun illumination. Reprinted from [47], with the permission of AIP Publishing. © 2017 Author(s) 262

Figure 6.8 (top) *k*-value in the un-degraded and degraded states plotted against the trap level relative to the valence band, showing *k*-values of approximately 1.4 and 28.5, respectively, between 0.3 eV and 0.8 eV above E_V. Adapted from [78]. (bottom) *k*-value plotted against the trap level relative to the intrinsic energy level for LeTID in a p-type mc-Si wafer with lifetimes measured at temperature (T) from–25 °C to 75 °C with 25 °C increments. Dots are defect parameters from the literature. Adapted from [79] 265

Figure 6.9 A depiction of a proposed model for LeTID involving metallic impurities by Bredemeier et al. [77] and a more recent adaption that includes dissociation of a M_i-H complex as in [100]. Reproduced from [101] 266

Figure 6.10 (top) LeTID degradation extent (maximum NDD) plotted against the hydrogen fraction released from SiN_x:H layers on two sets of mc-Si wafers. © 2019 IEEE. Reprinted, with permission, from [96]. (bottom) NDD with degradation time for mc-Si samples with varied SiN_x:H thicknesses. © 2019 IEEE. Reprinted, with permission, from [116] 270

Figure 6.11 Maximum effective defect concentrations, N_{max}^* on mc-Si sister wafers, plotted against total the hydrogen content $[H]_{tot}$ in corresponding boron-doped FZ-Si wafers processed in parallel; both surfaces were coated with SiN_x:H films with different Si:N ratios before firing at 900 °C. © 2019 IEEE. Reprinted, with permission, from [100] 272

Figure 6.12 Relative degradation in maximum power output of Cz-Si and mc-Si undergoing LeTID and B-O LID stability testing. Reprinted from [73], Copyright (2017), with permission from the Authors 276

Figure 6.13 (left) Proposed three-state LeTID defect model based on observations in boron-doped FZ-Si [81]; (right) Effective lifetime during LeTID testing at 80 °C under approximately 1-sun illumination (a) on a fresh sample, (b) after LeTID testing and 5 min, 200 °C dark anneal, (c) after further LeTID testing and another 5 min, 200 °C dark anneal. Reproduced from [81] with permission from John Wiley and Sons. © 2016 WILEY-VCH Verlag GmbH &Co. KGaA, Weinheim 276

Figure 6.14 Effective lifetime measured during successive LeTID degradation and regeneration cycles at 135 °C under 25.7 kW/m^2 illumination (squares) and after 8 min 232 °C dark anneals (triangles) preceding each degradation cycle. Reprinted from [82] Copyright (2018) with permission from Elsevier 277

Figure 6.15 A four-state model for LeTID. This model was based on the three-state B-O LID model by Herguth et al. [145]. State R represents a reservoir of precursors that feed into state A. Adapted from [82] 278

Figure 6.16 (left) NDD of LeTID and B-O regeneration rate following different temperature dark anneals, with the B-H integrated absorption coefficient during dark anneal extracted from McQuaid et al. [146]. (right) A proposed LeTID defect model including a reservoir state R (H_2A molecules), two precursor states A1 (B–H pairs) and unspecified A, defect state B and a dark-annealed state of H_2C molecules, with hydrogen transitioning through different interstitial charge states. Reprinted from [143] with the permission of AIP Publishing. © 2018 Author(s) 278

Figure 6.17 (top) Effective lifetime (extracted at a minority carrier density of 1×10^{15} cm^{-3}) of a p-type mc-si wafer at various stages of the SHJ fabrication process. Stage I refers to the fired wafer with SiN_x surface passivation, stage II is after the intrinsic a-Si:H deposition and stages III–VI represent a simulated thermal treatment representing subsequent process in the SHJ fabrication sequence. Adapted from [120]. (bottom) Injection resolved lifetime of a single sample undergoing the same treatment 281

Figure 6.18 FF of industrial silicon solar cells before and after a post-processing belt furnace anneal for set temperatures ranging from 400 °C to 700 °C. Reproduced from [157] with permission of John Wiley and Sons. © 2017 WILEY-VCH Verlag GmbH & Co. KGaA, Weinheim 282

Figure 6.19 (top) TLMs after firing (before) and after post-fire annealing at 353 °C for 2 hours after subsequent application of a variety of bias conditions. (bottom) Resistance measurements during the application of various biases. FB and RB stand for FB and RB, respectively. Reprinted from [129] Copyright (2018) with permission from Elsevier 284

Figure 6.20 Change in area normalised series resistance as the applied bias is switched from +0.2 V to –0.4 V during annealing at 400 °C for (A) multi PERC and (B) mono PERC solar cells. The open symbols represent samples that were annealed at a fixed bias of either +0.2 V (circle) or –0.2 V (square). © 2018 IEEE. Reprinted, with permission, from [158] 285

Figure 6.21 A p-n junction formed on p-type Cz wafers using a 30-minute hydrogen plasma treatment at 450 °C. Reprinted from [160] Copyright (2000) with permission from Elsevier 286

Figure 6.22 (left) $[H2^{*}]^{D}n$ preferred when no lattice dilation, early stages. Transforms into (right) 2Si-H with H_2 molecules. Reprinted with permission from [174], Copyright 2001 by the American Physical Society. https://doi.org/10.1103/PhysRevLett.86.1773 288

Figure 6.23 {100} platelet formation and progression into H_2 gas bubbles. Reprinted from [179], with permission from AIP Publishing. https://doi.org/10.1063/1.1494844 289

Figure 7.1 Diffusivity of dopants in silicon of key relevance for silicon solar cells as extracted from figures 12.11 and 12.12 in [10]. Reprinted by permission from Springer Nature: Springer Nature, Thermochemical and Kinetic Databases for the Solar Cell Silicon Materials. Kai Tang et al. Copyright (2009) 308

Figure 7.2 Phosphorus concentration profiles from gas-phase dopant diffused c-Si as a function of $POCl_3$:N_2 ratios. (a) SIMS spectrum showing the total phosphorus concentration. (b) Electrically active phosphorus measured with the ECV method. Reprinted from [14], Copyright (2003), with permission from Elsevier 309

Figure 7.3 SIMS profiles of phosphorus- and boron-implanted silicon before and after laser annealing. Data from [24] and [25] 311

Figure 7.4 Illustration of a laser annealing post-implantation process and recrystallisation 312

Figure 7.5 Cross section of silicon device showing the adsorption of the dopant species after dopant gas exposure (GILD process). The dopant is incorporated into the silicon upon laser annealing 313

Figure 7.6 Illustration of LCP for junction formation on a silicon device 313

Figure 7.7 Solar cell manufacturing sequence including laser doping (top). Cross section illustration of a LIMPID process using $POCL_3$ emitter diffusion and laser doping from PSG layer (bottom) 314

Figure 7.8 Illustration of a LIMPID process using a SOD method for deposition 315

Figure 7.9 Illustration of a LIMPID process using ultrasonic spray deposition 315

Figure 7.10 ECV profiles of areas doped using a Q-switched frequency-doubled solid-state laser. The profile for four different regions, with varying sheet resistances, is shown. The sheet resistance was reduced by increasing the pulse energy density of the laser-doped region. Reproduced (digitised) from [58] under Creative Commons CC BY-NC-ND. Original image © 2011. Published by Elsevier Ltd. https://doi.org/10.1016/j.egypro.2011.06.129 318

Figure 7.11 Combined SEM/EBIC image *with* ×2000 magnification showing a laser-doped region. The bright EBIC signal shows the location of the p-n junction. Reprinted from (60), Copyright (2013), with permission from Elsevier 320

Figure 7.12 Simulated depth of the molten silicon as a function of the pulse energy density. The laser pulse duration in the model is varied from 10 ns to 200 ns. Reproduced with permission from [61] 321

Figure 7.13 Junction depth profiles measured using SIMS, for laser-doped lines formed using a Q-switched Nd:YVO_4 diode laser (l = 532 nm) with a repetition rate of 100 kHz and scanning speed of 10 mm/s. Three different diode currents were used, indicating the relationship between laser energy density and junction depth. The dashed line represents the background boron-doping level. Reprinted from [62], Copyright (2011), with permission from Elsevier 321

Figure 7.14 Box-like diffusion profile formed using gas-immersed laser doping. Reprinted from [48]. Copyright (1993), with permission from Elsevier 323

Figure 7.15 Comparison of SIMS doping profile with Gaussian fit for laser-doped regions formed using a 532 nm CW laser. Reproduced from [68] 324

Figure 7.16 Combined SEM/EBIC images of n-type laser doping performed at a laser speed of (a) 1 m/s, (b) 2 m/s, (c) 5 m/s and () 10 m/s. A 532 nm CW laser was used. The bright line in the image corresponds to the p-n junction. (right) SIMS-doping profiles of n-type laser doping performed at various scanning speeds. Reprinted from [69], Copyright (2013), with permission from Elsevier 325

Figure 7.17 Comparison of laser emission for CW and Q-CW. The CW laser output remains constant over time, as indicated by horizontal line. Q-CW lasers emit high-energy pulses over a short period of time, with a very high switching frequency. As the switching frequency is very fast, the average power emitted from a Q-CW laser can be comparable to that of a CW laser. This is illustrated schematically in the figure; the amount of power emitted by the Q-CW shown in the dark grey regions is equal to the continuous output of the CW laser at lower power, shown by the cross hatch. The output of the laser is drawn as a square pulse for simplicity 326

Figure 7.18 Sheet resistance caused by dopant distribution during laser doping, plotted as a function of laser-processing speed reprinted from [71], Copyright (2015), with permission from Elsevier 327

Figure 7.19 Absorption coefficient of silicon at different temperatures. Data from [73, 77] 328

Figure 7.20 Comparison of the beam profile of a Nd:YAG laser. The intensity map and 2D intensity distribution on the left side shows the original Gaussian beam profile, the profile on the right is after

	the top hat beam-shaping optics, illustrating the more uniform distribution. Reproduced from [79] with permission. © (2016) COPYRIGHT Society of Photo-Optical Instrumentation Engineers (SPIE)	330
Figure 8.1	(top) PERL solar cell architecture using front-side selective emitter and rear locally diffused contacts. Reprinted from [6]. Copyright (1996), with permission from Elsevier. (bottom) Screen-printed Al-BSF solar cell with a front heavily doped homogeneous emitter	338
Figure 8.2	Improvement in the IQE at short wavelengths of a silicon solar cell with a selective emitter compared to a cell incorporating a homogeneous emitter. Reproduced from [9] with permission from John Wiley and Sons. Copyright © 2010 John Wiley & Sons, Ltd	341
Figure 8.3	(a) Modelled n-type emitter dopant profiles on a 1.5 Ω·cm p-type wafer. The doping profiles were created using a ERFC with different depths. (b) Simulated total extracted photo-generated current (filled symbols) for monochromatic 400 nm incident light with an illumination intensity corresponding to 40 mA/cm^2 short-circuit current generation under 100% EQE. Also shown are the corresponding Auger recombination current density (hollow symbols). Simulations were performed using EDNA2 [10]	341
Figure 8.4	(left) Impact of surface dopant concentration on contact resistivity of phosphorous-doped profiles and evaporated aluminium. The diffused profiles were created using a high-temperature quartz furnace tube with $POCl_3$ as the dopant source. The temperatures indicate the temperature at which the sintering was performed for 30 minutes in forming gas. © 2015 IEEE. Reprinted, with permission, from [11]. (right) Example transmission line measurement (TLM) data from a screen-printed Ag finger on an n-type LDSE region for PERC	342
Figure 8.5	The calculated total emitter recombination current density, J_{0e}, as a function of minority carrier SRV for different phosphorous dopant density profiles provided in panel (a) of Figure 8.3. Simulations performed with EDNA2 [10]	343
Figure 8.6	Contour map illustrating the total power loss due to combined effects of emitter resistive losses and finger shading as a function of emitter sheet resistance and spacing between the fingers. The modelling was performed using Sentaurus TCAD assuming a maximum power point voltage (V_{MP}) and current (J_{MP}) of 583 mV and 38.56 mA/cm^2, respectively	345
Figure 8.7	Processing sequence for the UNSW PERL cell. Adapted from [19]	346
Figure 8.8	Schematic of the UNSW BCSC architecture. Image reproduced from [15] with permission. © 1993 by John Wiley & Sons, Ltd	347

Figure 8.9 (a) Schematic of an industrial LDSE solar cell with aligned screen-printed contacts and (b) the corresponding process sequence. (c) SEM image, (d) EBIC image, (e) combined SEM/EBIC image at low magnification and (f) combined SEM/EBIC image at high magnification in the vicinity of a Ag screen-printed laser-doped line. (g) Histogram of the current collected from the undamaged region and the regions damaged due to laser doping. (c–g) Reprinted from [40], Copyright (2020), with permission from the International Solar Energy Society 352

Figure 8.10 (left) Efficiency distribution for 40 228 cells from an industrial LDSE PERC production line with aligned screen-printed contacts from LONGi, treated with advanced hydrogenation. (right) Certified *J–V* and power–voltage curves for high-efficiency LDSE PERC champion cell. Figures reproduced from [2] under Creative Commons CC BY. © 2020 The Authors. Progress in Photovoltaics: Research and Applications published by John Wiley & Sons Ltd. https://doi.org/10.1002/pip.3243 353

Figure 8.11 (top left) Schematic of an LDSE Al-BSF solar cell with self-aligned Ni/Cu/Ag-plated contacts. Adapted from [46]. (bottom left) Contact focused ion beam image on a laser-doped finger with high aspect ratio self-aligned Ni/Cu/Ag light-induced plating. The image has been stretched in the vertical direction to correct for the angle. Reproduced from [48]. (right) Corresponding process flow diagram for an LDSE Al-BSF solar cell with self-aligned plated contacts 355

Figure 8.12 Cross-sectional SEM images of the laser-doped region for samples processed with a 532 nm CW laser and 13 W optical power to dope through H_3PO_4 SOD source using (a) 0.5 m/s, (b), 2 m/s and (c) 5 m/s laser scan speed. The images illustrate the voids created in the laser-doped region with clearly larger ones formed at low laser speeds. (d) Top SEM image of a crack formed along crystal planes after laser doping at 0.01 m/s. Images adapted from [42, 48] 357

Figure 8.13 (Top) TEM results of silicon with a (100)-oriented surface, irradiated with a laser beam focused to a line with a pulse energy density of 2.5 J/cm^2, 28 ns pulse length, 8 µm line width at full width at half maximum (FWHM). (A) Two surfaces are joined, allowing an analysis of two cross sections perpendicular to these surfaces at a time. The darker areas are bending contours evolving due to the furling of the thinned silicon. (B) Diffraction pattern. The positions of the maxima correspond to the maxima of a (100)-oriented crystal. The silicon is epitaxially regrown with no defects/dislocations. (Bottom) Corresponding TEM results of silicon with a (111)-oriented surface, irradiated with a pulse energy density of 3.7 J/cm^2, 62 ns pulse length, 15.5 µm line width at FWHM. (C) TEM image shows a high defect concentration in some areas. (D) Diffraction image indicating a (111)-oriented crystal. The silicon is epitaxially regrown. © 2011 IEEE. Reprinted, with permission, from [71] 359

Figure 8.14 (a) Injection-dependent inverse lifetime curves of a symmetrical p-type lifetime test structure with n-type emitter passivated with hydrogenated SiN_x in the as-deposited start (square symbols), after firing (triangle symbols) and after laser doping (circle symbols). The inset uses a linear scale to demonstrate the impact of each step on high-injection level slope associated with J_{0e}. (b) Schematic of possible recombination pathways in the laser-doped region and its vicinity 360

Figure 8.15 (a) Optical surface image of a severe laser damaged area, using high-fluence 248 nm KrF pulsed laser (25 ns pulse duration), with the three components of the laser-doped region labelled. (b) PL spectra of the doped region, excited with a 532 nm CW laser, measured at different temperatures. (c) PL spectra detected from the three regions identified in panel (a). The impact of excitation beam energy on the dislocation-induced PL signal is also illustrated (dashed line vs dash-dotted line). The emitted PL signal from the locally excited edge regions is very strong compared to the signal from the doped region only for 532 nm excitation wavelength and, therefore, its spectra are normalised. (a–c) Reprinted from [65], with the permission of AIP Publishing. (d) Schematic of the shallow and deep states with respect to the band-edges, which contribute to the PL emission from laser-created dislocations 363

Figure 8.16 Impact of laser-doped line width on the solar cell FF and V_{OC}. Laser-doped line width is considered here to be equal to the metal finger line width 365

Figure 8.17 Schematic of (a) wide laser-doped line with aligned screen-printed fire-through finger in the middle (desirable) depicting the current implementation for industrial LDSE PERC, and (b) misaligned screen-print finger touching the undoped silicon (undesirable) 367

Figure 8.18 Example cross-sectional SEM image, corresponding EBIC image and combined SEM/EBIC image of a typical laser-doped groove silicon wafer. Reproduced from [85] Copyright (2017) with permission of Elsevier 368

Figure 8.19 The impact of laser conditions in the presence (square data points) and absence (circle data points) of a thin silicon oxide layer underneath the silicon nitride (SiO_2/SiN, denoted "SON") on implied V_{OC}. Implied V_{OC} of laser doping performed on bare wafers is shown with triangle data points, with the least drop. Reproduced from [68]. Copyright © 2010 John Wiley & Sons, Ltd 370

Figure 8.20 Schematic illustrations of possible scenarios after laser doping for selective emitter formation. Panel (a) depicts a scenario in which implementation of optimised dielectric layers' types and thicknesses allow LDSE technology with preferred coverage of the laser-induced edges coverage (i.e. laser-doped region extends laterally under the dielectric opening) to reduce its detrimental effect. Panel (b) the exposed, un-doped and un-passivated regions indicated in the 'undesirable' case are prone to high recombination rates and possible shunting upon metallisation. Adapted from [63] 371

Figure 8.21 (a) Combined SEM/EBIC images at ×2000 magnification of an n-type laser-doped line processed at 0.2 m/s on a textured surface using a CW laser. Adapted from [91]. (b) Combined SEM/EBIC image of a laser-doped region processed with a CW laser highlighting the loss of pyramids in the laser-doped region, collected at 5 kV accelerating voltage. Adapted from [92] 372

Figure 8.22 The impact of furnace firing at 800 °C and long-pulsed laser annealing on (a) the increase in the band-to-band PL signal after laser processing a bare p-type FZ silicon wafer and (b) decrease in the dislocation-induced PL signal. Reprinted from [95], with the permission of AIP Publishing. © 2018 Author(s) 373

Figure 8.23 Injection-level-dependent effective minority carrier lifetime curves for p-type Cz wafers, 169 ± 1 μm thick, with a 102 ± 3 Ω/ϒ n-type emitter and laser doping at 4 m/s with a 532 nm wavelength CW laser with an optical power of 14 W. A fast-firing process in a belt furnace was performed at 680 °C before laser doping. Laser-induced defect passivation was performed through a second firing step under identical conditions as the first firing step 375

Figure 8.24 Injection-level-dependent effective minority carrier lifetime curves for p-type Cz wafers, 167±2 μm thick, with a 102±3 Ω/ϒ n-type emitter and laser doping at 4 m/s with a 532 nm wavelength CW laser at an optical power of 14 W. Firing for a duration of several seconds in a firing furnace at 680 °C was performed before laser doping to isolate the effects of the passivation of laser-induced defects from that of other bulk/surface defects. Laser-induced defect passivation was performed at approximately 400 °C for approximately 4 minutes in the same belt furnace. No SiO_2 passivation was used for the lifetime test structure 377

Figure 8.25 Implied V_{OC} as a function of laser scan speed for p-type laser doping from AlO_x layers directly after laser doping, and after a subsequent hydrogenation process for the passivation of laser-induced defects (annealed). ©2014 IEEE. Reprinted, with permission, from [94] 379

Figure 8.26 Implied V_{OC} images of a laser-doped Al-BSF cell (a) after firing for Al-BSF formation, (b) after laser doping and (c) after laser-induced defect passivation by low-temperature BFA. ©2014 IEEE. Reprinted, with permission, from [94] 380
Figure 8.27 Injection-level-dependent effective minority carrier lifetime curves for p-type Cz wafers, 170 ± 2 µm thick, with a 102 ± 3 Ω/□ n-type emitter and laser doping at 4 m/s with a 532 nm wavelength CW laser at an optical power of 14 W. A firing for a duration of several seconds in a firing furnace at 680 °C (filled triangles) was performed before laser doping (filled circles). Laser-induced defect passivation was performed through a rapid 10-second laser-based AHP (filled diamonds). To test for LID (open circles), an accelerated degradation test was performed using the same commercial laser system (DR laser) as for AHP with the exception that the temperature of the sample stage was held at 150 °C during the 10 seconds laser illumination 383
Figure 8.28 (A) NDD values, reported at minority carrier density of $10^{15}/cm^3$ of the laser-doped samples calculated for the passivated state with respect to the initial fired state for various CW laser speeds. The values reported at 0 m/s are from the non-laser-doped control samples. (B) NDD values for the control and laser-doped samples calculated for the LID state with respect to the passivated state. The LID test was performed using the same conditions as in Figure 8.27 385
Figure 9.1 (a) Solid solubilities of Al and B in crystalline Si as a function of temperature under thermal equilibrium conditions from Rauer et al. [24] and Tang et al. [25]. (b) Electrochemical capacitance-voltage (ECV) measurements of the doping profiles formed by the Al-Si alloying process (at 850 °C) with standard Al paste and Al paste with B additive from Lölgen et al. [23]. Sheet resistance and J_0 values simulated using EDNA2 [25] using assumed values of S_{n0}, S_{p0} = 1 × 10^7 cm/s, τ_{n0}, τ_{p0} = 100 µs 398
Figure 9.2 SEM cross-sectional images of (a) an Al/Si contact with LBSF; (b) a Kirkendall void with no LBSF; (c) an Al/Si contact with negligible LBSF; and (d) an Al/Si contact with LBSF protecting against a void. Reproduced from reference [29] 399
Figure 9.3 (a) Boron ECV profiles for laser-doped regions formed using a boron SOD source and a 355 nm Q-switched laser with a scan speed of 0.5 m/s and power of 5 W. Adapted from Li et al. [7]. (b) Simulated value of J_0 as a function of JD (z_f) and peak concentration (N_{peak}) with a complementary error function profile using EDNA2 [25] 400
Figure 9.4 SEM cross-sectional images of (a) a boron laser-doped LBSF region and (b) an ideal Al/Si contact to a boron laser-doped LBSF where the eutectic region is protected by a thick LBSF.

	Reprinted from [8], Copyright (2014), with permission from Elsevier	402
Figure 9.5	(a) Focused ion beam (FIB) image of the surface of a laser-doped point contact on a textured silicon surface using a 532 nm CW laser; (b) combined SEM/electron beam-induced current (EBIC) image of the cross section of the laser-doped point contact. Reprinted from [36], Copyright (2013), with permission from Elsevier	403
Figure 9.6	(a) SIMS profiles of aluminium dopants for laser-doped regions formed with a 1064 nm Nd:YAG laser for various laser powers. Data reproduced from [43] along with estimated electronically active dopant concentrations based on sheet resistance measurements and EDNA2 simulations [26]. (b) The measured four-point-probe sheet resistance of the profiles in (a) reproduced from [43] (closed squares) as well as simulated sheet resistance if all dopants were electrically active as performed using EDNA2 (open triangles) [26]	404
Figure 9.7	(a) SIMS profiles of Al concentration for laser-doped regions formed through a dielectric stack of 10 nm ALD Al_2O_3 capped with 120 nm PECVD SiN_X using a 355 nm Q-CW laser with speeds of 8 m/s and 1 m/s. Reprinted from [12], Copyright (2015), with permission from Elsevier. (b) Sheet resistance of laser-doped regions as a function of laser speed for the same laser as (a) when performed through the 10 nm ALD Al_2O_3 layer with and without a capping layer of 120 nm PECVD SiN_X:H	405
Figure 9.8	Combined SEM/EBIC images of Al p^{++} laser doping on an n-type wafer using PECVD AlO_X as a dopant source. The p-n junction appears as a bright line, revealing a deep laser-doped region directly beneath the opening in the AlO_X layer, as well as a shallow-doped region extending laterally beneath the AlO_X layer. Image reproduced from [34]	406
Figure 9.9	Schematic of the SFSC structure with heavily doped channels running perpendicularly to the front metal fingers. The metal fingers only contact the Si at the intersections with the heavily doped channels and are insulated from the silicon by the front dielectric layer elsewhere	407
Figure 9.10	Processing sequence of (left) a standard conventional screen-printed homogeneously diffused solar cell, (middle) a standard SFSC, and (right) a laser-doped SFSC. The additional processing steps required for producing the semiconductor fingers compared to conventional SP technology are italicised. Figure adapted from [50]	409
Figure 9.11	Schematic of a typical IBC solar cell, where both polarities of contacts are on the rear of the device. Figure adapted from [56]	413

Figure 9.12 Processing sequence of an IBBC solar cell. Figure adapted from [70] 415

Figure 9.13 Process flow for the University of Stuttgart's laser-doped IBC cell adapted [56] 417

Figure 9.14 Schematic of the process for forming contacts through an emitter of opposite polarity using laser doping by (a) coating the silicon wafer with a SOD source, (b) illuminating the surface with a laser, ablating the dielectric layer and melting the silicon surface, (c) continuing the laser illumination to expand the molten region and incorporate dopants and (d) the finished structure. Figure adapted from [34, 78] 419

Figure 9.15 Combined SEM/EBIC image of a cross section of a p-type laser-doped line formed through a 120 Ω/□ n-type emitter, making contact to the underlying p-type wafer. The bright EBIC signal indicates the p-n junction near the surface, with no p-n junction visible underneath the laser-doped region indicating full compensation of the n-type dopants in the laser-doped region. Image reproduced from [34] 420

Figure 9.16 EBIC image of a cross section of a p-type laser-doped line formed through a 100 Ω/□ n-type emitter. The n-type emitter was not fully compensated by the p-type dopants, leaving behind a semi-circular p-n-p transistor structure indicated by the label 'PNP'. Image adapted from [80] 421

Figure 9.17 Schematic of simplified laser-doped IBC cell. Reprinted from [77], Copyright (2012), with permission from Elsevier 421

Figure 9.18 Effective lifetime as a function of excess carrier density as measured by QSS-PL of a symmetrical phosphorus diffused p-type lifetime test structure after an increasing amount of damage is introduced outside the measurement test area. © 2014 IEEE. Reprinted, with permission, from [87] 422

Figure 9.19 Schematic of edge isolation of a solar cell using laser doping. Reprinted from [91], Copyright (2015), with permission from Elsevier 423

Figure 9.20 Calibrated PL images showing a spatial map of iV_{OC} of (left) a solar cell with laser cleaved edges and (right) a solar cell within a larger substrate that has been isolated using laser doping. The edge of the host wafer (indicated by the dotted line) is not visible due to the edge shunt. Reprinted from [91], Copyright (2015), with permission from Elsevier 424

Figure 9.21 Plot of τ_{bulk} vs. various laser doping or melting conditions with or without a thermal diffusion ($POCl_3$). Image reproduced from [99] 425

Figure 9.22 τ_{bulk} comparison for Cz-Si and mc-Si wafers with and without rear side laser doping, undergoing various thermal annealing or $POCl_3$ diffusion processes. Reprinted by permission from Springer

Nature Customer Service Centre GmbH: Springer Nature [98], © (2016). https://www.springer.com/journal/11708 426

Figure 9.23 [Fe_i] measured on Cz-Si and mc-Si wafers with and without rear side laser damage either undergoing phosphorus diffusion processing, an anneal in N2 ambient, or not receiving thermal treatment. Reprinted by permission from Springer Nature Customer Service Centre GmbH: Springer Nature [98], © (2016). https://www.springer.com/journal/11708 427

Figure 9.24 iV_{OC} calibrated photoluminescence image (left) and[Fe_i] map (right) of mc-Si wafers (a–b) without and (c–d) with a LEG process. Reprinted by permission from Springer Nature Customer Service Centre GmbH: Springer Nature [98], © (2016). https://www.springer.com/journal/11708 428

Figure 10.1 Sentaurus simulations highlighting the gain in efficiency that can be achieved through (left) a reduction in laser-doped line width for screen-printed solar cells with 40 μm or 20 μm wide fingers. (right) Impact of contact resistance and coverage area on device efficiency 446

Figure 10.2 Normalised effective lifetime (a) and apparent J_0 for n-type Cz silicon subjected to annealing at 160 °C under a variety of illumination intensities. Figure reproduced from [17] 447

Figure 10.3 Normalised PL intensity during light soaking at 75 °C, 1-sun for (a) boron-doped and (b) gallium-doped PERC solar cells. A current-injection stabilisation process was applied to the boron-doped cell, but not to the gallium-doped cell. Reproduced from [23] under Creative Commons CC-BY. © 2019 The Authors. Published by Elsevier B.V. https://doi.org/10.1016/j.solmat.2019.110299 448

Figure 10.4 Schematic diagram displaying the device architecture for (left) SHJ and (right) TOPCon solar cells 449

Figure 10.5 (a) Schematic diagram displaying the selective emitter TOPCon device. The inset shows (b) optical microscope image of the front selective emitter metallisation and (c) TEM image of the rear passivating contact. Reprinted from [29], with the permission of AIP Publishing. © 2017 Author(s) 451

Figure 10.6 Evolution of effective lifetime (a) and J_0 (b) for symmetrical TOPCon lifetime test structures with symmetrical tunnelling oxide/n-doped polysilicon passivation, capped with SiN_x. The light soaking was performed at 140 °C under 1-sun illumination. The structure of the samples is shown in the inset of (a). The changes in lifetime for a sample passivated with a superacid passivation scheme is also shown. Reprinted from [35], Copyright (2012) with permission from Elsevier 453

Figure 10.7 (a) Influence of post-cell treatment on the efficiency of n-type solar cells (b) current–voltage characteristic curves before and after the treatment. The inset shows the reduction in series resistance (c) injection-dependent R_s. Reprinted from [37], with the permission of AIP Publishing. © 2019 Author(s) 454

Figure 10.8 Technology road map for various industrial silicon solar cell technologies. Reprinted from [41], with the permission of AIP Publishing. © 2020 Author(s) 455

List of tables

Table 4.1	I–V properties of an Al-BSF solar cell fabricated on p-type cast-mono material before and after an advanced hydrogenation process (AHP)	130
Table 4.2	Influence of hydrogenation process on the current–voltage characteristics of industrial n-PERT silicon solar cells	141
Table 4.3	Current–voltage (I–V) results for SHJ solar cells highlighting the impact of defect engineering processes	161
Table 5.1	Basic reaction rates in the B-O defect system as reported in the literature	205
Table 5.2	Impact of the composition of PECVD SiN_X:H on the passivation rate of B-O defects. Data extracted from [149]	214
Table 5.3	Impact of dielectric stack composition for ALD AL_2O_3 and PECVD SiN_X:H on the passivation rate of B-O defects suggesting the blocking of hydrogen diffusion for sufficiently thick Al_2O_3 layers. Data extracted from [149]	215
Table 5.4	J–V characteristics of the defect-engineered p-type SHJ solar cells with and without an AHP for passivation of B-O defects before and after LID (degraded). Data from [200]	231
Table 6.1	Reported relative power and efficiency losses on p-type cells and modules	259
Table 6.2	LeTID capture cross section ratios (k-values) and apparent defect energy levels in various silicon materials as reported in the literature in chronological order	263
Table 6.3	Summary of various suppression or mitigation strategies at a wafer or cell level as highlighted throughout the literature	274
Table 8.1	Effect of front homogeneous emitter doping on solar cell performance: desirable =√ undesirable =×	340
Table 8.2	Baseline parameters used as an input to the software Sentaurus TCAD to simulate the impact of laser-doped line width/finger width on LDSE PERC solar cells as presented in Figure 8.16	366
Table 8.3	Lifetime parameters for the lifetime curves measured on the samples that went through similar process sequences as those in Figure 8.23. The iV_{OC} is extracted at 1-sun and the τ_{eff} value is extracted at a minority carrier density $10^{15}/cm^3$. J_{0e} is the total emitter recombination current density of both surfaces	376

Table 8.4	Lifetime parameters for the lifetime curves measured on the samples that went through similar process sequences as those in Figure 8.24. The iV_{OC} is extracted at 1-sun and the τ_{eff} value is extracted at a minority carrier density 10^{15}/cm^3. J_{0e} is the total emitter recombination current density of both surfaces	378
Table 8.5	Impact of hydrogenation processes on the passivation of laser-induced defects for LDSE solar cells. P-type Al-BSF and n-type PERT data from [94]. P-type PERC data from [57]	381
Table 8.6	Lifetime parameters for the lifetime curves measured on the samples that went through similar process sequences as those in Figure 8.27. The iV_{OC} is extracted at 1-sun and the τ_{eff} value is extracted at a minority carrier density 10^{15}/cm^3. J_{0e} is the total emitter recombination current density of both surfaces	384
Table 9.1	Efficiencies of various PERL cells utilising laser-doped rear contacts	397

About the editors

Brett Hallam is an associate professor and the research director for Advanced Hydrogenation at the University of New South Wales, Sydney, Australia, working on a large government-funded project that is also supported by 15 key solar cell manufacturers. He has transferred patented hydrogen passivation and laser-doping processes to several companies and research institutes. He received the 2020 IEEE PVSC Stuart Wenham Young Professional Award for his contributions to the photovoltaic industry.

Catherine Chan is a senior researcher with the Australian Centre for Advanced Photovoltaics at the University of New South Wales, Sydney, Australia. She completed her PhD on laser doping and her research currently focuses on addressing the negative impacts of hydrogen with light- and elevated-temperature-induced degradation (LeTID) in PERC cells. She has demonstrated rapid approaches to eliminate such degradation and her work has led to the development of industrial tools to mitigate LeTID.

Chapter 1
Industrial silicon solar cells

Matthew Wright[1], Brett Hallam[1], Catherine Chan[1], and Martin Green[1]

Climate change is the biggest threat facing humanity. In order to curtail the emission of greenhouse gases into the atmosphere to address climate change, it is essential to rapidly shift global energy systems away from a reliance on fossil fuels. The abundance of the solar resource and the recent reductions in the cost of solar panels makes solar photovoltaics (PV) an ideal candidate for this purpose. The performance of industrial silicon solar cells has improved significantly over the past ten years, with multiple manufacturers mass-producing solar cells with efficiencies exceeding 23%. This is related to both incremental improvements of the industrial processes on the production line and improved device architectures. Passivated emitter and rear cell (PERC) solar cells are currently dominating the market of commercial silicon solar cells because of the high-efficiency design compared with conventional cell structures. Two key factors will help further drive this technology's success: the passivation of bulk defects in silicon using hydrogen to improve the cell reliability and the use of lasers to effectively produce a 'selective emitter', which is crucial to help improve the cell efficiency. Suntech used these technologies to achieve the world's first commercial p-type solar cell with an efficiency above 20% [1] and the same technologies were used by LONGi Solar to recently reach over 23.8% efficiency [2]. The purpose of this textbook is to provide a comprehensive overview of the current state of the art for both hydrogen passivation and laser doping of silicon solar cells.

1.1 Climate change and the use of fossil fuels

The effects of climate change on our world are already being observed. The average global temperature has increased above pre-industrial levels by approximately 1.1 °C [3, 4]. Figure 1.1 shows the change in average global temperature in the range spanning from 1850 to 2019. The values are based on the HadCRUT4 land sea data set [3].

[1]The School of Photovoltaic and Renewable Energy Engineering, The University of New South Wales, Australia

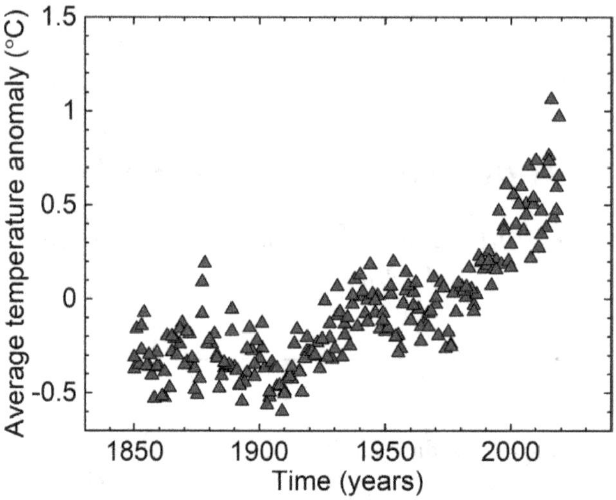

Figure 1.1 *Global average land-sea temperature anomaly, plotted over the time period from 1850 to 2019. The values are plotted relative to the average temperature during the time spanning from 1961 to 1990. The data has been replotted, original data from [3]*

The 2015 Paris Agreement called for an international effort to limit this global mean temperature rise to 1.5 °C [5]. As outlined in the IPCC 5th Assessment Report, failure to do so will lead to significant ecological and health impacts, including [6]

- an increased number of extreme weather events, such as floods, droughts and severe storms
- rising sea levels and ocean acidification
- altered crop growth.

This increase in global temperature is related to increases in the concentration of atmospheric carbon dioxide (CO_2), one of the primary greenhouse gases. For the first time in 800 000 years, the atmospheric concentration of CO_2 is now above 400 parts per million (ppm) [7, 8]. The rising concentration of greenhouse gases, as a result of human activity, is by far the biggest cause for the observed increase in global temperature [9]. Figure 1.2 displays the historical CO_2 emissions between the years 1751 and 2017. These emissions have risen steeply since the beginning of the industrial revolution in approximately 1850, with the data indicating that the global emissions in 2017 were 37 Gt/year [11]. At the beginning of the industrial revolution, the emissions can be attributed primarily to Europe and the USA; however, over the past 30 years, emissions have shifted to Asia, with China contributing the highest value in 2017.

The amount of CO_2 emissions in a country is related to economic prosperity, and there is currently a large inequity in global emissions. The richest countries

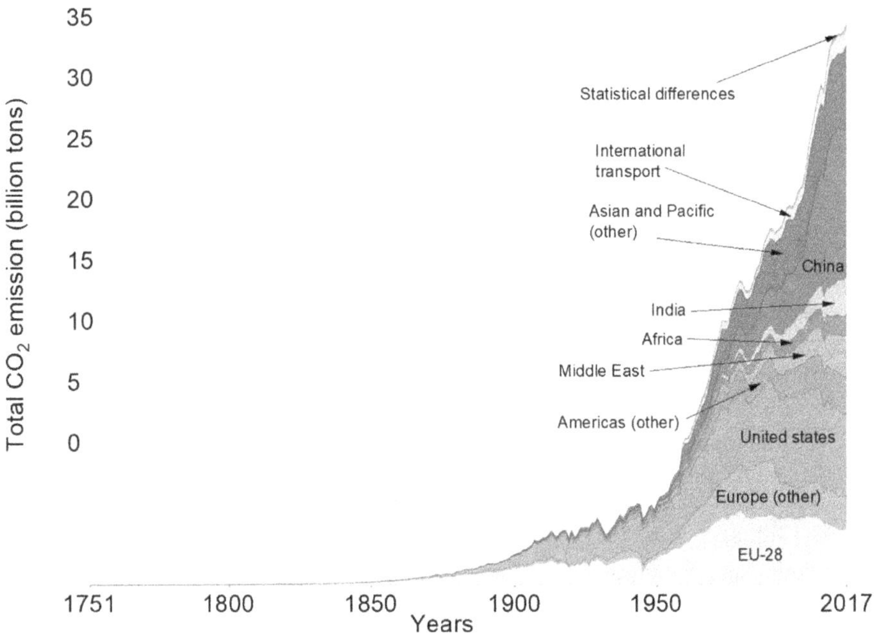

Figure 1.2 *Historical global CO_2 emissions shown by region. Data is shown in the range from 1751 to 2017. The original graph was produced by Our World in Data from [4] and is reproduced according to the Creative Commons CC BY 4.0 license. Original data from [7] and [10]*

in the world, which account for 50% of the human population, are responsible for 86% of CO_2 emissions. On the other hand, the poorest 9% of the population are responsible for only 0.5% of global emissions [4]. This link between emissions and economic growth causes contention in the pursuit of an international approach to overcome the problem.

Over the past decade, about 45% of global CO_2 emissions can be attributed to the energy sector, with electricity and heat being the primary causes [12]. In order to overcome this rapidly rising increase in CO_2 emissions, priority must be given to policies that both reduce the use of fossil fuel generation and promote the rapid uptake of renewable energy [12].

To understand the action required to limit warming to 1.5 °C, Rogelj *et al.* provided a comprehensive comparison of various models [13]. In their analysis, they use the metric of 'radiative forcing', which describes the balance between energy from the Sun that is either: (i) absorbed in the atmosphere; or (ii) radiated back to space. In order to reach the target of 1.5 °C by the end of the century, the radiative forcing must be as low as 1.9 W/m^2. The authors compare the viability of five different scenarios [13] that use different approaches to supplying the world's energy, which are:

1. SSP1. development under a green growth paradigm
2. SSP2. a middle of the road development along historical patterns
3. SSP3. a regionally heterogeneous development
4. SSP4. a development that results in both geographical and social inequalities
5. SSP5. a development path dominated by high energy demand supplied by extensive fossil fuel use.

The results showed that multiple pathways to limiting warming to 1.5 °C are possible; however, the risk of failure is much higher for scenarios with high growth, high levels of inequality or high demand, represented by SSP3, SSP4 and SSP5, respectively. Common amongst the successful scenarios are various stringent requirements, including:

1. The emission of CO_2 (and other greenhouse gases) peaks before 2030 and declines rapidly in the following decades
2. Rapid upscaling of carbon capture and storage as well as CO_2 removal technologies
3. Significant limiting of energy demand growth
4. Rapid upscaling of renewable energy, in particular solar PV and wind.

The success of each pathway is heavily dependent on the level of global CO_2 emissions by 2030. The probability of achieving the target of 1.5 °C is halved if the emissions are high (>45 Gt/year) rather than low (<30 Gt/year). Amongst the requirements, it is critical that the amount of energy produced by renewable sources is rapidly increased. While there are multiple approaches to do this, experts suggest that solar PV is the forerunner technology to meet this demand [14].

1.2 Solar photovoltaics for future energy generation

Although solar PV is not the only solution to all energy-related problems, it is a vital part of a transition to a sustainable future. Solar energy is the most abundant energy resource on earth. Every moment, 170 000 TW of solar power strikes the earth; the amount of energy striking the earth's surface every hour is more than enough to meet the current annual global energy demand [15]. The abundance of this energy far exceeds that of traditional fossil fuel reserves and all other forms of renewable energy. Figure 1.3 shows a map of the world, which outlines the strength of the solar resource. Shown on the right-hand side is a map of Australia. The shaded square area represents the area of solar panels required to meet the global energy demand. This area assumes there is no contribution from rooftop solar. This indicates that solar PV has vast potential to provide renewable electricity on a terawatt (TW) scale, as required to mitigate the impact of increased CO_2 emissions.

Over the past few years, we have begun to see a rapid shift towards solar energy as the choice for new electricity installations. The year 2016 was a landmark for renewable energy; for the first time in history, solar PV was the most installed

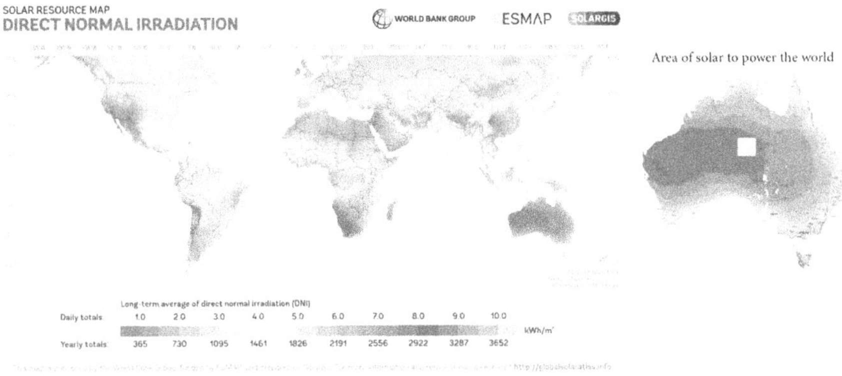

Figure 1.3 *World map indicating the direct normal irradiation (left). The area of Australia required to power the world using solar exclusively (right). Reproduced from [16]. © 2019 The World Bank, Source: Global Solar Atlas 2.0, Solar resource data: Solargis. Creative Commons CC BY 4.0*

electricity-generation technology, exceeding that of coal by 20 GW_p [17]. This trend continued in the following years, with the difference between net additions of solar compared to that of coal of 60 GW_p in 2017 and 80 GW_p in 2018 [17]. These milestones indicate that the shift towards solar PV as a major contributor to the energy system has already begun. This rapid uptake of solar-generated electricity, specifically crystalline silicon-based solar panels, will help to offset emissions from fossil fuels. However, the trajectory needed to curtail increases in global average temperature will require the installation of solar PV on a TW scale [18]. Breyer *et al.* built a new energy transition model to investigate the role of solar PV in the transition to world renewable energy generation. This comprehensive model, which analyses world energy consumption with a time resolution of one hour, predicts that solar PV will have the largest share of electricity generation in the mid- to long-term [14]. Their analysis suggests that a total of 27 TW_p of electricity will be generated by solar PV by 2050 [14]. By the end of 2018, the global installed solar electricity capacity was approximately 500 GW_p [19], which indicates that the growth of future solar PV installations must be very large. According to the ITRPV 11th edition report, a broad electrification scenario, corresponding to a total PV module installation of 63.4 TW_p in 2050, would require an installation rate of 1 TW_p per year is reached as soon as 2028 [19]. These views are echoed by the Shell Sky Scenario, which outlines the possibility of a huge transformation of the energy sector. Their analysis indicates that by 2070, solar will be the biggest contributor to global primary energy supply by far, with a share of 32% of the energy supply [20]. This will be more than double that of any other technology.

When discussing reductions in CO_2 emissions, it is important to clarify the difference between electricity and energy. As discussed above, electricity is the primary contributor to global CO_2 emissions. Other major contributors to emissions include the energy required for transport and industrial processes [18]. A recent report from Haegel *et al.* indicates that solar PV not only has the capacity to contribute to electricity generation, but will also play a critical role in all sectors of global energy production, including transportation and industrial processes. It was predicted that the required amount of total solar PV in 2050 will be in the range of 30–70 TW_p. However, some shifts in the global energy system will be required to facilitate this [18], including:

1. significantly increased electricity storage capabilities
2. widespread electrification of transport
3. increased research and development to support the production of hydrogen or complex hydrocarbons using solar PV.

The past few years have seen rapid advancements in electric vehicles. Reductions in the price of electric vehicles has been driven by falling lithium-ion battery pack costs, which are reducing by 18% for every doubling of the cumulative production volume. The 2020 Bloomberg New Energy Outlook report indicates that for many vehicle types and geographies, electric vehicles will reach price parity with equivalent internal combustion vehicles by 2025. It is likely that the rapid reduction in price will lead to the rapid adoption of the technology [21]. In line with this argument, Blakers *et al.* suggest that a global pathway to 100% renewable electricity and zero energy-related emissions is achievable using technologies available today [22]. Solar PV plays a critical role in this view. The authors suggest that TW-scale solar PV and wind must be coupled with hydro-electricity storage and high-voltage DC transmission to achieve these three objectives.

Predictions for future global energy consumption vary widely, depending on the input parameters used. When analysing future energy consumption, Green and Newman suggest that solar PV, coupled with energy storage, displays the characteristics of a disruptive technology [23]. The five key factors that could lead to solar PV becoming a disruptive technology include:

1. significantly reduced cost of solar panels
2. significantly reducing the cost of battery storage
3. policy constraints limiting the release of CO_2
4. improvements in efficiency driven by technological innovation
5. trends towards ‘knowledge-based’ economies.

If this technology is truly disruptive, the uptake could occur on a much faster time frame than previously projected scenarios, which only consider small technological changes. The most important factor, which has driven the recent rapid uptake of solar PV, has been the significant reductions in cost, particularly within the last decade.

1.3 The reducing cost of solar PV

The cost of solar modules has reduced at a substantial rate over the past 40 years. While the technology has been viewed as prohibitively expensive for the majority

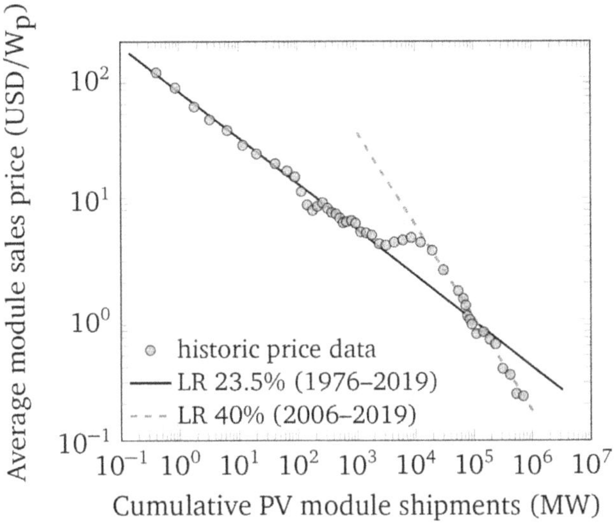

Figure 1.4 *Learning curve for average PV module prices, shown as a function of cumulative PV module shipments. Reproduced with permission from [19]. Copyright 2020 VDMA Photovoltaic Equipment*

of its time as a commercial product, the rapid cost reductions over the last five to ten years means solar electricity is becoming an economically viable alternative to traditional fossil fuel-based generation. Multiple recent power purchase agreement bids have seen lower prices for electricity generated using solar power compared to that of coal-fired power stations, with several large-scale solar plants now bidding in at US$ 19–24/MWh [24]. These significant cost reductions imply that solar PV will make up the competitive backbone of the future global energy sector. Figure 1.4 displays the historical learning curve for average solar module selling prices, plotted as a function of cumulative PV module shipments, adapted from the data presented in the ITRPV 11th report [19]. The curve is plotted on a log-log scale, which displays a linear trend up until a cumulative module shipment of about 3 GW_p. This corresponds to the time period spanning 1976–2003. During this time, the learning curve was stable. Beyond this, significant variability in the sales price is observed. These deviations, described by an increase in price (reduction in learning rate), followed by a significant reduction in price (increase in learning rate), can be broadly attributed to market events. In particular, recent cost reductions have been so substantial that they have forced adjustments to the historical learning rate. As displayed on the graph, the learning rate in the time period from 2006 to 2019 is approximately 40%.

Many factors have caused the price of solar modules to reduce over the past ten years [25]. These include:

- the cost of raw polysilicon
- wafering costs

- cost of non-silicon components
- the efficiency of the cell
- production yield
- production economies of scale.

Each of these factors is subject to a variety of technical and economic forces, which means the cost reduction has been created by an array of elements, each changing at a different rate.

Rather than merely analysing historical changes in each variable, Green provided a nuanced analysis, which highlighted crucial 'serendipitous' events and interactions between people that each had a significant bearing on the pathway to low cost PV [26]. The first event described was the oil embargoes in the USA in 1973/1974, which prompted the USA to strive for energy independence. The policy reaction to this crisis resulted in the development of the 'Flat Plate Solar Array Project', which led to one crucial development in the fabrication of solar cells, the robust fabrication of solar modules using screen printing to form the metal contacts. This technique, first developed by Spectrolab [27], provided significant cost advantages compared to the more traditional vacuum-based processes and is a mainstay in modern low-cost solar PV modules.

Progressive policies in various countries also played a role in producing a growing market for solar modules in the 1990s and 2000s. In particular, the extremely ambitious 'Million Roof' program in Japan, announced in 1993, which boosted Japan as the world leader for installing rooftop solar. Following the reduction in the subsidies provided by this scheme was the strong feed-in tariffs offered throughout Europe, which successfully provided a steady market for this emerging industry.

Perhaps the most important development in the evolution of industrial silicon solar cells was the transition of solar module fabrication to China. This transition was primarily engineered by UNSW (University of New South Wales) alumni, Dr Zhengrong Shi, who started the first production line in China in 2002 with the company Suntech. The success of Suntech led to the rapid development of the Chinese module market, which ultimately led to the establishment of many Chinese companies in the following years.

Over the past ten years, the involvement of Chinese manufacturers in the PV manufacturing industry has led to multiple market forces, which have driven unprecedented cost reductions. To provide a clearer picture of the cost during this time, Figure 1.5 shows the average quarterly price for multi-crystalline PV modules in the period 2008–2018 [26]. This shows that the price reduced by a factor of 15 during the 10-year time frame, caused by two periods of rapid decline. The first rapid reduction in price was closely linked to a sharp decline in the cost of polysilicon, due to a market overcorrection to accommodate for the large increase in Chinese companies requiring the material. The second sharp decline, beginning in 2010, was caused by an oversupply of modules on the market, as many companies became established and started producing modules on a large scale. The significant competition in the Chinese market has reduced the margins on modules to almost zero. This extremely competitive environment currently present in the market has pushed further steady

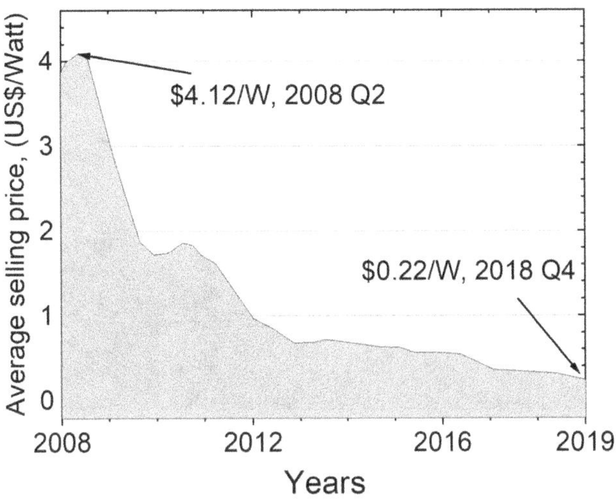

Figure 1.5 *Average quarterly selling price for multi-crystalline silicon photovoltaic modules. Reprinted from (25) Copyright (2019) with permission from Elsevier*

reductions, with the average selling price dropping to US$ 0.21/W in the last quarter of 2019 [28].

The rapid learning curve for solar modules has not only reduced the cost of modules but also consequently shrunk the fractional cost of the system attributed to modules [24]. This means that the other cost components now have a larger bearing on the overall cost of photovoltaic-generated electricity. In their analysis, Aposteleris *et al.* suggest that 'soft costs' related to taxes, regulation and installation will play a large role in determining the price of solar electricity [24]. These costs are related to the country or region where the solar modules are being installed. In light of this, very low costs have been observed in some unexpected countries, including the United Arab Emirates and Saudi Arabia. The analysis by Aposteleris *et al.* suggests that unsubsidised prices below $US 30/MWh are attainable given favourable local conditions. A consequence of the increased competition in the industrial silicon solar cell market has led to rapid improvements in the efficiency of solar PV modules. The current state of the art of the technologies driving industrial silicon solar cells are considered in Section 1.4.

1.4 Screen-printed silicon solar cells – the workhorse of the photovoltaics industry

The current PV market is completely dominated by crystalline silicon solar cells, which account for ~95% market share, while the other ~5% is made up of thin-film technologies [19]. A historical overview of the breakdown of these technologies is shown in Figure 1.6, with data kindly provided by Paula Mints [29].

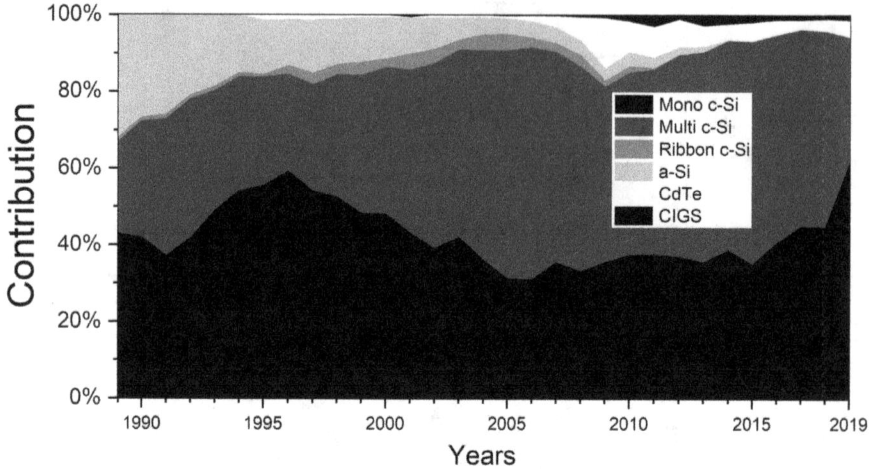

Figure 1.6 Historical overview of the market share of solar cell technologies, spanning from 1980 to 2019. Data provided by Paula Mints [29]

During the 1980s and 1990s, there was a belief that the economic viability of crystalline silicon PV would be surpassed by thin-film technologies, which were dubbed as the 'second generation' of solar cells [30, 31]. This is reflected by the large market share for thin-film technologies in this era, shown in Figure 1.6. Note that the graph represents the fractional market share, but the actual production volume of thin-film technologies in the 1980s was small by today's standards. The perceived promise of second-generation solar cells was the reduction in material costs, facilitated by the use of thinner films, and the avoidance of costly purification techniques. Early research into thin-film solar cells focused on compound semiconductors, such as cadmium CdTe and $CuInSe_2$. This led to impressive results in a laboratory setting. However, there are several barriers to market acceptance [30]. These include:

1. The use of toxic materials such as cadmium does not fit with the environmentally friendly promise of solar energy.
2. Problems with instabilities during damp heat testing reduced investor confidence.
3. The scarcity of some elements, namely indium, cast dispersions on the scalability of these technologies.

These issues for compound semiconductor thin film PV sparked renewed research interest in silicon thin films. Device architectures in this class of second-generation devices include amorphous silicon (a-Si) p-i-n devices and thin-film polycrystalline silicon devices on glass films, referred to as crystalline silicon on glass. The peak of the thin-film market share occurred in 2009 during the silicon shortage. However, the rapid reduction in silicon wafer and processing cost and consistent efficiency improvements in industrial crystalline silicon solar cells removed the cost advantage of thin-film approaches. These factors caused a significant decrease in market share for thin-film technologies. In fact, the diminished market share for thin-film approaches shown in Figure 1.6 closely correlates with the reduction in the cost of crystalline silicon PV displayed in Figure 1.5.

In 2017, the 4.5% of annual production attributed to thin films materials comprised of cadmium telluride (CdTe) (2.3%), copper indium gallium diselenide ($CuInSe_2$) (1.9%) and a-Si (0.3%) [29]. Since 2014, another thin-film material – referred to as 'perovskites' due to the crystal structure – has garnered a lot of interest in the PV community [32–34]. In a few years, perovskite solar cells have improved remarkably, with efficiencies now reaching 25% [35]. One significant drawback of this technology is the poor environmental stability of the material [36]. It seems the most likely pathway to market for perovskites would be for them to be coupled with a crystalline silicon (c-Si) solar cell to form a tandem structure [37]. The company Oxford PV is heavily investigating the possibility of tandem perovskite on silicon solar cells and have achieved an impressive efficiency of 28% for a large area device [35]. It seems as though silicon solar cells will play a dominant role in the near and mid-range future of the solar cell market. As such, this book focuses exclusively on silicon solar cell technologies.

Over the past 30 years, industrial solar cells have been overwhelmingly made using a standard processing approach, a phosphorus diffusion on boron-doped p-type silicon wafers, metallised by screen-printed silver on the front, and aluminium on the rear [38]. Subsequent sintering of the aluminium rear contacts leads to the formation of an aluminium back-surface field (Al-BSF) at the rear, through the alloying of aluminium and silicon above the eutectic temperature of 577 °C, which increases the open circuit voltage (V_{OC}) [39]. As such, these cells are typically referred to as Al-BSF solar cells [40]. One of the key advantages of this approach is the simple production methods required. Although the efficiency of industrial Al-BSF cells has improved significantly over the past 20 years, the efficiency of cells using this structure is limited to about 20%, due to the large contact area between the rear metal contact and the silicon substrate [39]. While the formation of the p^+/p back-surface field does provide some shielding from recombination at the metal–Si interface, the strength of the field-effect passivation is limited compared to that of some dielectric layers containing fixed charge. This is caused by the relatively low solid solubility of aluminium in silicon, which limits the peak doping concentration from the aluminium alloy to about $3\times10^{18}/cm^3$ [41]. The achievable dark saturation current density (J_0) on the rear surface depends on the thickness of the p^+ layer region and the temperature of the firing step; however, given the processing constraints of a commercial manufacturing line, it is difficult to achieve J_0 values lower than 400 fA/cm^2 [41]. This places severe limitations on the achievable V_{OC}. To overcome this, it is desirable to remove the impact of the area metal–Si interface by the formation of local contact points.

Increasing the efficiency of the cell and module is an effective way to reduce the overall cost of a PV system. This is particularly true when considering how the cost components that make up the total PV system costs are changing over time. Ten years ago, when the price of modules was higher, the PV module made up a large fraction of the total system cost. As the module price has reduced rapidly (shown in Figure 1.5), the fractional cost of the PV system attributed to the module is shrinking [42, 43]. This means that the balance of systems (BOS) costs, which includes the inverter, mounting structure and cabling, have a larger impact on the levelised cost of electricity (LCOE). One simple way to reduce the LCOE is to increase the efficiency of the cell, which will reduce the area-dependent BOS costs.

One popular high-efficiency technology, developed by UNSW in the 1980s, is the PERC structure [44, 45]. PERC cells demonstrate higher efficiency than Al-BSF cells due to two main reasons [46]:

1. The surface recombination on the rear surface is significantly reduced, due to the introduction of a passivating dielectric layer and reduced metal–Si interface.
2. Increased reflection at the rear due to the rear dielectric reflector.

A comparison of the PERC cell structure with the traditional Al-BSF solar cell structure is shown in Figure 1.7.

Although the PERC solar cell technology involves extra processing steps, the process flow is very similar to that of the conventional Al-BSF structure (this is described in detail in Chapter 2). This means the cost of production is kept relatively low, and importantly, widespread processing expertise gained from the production of Al-BSF solar cells can be directly translated to the manufacture of PERC solar cells. Due to these factors, the market share of PERC solar cells is increasing rapidly. Figure 1.8 shows an overview of the projected market share of different cell technologies, projected forward to 2030 [19]. This indicates that the capacity of PERC surpassed that of Al-BSF cells in 2019, with a market share of more than 70% in 2020. This market domination is expected to continue for the next decade, with approximately 70% of the market expected to be made up of PERC cells in 2030.

As the global production of PERC solar cells increases, incremental improvements in various aspects of cell processing are leading to steady improvement in cell efficiency, with current production values for multiple companies, including Trina and Hanwha Q-Cells (as presented at [47], now exceeding 23% [2, 47]. Figure 1.9 displays the average cell efficiency data from production at Hanwha Q-Cells. The graph displays both Al-BSF (labelled 'BSF') and the Q.ANTUM technology (labelled 'QNT', representing Hanwha Q-Cells' implementation of PERC). This representative example excellently illustrates the incremental improvements in both the Al-BSF and PERC technologies, for both multi-crystalline and mono-crystalline p-type silicon wafers. Extrapolating across the more than ten years of data indicates a cell efficiency learning rate of $0.5\%_{abs}$ per year for both multi- and mono-crystalline solar cell technologies. This learning rate has reliably remained in place as the cell technology switched from Al-BSF to PERC. The transition period is highlighted in the graph in the dark grey region. This data was kindly provided for reproduction by Markus Fischer of Hanwha Q-Cells.

It is expected that, as with the Al-BSF technology, industrial cells will approach the efficiency values achieved on a laboratory scale [46]. Although PERC solar cells obtain significant efficiency benefits, there are still several limitations that must be

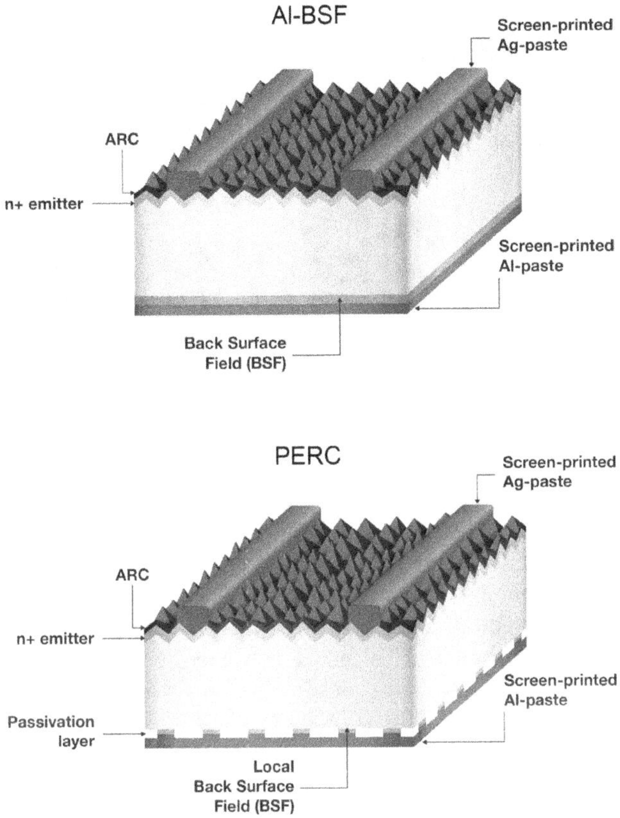

Figure 1.7 *Schematic comparison of the cell architecture for Al-BSF (top) and PERC solar cells (bottom)*

overcome to unlock the full potential of the technology. These can be characterised by two key areas:

1. overcoming performance-limiting defects in the solar cells and light-induced degradation (LID)
2. design of the emitter to reduce front surface recombination.

These aspects will be discussed in Section 1.5.

1.5 Limitations to current commercial silicon solar cells

1.5.1 Performance-limiting defects and light-induced degradation

There are many types of defects that can limit the performance of silicon solar cells. All silicon solar cells are susceptible to recombination at the surface. At any bare

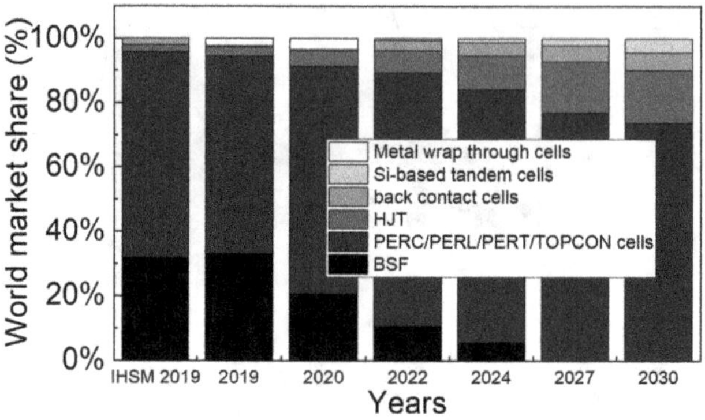

Figure 1.8 *Projected worldwide market share for different silicon solar cell technologies. Reproduced with permission from [19]. Copyright 2020 VDMA Photovoltaic Equipment*

silicon surface, the abrupt discontinuity in silicon atom bonding leads to the formation of dangling bonds, which form surface energy traps. Hydrogen passivation of the surface states, whereby hydrogen atoms are introduced to the surface so that they can bond to and neutralise these surface states, is critical to increasing the performance of silicon solar cells. This can occur via hydrogen-containing dielectrics

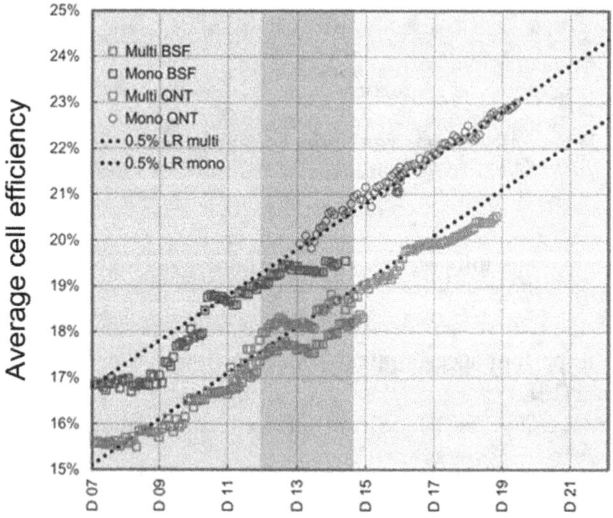

Figure 1.9 *Progression of Hanwha Q-Cells production cell efficiency data for both Al-BSF (labelled 'BSF') and Q.ANTUM technology (labelled 'QNT', representing Hanwha Q-Cells' implementation of PERC). The labels on the x-axis represent December in the allotted calendar year. Graph kindly provided by Markus Fischer from Hanwha Q-Cells*

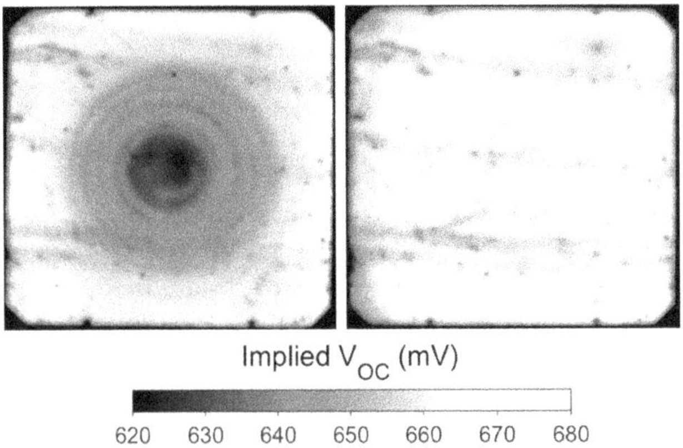

Figure 1.10 *1-sun calibrated iV_{OC} maps of a boron-doped p-type Cz wafer exhibiting oxygen precipitates (left) directly after deposition of SiN_x:H, and (right) following a subsequent high-temperature firing process for hydrogenation. Image reproduced with permission from [49] under Creative Commons CC-BY-NC-ND. © CSIRO 2018 Open Access*

such as hydrogenated amorphous silicon (a-Si:H) films for world record silicon heterojunction solar cells, or by hydrogenated silicon nitride (SiN_X:H) in Al-BSF and PERC solar cells. The hydrogen passivation of surfaces is covered in detail in Chapter 3.

A variety of defects can also form in the bulk of the silicon. One prominent example of this is crystallographic defects, such as dislocation clusters and grain boundaries [48]. Such defects are particularly prominent in lower-cost silicon formation methods, such as multi-crystalline and cast-mono silicon wafers. In addition to the crystallographic defects formed during the wafer growth, defects can form during the processing steps involved in making the solar cell. Examples of these include laser-induced defects that can form via laser doping or laser ablation, or oxygen precipitate rings, which can form during high-temperature processing such as emitter formation [49]. Hydrogen can be used to passivate many defects in the bulk of silicon. Figure 1.10 shows an example of a p-type silicon wafer that contains oxygen precipitate rings. The dark circles present in the implied V_{OC} (iV_{OC}) map in the left-hand image represent regions of increased recombination activity and thus reduced material quality. The image on the right displays the same wafer after it has undergone a hydrogenation process, indicating that hydrogen can be used to successfully passivate the defect. The role of hydrogen in passivating bulk defects is discussed in detail in Chapter 4.

With the exception of n-PERT (n-type passivated emitter and rear, totally diffused), passivated emitter solar cells are generally fabricated using p-type substrates, which are primarily doped with boron impurities to enable relatively uniform doping profiles throughout ingots during crystallisation. Due to the predicted dominance of

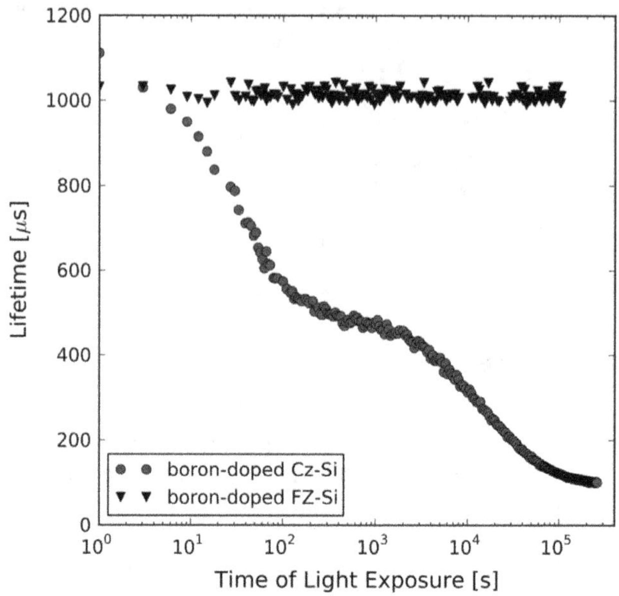

Figure 1.11 *Effective minority carrier lifetime as a function of light soaking time for both Cz and FZ boron-doped p-type silicon wafers, illustrating the significant LID in p-type Cz silicon. Reproduced with permission from Tine Nærland [53]*

PERC technologies in the future, p-type Czochralski (Cz) silicon wafers will play a substantial role in the solar industry over the coming ten years [19]. Unfortunately, boron-doped Cz material is known to be susceptible to LID, which represents one of the key challenges faced by the industry at the present moment [50], and has plagued the industry since its discovery in p-type Cz silicon in 1973 [51]. The primary cause of this defect is linked to the formation of recombination active boron-oxygen (B-O) complexes [52]. The result of the degradation is observed as a significant reduction in the effective minority carrier lifetime under the presence of light soaking. An example of this LID is shown in Figure 1.11. This shows the effective minority carrier lifetime as a function of light soaking time for float-zone (FZ) and Cz grown p-type wafers, with a degradation occurring in two stages. These two stages are referred to as the fast- and slow-stages. In a finished cell, this primarily manifests as a reduction in V_{oc}. Although the term used to describe this degradation is 'light-induced degradation', this phenomenon has also been observed in the dark in the presence of an applied bias, indicating that the cause is related to the generation of excess minority carriers, not solely an interaction with photons [54].

In order to avoid this detrimental process from occurring, the responsible defect must be permanently deactivated such that the complex is no longer recombination active. Herguth *et al.* first demonstrated such a process in 2006 by using an illuminated anneal [55]. While some uncertainty has surrounded the exact physical mechanisms underpinning deactivation, it is now clear that hydrogen is crucial for the passivation of the B-O defect. This is explored in detail in Chapter 5.

1.5.2 Selective emitter design

The laboratory design of the best passivated emitter and rear, locally diffused (PERL) cell, which has the highest efficiency of the passivated emitter cell series, exhibits an efficiency of 25% [56, 57]. One of the key pathways in bridging the gap between current industrial cells and the high-efficiency laboratory cells is increasing the V_{OC}. In the 25% efficiency PERL cell, the total J_0 is only about 50 fA/cm^2, with an emitter J_0 (J_{0e}) contribution as low as J_{0e} = 15 fA/cm^2 [46]. In current industrial p-type silicon cells, with total J_0 of ~ 260 fA/cm^2 [19], one of the main contributors to these recombination losses is from the emitter. As such, emitter design is one of the key areas that will drive future improvements, which will be discussed in detail in Chapter 8. Furthermore, hydrogen played a key role in surface passivation during the development of the PERL cell, as will be discussed in detail in Chapters 2 and 3.

In 2017, Min *et al.* laid out a road map to achieve industrial PERC cells with efficiencies exceeding 24% [58]. This comprehensive loss analysis indicates that recombination losses are dominated by recombination in the emitter, and that reducing the J_{0e} component is critical to increasing the efficiency. LONGi recently demonstrated a peak efficiency of 23.83% for an industrial p-type PERC solar cell [2]. Other companies, such as TRINA Solar and Hanwha Q-Cells, have also demonstrated p-type PERC efficiencies exceeding 23%. These impressive efficiencies are facilitated by reductions in emitter recombination, driven by a 'selective emitter' design.

The term selective emitter refers to an approach where the doping profile varies across the surface of the solar cell. This emitter design requires careful consideration of the interplay between multiple competing effects and is closely linked to the metallisation. The ability for metal fingers to collect carriers is dependent on the relationship between the lateral conductivity of the emitter and the spacing between the metal fingers. It is desirable to increase the spacing of the fingers, to reduce shading losses; however, larger spacing requires heavier doping. Heavy doping is also required under the contacts to reduce the contact resistance. However, heavy doping in the emitter leads to the formation of a 'dead layer', where the response to short wavelengths is poor, due to losses at the surface. The heavy doping also increases the J_{0e} component, which limits the V_{OC}. Using a homogeneous emitter, whereby the emitter doping is constant in all areas, requires a trade-off to be made between these competing factors. A selective emitter approach avoids losses imposed by this trade-off. Practically, this requires a two-step process, so that the emitter is lighter in between the fingers and heavy under the contacts. The first step involves a typical thermal diffusion in a furnace; however, a lighter emitter profile is produced by varying the processing conditions. Laser doping can then be employed to form the heavy diffusion in selected areas [59]. One key advantage of laser doping is the highly localised nature of the diffusion. As a result, substantially higher temperatures can be used well above 1 000 °C, greatly increasing diffusivity. This means that

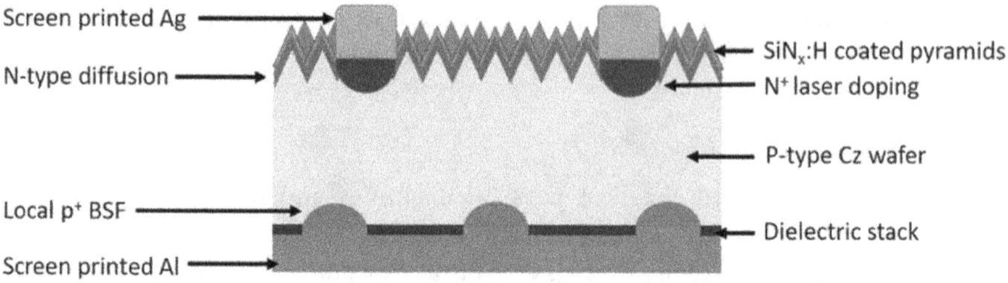

Figure 1.12 *Schematic diagram illustrating the selective emitter design in a PERC solar cell. Image reproduced with permission from [2] under Creative Commons CC BY. © 2020 The Authors. Progress in Photovoltaics: Research and Applications published by John Wiley & Sons Ltd. https://doi.org/10.1002/pip.3243*

diffusion properties that would require multiple hours of processing in a standard furnace can occur on the timescale of microseconds, particularly when doping takes place in the liquid state (above 1414 °C), rather than the solid-state. Figure 1.12 displays a schematic figure from LONGi that illustrates the selective emitter design of their high-efficiency p-type PERC solar cells, as formed using a pulsed 532-nm laser. The doping under the contacts is much heavier than between the fingers as will be discussed in Chapter 8.

Efficiency enhancements to push industrial cells closer to the record PERL cell will require further optimisation of the delicate balance in the selective emitter. Laser doping processes will be crucial in unlocking this pathway to higher efficiency industrial PERC solar cells. In line with this argument, the recent ITRPV report predicts that selective emitter design, facilitated by laser doping, will make up almost 60% of industrial p-type silicon solar cell production by 2030 [19]. This is illustrated in Figure 1.13.

1.6 Overview of chapter structure

The purpose of this book is to provide a comprehensive overview of the current state of the art of hydrogenation and laser doping processes for silicon solar cells and to outline the necessary pathways to drive further improvements in industrial silicon solar cells.

In Chapter 2, the mechanisms of hydrogen passivation are outlined. A historical overview of hydrogen passivation in silicon solar cells is provided, and the importance of the charge state of hydrogen in silicon is discussed, leading to the development of advanced hydrogen passivation processes incorporating illumination.

Chapter 3 focuses on hydrogen passivation of the silicon surface. It discusses the role of hydrogen-containing dielectric layers, with particular regard to the hydrogen content of these layers. It outlines current progress and future directions for surface passivation using hydrogen.

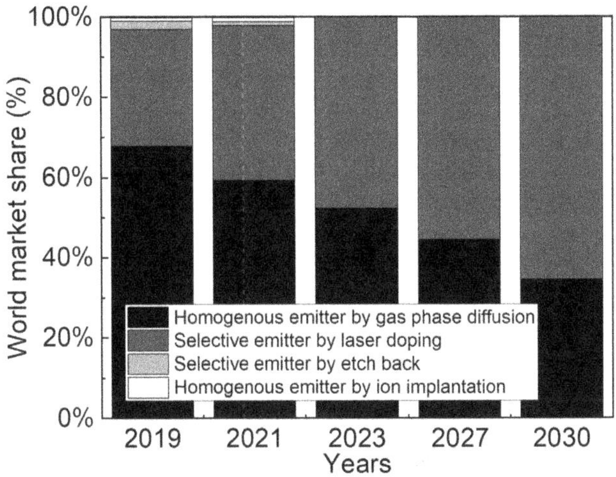

Figure 1.13 *Projected market share for the different phosphorous emitter technologies for p-type silicon solar cells. This indicates the expected importance of laser doping for selective emitter formation. Reproduced with permission from [19]. Copyright 2020, VDMA Photovoltaic Equipment*

Chapter 4 explores the hydrogen passivation of bulk defects. These include structural defects, such as grain boundaries, impurities, such as dopants, and process-induced defects, such as oxygen precipitates and laser-induced defects. The discussion is extended beyond p-type multi-crystalline and Cz wafers to include both n-type Cz and upgraded metallurgical grade (UMG) silicon, and the integration of hydrogenation processes for bulk defect passivation in p-type silicon heterojunction solar cells to enable p-type multi-crystalline silicon solar cells with V_{OC} values over 700 mV.

Chapter 5 details the current understanding of B-O-related LID in p-type Cz silicon solar cells. It includes the multiple pathways for eliminating B-O-related LID that has caused significant controversy in the literature, which are the thermal annihilation of defect precursors with high-temperature processing, as well as hydrogen passivation during illuminated annealing. It also highlights the importance of accelerating defect formation to enable rapid passivation of the B-O defects. Recent results are also presented demonstrating boron-doped p-type Cz silicon heterojunction solar cells with stable V_{OC} values over 735 mV after hydrogenation of B-O defects.

In addition to passivating defects, the presence of hydrogen in silicon can also induce recombination-active defects, which have negative impacts on solar cell performance. Chapter 6 outlines the negative impact of hydrogen in silicon solar cells. This includes different hydrogen-containing complexes that can form in silicon and the evolving understanding of hydrogen's role in light and elevated temperature degradation (LeTID). It also highlights the phenomenon of increased contact resistance due to an accumulation of hydrogen at the silicon surface, which can occur during attempts to solve LeTID.

Chapter 7 introduces the concept of laser doping for rapid diffusion in silicon solar cells. It provides a fundamental understanding of the movement of dopants in silicon and outlines the practical approaches to laser doping. It also discusses the importance of the laser processing parameters on dopant profiles.

Chapter 8 explores the use of laser doping to form a 'selective emitter', which is an integral design approach to drive further efficiency improvements at the present moment. It outlines and assesses the various methods to achieve selective doping and discusses some of the pressing challenges in this field, such as the mitigation of laser-induced defects. This highlights the complementary nature of laser doping and hydrogen passivation.

Chapter 9 extends this discussion to various other applications of laser doping in silicon solar cells. Examples of these applications include the semiconductor finger technology, the laser edge isolation to avoid edge recombination and the use of the rear surface aluminium oxide passivation layer as a dopant source for forming heavily doped p^+ regions such as the rear contact of PERC/PERL.

Chapter 10 provides a concise overview of the current status of hydrogenation and laser doping in silicon solar cells. It then provides insights into future directions in this field, which will be at the forefront of improvements for industrial silicon solar cells over the next ten years. The technological advancements discussed in this book will be crucial in progressing the rapid uptake of solar PV, which is required to tackle the pressing challenge of global CO_2 emissions.

1.7 Acknowledgements

The authors would like to thank Dr Moonyong Kim and Dr Daniel Chen for assistance with figure formatting. We acknowledge Paula Mints, Markus Fischer of Hanwha Q-Cells and Tine Nærland for providing data for Figure 1.6, Figure 1.9 and Figure 1.11, respectively. Thanks to Aaron Kuswara for the graphic design of Figure 1.7. Finally, thanks to Bruno Vicari Stefani for useful discussions regarding the link between economic prosperity and CO_2 emissions.

References

[1] Wang Z., Han P., Lu H., *et al.* 'Advanced PERC and PERL production cells with 20.3% record efficiency for standard commercial p-type silicon wafers'. *Progress in Photovoltaics: Research and Applications*. 2012;**20**(3):260–8.

[2] Chen R., Tong H., Zhu H., *et al.* '23.83% efficient mono-PERC incorporating advanced hydrogenation'. *Progress in Photovoltaics: Research and Applications*. 2020;**28**(12):1239–47.

[3] Morice C.P., Kennedy J.J., Rayner N.A., Jones P.D. 'Quantifying uncertainties in global and regional temperature change using an ensemble of

observational estimates: the HadCRUT4 data set'. *Journal of Geophysical Research: Atmospheres*. 2012;**117**(D8):n/a.

[4] Ritchie H., Roser M. CO_2 and Greenhouse Gas Emissions. 2017. Available from https://ourworldindata.org/co2-and-other-greenhouse-gas-emissions [Accessed 23 Feb 2021].

[5] Hulme M. '1.5 °C and climate research after the Paris agreement'. *Nature Climate Change*. 2016;**6**(3):222–4.

[6] Pachauri R.K., Meyer L.A., Change C. *Synthesis Report. Contribution of Working Groups I,II and III to the Fifth Assessment Report of the Intergovernmental Panel on Climate Change*. **2014**. Geneva, Switzerland; 2014.

[7] Friedlingstein P., Jones M.W., O'Sullivan M., *et al*. 'Global carbon budget 2019'. *Earth System Science Data*. 2019;**11**(4):1783–838.

[8] Dlugokencky E., Tans P. 'Trends in Atmospheric Carbon Dioxide. National Oceanic & Atmospheric Administration'. *Earth System Research Laboratory*. 2020.

[9] Haustein K., Allen M.R., Forster P.M., *et al*. 'A real-time global warming index'. *Scientific Reports*. 2017;**7**(1):1–6.

[10] Andrew R. Figures from the Global Carbon Budget. 2020. Available from https://folk.universitetetioslo.no/roberan/GCB2020.shtml [Accessed 23 Feb 2021].

[11] Global Carbon Atlas. Available from www.globalcarbonatlas.org/en/CO2-emissions [Accessed 23 Feb 2021].

[12] Peters G.P., Andrew R.M., Canadell J.G., *et al*. 'Carbon dioxide emissions continue to grow amidst slowly emerging climate policies'. *Nature Climate Change*. 2020;**10**(1):3–6.

[13] Rogelj J., Popp A., Calvin K.V., *et al*. 'Scenarios towards limiting global mean temperature increase below 1.5 °C'. *Nature Climate Change*. 2018;**8**(4):325–32.

[14] Breyer C., Bogdanov D., Gulagi A., *et al*. 'On the role of solar photovoltaics in global energy transition scenarios'. *Progress in Photovoltaics: Research and Applications*. 2017;**25**(8):727–45.

[15] Service R.F. 'Solar energy. is it time to shoot for the sun?' *Science*. 2005;**309**(5734):548–51.

[16] Solargis. Solar resource map of Australia. Available from https://solargis.com/maps-and-gis-data/download/australia [Accessed 23 Feb 2021].

[17] Whiteman A., Rueda S., Akande D., Elhassan N., Escamilla G., Arkhipova I. 'Renewable capacity statistics'. *International Renewable Energy Agency*. 2020;**2020**.

[18] Haegel N.M., Atwater H., Barnes T., *et al*. 'Terawatt-scale photovoltaics: transform global energy'. *Science*. 2019;**364**(6443):836–8.

[19] 'International technology roadmap for Photovoltaics, 11th edition'. 2020.

[20] Shell. 'Sky - Meeting the Goals of The Paris Agreement'. 2019.

[21] 'New energy outlook 2020, Bloomberg new energy finance'. 2020.

[22] Blakers A., Stocks M., Lu B., Cheng C., Stocks R. 'Pathway to 100% Renewable Electricity'. *IEEE Journal of Photovoltaics*. 2019;**9**(6):1828–33.

[23] Green J., Newman P. 'Disruptive innovation, stranded assets and forecasting: the rise and rise of renewable energy'. *Journal of Sustainable Finance & Investment*. 2017;7(2):169–87.

[24] Apostoleris H., Sgouridis S., Stefancich M., Chiesa M. 'Evaluating the factors that led to low-priced solar electricity projects in the middle East'. *Nature Energy*. 2018;**3**(12):1109–14.

[25] Kavlak G., McNerney J., Trancik J.E. 'Evaluating the causes of cost reduction in photovoltaic modules'. *Energy Policy*. 2018;**123**:700–10.

[26] Green M.A. 'How did solar cells get so cheap?' *Joule*. 2019;**3**(3):631–3.

[27] Ralph E.L. Recent advancements in low cost solar cell processing. 11th IEEE Photovoltaic Specialists Conference; Scotsdale; 1975. pp. 315–6.

[28] PV insights. Solar PV Module Weekly Spot Price. Available from http://pvinsights.com [Accessed 23 Feb 2021].

[29] Mints P. 'Photovoltaic Manufacturer Shipments: Capacity Price & Revenues 2019/2020'. 2020.

[30] Green M.A. 'Crystalline and thin-film silicon solar cells: state of the art and future potential'. *Solar Energy*. 2003;**74**(3):181–92.

[31] Andersson B.A. 'Materials availability for large-scale thin-film photovoltaics'. *Progress in Photovoltaics: Research and Applications*. 2000;**8**(1):61–76.

[32] Green M.A., Ho-Baillie A., Snaith H.J. 'The emergence of perovskite solar cells'. *Nature Photonics*. 2014;**8**(7):506–14.

[33] Snaith H.J. 'Perovskites: the emergence of a new era for low-cost, high-efficiency solar cells'. *The Journal of Physical Chemistry Letters*. 2013;**4**(21):3623–30.

[34] Liu M., Johnston M.B., Snaith H.J. 'Efficient planar heterojunction perovskite solar cells by vapour deposition'. *Nature*. 2013;**501**(7467):395–8.

[35] Green M.A., Dunlop E.D., Hohl-Ebinger J., Yoshita M., Kopidakis N., Hao X. 'Solar cell efficiency tables (version 56'. *Progress in Photovoltaics: Research and Applications*. 2020;**28**(7):629–38.

[36] Wang D., Wright M., Elumalai N.K., Uddin A. 'Stability of perovskite solar cells'. *Solar Energy Materials and Solar Cells*. 2016;**147**:255–75.

[37] Shen H., Walter D., Wu Y., *et al.* 'Monolithic Perovskite/Si Tandem Solar Cells: Pathways to Over 30% Efficiency'. *Advanced Energy Materials*. 2020;**10**(13):1902840–18.

[38] Green M.A. 'Silicon photovoltaic modules: a brief history of the first 50 years'. *Progress in Photovoltaics: Research and Applications*. 2005;**13**(5):447–55.

[39] Dullweber T., Schmidt J. 'Industrial silicon solar cells applying the Passivated emitter and rear cell (PERC) Concept—A review'. *IEEE Journal of Photovoltaics*. 2016;**6**(5):1366–81.

[40] Fellmeth T., Mack S., Bartsch J., *et al.* '20.1% efficient silicon solar cell with aluminum surface Field'. *IEEE Electron Device Letters*. 2011;**32**(8):1101–3.

[41] Rauer M., Schmiga C., Glatthaar M., Glunz S.W. 'Alloying from Screen-Printed aluminum pastes containing boron additives'. *IEEE Journal of Photovoltaics*. 2013;**3**(1):206–11.

[42] Werner J., Niesen B., Ballif C. 'Perovskite/Silicon tandem solar cells: marriage of convenience or true love story? – an overview'. *Advanced Materials Interfaces*. 2018;**5**(1):1700731.

[43] Peters I.M., Rodriguez Gallegos C.D., Sofia S.E., Buonassisi T. 'The value of efficiency in Photovoltaics'. *Joule*. 2019;**3**(11):2732–47.

[44] Blakers A.W., Wang A., Milne A.M., Zhao J., Green M.A. '22.8% efficient silicon solar cell'. *Applied Physics Letters*. 1989;**55**(13):1363–5.

[45] Blakers A.W., Green M.A. '20% efficiency silicon solar cells'. *Applied Physics Letters*. 1986;**48**(3):215–7.

[46] Green M.A. 'The Passivated emitter and rear cell (PERC): from conception to mass production'. *Solar Energy Materials and Solar Cells*. 2015;**143**(2):190–7.

[47] Müller J. Approaching 24% efficiency. Oral presentation by Hanwha Q-Cells at PV Cell Tech; 2020.

[48] Chen D., Kim M., Shi J., Vicari Stefani B., Yu Z., Liu S. 'Defect engineering of P-type silicon heterojunction solar cells fabricated using commercial-grade low-lifetime silicon wafers'. *Prog Photovoltaics Res Appl*. 2018;**2019**:1–15.

[49] Hallam B.J., Ciesla A.M., Chan C.C., *et al*. 'Overcoming the challenges of hydrogenation in silicon solar cells'. *Australian Journal of Chemistry*. 2018;**71**(10):743–52.

[50] Lindroos J., Savin H. 'Review of light-induced degradation in crystalline silicon solar cells'. *Solar Energy Materials and Solar Cells*. 2016;**147**:115–26.

[51] Fischer H., Pschunder W. Investigation of photon and thermal induced changes in silicon solar cells. Proceedings of the 10th IEEE Photovoltaic Specialists Conference. IEEE; 1973. pp. 404–11.

[52] Bothe K., Schmidt J. 'Electronically activated boron-oxygen-related recombination centers in crystalline silicon'. *Journal of Applied Physics*. 2006;**99**(1):13701.

[53] Nærland T.U. *Characterization of light induced degradation in crystalline silicon. Norwegian*. Trondheim, Norway: University of Science and Technology; 2013.

[54] Niewelt T., Schön J., Warta W., Glunz S.W., Schubert M.C. 'Degradation of crystalline silicon due to Boron–Oxygen defects'. *IEEE Journal of Photovoltaics*. 2017;**7**(1):383–98.

[55] Herguth A., Schubert G., Käs M., Hahn G., Kaes M., Hahn G. A new approach to prevent the negative impact of the metastable defect in boron doped cz silicon solar cells. 4th IEEE World Conference on Photovoltaic Energy Conversion. IEEE; 2006. pp. 940–3.

[56] Green M.A. 'The path to 25% silicon solar cell efficiency: History of silicon cell evolution'. *Progress in Photovoltaics: Research and Applications*. 2009;**17**(3):183–9.

[57] Zhao J., Wang A., Altermatt P., Wenham S., Green M. '24% efficient PERL silicon solar cell: Recent improvements in high efficiency silicon cell research'. *Solar Energy Materials and Solar Cells*. 1996;**41-42**:87–99.

[58] Min B., Müller M., Wagner H., *et al*. 'A roadmap toward 24% efficient PERC solar cells in industrial mass production'. *IEEE Journal of Photovoltaics*. 2017;**7**(6):1541–50.

[59] Hallam B., Urueña A., Russell R., *et al*. 'Efficiency enhancement of i-PERC solar cells by implementation of a laser doped selective emitter'. *Solar Energy Materials and Solar Cells*. 2015;**134**(4):89–98.

Chapter 2
Hydrogen passivation mechanisms

Phillip Hamer[1], Chang Sun[2], and Brett Hallam[1]

Hydrogen passivation is widely used for silicon solar cells to reduce the recombination activity of performance-limiting defects. Silicon solar cells may contain many different types of recombination active defects that trap charges and thus reduce the output of the solar cell. These defects can be present at the silicon surface, where the abrupt discontinuity of silicon bonds leads to the formation of many dangling bonds, or they may form within the bulk of the silicon via a variety of mechanisms. The presence of recombination active defects causes a reduction in the lifetime of separated charge carriers in silicon, which ultimately reduces the cell efficiency. The concept of hydrogen passivation refers to the idea that hydrogen atoms present in silicon can bond with these recombination centres to reduce or remove their electrical activity. Hydrogen is a very powerful atom to promote passivation of defects because it is highly mobile and very electrically active in silicon. As a result, hydrogen passivation at the surface or in the bulk of silicon can be used to significantly enhance the performance and reliability of commercial silicon solar cells. Virtually all laboratory-based and commercial silicon solar cell technologies incorporate hydrogen passivation in one form or another. As a result, decades of research on silicon solar cells have generated an immense number of publications on hydrogen passivation. A search for publications with all of the terms 'hydrogen passivation', 'silicon' and 'solar cell', using Google Scholar generates 'about 3 380 results' (search date 31st January 2021). Figure 2.1 shows the number of papers published each year since 1980 featuring such terms.

2.1 The role of hydrogen passivation in high-efficiency silicon solar cells

Even in the early development of high efficiency silicon solar cells in the 1980s and 1990s, the incredible importance of hydrogen had been noted, leading to multiple

[1]The School of Photovoltaic and Renewable Energy Engineering, The University of New South Wales, Australia
[2]School of Engineering, The Australian National University, Australia

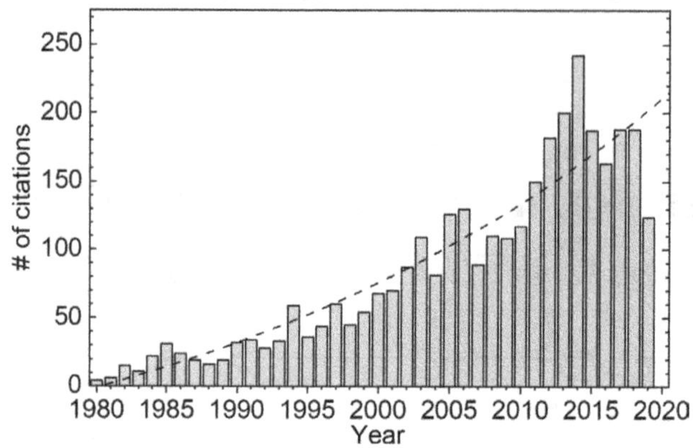

Figure 2.1 *Google scholar results per year of publication for search containing the terms 'hydrogen passivation', 'silicon' and 'solar cell'*

world records. In particular, hydrogen was crucial for the development of back contact solar cells [1–4] at Stanford University, and the passivated emitter and rear cell (PERC) family at the University of New South Wales (UNSW), including the passivated emitter and rear, locally diffused (PERL) solar cell [5–7].

The successful development of these technologies capitalised on work from investigations of hydrogen passivation in the semiconductor industry during the 1960s. In particular, forming gas annealing (FGA) and the 'alneal' process developed by Balk *et al.* and Deal *et al.* was used for metal-oxide-semiconductor (MOS) structures, where interactions between SiO_2 and active metals such as aluminium can generate atomic hydrogen to passivate defects at the Si/SiO_2 interface [8, 9].

For silicon solar cells, this was primarily used to improve surface passivation. This was particularly important for the non-diffused p-type rear surface for PERC and PERL solar cells, which are highly susceptible to surface recombination due to the relatively high concentration of minority carriers (electrons) and their high mobility. However, an alneal was also used on the front surface to improve surface passivation, leading to a 30 mV increase in the open circuit voltage (V_{OC}) [6]. In this instance, aluminium was evaporated over the entire front surface, followed by FGA for the alneal process. Subsequently, regions of the aluminium were masked, followed by a patterned removal of aluminium to define the active cell area. Effective hydrogen passivation of the surfaces enhanced the effective minority carrier lifetime of the material from 14 μs after thermal oxidation to 400 μs, essentially making the difference between a 600 mV device and a 700 mV device. The inclusion of the alneal process, among other cell improvements ultimately enabled a world-record efficiency of 25% [10]. Since then, as will be discussed in Chapter 3, surface passivation has greatly improved with the development of hydrogen-containing dielectric layers such as hydrogenated silicon nitride (SiN_X:H), resulting in lower dark saturation current density (J_0) values (see Figure 2.2)

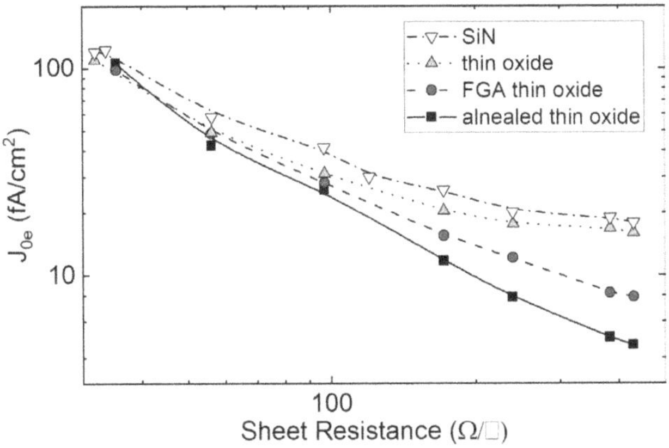

Figure 2.2 Emitter saturation current density (J_{0e}) as a function of emitter sheet resistance for n^+pn^+ lifetime structures. Reprinted from [11] with the permission of AIP Publishing.

The current world-record-holding silicon solar cell with 26.7% efficiency again uses hydrogen to improve surface passivation [12]. This is achieved through the use of hydrogenated amorphous silicon (a-Si:H) deposited by PECVD (plasma-enhanced chemical vapour deposition) [13]. High-efficiency cells using polysilicon-based passivating contacts on n-type silicon substrates use high-temperature annealing and hydrogen passivation to tune the morphology and band-gap of the doped polysilicon thin film layers on the top of tunneling oxides [14]. Using this approach, efficiencies as high as 26.1% have been reached [15–17].

In industry, the dominant PERC cell technology relies heavily on hydrogen passivation of surface and bulk defects, primarily supplied by silicon nitride dielectrics and released during metal contact co-firing. The use of hydrogen passivation in industrial high-efficiency (>24%) silicon solar cells is more difficult to ascertain due to the commercial sensitivity, with few details published in the literature. However, evidence for the use of hydrogen passivation can still be found. For example, several patents and publications relating to the development of Sunpower's rear interdigitated back-contact (IBC) solar cells and passivated contact structures by Silevo and Tetrasun have alluded to the use of hydrogen-containing dielectric layers and FGA throughout the fabrication sequence [18–23]. New emerging results with record cell efficiencies for cast silicon material on both n-type and p-type wafers using tunnel-oxide passivated contacts (TOPCon) also use a process sequence that enables hydrogen passivation of both the bulk and surface [24, 25].

However, despite the use of hydrogen passivation in virtually all silicon solar cells, in many papers, hydrogen passivation has often been considered somewhat of a 'black box'. Reading the literature one can easily get the impression that as long as hydrogen is incorporated into a dielectric layer, and a subsequent thermal process is used, effective hydrogen passivation will be achieved. However, this is not the case. Conventional solar cell structures, process sequences and attempts of hydrogen passivation do not necessarily result in effective passivation of all recombination active defects in the device. The difficulty in passivation is complicated by many factors including the defect type, whether it is structural, process-induced or carrier-induced, and at what point during the production sequence the defects are introduced into the device [26].

2.2 Methods of introducing hydrogen in silicon solar cells

There are a wide variety of ways hydrogen can be introduced into silicon. Indeed, given the wide spread use of wet chemical processing it is nearly impossible to fabricate a silicon solar cell without incorporating *some* hydrogen. However, defect passivation typically requires high concentrations of atomic hydrogen to be available in the vicinity of the defect. The techniques that have seen widespread use for intentionally introducing such hydrogen to silicon solar cells can be divided into three main classes:

1. Annealing in a molecular hydrogen ambient;
2. Exposure to a hydrogen containing plasma: and
3. Introduction of hydrogen from a hydrogen-containing dielectric

2.2.1 Annealing in a molecular hydrogen ambient

FGA is a common method to introduce H into Si. Forming gas comprises a small concentration of H_2 (2–6%), with the balance made up by an inert gas (Ar_2 or N_2). Samples are annealed in a forming gas ambient, typically at temperatures between 400 and 450 °C. FGA has been demonstrated to be capable of significantly improving surface passivation of Si/SiO_2 interfaces. For example, in the development of the PERL cells, forming gas improved the effective minority carrier lifetime from 14 to 40 μs.

The effectiveness of this process can be increased through the use of a metal layer deposited on top of an oxide. This has been exploited for decades in the MOS industry [8, 9, 27]. This technique is the basis of the 'alneal' process, which has been widely used to achieve excellent passivation of dangling bonds at the Si/SiO_2 interface. This process consists of depositing a thin layer of aluminium on an oxide passivated silicon surface and subjecting this structure to an FGA (although similar results can be obtained by annealing in N_2). During this process the oxidation of the Al by hydroxyl ions in the oxide leads to the release of atomic hydrogen, which can then migrate to the Si/SiO_2 interface (see Figure 2.3). Studies have shown remarkable improvements in surface passivation [11, 29], leading to the world's first 25% efficient silicon solar cell [30]. However, due to the numerous processing steps involved, and the improvement in surface passivation by hydrogenated dielectric layers such as silicon nitride (SiN_X:H) and aluminium oxide (AlO_X:H) layers, the alneal has not seen widespread commercial use.

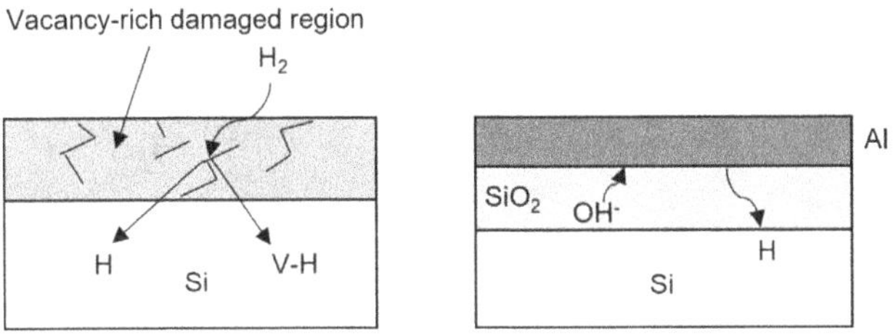

Figure 2.3 *(Left) Proposed mechanisms for splitting of hydrogen molecules by vacancies and surface damage adapted from [28]; (right) proposed alneal mechanism highlighting interactions between hydroxyl ions and Al layers*

An alternate approach is to introduce damage to the silicon surface, via such techniques as plasma exposure and ion implantation. Vacancies associated with surface damage can spontaneously dissociate H_2 [28] as also shown in Figure 2.3. Once atomic hydrogen enters the silicon bulk it can diffuse either by itself, or while attached to a vacancy in order to passivate defects. Given that surface damage is highly undesirable in a solar cell, this approach has not been widely used in industry.

Although annealing in a molecular hydrogen ambient has been shown to be suitable for surface passivation, in order to effectively passivate defects in the silicon bulk higher concentrations of atomic hydrogen are required. In addition, a sufficient thermal budget is required to allow diffusion of hydrogen throughout the bulk of the wafer, with a thickness in the range of 150–200 μm.

2.2.2 Exposure to a hydrogen containing plasma

The ability to introduce hydrogen for defect passivation can be drastically improved when hydrogen is introduced in its atomic form. This removes the dependence on surface conditions to split H_2 and can be performed over a wide range of temperatures. The most common approach for generating atomic hydrogen is to strike a plasma. This is typically generated through an inductively or capacitively coupled plasma driven with a 2.4 GHz (microwave) or 13.56 MHz (radio frequency [RF]) power supply. Along with pure molecular hydrogen, reactive ammonia (NH_3), or a mix of hydrogen and an inert gas (He or Ar) can be used as the process gas.

2.2.2.1 Direct hydrogen plasma

The simplest, and most efficient configuration for introducing atomic hydrogen is a direct hydrogen plasma, where the silicon sample is located within the plasma itself, typically on top of the grounded electrode (see Figure 2.4). However, this technique produces highly energetic H ions (100–500 eV) and hard short wavelength ultraviolet (UV) photons with high energy, which damage the Si sample surface [32]. Furthermore, the high concentrations of atomic hydrogen introduced often interacts with this damage to form extended defects. This is discussed in more detail in Chapter 6.

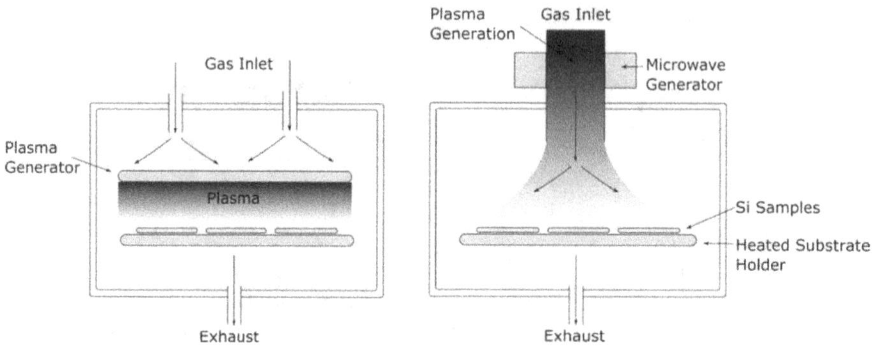

Figure 2.4 *Schematic of PECVD equipment used for (left) direct and (right) remote plasma hydrogenation. Image reproduced with permission from PV-Manufacturing.org [31]*

2.2.2.2 Remote plasma hydrogenation

An alternative for plasma-based hydrogenation is the remote plasma system (see Figure 2.4). This technique is similar in apparatus to the direct plasma approach but is characterised by the placement of the Si sample outside of the plasma region by means of a remote chamber [33, 34]. This prevents the sample from being exposed to UV radiation and allows for atomic H to lose some of its energy before incidence at the sample. Some damage has still been observed at temperatures below 300 °C. However it still remains unclear whether this is due to the impact of hydrogen radicals or the formation of hydrogen related defects within the silicon [35].

This approach has been used for the passivation of polycrystalline thin film solar cells. High concentrations of hydrogen are needed to passivate the sheer quantity of defects in these devices and industrial scale remote plasma systems have been demonstrated to be capable of delivering more than 3×10^{13} hydrogen atoms/cm^2/s to a silicon sample [35].

More recently remote hydrogen plasmas have been used in the fabrication of 25.6% efficient TOPCon solar cells [14] to introduce hydrogen that can diffuse across a thin polysilicon layer and passivate defects at the tunnel oxide/silicon interface.

2.2.3 Dielectric layer deposition

The standard industrial method used to introduce H into Si solar cells is through release from dielectric layers during belt furnace annealing. The most common dielectric sources of hydrogen for silicon solar cells are hydrogenated silicon nitride (SiN_X:H) and aluminium oxide (AlO_X:H) layers [36–40]. These may be deposited using chemical vapour deposition, in particular, PECVD. Stoichiometric hydrogenated aluminum oxide Al_2O_3:H can also be grown by atomic layer deposition (ALD) . The amount of hydrogen incorporated into these layers has been reported at up to 30% atomic for SiN_X:H and 12% for AlO_X:H, depending upon the deposition techniques and processing parameters [41]. This will be discussed more in Chapter 3.

While it is tempting to assume that a higher concentration of hydrogen within the dielectric layer will lead to a higher concentration of hydrogen in the silicon bulk after firing there are a number of other factors to consider. The first is the concentration of hydrogen in various binding configurations. In the case of SiN_X:H, hydrogen is primarily detected as either Si–H or N–H bonds. The concentration in each configuration is heavily dependent upon the Si:N ratio of the film. This is important because Si-H dissociates at lower temperatures (with an activation energy (E_a) ~ 3.1 eV) than N–H (E_a ~ 4.1 eV) [42]. This is a large part of the reason why 'N-rich' films with [Si]/[N] ratios below 1 are found to be more stable during firing.

The effusion of hydrogen into the atmosphere also requires consideration. In general, far more of the hydrogen released by dielectric layers escapes into the atmosphere than enters the silicon substrate. Two main mechanisms have been observed for this process in SiN_X:H. The first is via the formation and desorption of molecular hydrogen, which is accelerated in low density films and by the presence of voids [42–44]. This process is dominant at temperatures up to 800 °C. At higher temperatures or in dense films, the effusion of hydrogen is driven by the diffusion of atomic hydrogen through the film and desorption at the surface. The diffusivity of atomic hydrogen also decreases with film density. This diffusion of atomic hydrogen is required for introduction of hydrogen to the silicon substrate.

In order for effective passivation of bulk and surface defects by hydrogen, it is therefore advantageous to use a SiN_X:H layer that is sufficiently dense to impede the release of hydrogen as H_2 (typically achieved with a refractive index $n < 2.4$), while not so dense as to reduce the ability of atomic hydrogen to diffuse through the film thickness to the surface, and hence, preferably $n > 2.1$ [45], which is approximately the value required for antireflection coating purposes. For effective passivation it is also important to fire the wafers at temperatures above 600 °C to break N–H bonds and to allow atomic hydrogen to diffuse through the dielectric and into the wafer, which is readily achieved during the fast co-firing process used for metal/Si interface formation with screen-printed solar cells.

If it is desired to further increase the amount of hydrogen introduced to the silicon, the use of a density gradient or a higher density capping layer on top of the SiN_X:H can be used to impede the out-diffusion of atomic hydrogen, increasing the concentration in the dielectric film and hence the amount available to enter the silicon bulk [46]. Conversely a higher density layer in between the nitride and the silicon can impede in-diffusion to the substrate [47].

There have been fewer studies on hydrogen introduction from AlO_X:H layers, however, similar trends for the dependence on density have been observed. For ALD, films deposited at lower temperatures (<100 °C) had lower densities and released significant hydrogen at temperatures below 450 °C (which may play a role in the improved passivation seen after 'activation' anneals of AlO_X:H, while higher density films deposited at higher temperatures did not release significant hydrogen at temperatures below 600 °C. The effusion of hydrogen molecules was also only observed for low density films with a substrate deposition temperature of 150 °C, which was sufficient to almost completely suppress H_2 diffusion in the films [48].

2.2.4 *Unintentional incorporation of hydrogen*

It is nearly impossible to prevent the incorporation of hydrogen during silicon processing, whether it be for solar cells or MOS devices. The most common source of this unintentional hydrogen is via wet chemical processing. Hydrogen incorporation up to a depth of 1 μm has been observed as a result of heating silicon in water or acidic solutions [49, 50]. The actual amount introduced is strongly dependent upon the number of surface and near-surface sites for hydrogen, with the removal of native oxides and the presence of high concentrations of dopants both observed to increase hydrogen incorporation [51]. Subsequent thermal processing can potentially distribute hydrogen throughout the device. This unintentional incorporation of hydrogen can create difficulties in determining the role of hydrogen for passivation of a given defect. This is discussed in more detail in Chapter 5.

2.3 Behaviour of interstitial hydrogen in silicon

Once atomic hydrogen has been introduced into a silicon solar cell it is capable of interacting with nearly all impurities and crystal defects in the device. In order to achieve optimum electrical performance, it is essential to understand the behaviour of hydrogen in silicon and how this can be manipulated. In particular, the charge state of interstitial hydrogen plays a major role in both its diffusivity and ability to passivate defects.

2.3.1 *Donor and acceptor levels of hydrogen in silicon*

An often overlooked property of interstitial hydrogen in silicon is its amphoteric nature. In silicon, hydrogen can exist as a positively charged (H^+), neutral (H^0) or negatively charged (H^-) species [52–54]. Hydrogen is a negative-potential (negative-U) impurity with a donor level (E_D) above the acceptor level (E_A), meaning that the neutral charge state is unstable. The donor and acceptor levels of hydrogen, as determined by Herring *et al.* [55] are given by Equations 2.1 and 2.2, respectively, where E_C is the energy level of the conduction band and E_V is the energy level of the valence band. A diagrammatic representation of the energy levels of hydrogen within silicon is shown in Figure 2.5.

$$E_D = E_C - 0.16\text{ eV} \tag{2.1}$$

$$E_A = E_V + 0.48\text{ eV} \tag{2.2}$$

It should be noted that there is some uncertainty over the position of the acceptor level due to the interaction of hydrogen in silicon with oxygen, which will be discussed in section 2.3.4. Bonde Nielsen *et al.* [56] provided evidence that (2.2) above may actually be the acceptor level for the process:

$$\mathrm{H}^0(BC) - \mathrm{O}_i + e^- \leftrightarrow \mathrm{H}^-(T) - \mathrm{O}_i \tag{2.3}$$

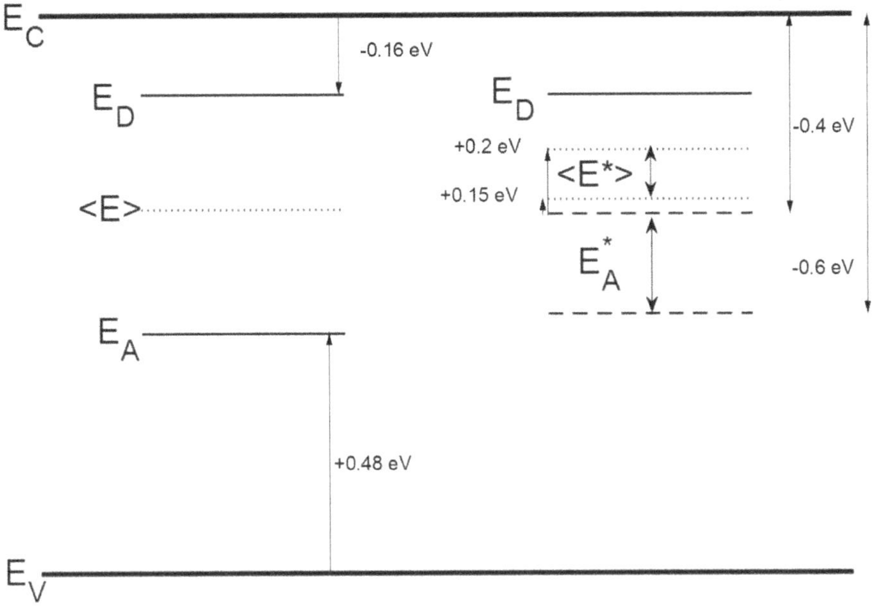

Figure 2.5 *Energy levels of interstitial hydrogen according to work of both (left) Herring et al. [55] and (right) Bonde Nielsen et al. [56]. <E> denotes the energy level at which concentrations of* H^+ *and* H^- *are equal for each set of values*

However, they could not accurately determine the true acceptor level for isolated hydrogen in silicon, with their best estimate being between 0.4 and 0.6 eV below the conduction band edge.

The practical upshot of these energy levels is that H^+ is the dominant form of hydrogen in p-type silicon, with significant concentrations of the other charge states requiring high temperatures well beyond silicon solar cell operating temperatures in the field, or as discussed later, a significant injection of minority carriers. In n-type silicon the situation is little more complex. In heavily doped n-type, H^- will be dominant. However, in lightly doped material the relative concentration of the charge states will be heavily temperature dependent. In many cases it will switch from being predominantly H^- at low temperatures to H^+ being dominant at elevated temperatures, such as those used in belt firing in the range of 700–900 °C. As a negative-U impurity, H^0 is always a minority species, but its concentration is generally maximised when the Fermi level is near the midpoint of the donor and acceptor energies (approximately 0.15 to 0.2 eV above mid-bandgap).

First principle calculations have been used to identify the minimum energy locations for each of the charge states [57]. In the positive charge state (H^+), hydrogen acts as a proton, and is Coulombically attracted to the region of highest electronic charge density. In the silicon lattice (diamond cubic), this occurs at the bond-centred (BC) site, which is half-way between two silicon atoms. However, in order to accommodate the hydrogen atom, the silicon atoms are displaced approximately 0.41 Å. In the process, the strain energy required to move the silicon atoms is sufficiently compensated by the energy gained in forming the Si-H-Si structure. Similarly for H^+, the site of lowest energy for H^0 occurs at the BC site, this time with a slightly larger displacement of the silicon atoms (0.45 Å) to account for the extra electron.

On the other hand, interstitial hydrogen in the negative charge state (H^-) takes on a completely different location in the lattice compared to H^+, driven by Coulombic repulsion away from regions of high electronic charge density in the lattice and hence towards the region of the lowest electronic density. In silicon, the region of lowest electronic charge density occurs at the tetrahedral site. For hydrogen in this configuration, the relaxation of the silicon atoms is negligible. Figure 2.6 shows the calculated positions of the different hydrogen charge states in the lattice [55].

Transitions between charge states occurs through the neutral charge state, indicating a shift of the hydrogen location as hydrogen transitions from H^+ to H^- or vice versa. The transition from H^0 to H^+ with the emission of an electron (e^-) is a phonon mediated process and occurs rapidly (on the order of one nanosecond). On the other hand, the transition from H^- to H^0 with the emission of e^- has a relatively high activation enthalpy of 0.84 eV (0.65 eV according to Bonde Nielsen *et al.*). This is not only due to the acceptor level being deeper in the bandgap but also the need for H^0 to transition between the bond-centred and tetrahedral sites in the lattice. The activation energy for these transitions has been measured to be 0.295 eV for H^0 (BC→T) and estimated at 0.2 eV for H^0 (T→BC) [58].

Given that the migration barrier for T→T diffusion of H^0 has been calculated to be only 0.12 eV, it is possible for H^0(T) to diffuse a significant distance through the lattice before moving to the overall energy minima at the BC site.

2.3.2 *Neutralisation of dopant atoms*

The charge state of hydrogen has important implications for the passivation of defects. Certain defects can only be passivated by a certain hydrogen charge state and require the presence of electrons. For example, hydrogen can bind with dopant atoms to form neutral hydrogen-dopant complexes [59–63] such as hydrogen-boron (HB) or hydrogen-phosphorus (HP) pairs. In general, it is believed that the bonds form due to an interaction between the ionised dopant and the opposite charge of interstitial hydrogen. That is, H^+ can interact with negatively charged acceptor dopants such as boron (B^-) or gallium (Ga^-), and H^- can interact with positively charged donor dopants such as phosphorus (P^+). For the case of HB pairs, hydrogen is located near the BC site between B and Si [64]. Equations 2.4 and 2.5 show the reversible reactions for the neutralisation of ionised boron and phosphorus dopants, respectively.

$$\mathrm{H}^+ + \mathrm{B}^- \leftrightarrow \mathrm{HB} \tag{2.4}$$

$$H^- + P^+ \leftrightarrow HP \tag{2.5}$$

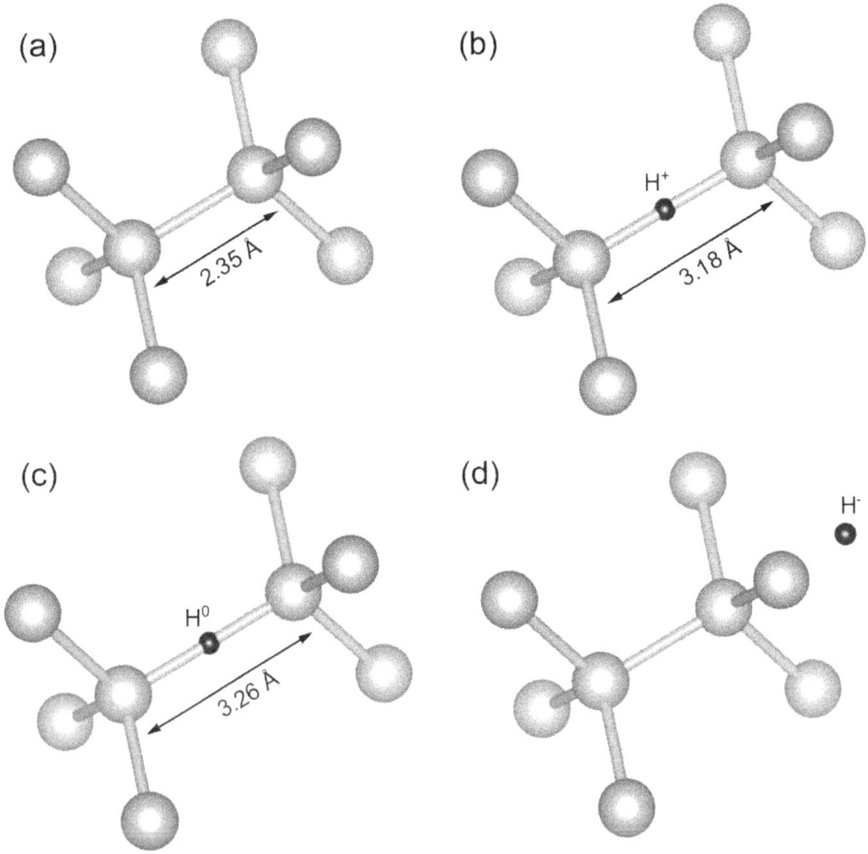

Figure 2.6 *Minimum energy configurations of hydrogen charge states in silicon. (a) shows two silicon atoms with no hydrogen in the vicinity. (b) Shows H^+ at the BC site with stretching of the Si–Si bond. (c) Shows H^0, also at the BC site, also with stretching of the Si–Si bond. (d) Shows H^- at the tetrahedral site with negligible distortion of the silicon lattice. Figure adapted from [55]*

The binding of hydrogen with boron has been exploited for the formation of selective emitters. For example, Shumate *et al.* used a hydrogen plasma to passivate boron atoms in non-metallised regions, hence locally increasing the sheet resistance in a self-aligned manner, from values of 35 Ω·cm to approximately 100 Ω·cm [65]. Other studies using shielded hydrogen plasma have also demonstrated significant reductions in the electronically active boron-doping concentration in the vicinity of the surface [66, 67]. A recent study also suggested that after an initial high-temperature firing process, during subsequent thermal processing, hydrogen within

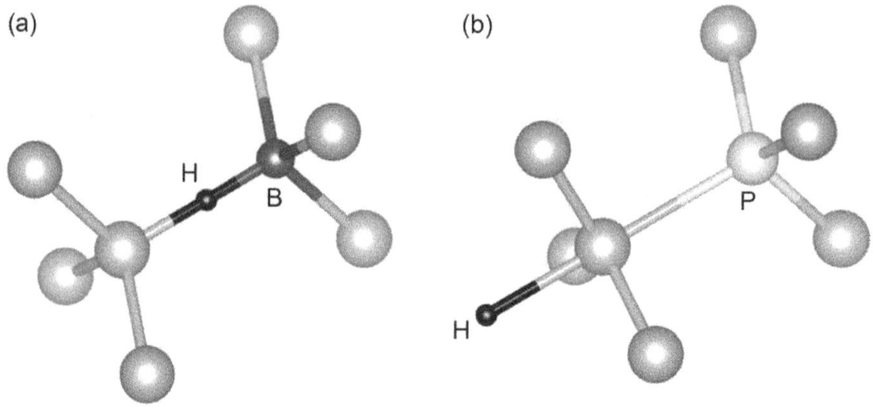

Figure 2.7 *Minimum energy configurations of hydrogen-dopant pairs in silicon. (a) Shows H bound to substitutional Boron while (b) shows H bound to a Si atom neighboring substitutional P. Figure adapted from [70]*

the p-type bulk of a solar cell tends to migrate towards heavily diffused phosphorus-doped regions and can passivate a significant fraction of the phosphorus dopants [68].

In the case of HP pairs, rather than a direct bonding of H^- to ionised P^+ to form a neutral $(HP)^0$ complex, H^- is still located at the tetrahedral site, forming a P–Si–H bond [69] (see Figure 2.7). The hydrogen atom is located at an antibonding site of a silicon atom that is adjacent to the phosphorus atom. In this instance, both the silicon and phosphorus atom undergo relaxations along the bond axis to accommodate the hydrogen atom.

However, hydrogen-dopant pairs have relatively low binding energies. For HB pairs, the binding energy is in the range of 0.7–0.85 eV [62, 71, 72], while the binding energy for HP is slightly lower at 0.5–0.71 eV [52, 55, 73]. As a result, for dopant concentrations of up to $1 \times 10^{16}/cm^3$, concentrations that are typical for bulk doping in silicon solar cells, these pairs can readily dissociate at temperatures above 200 °C. Typically, HP pairs are thought to dissociate more readily than HB pairs. However, this may be due to the asymmetry in the hydrogen charge states rather than a significant difference in binding energies [55]. The dissociation rates of hydrogen-dopant complexes are substantially increased in the presence of minority carrier injection from either illumination or an applied bias [55, 63, 71, 74], and hence less stable under normal solar cell operation than in thermal equilibrium at the same temperature.

2.3.3 Hydrogen dimer formation

In addition, interstitial hydrogen can interact with itself and other forms of hydrogen to form hydrogen dimers, or extended defects [75–77]. A recent study by Voronkov *et al.* identified at least three hydrogen dimer species at low temperatures in silicon

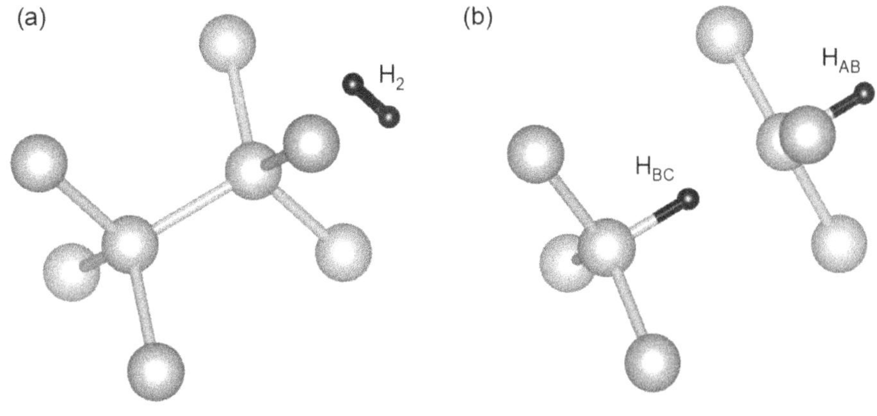

Figure 2.8 *Known stable H dimer formations in Si. (a) depicts isolated H_2 molecules at the tetrahedral sites. (a) shows with hydrogen bound to neighbouring Si atoms at the BC and AB sites*

[78], which are predicted to be the dominant hydrogen species at temperatures of up to 650 °C under equilibrium conditions [78, 79].

As shown in Figure 2.8., two species of hydrogen dimers have been identified using Raman/IR spectroscopy [80–82]. The first are interstitial H_2 molecules, while the second, H_2^* , consists of two hydrogen atoms, bound to a silicon atom in the lattice at the bond centred and anti-bonding sites. H_2 is the dominant form of hydrogen after high temperature processing in silicon free of significant defect or impurity concentrations, while H_2^* formation is thought to be the result of hydrogen interaction with vacancies and self-interstitials in silicon [83]. No electrical activity has been assigned to either of these dimers, however, they play an important role as traps to reduce the transport of hydrogen, and as 'hidden' sources that can release atomic hydrogen during solar cell processing.

The equilibrium concentrations of the various monomeric species, dimeric species, and hydrogen-dopant complexes are dependent on temperature, effective doping concentration and the diffusivity of the different hydrogen species. In particular, the dependence on temperature and effective doping results in spatial variations of the localised concentration of the various species of hydrogen throughout a silicon solar cell, which can also be modulated by thermal processing. The mechanisms and underlying equations between the formation/dissociation of dimeric and hydrogen-dopant complexes are discussed in detail in work by Voronkov *et al.* and Hamer *et al.* [68, 78].

The ability to bind and release interstitial hydrogen from dopant atoms and hydrogen dimers can allow such species to act as a storage mechanism during cell fabrication, whereby interstitial hydrogen can be later released during thermal processing and used for defect passivation [84] as will be discussed in Chapters 4 and 5. Furthermore, the passivation of defects leads to a depletion of interstitial hydrogen species, and subsequent replenishment of the interstitial hydrogen species from other hydrogen species to maintain equilibrium. Undesirably, the release of hydrogen from these sites also poses issues for hydrogen-related defects as will be discussed in Chapter 6.

2.3.4 Hydrogen interactions with other defects

The interactions of hydrogen with a range of impurities, as well as crystal defects such as vacancies, can greatly alter the behaviour of hydrogen in silicon and have been intensively studied. For the purposes of this chapter we will consider three common cases:

1. Oxygen
2. Carbon, and
3. Silicon vacancies

Hydrogen is well known to be highly effective at passivating dangling silicon bonds. Therefore, the interaction of hydrogen with vacancies comes as no surprise, with the existence of VH_1, VH_2, VH_3 and VH_4 complexes in silicon calculated and observed [85–88]. Complexes with vacancy clusters and higher numbers of hydrogen atoms V_xH_n have also been investigated [87, 89]. The $VH_{n=1.4}$ complexes are quite stable in silicon, with dissociation energies between 2.2 and 2.6 eV [85], significantly higher than most dopant-hydrogen complexes.

While VH do possess levels within the silicon bandgap, they have not to date been reported as a significant source of recombination within silicon solar cells. However, the interaction of hydrogen with vacancies can have a significant impact on hydrogen diffusion. The first impact is relatively straightforward, with vacancies being highly effective traps for isolated hydrogen atoms. However, the impact on hydrogen molecules is more interesting. It is well established that vacancies can split hydrogen molecules, resulting in Si–H bonds [28, 90]. This can be exploited to increase the incorporation of hydrogen from a molecular ambient through the presence of a damaged surface layer [28]. It has also been reported as a possible method to improve bulk passivation by hydrogen, whereby vacancies introduced from a surface of the wafer interact with hydrogen molecules [28]. There are two mechanisms by which this might take place; firstly it has been suggested that $VH_{n<4}$ complexes are mobile, with activation energies between 1.0 and 3.4 eV [58] and under circumstances where diffusion of isolated hydrogen is trap-limited may actually increase the effective diffusivity of hydrogen [28]. A, perhaps more likely, explanation is that vacancies can act to split hydrogen molecules in the bulk, and that either the resulting vacancy-hydrogen complex or resulting atomic hydrogen can then passivate nearby defects [90, 91].

Two main hydrogen-carbon complexes in silicon have been observed via IR spectroscopy [92–94]. The first is H_2^*, which is formed when the H_2^* dimer (one H atom at a bond-centred site and one at the anti-bonding site) is trapped by substitutional carbon. The second is the $(CH)_2$ complex, which consists of bound substitutional carbon-hydrogen pairs (C-H-H-C). Where the carbon concentration [C] > 2

$\times$ 10^{17}/cm^3, such as in multi-crystalline silicon, the H_2*C complex can become the dominant form of hydrogen within a silicon wafer after thermal treatment [94, 95], replacing the hydrogen dimers which are most prevalent in defect-free material. This complex is electrically inactive and more thermally stable than hydrogen-dopant pairs, however, it can still act as a reservoir for hydrogen during subsequent thermal processing. Because $(CH)_2$ requires two adjacent substitutional carbon atoms it is only prevalent in silicon with carbon concentrations in excess of 5 $\times$ 10^{17}/cm^3. It has been observed that the presence of carbon does not materially affect hydrogen diffusion at temperatures in excess of 750 °C [94].

Hydrogen can also form complexes with interstitial oxygen, O_i. However, none of the complexes observed are stable above 70 °C, or have any significant electrical activity [96]. The main interaction of interest is the increase in oxygen diffusivity observed in the presence of hydrogen [97, 98], with a reduction in activation from 2.5 to 2 eV observed. It has been proposed that this is a result of the capture of hydrogen at O_i, followed by dissociation of the unstable complex. On dissociation a diffusion jump of O_i may occur. While initial work primarily considered the interaction of isolated hydrogen with O_i, further studies suggested that O_i could also interact with interstitial hydrogen molecules [96, 98]. This effect is of particular interest because it accelerates the formation rate of oxygen-related thermal donors [97, 99]. This has been used to form deep (tens of microns) p-n junctions through the diffusion of hydrogen into lightly doped p-type material at temperatures 250–400 °C [100]. It should be noted that the introduction of hydrogen can also lead to the generation of another, less studied, class of shallow thermal donors [99].

To complicate things further, hydrogen forms vacancy-carbon-hydrogen [58] and vacancy-oxygen-hydrogen [89] complexes. These are typically more stable than their counterparts that do not include a vacancy and can be significant reservoirs of hydrogen in silicon. Hydrogen can also form recombination active oxygen-carbon-hydrogen complexes, which will be discussed in detail in Chapter 6.

The complex behaviour of hydrogen in silicon has led to significant challenges in interpreting experimental results. Several key complications that arise are:

1. atomic hydrogen interacts with the silicon lattice and virtually all impurities and defects within the silicon [101]
2. hydrogen is difficult to measure and detect [50, 102]
3. hydrogen can be unintentionally incorporated into silicon during a variety of processing steps such as ion beam or chemical processing [50, 79, 103]

The third aspect is a key challenge. Even a dilute HF dip at room temperature can result in a significant concentration of atomic hydrogen diffusing several microns into the bulk of the silicon [103]. Although high temperatures are often considered critical for introducing hydrogen to silicon solar cells, with sufficient time and appropriate processing conditions, hydrogen can diffuse entirely throughout a silicon wafer at temperatures below 100 °C [104], hence creating confusion in the literature whereby attempts to avoid hydrogen injection may still allow for a slow passivation of defects.

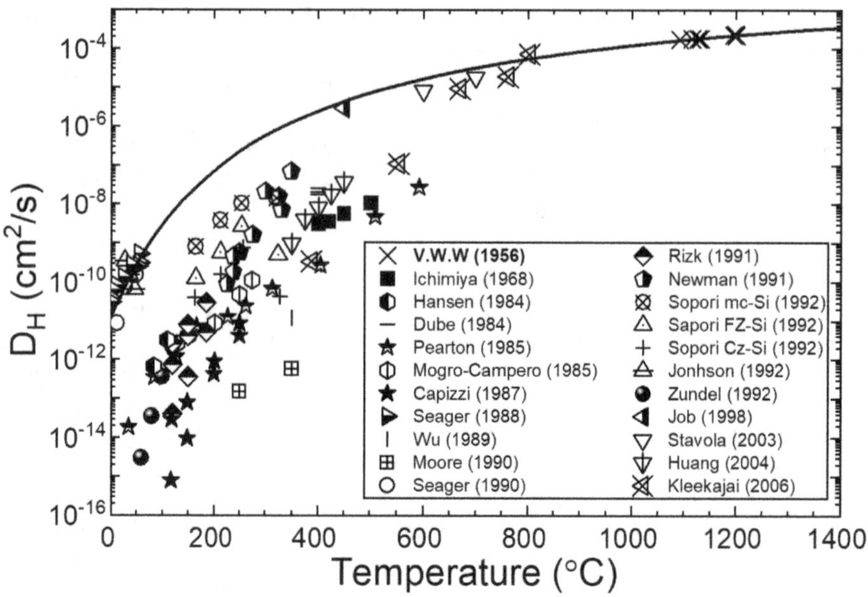

Figure 2.9 Hydrogen diffusivity data taken from the literature [54, 73, 97, 100, 104–119]. The solid line represents the extrapolation of the hydrogen diffusivity from the work of Van Wieringen et al. (V.W.W.) [104]. Adapted from [120]

2.4 Hydrogen migration in silicon

2.4.1 Hydrogen diffusivity

While the benefits of hydrogen in silicon-based semiconductors are widely accepted, the diffusion and passivation mechanisms and dependence on material parameters are poorly understood. For example, extensive studies by multiple groups have reported on hydrogen diffusivity data, with differences of up to five orders of magnitude in the diffusion coefficient reported for a given temperature (see Figure 2.9.). A consensus on exact values for hydrogen diffusivity has not been obtained. However, it is generally agreed that hydrogen transport mostly occurs in the interstitial form [55, 68].

Work by Van Wieringen and Warmoltz in the temperature range of 1000–1200 °C is often cited as an upper limit of hydrogen diffusivity in defect-free silicon [104]. At lower temperatures, a broad range of diffusivities has been reported and is heavily dependent on the background doping concentration [53, 73, 103, 105]. The astonishing discrepancies are primarily due to the interaction of hydrogen with electric fields and numerous impurities and defects within crystalline silicon, including interactions with the silicon lattice [28, 103, 106, 121–125]. In many cases, what is measured is the trap-limited diffusion of hydrogen in silicon [103, 107].

For silicon solar cells, industrial processing times are in the vicinity of one second per wafer, to enable high throughput production. The typical thickness of silicon solar cells is in the range of 150–200 μm. Therefore, after releasing hydrogen from bound states in the dielectric layers, to have hydrogen distributed throughout the wafer during thermal processing, a diffusion length of hydrogen in the order of 100 μm is required. Hence, to diffuse hydrogen throughout a wafer in a timescale of approximately one second, a temperature of more than 700 °C is needed, highlighting the importance of the fast-firing process for metallisation pastes. On the other hand, for a process of 200 °C, as is used for silicon heterojunction (SHJ) solar cells, a processing time of approximately 22 minutes is required, assuming an upper limit of diffusivity as extrapolated from [104]. This duration is within that expected for SHJ solar cells during deposition of PECVD a-Si:H and curing of metal pastes.

2.4.2 Impact of charge state on hydrogen diffusivity

Due to a variety of interactions in the lattice, large differences have been observed for the mobility of the various charge states within crystalline silicon. For example, being neutral, H^0 is not affected by electric fields and charge effects within the silicon. On the other hand, as a result of the high charge to mass ratio, for H^+ and H^-, motion can be greatly affected by electric fields. For example, an applied bias across a Schottky junction and internal electric fields due to doping profiles can have a significant influence on the motion of hydrogen [68, 73, 79, 126–128].

Similar to many properties of hydrogen, there is a staggering level of uncertainty over the diffusivity of each hydrogen charge state. There is even uncertainty over which charge state has the highest diffusivity. A study by Rizk *et al.* investigating hydrogen diffusivity in the range of 120 to 185 °C using deuterium concluded that H^0 had the lowest diffusivity [105]. In contrast, a study by Mathiot *et al.* calculated that H^0 had the highest diffusivity by several orders of magnitude, being three orders of magnitude higher than that of H^- and five orders of magnitude higher than that of H^+ [100, 105].

While there is a consensus on the activation energy for the diffusivity of H^+ between 0.43 and 0.48 eV [58, 104, 129], there are significant uncertainties for the activation energy reported for the diffusivity of H^-. For example, between 27 °Cand 47 °C, Johnson *et al.* determined an activation energy of 0.7 eV [73]. However, a recent theoretical investigation by Estreicher *et al.* identified an energy barrier of only 0.39 eV [58]. Due to interactions in the lattice, an unambiguous determination of the diffusion of H^- at elevated temperatures has proven challenging.

An accurate determination of the diffusivity of H^0 is even more difficult. In addition to the proportion of H^0 being much lower than that in other charge states, H^0 can be found at both the bond-centred and tetrahedral sites within the silicon lattice. Although the bond-centred site is the minimum energy location for both H^+ and H^0, H^- is usually found at the tetrahedral site. As a result, the transitions between H^0 and H^- occur at the tetrahedral site [55, 58]. However, the fraction of H^0 located at the tetrahedral ($H^{0,T}$) and bond-centred ($H^{0,BC}$) site has not yet been reported. Theoretical calculations for $H^{0,T}$ have predicted that it should be metastable and migrate rapidly through the lattice with an activation energy of 0.14 eV [58]. The high diffusivity of $H^{0,T}$ has also been inferred from deep-level transient spectroscopy (DLTS) measurements at low temperatures [130]. In contrast, the diffusivity of $H^{0,BC}$ is expected to be closer to that of H^+ with a theoretical activation energy for diffusion of 0.38 eV [58]. Given the different diffusivities of the different charge states, the effective diffusion coefficient for interstitial hydrogen will greatly depend on the fractional concentration of hydrogen in each of the charge states.

2.4.3 Uncertainties around hydrogen passivation

Due to the challenges of measuring the properties of hydrogen, and the large uncertainties of diffusivity reported in the literature, there are large discrepancies in the conclusions drawn from experiments related to the possible involvement of hydrogen. Numerous authors have offered interpretations of hydrogen passivation mechanisms in silicon. However, as stated by Van de Walle *et al.*, attempts to explain observed results have led to several contradictory assumptions about the hydrogen charge state and hence the passivation reactions which can occur [57, 59–61, 131]. Questions have even been raised over the possibility and effectiveness of hydrogen to passivate certain defects [132, 133]. The likelihood of hydrogen passivating bulk defects has also been questioned due to the majority of the hydrogen in the dielectric layers leaving the solar cell rather than diffusing into the bulk of the silicon [134].

For solar grade silicon, interstitial iron is a common impurity. Several publications have investigated the impact of hydrogen passivation on iron with very different conclusions. Some studies have suggested that significant concentrations of iron can be hydrogen passivated [135–137]. Other studies have indicated that during such thermal processes iron can be gettered to the surface dielectric layers, or many other possibilities [138, 139].

Another defect of key importance for silicon solar cells with contradictions in the literature is the boron-oxygen (B-O) complex [26, 140–142], responsible for carrier-induced degradation in p-type Czochralski (Cz) silicon [143, 144]. A solution for the permanent deactivation of the B-O defect based on annealing in the presence of carrier injection was first proposed in 2006 [145]. In 2009, Münzer identified the critical role of hydrogen whereby permanent deactivation only occurred in solar cells fabricated using hydrogenated silicon nitride (SiN_X:H) grown by PECVD, and not those using hydrogen-lean SiN_X:H layers grown by low-pressure chemical vapour deposition (LPCVD) [146]. Subsequent work has provided further evidence for the role of hydrogen, by identifying the influence of the hydrogen concentration and thickness of dielectric layers in the passivation reaction [47, 147–150]. Despite the strong evidence presented for the role of hydrogen in the permanent deactivation of B-O defects, controversy has surrounded its involvement. For example, one study showed an alternative theory for permanent deactivation involving boron nanoprecipitates and a purely thermal mechanism [151]. A subsequent study attempted to avoid the injection of hydrogen into the wafer, yet still achieved permanent deactivation [152]. From this, along with assumptions about hydrogen diffusivity using values more than two orders of magnitude lower than experientially observed values at the same temperature shown in Figure 2.9, concluded that hydrogen could not be involved. The passivation of B-O defects will be discussed in detail in Chapter 5.

2.5 Detection of hydrogen in silicon

The primary reason so little is known about hydrogen passivation and migration processes in silicon solar cells is that the bulk hydrogen concentrations involved in these processes are well below the detection limits of most available techniques. There are three main classes of detection methods for the detection of hydrogen in crystalline silicon samples. The first is vacuum-based techniques to detect the total hydrogen concentration (including Secondary Ion Mass Spectroscopy (SIMS) [105, 153], Atom Probe Tomography (APT) [154, 155], and Glow Discharge Optical Emission Spectroscopy (GDOES) [156]).

The second class is the detection of hydrogen complexes such as hydrogen-metal complexes [132, 157–161], hydrogen-defect complexes [87, 89, 162] or hydrogen dimers [81, 163–165], typically using Infrared or Raman Spectroscopy [81, 163–165], Deep Level Transient Spectroscopy (DLTS) [56, 96, 157, 158, 162, 166] or electron paramagnetic resonance (EPR) [87, 161]. A third is the observation of hydrogen deactivation of dopant impurities [59, 60, 126, 167] via changes in resistivity, either locally using capacitance voltage methods [59, 126], or throughout the entire silicon bulk [165, 167].

The main problem for the first two approaches is sensitivity. Typical hydrogen concentrations in the bulk of silicon solar cells have been estimated between 10^{14} and 10^{16} atoms/cm^3 [159, 167], with at most a few per cent of this being mobile and able to participate in reactions below 200 °C. In a vacuum system it is practically impossible to evacuate enough atmospheric hydrogen to achieve detection limits below 10^{18} atoms/cm^3, which is only capable of profiling the relatively enormous concentrations of hydrogen trapped in amorphous and micro-crystalline silicon layers. The use of the ^{2}H isotope greatly improves this detection limit, allowing for SIMS profiling of near-surface hydrogen concentrations in silicon [105, 153] and more recently identification of hydrogen segregation to interfaces and crystallographic defects using APT [154, 155]. However, very few studies have been carried out using deuterium substitution in dielectric layers, such as are typical for hydrogen introduction to solar cells [154], and both the detection limit and practical maximum profiling depth of these techniques are not compatible with accurately determining hydrogen concentrations in the bulk of a crystalline silicon solar cell. Furthermore, such techniques provide no information on the configuration of hydrogen or what it might be bound to. An example from a study by Tweddle *et al.* showing the direct measurement of hydrogen at a grain boundary is shown in Figure 2.10.

Both DLTS and vibrational spectroscopy techniques have been used to observe the behaviour and properties of a range of hydrogen configurations, most notably hydrogen dimers [165], hydrogen-vacancy complexes [87, 89, 162] and hydrogen-oxygen complexes [56, 96]. The limitation is that these studies typically only look

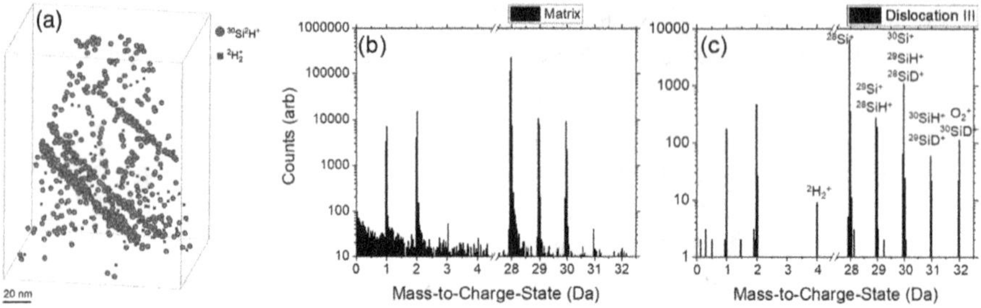

Figure 2.10 *APT datasets showing the enrichment of $^{2}H^{2+}$ and $^{29}Si^{2}H^{+}$ at three individual dislocations. (a) Viewing perpendicular to the dislocation. (b) Corresponding mass spectra of the matrix of the APT specimen. (c) Mass spectra of dislocation III, showing clear enrichment in ^{2}H at 4 Da, 31 Da and 32 Da. Figure adapted from [155]. Please see Tweddle et al. [155] for full color version and details of deconvolution*

at one or two specific forms of hydrogen and use tailored structures and processes to maximise the concentrations of these forms in the region of interest.

Similarly, studies of hydrogen-metal interactions [157, 158, 160] use elevated metal concentrations to maximise the signal. This means that it is problematic to correlate behaviour observed in these experiments with the behaviour of hydrogen introduced to a silicon solar cell. Not only is this behaviour significantly impacted by the presence of high concentrations of precursor impurities/defects used in such studies, but hydrogen is introduced at temperatures much lower than those used in solar cell manufacturing. An exception to this generalisation was the use of intentional Pt contamination at the rear of a typical solar cell structure by Kleekajai *et al.* [159]. Since the binding energy of the Pt-H complex is quite high it was possible to trap hydrogen at the rear, which had been released from a nitride layer on the front surface and diffused through the silicon bulk. This provided a valuable estimation of the quantity of hydrogen entering the silicon during a firing process, albeit the suppression of hydrogen effusion via the Pt traps likely introduced some error. Figure 2.11 shows an example of Raman spectroscopy measurements after exposure to a hydrogen or deuterium plasma [81].

A number of recent studies have used techniques such as Nuclear Resonant Reaction Analysis (NRRA) [149, 168–170] or Elastic Recoil Detection Analysis (ERDA) [171–173] to observe hydrogen concentrations of the order of an atomic per cent in the first few tens of nm of crystalline silicon samples using high energy MeV ion beams. Thorough reviews on NRRA and ERDA are presented by Wilde and Fukutani [168] and Lanford [174]. These techniques can provide a qualitative comparison of the amount of hydrogen introduced, e.g. from different dielectric layers [149, 171, 175], provided that the temperature and duration of hydrogen introduction remain constant. However, caution must be exercised when comparing

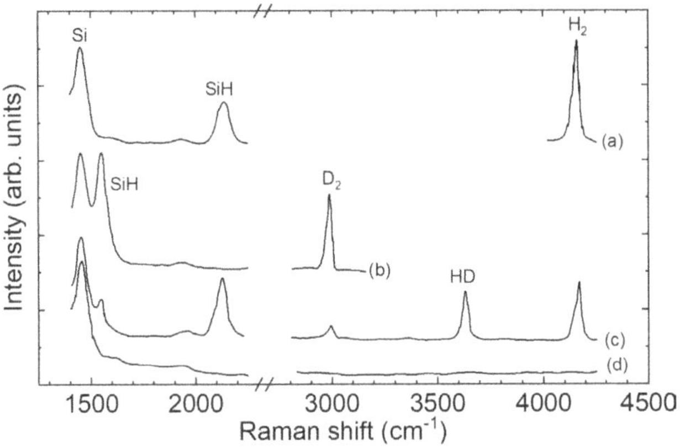

Figure 2.11 *Room temperature Raman spectra of p-type FZ silicon after (a) exposure an 8 hour H_2 plasma treatment at 250 °C, (b) D_2 plasma, (c) a H_2:D_2 (50:50) plasma, and d) no plasma treatment. Excitation wavelength: 488 nm. Reprinted figure with permission from [81] Copyright (1998) by the American Physical Society. http://dx.doi.org/10.1103/PhysRevLett.81.421*

different processing conditions as this will affect how much hydrogen in trapped in immobile forms near the surface and hence would be detected by these techniques.

The final class of techniques, that involving the deactivation of dopant impurities in silicon is both the most straightforward and amongst the most sensitive. Hydrogen-dopant pairs at concentrations of a few $10^{14}/cm^3$ have been accurately measured using both capacitance-voltage [176] and spreading resistance techniques [60, 61]. Furthermore, dopant atoms are present in high concentrations in virtually all crystalline silicon solar cells. However, due to the low binding energies of most hydrogen dopant pairs (0.7–0.85 eV for H-B and 0.5-0.71 eV for H-P [68]) very little hydrogen is present in this form at the end of solar cell processing. It is possible to release hydrogen trapped as dimers during processing to increase deactivation of dopants in the silicon bulk [165, 167], which gives a good estimation of the amount of hydrogen present in boron-doped float-zone samples rapidly cooled after firing, but it is unclear how effective this approach is on the more common commercial Cz and multi-crystalline materials where there are a large number of competing traps for hydrogen. Figure 2.12 shows an example of changes in the spreading resistance to due hydrogen passivation of dopants and deuterium near the silicon surface measured by SIMS [177].

Unfortunately for most hydrogen interactions critical to the bulk lifetime of solar silicon materials the hydrogen configurations are either unknown or present in concentrations below $10^{12}/cm^3$ and so, to our knowledge, direct observation of the involvement of hydrogen has not been reported to date. Instead, almost all papers rely on correlation between the behaviour observed and the expected behaviour of hydrogen from fundamental studies.

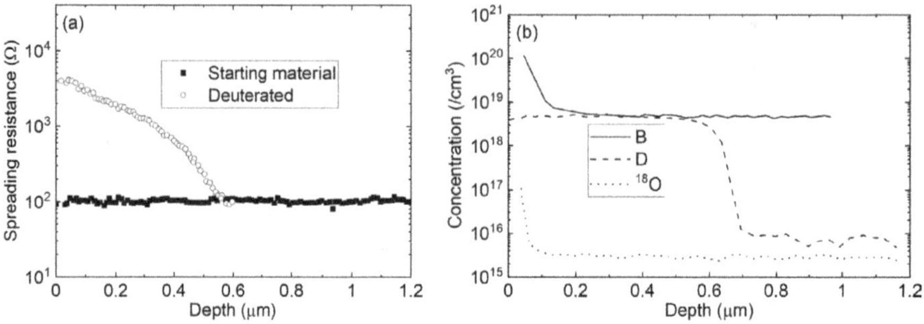

Figure 2.12 *(a) Spreading resistance and (b) SIMS profiles in boron-doped Czochralski-grown silicon with an initial uniform boron concentration of 5 × 10^{18}/cm^3. Reprinted figure with permission from [177] Copyright (1985) by the American Physical Society. http://dx.doi.org/10.1103/PhysRevB.31.5525*

2.6 Fractional charge state concentrations of interstitial hydrogen in silicon in thermal equilibrium

For defect passivation, it is assumed that the key species responsible is a form of interstitial hydrogen, due to the mobility and reactivity of interstitial hydrogen, and hence availability to bind with a range of defects in silicon. While species such as hydrogen-boron pairs, hydrogen-phosphorus pairs and hydrogen dimers are also likely to play a significant role in modulating kinetics of hydrogen passivation, here we only consider hydrogen in interstitial form.

A comprehensive study on the behaviour of hydrogen in silicon is done by by Herring *et al.* [55]. This was long called *The Hydrogen Bible* by the late Professor Wenham for its invaluable contribution in developing the advanced hydrogenation technology, in highlighting the ability to modulate fractional charge state concentrations using minority carrier injection. In thermal equilibrium, the dominant hydrogen charge state depends on the material's dominant conductivity. Undesirably, interstitial hydrogen naturally goes into a low mobility charge state. For example, at low temperatures in p-type silicon, due to a deficiency of electrons, essentially all hydrogen is H^+. For n-type silicon, due to an abundance of electrons, essentially all hydrogen is H^-. This greatly impacts hydrogen diffusivity, being the opposite charge to the ionised dopant atoms, and also has implications for reactivity.

Various models have been used to calculate the fractional charge state concentrations of interstitial hydrogen in silicon [55, 178–181]. In *The Hydrogen Bible*, the the fractional charge state concentrations in thermal equilibrium were calculated using the Fermi level [55]. For this chapter, Sah-Shockley statistics [28, 121] and a model developed in Matlab for hydrogen charge state occupation are used, based on the work by Sun *et al.* [181]. These simulations were reported in [120]. This assumes a hydrogen acceptor energy level of 0.07 eV below mid-gap and a donor energy level of 0.16 eV below the conduction band. The capture cross-section ratios

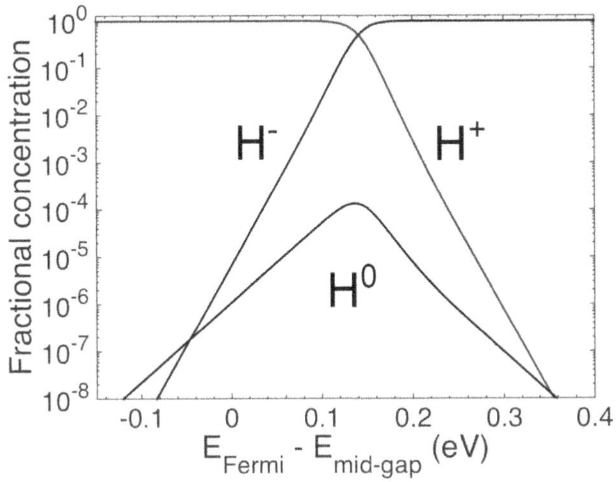

Figure 2.13 *Fractional hydrogen charge state concentrations of interstitial hydrogen in silicon in thermal equilibrium at room temperature as a function of the Fermi level (E_{Fermi}) relative to the mid-gap energy level ($E_{mid\text{-}gap}$). Figure adapted from [120]*

assumed for the acceptor and donor levels are 0.05 and 2, respectively [181]. Band-gap narrowing is used and the model assumes that the energy levels maintain their relative positions within the band-gap.

Figure 2.13 shows the fractional charge state concentrations at room temperature as a function of the position of the Fermi level (E_{Fermi}) relative to mid-gap ($E_{mid\text{-}gap}$). As shown, there is a crossover point for an E_{Fermi} at 0.14 eV above mid-gap, which is approximately halfway between the E_D and E_A for hydrogen. At this energy level, there are equal probabilities of H^+ and H^-. Hence, this is often termed the energy level for the H^+/H^- transition ($E_{+/-}$). At this E_{Fermi} level, the concentration of H^0 is maximised. Despite this, it is still orders of magnitude lower than that of H^+ and H^-, with a fractional concentration of 1.3×10^{-4}. This E_{Fermi} level corresponds to an effective donor doping concentration of $2.2 \times 10^{12}/cm^3$ at room temperature, which is essentially very lightly doped n-type material with a bulk resistivity 2 000 Ω·cm. This value is much higher than that typically used for wafers in the industry, with a value on the order of 2–5 Ω·cm for n-type. For silicon with an E_{Fermi} level above this value, which represents more heavily doped n-type material, essentially all hydrogen is H^-, with the concentrations of H^+ and H^0 rapidly falling off. For 1-Ω·cm n-type material with a donor doping concentration of $N_D = 5 \times 10^{15}/cm^3$, the E_{Fermi} level is 0.34 eV above mid-gap. Hence, this silicon is completely dominated by H^-. On the other hand, for E_{Fermi} levels below the $E_{+/-}$ level, representing p-type silicon or extremely lightly doped n-type silicon, essentially all hydrogen is H^+. For 1-Ω·cm p-type material with an acceptor doping concentration of $N_A = 1.5 \times 10^{16}/cm^3$, the E_{Fermi} level is 0.37 eV below mid-gap, and H^+ completely dominates.

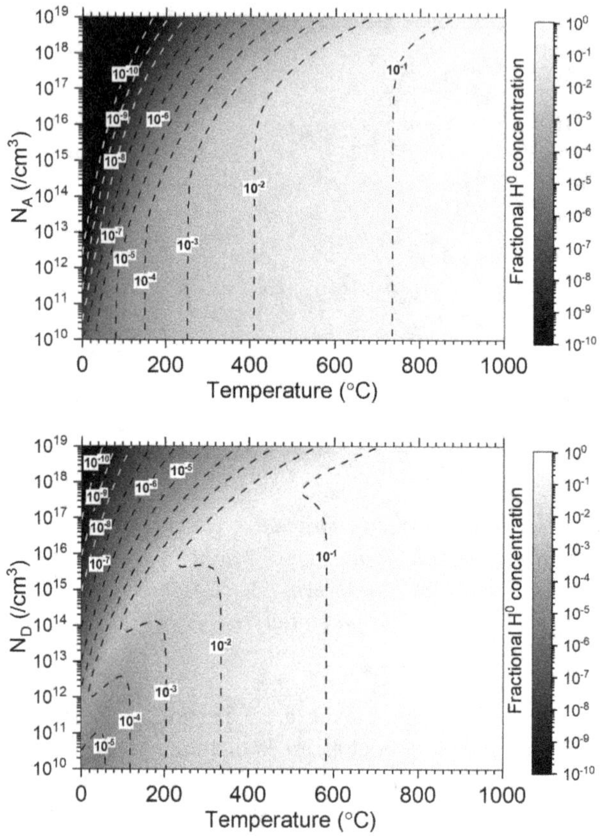

Figure 2.14 *Fractional H^0 concentration in silicon in thermal equilibrium as a function of temperature for (top) p-type silicon as a function of acceptor dopant concentration (N_A), and (bottom) n-type silicon as a function of donor dopant concentration (N_D). It should be noted that the lower limit of the colour bar is truncated at 1×10^{-10}. Figure adapted from [120]*

As a result, for practical solar cell applications, the background doping alone cannot be used to significantly enhance H^0 concentrations. However, while still under thermal equilibrium conditions, increasing the temperature is a very effective tool to control hydrogen charge state concentrations. Increasing the temperature makes the silicon behave more like intrinsic silicon as the E_{Fermi} level moves closer to mid-gap, and hence closer to the $E_{+/-}$ transition (apart from very lightly doped n-type material). This acts to increase the H^0 concentration. Figure 2.14 highlights the impact of both the background doping concentration and temperature on the fractional H^0 concentration in both p-type and n-type. silicon. As shown, increasing the temperature can greatly enhance the fractional H^0 concentration. For example, at room temperature, the fractional H^0 concentrations in p-type and n-type silicon are

4.1×10^{-13} and 2.3×10^{-8}, respectively. At a modest 400 °C, a temperature typically used for the deposition of SiN_X:H, the fractional charge state concentrations increase dramatically, to 7.5×10^{-3} and 2.4×10^{-2}, respectively. For a temperature of 700 °C, commonly achieved during the high-temperature fast-firing process, the fractional concentrations increase further to 8.6×10^{-2} and 1.7×10^{-1}, respectively. The use of higher temperatures also greatly increases the diffusivity of hydrogen as was shown in Figure 2.9.

In general, as the doping concentration increases for both p-type and n-type silicon, the H^0 concentration rapidly falls. Hence, as the bulk doping of the silicon is increased, higher temperatures are required to generate the same fractional H^0 concentration. Furthermore, it would appear easier to generate H^0 in n-type than in p-type silicon, presumably due to the asymmetry in the donor and acceptor levels of hydrogen in the band-gap, with the $E^{+/-}$ level in the top half of the band-gap.

However, despite the ability to generate large quantities of H^0 at high temperatures, the fast-firing process typically used for metallisation in industrial solar cells is not effective for the passivation of performance limiting defects such has the boron-oxygen complex. This is due to the thermal dissociation of hydrogen-defect complexes and in the case of the B-O defect, the dissociation of the defect complex, too [182]. A number of studies have investigated the activation energy for dehydrogenation, as the bond between hydrogen and the defect is broken [183, 184].

As a result, for certain defects, temperatures lower than that used during fast firing are required to reduce the reaction rates for dehydrogenation and defect dissociation. Unfortunately, the temperatures required to reduce these reactions sufficiently may be too low to thermally generate enough hydrogen in high mobility charge states to move the hydrogen within a capture cross-section of the defects and allow effective passivation. This leaves boron-doped solar cells susceptible to light-induced degradation after cell fabrication, and hence other methods are required to enhance the H^0 concentration and therefore defect passivation. This will be discussed in detail in Chapter 5.

2.7 Increased generation of H^0 using carrier injection

The key to increasing the fractional H^0 concentration in silicon solar cells is to change the availability of electrons and holes in the material.

Manipulating fractional hydrogen charge state concentrations

There are several key factors that can manipulate fractional hydrogen charge state concentrations in silicon solar cells, and therefore be used to increase the fractional concentration of the highly mobile and reaction H^0 charge state:

- Effective doping concentration
- Temperature
- Excess carrier concentration
- Effective lifetime of the material

Manipulating fractional hydrogen charge state concentrations

A variety of techniques can be used to provide excess carriers such as current injection, an applied bias or illumination [185, 186]. This forms the basis of techniques for the advanced hydrogenation technology to increase the concentration of H^0 [187].

When modelling fractional charges state concentrations under non-equilibrium conditions, particularly for medium- to high-injection conditions, as noted by. Sun *et al.* [181, 188]. As shown in Figure 2.15, the fractional concentration of H^0 is greatly enhanced with carrier injection. Even with a modest minority carrier concentration of $\Delta n = 1 \times 10^{15}/\text{cm}^3$, which can easily be achieved with 1 sun illumination, can dramatically increase the fractional H^0 concentration. It is noted that a different scale has been used compared to that in Figure 2.14. As shown, increases in the H^0 concentration by orders of magnitude are realised for both p-type and n-type silicon for low temperatures below 200 °C, even in relatively heavily doped silicon. However, for p-type silicon with $N_A > 1 \times 10^{19}/\text{cm}^3$, typical of boron emitters, the concentration begins fall. Heavily doped n-type silicon is somewhat less impacted, still maintaining relatively high fractional H^0 concentrations up to a doping density of $N_D = 1 \times 10^{20}/\text{cm}^3$.

Importantly, the use of minority carrier injection overcomes a fundamental limitation of extremely low fractional H^0 concentrations in silicon solar cells, which are naturally imposed when performing hydrogenation processes at low temperatures in the dark, such as during cooling after fast firing, and hence, in thermal equilibrium. This allows moderately high fractional H^0 concentrations to be realised both in the bulk silicon region as well as diffused emitter regions, even at room temperature, orders of magnitude higher than that in thermal equilibrium. It is again noted that higher H^0 concentrations can be realised in n-type silicon than that in p-type silicon at low temperatures.

Given the challenges of dehydrogenation for certain defects at high temperatures, or low fractional H^0 concentrations in the dark while cooling, after the high-temperature firing process distributes hydrogen throughout the bulk of the silicon during solar cell fabrication, a subsequent advanced hydrogenation process incorporating minority carrier injection may be required to enable rapid hydrogen diffusion

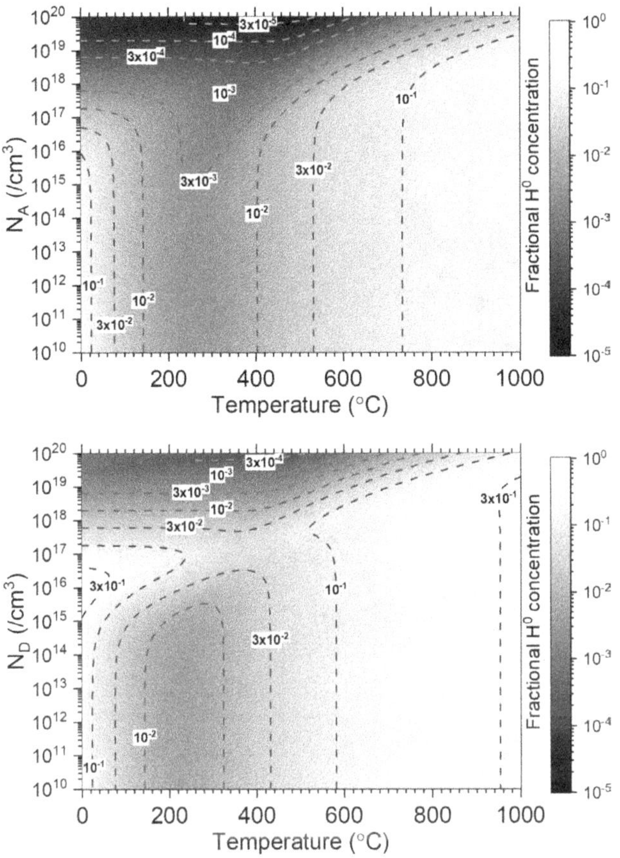

Figure 2.15 *Fractional H^0 concentration in silicon with an excess minority carrier concentration of $\Delta n = 1 \times 10^{15}/cm^3$ as a function of temperature for (top) p-type silicon as a function of acceptor dopant concentration (N_A), and (bottom) n-type silicon as a function of donor dopant concentration (N_D). Note, a different scale has been used compared to that in Figure 2.14. Figure adapted from [120]*

to the defect site to passivate difficult defects. Alternatively, carrier injection, most likely in the form of illumination, could be provided during the cooling within the belt furnace for fast-firing. Hence, it is important to understand the impact of minority carrier injection of the fractional H^0 concentration in the bulk of the device.

Figure 2.16 highlights the fractional H^0 concentration as a function of excess carrier concentration and temperature for 1 Ω·cm silicon material. Particularly for p-type silicon, the generation of H^0 is highly sensitive to the injection level for temperatures below 200 °C. Regarding how much carrier injection should be provided

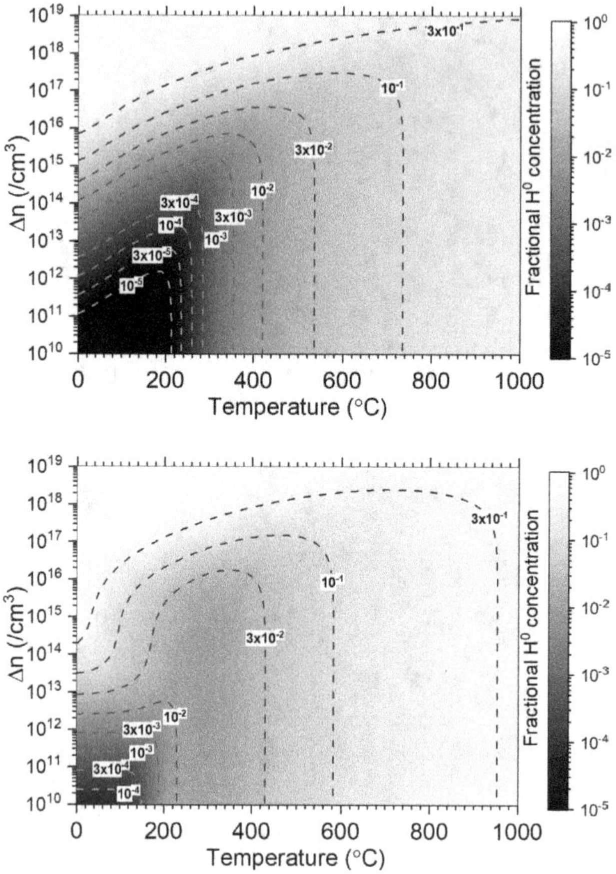

Figure 2.16 *Fractional H^0 concentrations in 1 Ω·cm silicon material as a function of temperature and excess carrier concentration (Δn) for (top) p-type silicon and (bottom) n-type silicon using Sah-Shockley occupation statistics. Note, a different scale has been used compared to that in Figure 2.14, but the same scale as for Figure 2.15. Figure adapted from [120]*

to increase the H^0 concentration, the more, the better! Ideally sufficient carrier-injection should be provided to enable an $\Delta n > 1 \times 10^{16}/cm^3$. Although n-type silicon is less sensitive, carrier injection is still beneficial, and relatively high concentrations of H^0 can be achieved provided at these low temperatures provided an excess carrier concentration over $\Delta n = 1 \times 10^{12}/cm^3$ is used, which is readily achieved using a low illumination intensity of 0.1 sun.

Particularly in low-lifetime samples, high illumination intensities are essential to enhance the H^0 concentration. Furthermore, using high illumination intensities can reduce the impact of spatial non-uniformities in the minority carrier lifetime, which tend to affect the lifetime in low-injection. This is particularly important for multi-crystalline silicon, where lifetimes can spatially vary by more than an order of magnitude in low-injection. By operating in high-injection where the dark saturation current density and intrinsic recombination dominates the lifetime, and hence minority carrier concentration, greater control over the passivation reactions can be achieved.

Due to resistive effects, such conditions are hard to achieve using current injection or an applied bias. In contrast, illuminated annealing in open circuit does not suffer from resistive effects.

Impact of solar cell structure and processing on fractional hydrogen charge state concentrations

Due to the impact of the excess carrier concentration, doping density and lifetime on the H^0 concentration, further conclusions of relevance for silicon solar cells can be drawn for device aspects that can also impact fractional hydrogen charge state concentrations:

- The solar cell structure
- The background lifetime of the material/structure
- Any process that changes the lifetime of the material
- The evolution of light-induced defects

As such, changes in the above must be considered when studying hydrogen passivation for silicon solar cells. Care must be taken when making assumptions about the behavior of hydrogen in silicon solar cells based on conclusions drawn from test structures or when using different processing conditions.

The solar cell structure, and processes used to fabricate the solar cell, can greatly impact aspects of the device including doping concentration and lifetime, either globally, or in localised regions. For example, a higher H^0 concentration will be expected in PERC solar cells than aluminium back surface field (Al-BSF) solar cells, and similar for TOPCon or SHJ solar cells compared to PERC due to the improved surface passivation or bulk lifetime. As time goes by, with V_{OC} values of silicon solar cells increasing and the transition to high efficiency solar cell architectures, the Δn will naturally increase, thereby relaxing the requirement for the use of high illumination intensities to provide sufficient excess carrier concentrations to manipulate hydrogen charge state concentrations. In other words, advanced hydrogen passivation processes will become easier. In addition, different regions in the device may respond differently due to differing local lifetimes and doping densities. Any process that changes lifetime in the material, whether that be a change in bulk lifetime, or in surface passivation, can also impact hydrogen passivation processes. This will be discussed more in Chapters 4 and 5.

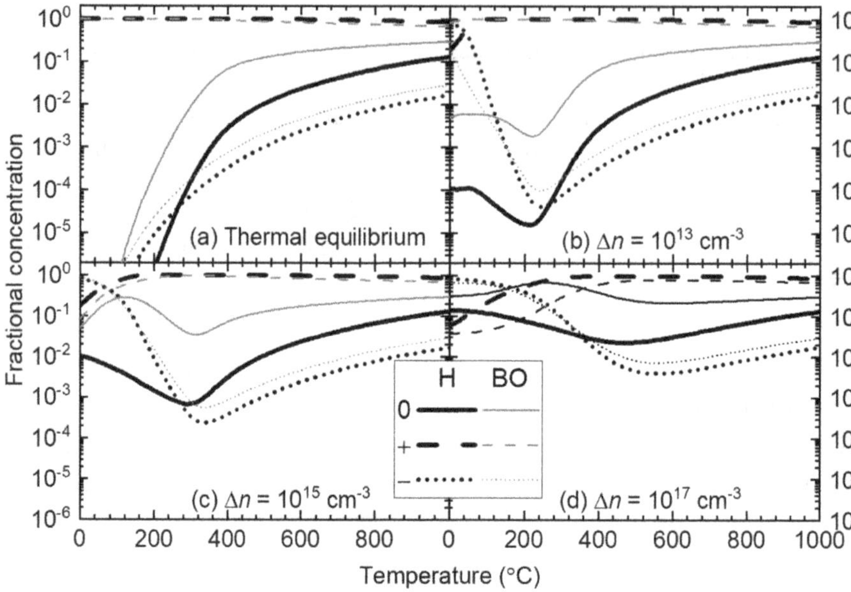

Figure 2.17 *Fractional hydrogen (H) and boron-oxygen (B-O) charge state concentrations in 1 Ω.cm p-type silicon as a function of temperature for (a) thermal equilibrium and excess carrier densities of (b) 1 × 10^{13}/cm^3, (c) 1 × 10^{15}/cm^3 and (d) 1 × 10^{17}/cm^3. Figure based on modelling presented in [181]*

2.8 Unified modelling of hydrogen and defects

Silicon solar cells often contain a range of defects and impurities. In order to passivate the recombination due to the defects, it is often advisable to maximise the available H^0 concentration. However, as discussed by Sun *et al.* [181], in order to get the best passivation of a single known defect it is important to consider the charge state, not only of hydrogen, but also of the defect itself. Figure 2.17 presents the charge states of the B-O defect and interstitial hydrogen in 1 Ω.cm p-type silicon as a function of temperature and excess carrier density. Given that the defect itself is unstable below 250 °C, it may be seen that hydrogen passivation at low excess carrier densities will be inefficient. This is because both hydrogen and the defect will overwhelmingly be in the positive charge state.

This can be overcome by increasing the excess carrier density. Figure 2.18 shows the charge state products for the three most likely defect passivation reactions, and their dependence on excess carrier density. Due to Auger recombination achieving an excess carrier density in excess of 1 × 10^{17}/cm^3 is practically impossible, and so the best approach, and the one that has demonstrated success in industry, is to inject as many carriers as possible to improve passivation. Although H^0 may not be directly responsible for the passivation reaction, it is highly mobile, enabling

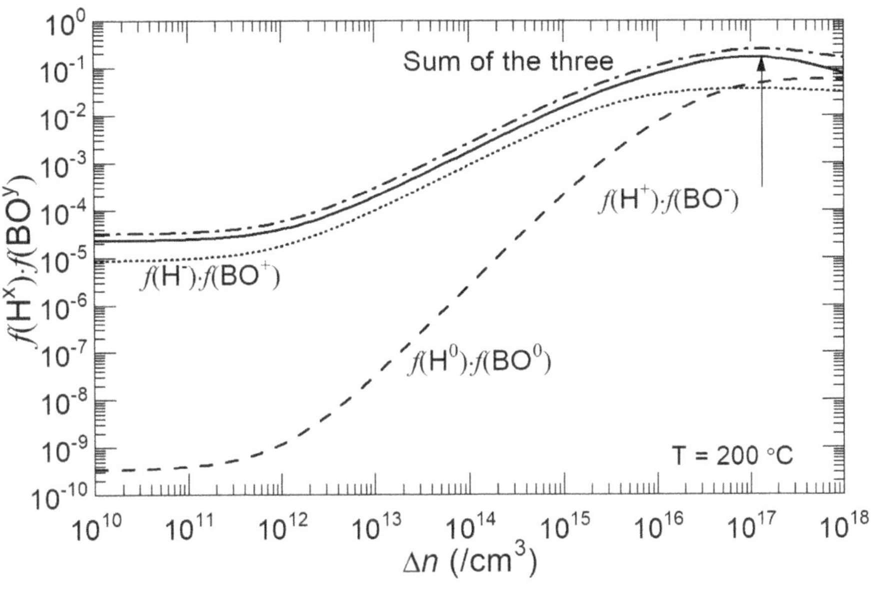

Figure 2.18 *Reaction products for H and B-O in 1 Ω.cm p-type silicon at temperature = 200 °C as a function of excess carrier density. Figure based on modelling presented in [181]*

rapid diffusion to the defect site. In addition, conditions that result in a high H^0 concentrations, ensure high concentrations of all charge state species. H^0 is also highly unstable and reactive, meaning that there is a constant toggling from H^0 into charged species of H^+ or H^-. This is also likely the case for defects.

While increasing excess carrier density works in many cases, different defects with different donor and acceptor levels may require alternate strategies to ensure complete passivation. This is an area of work still requiring investigation.

2.9 Production sequences for industrial silicon solar cells

2.9.1 Al-BSF solar cells

Over the past twenty years, the vast majority of silicon solar cells have been fabricated using an architecture known as Al-BSF. An overview of the cell structure and operation of Al-BSF solar cells was detailed in Chapter 1. The process flow for the fabrication of Al-BSF solar cells is relatively simple and has undergone progressive optimisation and efficiency improvements. Figure 2.19 displays the process flow for an industrial silicon Al-BSF solar cell.

The starting material is a silicon wafer, typically p-type wafers have been used. The first processes are chemical treatments of the wafer. When the wafer is cut from the ingot, the sawing process induces damage on the surface. This saw damage is detrimental for the solar cell performance, so it needs to be removed. This is

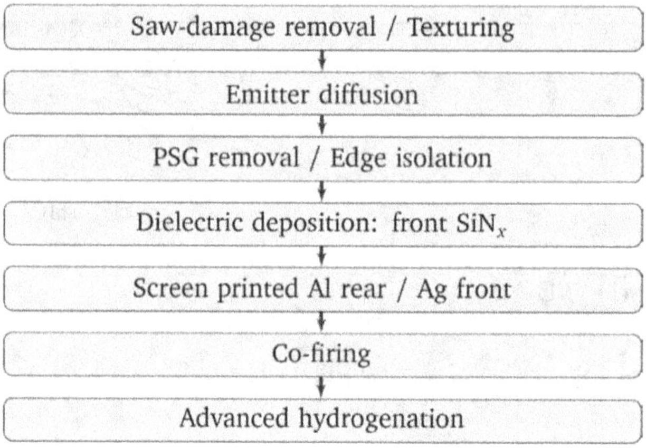

Figure 2.19 Process flow for industrial silicon Al-BSF solar cells

performed using an alkaline etching solution, most commonly potassium hydroxide (KOH). The etch rate is controlled by changing the concentration of the KOH solution and the temperature. The amount of saw damage removal depends on the sawing method. Traditional slurry-based sawing techniques can require the removal of up to 10 μm per side, while newer diamond wire sawing only produce about 5 μm of damage per side. Following saw damage removal is wafer texturing. Untextured wafers have high reflection properties, which limits the amount of light the solar cell can absorb. To reduce reflection, a chemical texturing on the surface is applied. For mono crystalline wafers, this is achieved using an alkaline solution (typically also KOH) with a much lower concentration than for saw damage etching, with the purpose of preferentially etching along the <111> crystallographic planes. This leads to the exposure of 'pyramids' on the front surface, which significantly reduce reflection. For multi-crystalline wafers, due to the varied grains formed at different angles, this approach cannot be used. Instead, multi wafers can be textured using an acidic solution, referred to as 'iso texturing'. However, this iso texturing approach relies on the surface damage induced by the sawing process. With diamond wiring sawing, the amount of damage is significantly reduced, which actually reduces the effectiveness of the iso texturing. An alternative texturing approach is the so called 'black silicon' approach, where texturing is achieved using either metal catalysed chemical etching (MCCE) or reactive ion etching (RIE). These techniques remove the reliance on surface damage, however, particularly for RIE, are yet to make a significant impact on the solar industry.

To form the p-n junction, the diffusion of an emitter region on the surface is required. For p-type wafers, this is achieved using a phosphorous dopant source. The standard industry diffusion process is performed in a high temperature furnace, with a liquid or gaseous dopant source. In the Al-BSF cell design a single homogenous emitter with constant doping is used. Newer technologies benefit from a 'selective emitter' approach, which optimises the dopant profile in contacted and non-contacted regions. Selective emitter design using laser doping is detailed in

Chapters 7 and 8. The diffusion process serves another purpose; it provides defect engineering in the form of gettering. The diffused region can act as a sink for impurities that diffuse out of the bulk during the high temperature process and change the effectiveness of hydrogenation processes. When the phosphosilicate glass (PSG) layer is removed and the surface is cleaned, these impurities are removed from the bulk. This diffusion gettering is one example of defect engineering that is naturally incorporated into the process flow. This will be discussed more in Chapter 4.

The next step is to deposit the dielectric antireflection layer. Of particular importance for industrial solar cells is SiN_X:H films deposited by PECVD. The PECVD process involves a chemical reaction of pre cursor gases, namely silane (SiH_4) and ammonia (NH_3), in a reaction chamber, the substrate temperature of the wafers is typically heated in the range from 350 to 450 °C. Using PECVD to deposit these films has been critical to the development of screen-printed silicon solar cells, as it allows for precise tuning of the film stoichiometry, and thus refractive index, and can form films with very high hydrogen content. SiN_X:H films are crucial to the operation of screen-printed solar cells as they perform multiple roles. First, SiN_X:H serves as an anti-reflection coating (ARC). This is achieved by destructive thin film interference on the front surface, which depends on the thickness and refractive index of the film. When coupled with alkaline texturing, the silicon reflectance can be reduced to less than 1% across a broad spectral range after application of the ARC. The PECVD process allows for precise tunability of the refractive index of the film, and thus, it gives a great deal of control over the optical properties of the solar cell. Additionally, the hydrogen in the film provides surface passivation. The hydrogen atoms can bond with dangling bonds at the silicon surface, thus reducing the number of recombination active sites. The charge held in the layer also provides field effect passivation. This is discussed at length in Chapter 3.

Metallisation is performed after the dielectric layer deposition. In Al-BSF solar cells, the metal contacts are formed by screen printing metallic pastes, silver on the front surface in a 'H-grid' pattern and aluminium on the full area of the rear surface, along with Ag-Al tabs for soldering. After the printing of the metal pastes through a specially designed screen, the paste is dried and then undergoes curing in a high temperature belt furnace. This process is referred to as firing. The primary function of the firing process is to drive the Ag front contact through the SiN_X:H layer to form electrical contact to the underlying silicon and form a heavily-doped aluminium BSF through the alloying of Al and Si. On the front surface, the Ag paste needs to penetrate the insulating SiN_X:H layer to form an ohmic contact. To achieve this, the paste chemistry includes glass frit.

Optimisation of the firing is critical for the performance of the front metallisation. If the Ag paste does not successfully fire through the SiN_X:H layer, the contact resistance will be very high, however, if the metal fires too far through the emitter to reach the p-doped base, the cell will be shunted. However more recent pastes are generally self-limiting and do not lead to shunting. On the rear surface, the high temperature treatment causes alloying of the Al and Si, this occurs at temperatures above the eutectic temperature of 577 °C. As Al is a group V element, it serves as a p-type dopant in the alloy. This leads to the formation of a p^+ layer on the rear surface, which forms a back-surface field, thus the name 'aluminium back surface field'. This field repels carriers from the rear metal/Si interface, which is a region of high recombination activity. Thus, this back-surface field reduces the J_0 on the rear, which leads to improved V_{OC}.

The firing process serves a secondary purpose, it also releases hydrogen from bonds in the SiN_X:H and promotes movement of hydrogen from the dielectric layer into the bulk of the silicon. This illustrates the multi-functional nature of the SiN_X:H layers, not only do they reduce reflection and provide surface passivation, they also act as a hydrogen source. The movement of hydrogen into the bulk can facilitate the passivation of bulk defects. Such defects could be structural, such as dislocations in multi crystalline silicon, or processing induced, such as laser-induced defects. Bulk defect passivation with hydrogen is explained in Chapter 4. Further to this, hydrogen passivation in the bulk can also be used to overcome light-induced degradation. In p-type silicon, during illumination, recombination active boron-oxygen defects can form, which can severely limit the cell performance. The role of hydrogen passivation in mitigating light-induced degradation (LID) in p-type silicon is discussed at length in Chapter 5.

2.9.2 PERC solar cells

The Al-BSF structure has been the workhorse of the solar cell industry over the past twenty years. The relative simplicity of the process and intense optimisation has seen the $/W for Al-BSF cells reduce drastically since 2008. However, this approach suffers from one fundamental problem that limits the obtainable efficiency. The full area metal contact on the rear means that the area of metal/silicon interface is very large. This metal/silicon interface contains many recombination active sites, which limits the V_{OC} of this approach to about 650 mV. In order to overcome this limitation, the PERC design incorporates a thin dielectric layer in between the silicon and metal on the rear surface. In this approach, the dielectric layer provides excellent surface passivation on the rear surface, which means that much higher V_{OC} and efficiency can be realised. This improved efficiency outlook has seen PERC solar cells overtake Al-BSF in terms of the largest market share for industrial silicon solar cells. The processing flow for industrial PERC solar cells is shown in Figure 2.20.

This shows that the process flow is very similar to that of Al-BSF, which has allowed for PERC solar cells to remain a relatively low-cost option, even though the efficiency potential is much greater than Al-BSF.

There are two main differences in the process flow for industrial PERC solar cells. Firstly, the dielectric deposition occurs both on the front and rear of the cell. Early reports on the development of PERC solar cells employed a SiO_2/SiN_X stack on the rear surface, which led to significantly reduced recombination compared to the full area metal/silicon interface. However, the development of Al_2O_3 as a passivation layer for p-type surfaces by Hoex *et al.* has been critical to the rapid progress of industrial PERC solar cells [189]. The development of ALD Al_2O_3 or PECVD AlO_X:H passivation is discussed at length in Chapter 3. The rear dielectric layer has implications for not only the surface passivation, the additional hydrogen containing

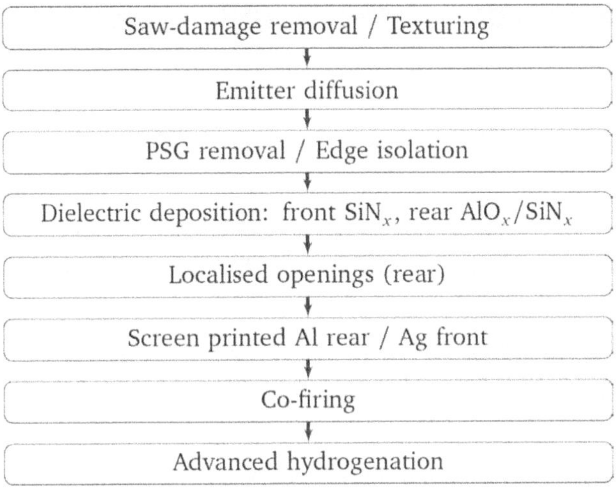

Figure 2.20 Process flow for PERC solar cells

layers also means there is much more hydrogen available to diffuse into the bulk during firing.

The second modification in the process flow is the localised openings on the rear for the formation of point contacts. For the local Al-BSF structure, the rear surface metallisation is full area, meaning that the aluminium covers the entire rear surface. However, in PERC solar cells, the contacts on the rear surface are localised and are defined by laser ablation of the rear dielectric when using a full area Al layer. This means that the metal paste for the rear contacts needs to have a different composition, such pastes are referred to as 'non-fire through' pastes. To form the local contacts, a high-powered laser is used to form a contact opening in the rear dielectric. This local contacting on the rear leads to multiple advantages. It significantly reduces recombination on the rear, which leads to higher cell V_{OC}. The use of laser processing through the rear AlO_x:H passivation layer leads to opportunities for laser doping with the AlO_x:H as a source (discussed in Chapter 9). The rear dielectric also opens up the possibility for optimised rear metallisation schemes to facilitate bifacial cell designs. For such solar cells, a fire-through Al paste may be printed in a 'H-pattern', similar to that used on the front surface.

The additional hydrogen available in PERC solar cells makes hydrogen management more important. A delicate balance is required in order to passivate unwanted bulk defects (Chapters 4 and 5), without introducing hydrogen-induced recombination if too much hydrogen is introduced (discussed in Chapter 6). As detailed above, the fractional population of hydrogen in different charge states determines the ability to passivate defects. This can be controlled by altering the temperature or carrier injection in the silicon. Within industry, this concept is widely employed in the form of post-cell advanced hydrogenation processing. These approaches commonly achieve carrier injection either by illumination or direct current injection. Current injection techniques have become ubiquitous within industry, as many cells can be processed simultaneously using a 'coin-stack' approach, thus the cost can be kept low. These techniques are used to promote hydrogen passivation to stabilise the cells to light and elevated temperatures.

2.10 Summary

This chapter has attempted to give an introduction to the behaviour of hydrogen in silicon, and how this relates to its ability to passivate defects. It has been shown that the use of hydrogen passivation can results in remarkable improvements in carrier lifetimes, and hence performance of silicon photovoltaic devices.

The key considerations for effective hydrogen passivation have also been outlined as follows:

1. Introduction of atomic hydrogen to the device
2. Diffusion of atomic hydrogen within the device to reach defects
3. Reaction of atomic hydrogen with defects

A range of techniques have been developed to introduce hydrogen into silicon, with initial approaches using a source of molecular hydrogen such as forming gas. Over the last 25 years this approach has been superseded by the release of atomic hydrogen from deposited dielectric films.

Both the diffusivity and reactivity of hydrogen in silicon are heavily influenced by the charge state, as well as interactions with dopants and other common impurities. Interstitial hydrogen in silicon can be in either the positive, neutral or negative state. Because hydrogen is a negative-U defect the neutral state is highly unstable and is always in the minority. Under thermal equilibrium interstitial hydrogen is primarily found as H^+ in p-type and high-resistivity silicon and as H^- in moderate to heavily doped n-type silicon (temperature dependent).

Each charge state possesses different properties and interacts with different impurities and defects in the silicon lattice. It is well known that H^+ can bond with and electrically de-activate shallow acceptors, while H^- is capable of de-activating donors. Different charge states of interstitial hydrogen can also affect the formation of hydrogen dimers, which in turn can interact with vacancies and carbon impurities.

If the charge states, and impurity interactions of hydrogen can be controlled the ability for effective passivation is increased. However, a major stumbling block for effectively manipulating hydrogen in silicon solar cells is the difficulty in accurately observing its location and activity. Most methods for unambiguous detection rely on vacuum processes and are impacted by hydrogen present in the atmosphere, resulting in detection limits that are too high for many studies. The detection limit can be reduced somewhat using hydrogen isotopes but at the cost of an increase in experimental complexity. In contrast, carrier lifetime and resistivity-based techniques can be very sensitive, but are based on inferred hydrogen behaviour and are subject to a number of assumptions.

Despite these uncertainties, it is shown that by controlling the excess carrier density and temperature of a silicon substrate significant alterations to the charge state distribution of hydrogen in silicon can be achieved, with associated changes in diffusivity and passivation effectiveness. This is particularly through the generation of the highly mobile and reactive H^0 charge state, to avoid the extremely low fractional concentrations at low temperatures in thermal equilibrium. As such, hydrogen passivation processes incorporating minority carrier injection such as illuminated annealing or current injection can be used to increase the concentration of H^0. With a dependence on the excess carrier concentration, the solar cell structure, any process that changes the lifetime in the material, and the illumination intensity, can all modulate the H^0 concentration.

The typical production process for Al-BSF and PERC silicon solar cells was then examined, along with how hydrogen passivation is incorporated. In particular, it is described how the current PERC process uses high temperature hydrogen introduction from dielectric layers during metal co-firing along with low temperature carrier injection to improve passivation of defects.

2.11 Acknowledgements

The authors would like to thank Matthew Wright for proof-reading and editing, Moonyong Kim for image editing and generation, Dale Philip for assistance with Figure 2.1, Bruno Vicari Stefani for providing hydrogen diffusivity data for Figure 2.9 and David Tweddle for providing atom probe images for Figure 2.10. P. Hamer and C. Sun acknowledge the support of the Australian Center for Advanced Photovoltaics (ACAP) through postdoctoral fellowships. P. Hamer and B. Hallam also acknowledge the support of the Australian Renewable Energy Agency (ARENA) as part of ARENA's Research and Development Program – Solar PV Research (Grant No. 1-A060). B. Hallam also acknowledges the support of the Australian Research Council (ARC) through a Discovery and Early Career Researcher Award (DE170100620). The views expressed herein are not necessarily the views of the Australian Government, and the Australian Government does not accept responsibility for any information or advice contained herein.

References

[1] Verlinden P., Lafontaine B., Jacquemin P., Goutain E., Crahay A. Super self-aligned technology for backside contact solar cells: a route to low cost and high efficiency. Conf Rec IEEE Photovolt Spec Conf; 1990. pp. 257–62.

[2] Verlinden P., Swanson R., Sinton R., Kane D. Multilevel metallisation for large area point-contact solar cells. Proc 8th Eur Photovolt Energy Conf; 1988. pp. 1466–71.

[3] King R.R., Sinton R.A., Swanson R.M. 'Studies of diffused phosphorus emitters: saturation current, surface recombination velocity, and quantum efficiency'. *IEEE Transactions on Electron Devices*. 1990;**37**(2):365–71.

[4] Swanson R.M. Method of fabricating back surface point contact solar cells. US Patent 4927770; 1990.

[5] Blakers A.W., Wang A., Milne A.M., Zhao J., Green M.A. '22.8% efficient silicon solar cell'. *Applied Physics Letters*. 1989;**55**(13):1363–5.

[6] Zhao J., Wang A., Altermatt P., Wenham S., Green M. '24% efficient perl silicon solar cell: Recent improvements in high efficiency silicon cell research'. *Solar Energy Materials and Solar Cells*. 1996;**41-42**:87–99.

[7] Zhao J., Wang A., Green M.A. '24·5% Efficiency silicon PERT cells on MCZ substrates and 24·7% efficiency PERL cells on FZ substrates'. *Progress in Photovoltaics: Research and Applications*. 1999;**7**(6):471–4.

[8] Balk P. 'Effects of hydrogen annealing on silicon surfaces'. *Electrochem Soc Spring Meet*. 1965;**14**(1):237–40.

[9] Deal B.E., MacKenna E.L., Castro P.L. 'Characteristics of Fast Surface States Associated with SiO2-Si and Si3N4-SiO2-Si Structures'. *Journal of The Electrochemical Society*. 1969;**116**(7):997–1005.

[10] Green M.A. 'The path to 25% silicon solar cell efficiency: History of silicon cell evolution'. *Progress in Photovoltaics: Research and Applications*. 2009;**17**(3):183–9.

[11] Kerr M.J., Schmidt J., Cuevas A., Bultman J.H. 'Surface recombination velocity of phosphorus-diffused silicon solar cell emitters passivated with plasma enhanced chemical vapor deposited silicon nitride and thermal silicon oxide'. *Journal of Applied Physics*. 2001;**89**(7):3821–6.

[12] Yoshikawa K., Kawasaki H., Yoshida W., *et al*. 'Silicon heterojunction solar cell with interdigitated back contacts for a photoconversion efficiency over 26%'. *Nature Energy*. 2017;**2**(5).

[13] Taguchi M., Yano A., Tohoda S., *et al*. '24.7% Record efficiency HIT solar cell on thin silicon wafer'. *IEEE Journal of Photovoltaics*. 2014;**4**(1):96–9.

[14] Glunz S.W., Feldmann F., Richter A., Bivour M., Reichel C., Steinkemper H. The Irresistible Charm of a Simple Current Flow Pattern – 25 % With a Solar Cell Featuring a Full-Area Back Contact. In: Proceedings of the 31st European Photovoltaic Solar Energy Conference and Exhibition [Internet]. 2015. Available from http://www.eupvsec-proceedings.com/proceedings?paper=34845.

[15] Richter A., Benick J., Feldmann F., Fell A., Hermle M., Glunz S.W. 'N-Type Si solar cells with passivating electron contact: identifying sources for efficiency limitations by wafer thickness and resistivity variation'. *Solar Energy Materials and Solar Cells*. 2017;**173**(6):96–105.

[16] Feldmann F., Bivour M., Reichel C., Hermle M., Glunz S.W. 'Passivated rear contacts for high-efficiency N-type Si solar cells providing high interface passivation quality and excellent transport characteristics'. *Solar Energy Materials and Solar Cells*. 2014;**120**:270–4.

[17] Haase F., Hollemann C., Schäfer S., *et al*. 'Laser contact openings for local poly-Si-metal contacts enabling 26.1%-efficient POLO-IBC solar cells'. *Solar Energy Materials and Solar Cells*. 2018;**186**(6):184–93.

[18] Cousins P.J., Smith D.D., Luan H.-C., Manning J., Dennis T.D., Waldhauer A. Generation 3: Improved performance at lower cost. 35th IEEE Photovolt Spec Conf; 2010. pp. 275–8.
[19] Mulligan W.P., Cudzinovic M.J., Pass T., Smith D., Swanson R.M. Metal contact structure for solar cell and method of manufacture. US Patent 7388147; 2008.
[20] Taguchi M., Tanaka M., Matsuyama T., Matsuoka T., Tsuda S., Nakano S. Improvement of the conversion efficiency of polycrystalline silicon thin film solar cell. *Proc Fifth Int Photovolt Sci Eng Conf*; 1990. pp. 689–92.
[21] De Wolf S., Descoeudres A., Holman Z.C., Ballif C. 'High-Efficiency silicon heterojunction solar cells: a review'. *Green*. 2012;**2**(1):7–24.
[22] Heng J.B., Yu C., Xu Z., Fu J. Solar cell with oxide tunneling junctions. US Patent 8686283; 2014.
[23] Turner A.B., Schultz-Wuttnann O., De Ceuster D., Crafts D.E. Selective transformation in functional films, and solar cell applications thereof. US Patent 20120186649; 2010.
[24] Green M., Dunlop E., Hohl-Ebinger J., Yoshita M., Kopidakis N., Hao X. 'Solar cell efficiency tables (version 57'. *Progress in Photovoltaics: Research and Applications*. 2021;**29**(1):3–15.
[25] Liu C., Chen D., Chen Y., *et al*. 'Industrial TOPCon solar cells on n-type quasi-mono Si wafers with efficiencies above 23%'. *Solar Energy Materials and Solar Cells*. 2020;**215**:110690.
[26] Hallam B.J., Ciesla A.M., Chan C.C., *et al*. 'Overcoming the challenges of hydrogenation in silicon solar cells'. *Australian Journal of Chemistry*. 2018;**71**(10):743–52.
[27] Reed M.L., Plummer J.D. 'Chemistry of Si-SiO $_2$ interface trap annealing'. *Journal of Applied Physics*. 1988;**63**(12):5776–93.
[28] Sopori B., Deng X., Benner J., *et al*. 'Hydrogen in silicon: a discussion of diffusion and passivation mechanisms'. *Solar Energy Materials and Solar Cells*. 1996;**41-42**:159–69.
[29] Collett K.A., Bonilla R.S., Hamer P., Bourret-Sicotte G., Lobo R., Kho T. 'An enhanced alneal process to produce SRV < 1 cm/s in 1 Ω cm n-type Si'. *Solar energy materials and solar cells : an international journal devoted to photovoltaic, photothermal, and photochemical solar energy conversion*. 2017;**173**.
[30] Zhao J., Wang A., Altermatt P., Green M.A. 'Twenty-four percent efficient silicon solar cells with double layer antireflection coatings and reduced resistance loss'. *Applied Physics Letters*. 1995;**66**(26):3636–8.
[31] P.V.-Manufacturing. PECVD of Silicon Nitride [Internet]. 2020. Available from https://pv-manufacturing.org/pecvd-of-silicon-nitride/.
[32] Verhoef L.A., Michiels P-P., Sinke W.C., Denisse C.M.M., Hendriks M., van Zolingen R.J.C. 'Combined impurity gettering and defect passivation in polycrystalline silicon solar cells'. *Applied Physics Letters*. 1990;**57**(25):2704–6.

[33] Hahn G., Spiegel J.M., Fath P., Wileke G., Bucher E. Improvement of MC Si solar cells by Al-gettering and hydrogen passivation. Conference Record of the Twenty Sixth IEEE Photovoltaic Specialists Conference; 1997. pp. 75–8.

[34] Elgamel H.E., Barnett A.M., Rohatgi A., *et al*. 'Efficient combination of surface and bulk passivation schemes of high-efficiency multicrystalline silicon solar cells'. *Journal of Applied Physics*. 1995;**78**(5):3457–61.

[35] Keevers M.J., Turner A., Schubert U., Basore P.A., Green M.A. Remarkably effective hydrogenation of crystalline silicon on glass modules. 20th European Photovoltaic Solar Energy Conference; 2005. pp. 1305–8.

[36] Duerinckx F., Szlufcik J. 'Defect passivation of industrial multicrystalline solar cells based on PECVD silicon nitride'. *Solar Energy Materials and Solar Cells*. 2002;**72**(1):231–46.

[37] Cai L., Rohatgi A., Yang D., El-Sayed M.A. 'Effects of rapid thermal anneal on refractive index and hydrogen content of plasma-enhanced chemical vapor deposited silicon nitride films'. *Journal of Applied Physics*. 1996;**80**(9):5384–8.

[38] Hoex B., Heil S.B.S., Langereis E., van de Sanden M.C.M., Kessels W.M.M. 'Ultralow surface recombination of c-Si substrates passivated by plasma-assisted atomic layer deposited Al2O3'. *Applied Physics Letters*. 2006;**89**(4):42112.

[39] Hoex B., Schmidt J., Pohl P., van de Sanden M.C.M., Kessels W.M.M. 'Silicon surface passivation by atomic layer deposited Al2O3'. *Journal of Applied Physics*. 2008;**104**(4):44903.

[40] Schmidt J., Merkle A., Brendel R., Hoex B., van de Sanden M.C.M., Kessels W.M.M. 'Surface passivation of high-efficiency silicon solar cells by atomic-layer-deposited Al2O3'. *Prog photovoltaics Res Appl*. 2008;**16**(6):461–6.

[41] Dingemans G., Einsele F., Beyer W., van de Sanden M.C.M., Kessels W.M.M. 'Influence of annealing and Al2O3 properties on the hydrogen-induced passivation of the Si/SiO2 interface'. *Journal of applied physics*. 2012;**111**(9):93713.

[42] Hong J., Kessels W.M.M., Soppe W.J., Rieffe H.C., Weeber A.W. Sanden MCM van de. Structural film characteristics related to the passivation properties of high-rate (>0.5 nm/s) plasma deposited a-SiNx:H. 3rd World Conference on Photovoltaic Energy Conversion, 2003 Proceedings; 2003. pp. 1158–61.

[43] Morin P.F., Pelletier B., Laffosse E., Plantier L. 'Modeling stress development and hydrogen diffusion in plasma enhanced chemical vapor deposition silicon nitride films submitted to thermal cycles'. *Journal of Applied Physics*. 2013;**114**(15):154113.

[44] Beyer W., Dekkers H.F.W. 'Microstructure characterization of amorphous Silicon-Nitride films by effusion measurements'. *MRS Proceedings*. 2006;**910**.

[45] Bredemeier D., Walter D.C., Heller R., Schmidt J. 'Impact of Hydrogen-Rich silicon nitride material properties on Light-Induced lifetime degradation

in multicrystalline silicon'. *Physica status solidi (RRL) – Rapid Research Letters*. 2019;**190020113**:1–5.

[46] Jafari S., Hirsch J., Lausch D., John M., Bernhard N., Meyer S. Composition limited hydrogen effusion rate of a-SiNx: H passivation stack. AIP Conference Proceedings. AIP Publishing LLC; 2019. p. 50004.

[47] Wilking S., Herguth A., Hahn G. 'Influence of hydrogen on the regeneration of boron-oxygen related defects in crystalline silicon'. *Journal of Applied Physics*. 2013;**113**(19):194503.

[48] Dingemans G., Beyer W., Kessels W.M.M. Thermal effusion measurements: Probing hydrogen in surface passivation schemes. 2012 38th IEEE Photovoltaic Specialists Conference. IEEE; 2012. pp. 1256–9.

[49] Tavendale A.J., Williams A.A., Pearton S.J. 'Hydrogen injection and neutralization of boron acceptors in silicon boiled in water'. *Applied Physics Letters*. 1986;**48**(9):590–2.

[50] Seager C.H., Anderson R.A., Panitz J.K.G. 'The diffusion of hydrogen in silicon and mechanisms for "unintentional" hydrogenation during ion beam processing'. *Journal of Materials Research*. 1987;**2**(1):96–106.

[51] Pearton S., Stavola M., Corbett J.W. 'Hydrogen in semiconductors: crystal growth and device processing'. *Advanced Materials*. 1992;**4**(5):332–40.

[52] Zhu J., Johnson N.M., Herring C. 'Negative-charge state of hydrogen in silicon'. *Physical Review B*. 1990;**41**(17):12354–7.

[53] Mathiot D. 'Modeling of hydrogen diffusion in N- and P-type silicon'. *Physical Review B*. 1989;**40**(8):5867–70.

[54] Seager C.H., Anderson R.A. 'Real-time observations of hydrogen drift and diffusion in silicon'. *Applied Physics Letters*. 1988;**53**(13):1181–3.

[55] Herring C., Johnson N.M., Van de Walle C.G. 'Energy levels of isolated interstitial hydrogen in silicon'. *Physical Review B*. 2001;**64**(12):1–27.

[56] Nielsen K.B., Dobaczewski L., Søgård S., Nielsen B.B. 'Acceptor state of monoatomic hydrogen in silicon and the role of oxygen'. *Physical Review B*. 2002;**65**(7):75205.

[57] Van de Walle CG., Denteneer P.J., Bar-Yam Y., Pantelides S.T. 'Theory of hydrogen diffusion and reactions in crystalline silicon'. *Physical Review B*. 1989;**39**(15):10791–808.

[58] Estreicher S.K., Docaj A., Bebek M.B., Backlund D.J., Stavola M. 'Hydrogen in C-rich Si and the diffusion of vacancy-H complexes'. *physica status solidi*. 2012;**209**(10):1872–9.

[59] Sah Chih-Tang., Sun Jack Yuan-Chen., Tzou Joseph Jeng-Tao. 'Deactivation of the boron acceptor in silicon by hydrogen'. *Applied physics letters*. 1983;**43**(2):204–6.

[60] Johnson N.M., Herring C., Chadi D.J. 'Interstitial hydrogen and neutralization of shallow-donor impurities in single-crystal silicon'. *Physical Review Letters*. 1986;**56**(7):769–72.

[61] Pankove J.I., Wance R.O., Berkeyheiser J.E. 'Neutralization of acceptors in silicon by atomic hydrogen'. *Applied Physics Letters*. 1984;**45**(10):1100–2.

[62] Zundel T., Mesli A., Muller J.C., Siffert P. 'Boron neutralization and hydrogen diffusion in silicon subjected to low-energy hydrogen implantation'. *Applied Physics A Solids and Surfaces*. 1989;**48**(1):31–40.

[63] Seager C.H., Anderson R.A. 'Two-step debonding of hydrogen from boron acceptors in silicon'. *Applied Physics Letters*. 1991;**59**(5):585–7.

[64] Denteneer P.J., Van de Walle CG., Pantelides S.T. 'Microscopic structure of the hydrogen-boron complex in crystalline silicon'. *Physical Review B*. 1989;**39**(15):10809–24.

[65] Shumate S.D., Hutchings D.A., Mohammed H., Beilke G., Newton B.S., Young M.G. Self aligned hydrogenated selective emitter for n-type solar cells. Proc 38th IEEE Photovolt Spec Conf; 2012. pp. 1110–4.

[66] Bourret-Sicotte G., Hamer P., Bonilla R.S., Collett K., Wilshaw P.R. 'Shielded hydrogen passivation – a novel method for introducing hydrogen into silicon'. *Energy Procedia*. 2017;**124**(1-4):267–74.

[67] Bourret-Sicotte G., Hamer P., Bonilla R.S., *et al.* 'Shielded hydrogen passivation – a potential in-line passivation process'. *Physica Status Solidi*. 2017;**214**(7):1700383.

[68] Hamer P., Hallam B., Bonilla R.S., Altermatt P.P., Wilshaw P., Wenham S. 'Modelling of hydrogen transport in silicon solar cell structures under equilibrium conditions'. *Journal of Applied Physics*. 2018;**123**(4):043108.

[69] Denteneer P.J., Van de Walle CG., Pantelides S.T. 'Microscopic structure of the hydrogen-phosphorus complex in crystalline silicon'. *Physical Review B*. 1990;**41**(6):3885–8.

[70] Chang K.J., Chadi D.J. 'Hydrogen bonding and diffusion in crystalline silicon'. *Physical Review B*. 1989;**40**(17):11644–53.

[71] Zundel T., Weber J. 'Boron reactivation kinetics in hydrogenated silicon after annealing in the dark or under illumination'. *Physical Review B*. 1991;**43**(5):4361–72.

[72] Estreicher S.K., Seager C.H., Anderson R.A. 'Bistability of donor-hydrogen complexes in silicon: a mechanism for debonding'. *Applied Physics Letters*. 1991;**59**(14):1773–5.

[73] Johnson N.M., Herring C. 'Diffusion of negatively charged hydrogen in silicon'. *Physical Review B*. 1992;**46**(23):15554–7.

[74] Yarykin N. 'Minority-carrier-enhanced dissociation of the boron–hydrogen pair in silicon'. *Physica B: Condensed Matter*. 2009;**404**(23–24):5093–5.

[75] Zhang S.B., Jackson W.B. 'Formation of extended hydrogen complexes in silicon'. *Physical Review B*. 1991;**43**(14):12142–5.

[76] Estreicher S.K. 'Hydrogen-related defects in crystalline semiconductors: a theorist's perspective'. *Materials Science and Engineering: R: Reports*. 1995;**14**(7–8):319–412.

[77] Matsubara M., Godet J., Pizzagalli L. 'Theoretical study of hydrogen stability and aggregation in dislocation cores in silicon'. *Physical Review B*. 2010;**82**(2):34–7.

[78] Voronkov V.V., Falster R. 'Formation, dissociation, and diffusion of various hydrogen dimers in silicon'. *Physica Status Solidi*. 2017;**254**(6):1600779.

[79] Pearton S.J., Corbett J.W., Shi T.S. 'Hydrogen in crystalline semiconductors'. *Applied Physics A Solids and Surfaces*. 1987;**43**(3):153–95.
[80] Murakami K., Fukata N., Sasaki S., *et al.* 'Hydrogen molecules in crystalline silicon treated with atomic hydrogen'. *Physical Review Letters*. 1996;**77**(15):3161–4.
[81] Leitch A.W.R., Alex V., Weber J. 'Raman spectroscopy of hydrogen molecules in crystalline silicon'. *Physical Review Letters*. 1998;**81**(2):421–4.
[82] Holbech J.D., Bech Nielsen B., Jones R., Sitch P., Öberg S. 'H2* defect in crystalline silicon'. *Physical review letters*. 1993;**71**(6):875.
[83] Estreicher S.K., Hastings J.L., Fedders P.A. 'Radiation-Induced formation of H2* in silicon'. *Physical review letters*. 1999;**82**(4):815.
[84] Wilking S., Beckh C., Ebert S., Herguth A., Hahn G. 'Influence of bound hydrogen states on BO-regeneration kinetics and consequences for high-speed regeneration processes'. *Solar Energy Materials and Solar Cells*. 2014;**131**(10):2–8.
[85] Roberson M.A., Estreicher S.K. 'Vacancy and vacancy-hydrogen complexes in silicon'. *Physical Review B*. 1994;**49**(24):17040–9.
[86] Bech Nielsen B., Hoffmann L., Budde M. 'Sih stretch modes of hydrogen — vacancy defects in silicon'. *Materials Science and Engineering: B*. 1996;**36**(1):259–63.
[87] Stallinga P., Johannesen P., Herstrøm S., Bonde Nielsen K., Bech Nielsen B., Byberg J.R. 'Electron paramagnetic resonance study of hydrogen-vacancy defects in crystalline silicon'. *Physical Review B*. 1998;**58**(7):3842–52.
[88] Park Y.K., Estreicher S.K., Myles C.W., Fedders P.A. 'Molecular-Dynamics study of the vacancy and vacancy-hydrogen interactions in silicon'. *Physical Review B*. 1995;**52**(3):1718–23.
[89] Kolevatov I.L., Weiser P.M., Monakhov E.V., Svensson B.G. 'Interaction between hydrogen and vacancy defects in crystalline silicon'. *physica status solidi*. 2019;**216**(10):1800670.
[90] Estreicher S.K., Hastings J.L., Fedders P.A. 'Defect-induced dissociation of H 2 in silicon'. *Physical review. B*. 1998;**57**(20):R12663.
[91] Yelundur V., Rohatgi A., Ebong A., Gabor A.M., Hanoka J., Wallace R.L. 'Al-enhanced PECVD SiNx induced hydrogen passivation in string ribbon silicon'. *Journal of Electronic Materials*. 2001;**30**(5):526–31.
[92] Hourahine B., Jones R., Öberg S., Briddon P.R., Markevich V.P., Newman R.C. 'Evidence for H2* trapped by carbon impurities in silicon'. *Physica. B, Condensed matter*. 2001;**308**:197–201.
[93] McAfee J.L., Estreicher S.K. 'Structural and vibrational properties of Cs and {Cs,H,H} complexes in Si'. *Physica B: Condensed Matter*. 2003;**340-342**:637–40.
[94] Peng C., Zhang H., Stavola M., *et al.* 'Interaction of hydrogen with carbon in multicrystalline Si solar-cell materials'. *Journal of Applied Physics*. 2011;**109**(5):53517.
[95] Weiser P.M., Monakhov E., Haug H., Wiig M.S., Søndenå R. 'Hydrogen-related defects measured by infrared spectroscopy in multicrystalline silicon

wafers throughout an illuminated annealing process'. *Journal of Applied Physics*. 2020;**127**(6):65703.

[96] Markevich V.P., Suezawa M. 'Hydrogen–oxygen interaction in silicon at around 50 °C'. *Journal of Applied Physics*. 1998;**83**(6):2988–93.

[97] Newman R.C., Tucker J.H., Brown A.R., McQuaid S.A. 'Hydrogen diffusion and the catalysis of enhanced oxygen diffusion in silicon at temperatures below 500 °C'. *Journal of Applied Physics*. 1991;**70**(6):3061–70.

[98] Huang Y.L., Ma Y., Job R., Fahrner W.R., Simoen E., Claeys C. 'The lower boundary of the hydrogen concentration required for enhancing oxygen diffusion and thermal donor formation in Czochralski silicon'. *Journal of Applied Physics*. 2005;**98**(3):33511.

[99] Simoen E., Huang Y.L., Ma Y., *et al.* 'What do we know about hydrogen-induced thermal donors in silicon?' *Journal of The Electrochemical Society*. 2009;**156**(6):H434.

[100] Job R., Fahrner W.R., Kazuchits N.M., Ulyashin A.G. 'A two-step low-temperature process for a p-n junction formation due to hydrogen enhanced thermal donor formation in P-type Czochralski silicon'. *MRS Proceedings*. 1998;**513**(1):337–42.

[101] Sopori B. 'Silicon solar-cell processing for minimizing the influence of impurities and defects'. *Journal of Electronic Materials*. 2002;**31**(10):972–80.

[102] Queisser H.J., Haller E.E. 'Defects in semiconductors: some fatal, some vital'. *Science*. 1998;**281**(5379):945–50.

[103] Sopori B., Zhang Y., Ravindra N.M. 'Silicon device processing in H-ambients: H-diffusion mechanisms and influence on electronic properties'. *Journal of Electronic Materials*. 2001;**30**(12):1616–27.

[104] Van Wieringen A., Warmoltz N. 'On the permeation of hydrogen and helium in single crystal silicon and germanium at elevated temperatures'. *Physica*. 1956;**22**(6):849–65.

[105] Rizk R., de Mierry P., Ballutaud D., Aucouturier M., Mathiot D. 'Hydroge*n* diffusio*n* and passivation processes in *p*- and *n*-type crystalline silicon'. *Physical Review B*. 1991;**44**(12):6141–51.

[106] Ichimiya T., Furuichi A. 'On the solubility and diffusion coefficient of tritium in single crystals of silicon'. *The International Journal of Applied Radiation and Isotopes*. 1968;**19**(7):573–8.

[107] Zundel T., Weber J. 'Trap-Limited hydrogen diffusion in boron-doped silicon'. *Physical Review B*. 1992;**46**(4):2071–7.

[108] Stavola M., Jiang F., Rohatgi A., Kim D., Holt J., Atwater H. Hydrogenation of Si from SiNx:H films: how much hydrogen is really in the Si? Proc 3rd 3rd World Conf Photovolt Energy Convers; 2003. pp. 909–12.

[109] Huang Y.L., Ma Y., Job R., Ulyashin A.G. 'Hydrogen diffusion at moderate temperatures in P-type Czochralski silicon'. *Journal of Applied Physics*. 2004;**96**(12):7080–6.

[110] Kleekajai S., Jiang F., Stavola M., *et al.* 'Concentration and penetration depth of H introduced into crystalline Si by hydrogenation methods used to fabricate solar cells'. *Journal of Applied Physics*. 2006;**100**(9):93517.

[111] Hansen W.L., Pearton S.J., Haller E.E. 'Bulk acceptor compensation produced in *p* -type silicon at near-ambient temperatures by a H $_2$ O plasma'. *Applied Physics Letters*. 1984;**44**(6):606–8.

[112] Dubé C., Hanoka J.I., Sandstrom D.B. 'Hydrogen diffusion along passivated grain boundaries in silicon ribbon'. *Applied Physics Letters*. 1984;**44**(4):425–7.

[113] Pearton S.J. '13Th Int. Conf. on defects in semiconductors'. *Metallurgical Society of AIME*. 1985:737–43.

[114] Capizzi M., Mittiga A. 'Hydrogen in crystalline silicon: a deep donor?' *Applied Physics Letters*. 1987;**50**(14):918–20.

[115] Wu I.W., Lewis A.G., Huang T.Y., Chiang A. 'Effects of Trap-State density reduction by plasma hydrogenation in low-temperature Polysilicon TFT'. *IEEE electron device letters : a publication of the IEEE Electron Devices Society*. 1989;**10**(3):123–5.

[116] Moore C.B., Ast D.G. 'Measurements of Grain Boundary Trap Density and Hydrogen Diffusivity in Polycrystalline Silicon Fet's'. *MRS Proceedings*. 1990;**182**:341–6.

[117] Seager C.H., Anderson R.A., Brice D.K. 'In situ measurements of hydrogen motion and bonding in silicon'. *Journal of applied physics*. 1990;**68**(7):3268–84.

[118] Mogro-Campero A., Love R.P., Schubert R. 'Drastic changes in the electrical resistance of Gold-Doped silicon produced by a hydrogen plasma'. *Journal of The Electrochemical Society*. 1985;**132**(8):2006–9.

[119] Sopori B.L., Jones K., Deng X.J. 'Observation of enhanced hydrogen diffusion in solar cell silicon'. *Applied Physics Letters*. 1992;**61**(21):2560–2.

[120] Hallam B.J., Hamer P.G., Wenham S., Vicari Stefani B., Chan C.E., Ciesla nee Wenham A.M. 'Development of advanced hydrogenation processes for silicon solar cells via an improved understanding of the behaviour of hydrogen in silicon'. *Progress in Photovoltaics: Research and Applications*. 2018;**2020**:1–22.

[121] Pearton S.J., Corbett J.W., Borenstein J.T. 'Hydrogen diffusion in crystalline semiconductors'. *Physica B: Condensed Matter*. 1991;**170**(1):85–97.

[122] Pankove J.I., Johnson N.M. *Hydrogen in Semiconductors: Hydrogen in Silicon*. **Vol. 34**. Academic Press; 1991.

[123] Corbett J.W., Lindström J.L., Pearton S.J., Tavendale A.J. 'Passivation in silicon'. *Solar Cells*. 1988;**24**(1):127–33.

[124] Sopori B. 'Silicon nitride processing for control of optical and electronic properties of silicon solar cells'. *Journal of Electronic Materials*. 2003;**32**(10):1034–42.

[125] Johnson N.M., Moyer M.D. 'Absence of oxygen diffusion during hydrogen passivation of shallow-acceptor impurities in single-crystal silicon'. *Applied Physics Letters*. 1985;**46**(8):787–9.

[126] Tavendale A.J., Alexiev D., Williams A.A. 'Field drift of the hydrogen-related, acceptor-neutralizing defect in diodes from hydrogenated silicon'. *Applied Physics Letters*. 1985;**47**(3):316–8.

[127] Chan C., Hamer P., Bourret-Sicotte G., *et al.* 'Instability of increased contact resistance in silicon solar cells following Post-Firing thermal processes'. *Solar RRL*. 2017;**1**(11):1700129.

[128] Hamer P., Chan C., Bonilla R.S., *et al.* 'Hydrogen induced contact resistance in PERC solar cells'. *Solar Energy Materials and Solar Cells*. 2018;**184**(5):91–7.

[129] Gorelkinskii Y.V., Nevinnyi N.N. 'Epr of interstitial hydrogen in silicon: uniaxial stress experiments'. *Materials Science and Engineering: B*. 1996;**36**(1–3):133–7.

[130] Bonde Nielsen K., Nielsen B.B., Hansen J., Andersen E., Andersen J.U. 'Bond-centered hydrogen in silicon studied by *in situ* deep-level transient spectroscopy'. *Physical Review B*. 1999;**60**(3):1716–28.

[131] Pankove J.I., Zanzucchi P.J., Magee C.W., Lucovsky G. 'Hydrogen localization near boron in silicon'. *Applied Physics Letters*. 1985;**46**(4):421–3.

[132] Jiang F., Stavola M., Rohatgi A., *et al.* 'Hydrogenation of Si from SiNx(H) films: Characterization of H introduced into the Si'. *Applied Physics Letters*. 2003;**83**(5):931–3.

[133] Glunz S.W., Rein S., Warta W., Knobloch J., Wettling W. 'Degradation of carrier lifetime in cz silicon solar cells'. *Solar Energy Materials and Solar Cells*. 2001;**65**(1):219–29.

[134] Boehme C., Lucovsky G. 'H loss mechanism during anneal of silicon nitride: chemical dissociation'. *Journal of Applied Physics*. 2000;**88**(10):6055–9.

[135] Geerligs L.J., Azzizi A., Macdonald D.H., Manshanden P. 'Hydrogen passivation of iron in multicrystalline silicon'. *Energy Research Centre of the Netherlands*. 2003.

[136] Liu A., Sun C., Macdonald D. 'Hydrogen passivation of interstitial iron in boron-doped multicrystalline silicon during annealing'. *Journal of Applied Physics*. 2014;**116**(19):194902.

[137] Sun C., Liu A., Phang S.P., Rougieux F.E., Macdonald D. 'Charge states of the reactants in the hydrogen passivation of interstitial iron in P-type crystalline silicon'. *Journal of Applied Physics*. 2015;**118**(8):085709.

[138] Tan J., Cuevas A., Macdonald D., Trupke T., Bardos R., Roth K. 'On the electronic improvement of multi-crystalline silicon via gettering and hydrogenation'. *Progress in Photovoltaics: Research and Applications*. 2008;**16**(2):129–34.

[139] Liu A.Y., Sun C., Markevich V.P., Peaker A.R., Murphy J.D., Macdonald D. 'Gettering of interstitial iron in silicon by plasma-enhanced chemical vapour deposited silicon nitride films'. *Journal of Applied Physics*. 2016;**120**(19):193103.

[140] Hallam B., Herguth A., Hamer P., *et al.* 'Eliminating light-induced degradation in commercial P-type Czochralski silicon solar cells'. *Applied Sciences*. 2017;**8**(1):10.

[141] Hallam B.J., Wenham S.R., Hamer P.G., *et al.* 'Hydrogen Passivation of B-O defects in Czochralski silicon'. *Energy Procedia*. 2013;**38**(8):561–70.

[142] Hallam B., Kim M., Abbott M., *et al.* 'Recent insights into boron-oxygen related degradation: evidence of a single defect'. *Solar Energy Materials and Solar Cells*. 2017;**173**:25–32.

[143] Fischer H., W P. Investigation of photon and thermal induced change in silicon solar cells. Proceedings of the 10th IEEE Photovoltaic Specialists Conference. IEEE; 1973. pp. 404–11.

[144] Knobloch J., Glunz S.W., Biro D., Warta W., Schaffer E., Wettling W. Solar cells with efficiencies above 21% processed from Czochroalski grown silicon. Proceedings of the 25th IEEE Photovoltaic Specialists Conference; 1996. pp. 405–8.

[145] Herguth A., Schubert G., Käs M., Hahn G., Kaes M., Hahn G. A new approach to prevent the negative impact of the metastable defect in boron doped Cz silicon solar cells. Proceedings of the 4th IEEE World Conference on Photovoltaic Energy Conversion; 2006. pp. 940–3.

[146] Münzer K. Hydrogenated Silicon Nitride for Regeneration of Light Induced Degradation. 24th European Photovoltaic Solar Energy Conference; 2009. pp. 1558–61.

[147] Krugel G., Wolke W., Geilker J., Rein S., Preu R. 'Impact of hydrogen concentration on the regeneration of light induced degradation'. *Energy Procedia*. 2011;**8**:47–51.

[148] Dubois S., Tanay F., Veirman J., Enjalbert N., Stendera J., Butté S. The BOLID project: suppression of the Boron-Oxygen related Light-Induced-Degradation. 27th Eur Photovolt Sol Energy Conf; 2012. pp. 749–54.

[149] Wilking S., Engelhardt J., Ebert S., Beckh C., Herguth A., Hahn G. 'High speed regeneration of BO-Defects: improving long-term solar cell performance within seconds'. *29th European Photovoltaic Solar Energy Conference and Exhibition*. 2014:366–72.

[150] Joos S., Schiele Y., Terheiden B., Becker H.-W., Rogalla D., Hahn G. 'Fundamental studies of hydrogen at the silicon / silicon nitride interface'. *Energy Procedia*. 2014;**55**(11):786–90.

[151] Walter D.C., Lim B., Bothe K., Voronkov V.V., Falster R., Schmidt J. 'Effect of rapid thermal annealing on recombination centres in boron-doped Czochralski-grown silicon'. *Applied Physics Letters*. 2014;**104**(4):42111.

[152] Walter D.C., Schmidt J. 'Impact of hydrogen on the permanent deactivation of the boron-oxygen-related recombination center in crystalline silicon'. *Solar Energy Materials and Solar Cells*. 2016;**158**:91–7.

[153] Pearton S.J. 'Diffusion of hydrogen in N-type Si'. *Materials Science and Engineering: B*. 1994;**23**(2):130–6.

[154] Schnabel M., Van De Loo B.W.H., Nemeth W., Macco B., Stradins P., Kessels W.M.M. 'Hydrogen passivation of poly-Si/SiOx contacts for Si solar cells using Al2O3 studied with deuterium'. *Applied physics letters*. 2018;**112**(20):203901.

[155] Tweddle D., Hamer P., Shen Z., Markevich V.P., Moody M.P., Wilshaw P.R. 'Direct observation of hydrogen at defects in multicrystalline silicon'. *Progress in Photovoltaics: Research and Applications*. 2019;**215**(2).

[156] Tiron V., Velicu I.-L., Pana I., *et al.* 'HiPIMS deposition of silicon nitride for solar cell application'. *Surface and Coatings Technology*. 2018;**344**(3):197–203.

[157] Yarykin N.A., Weber J. 'Identification of copper-copper and copper-hydrogen complexes in silicon'. *Semiconductors*. 2013;**47**(2):275–8.

[158] Mullins J., Leonard S., Markevich V.P., *et al.* 'Recombination via transition metals in solar silicon: the significance of hydrogen–metal reactions and lattice sites of metal atoms'. *Physica Status Solidi*. 2017;**214**(7):1700304.

[159] Kleekajai S., Jiang F., Stavola M., *et al.* 'Concentration and penetration depth of H introduced into crystalline Si by hydrogenation methods used to fabricate solar cells'. *Journal of Applied Physics*. 2006;**100**(9):93517.

[160] Sachse J.-U., Sveinbjörnsson E.Ö., Yarykin N., Weber J. 'Similarities in the electrical properties of transition metal–hydrogen complexes in silicon'. *Materials Science and Engineering: B*. 1999;**58**(1):134–40.

[161] Williams P.M., Watkins G.D., Uftring S., Stavola M. 'Structure-Sensitive spectroscopy of transition-metal-hydrogen complexes in silicon'. *Physical Review Letters*. 1993;**70**(24):3816–9.

[162] Bonde Nielsen K., Dobaczewski L., Goscinski K., Bendesen R., Andersen O., Bech Nielsen B. 'Deep levels of vacancy-hydrogen centers in silicon studied by Laplace DLTS'. *Physica B: Condensed Matter*. 1999;**273-274**:167–70.

[163] Weber J., Hiller M., Lavrov E.V. 'Hydrogen molecules in semiconductors'. *Physica B: Condensed Matter*. 2007;**401-402**:91–6.

[164] Lavrov E V., Weber J. 'Ortho and Para Interstitial H 2 in Silicon'. *Physical review letters*. 2002;**89**(21):215501.

[165] Pritchard R.E., Tucker J.H., Newman R.C., Lightowlers E.C. 'Hydrogen molecules in boron-doped crystalline silicon'. *Semiconductor Science and Technology*. 1999;**14**(1):77–80.

[166] Bonde Nielsen K., Nielsen B.B., Hansen J., Andersen E., Andersen J.U. 'Bond-centered hydrogen in silicon studied by *in situ* deep-level transient spectroscopy'. *Physical Review B*. 1999;**60**(3):1716–28.

[167] Walter D.C., Bredemeier D., Falster R., Voronkov V.V., Schmidt J. 'Easy-to-apply methodology to measure the hydrogen concentration in boron-doped crystalline silicon'. *Solar Energy Materials and Solar Cells*. 2019;**200**(1–4):109970.

[168] Wilde M., Fukutani K. 'Hydrogen detection near surfaces and shallow interfaces with resonant nuclear reaction analysis'. *Surface Science Reports*. 2014;**69**(4):196–295.

[169] Gerke S., Becker H.–W., Rogalla D., Hahn G., Job R., Terheiden B. 'About nuclear resonant reaction analysis for hydrogen investigations in amorphous silicon'. *Energy Procedia*. 2015;**84**:99–104.

[170] Catoir J., Wolke W., Griesshammer M., Preu R., Trassl R., Grambol D. Analysis of hydrogen passivation by sputtered silicon nitride. Proc 23rd Eur Photovolt Sol Energy Conf Proceedings, Val; 2008. pp. 1530–3.

[171] Kim K., Borojevic N., Winderbaum S., *et al.* 'Investigation of industrial PECVD AlOx films with very low surface recombination'. *Solar Energy*. 2019;**186**(6):94–105.

[172] Oura K., Naitoh M., Shoji F., Yamane J., Umezawa K., Hanawa T. 'Elastic recoil detection analysis of hydrogen adsorbed on solid surfaces'. *Nuclear Instruments and Methods in Physics Research Section B: Beam Interactions with Materials and Atoms*. 1990;**45**(1–4):199–202.

[173] Dingemans G., Kessels W.M.M. 'Status and prospects of Al2O3-based surface passivation schemes for silicon solar cells'. 2016;**2015**(040802):1–27.

[174] Lanford W.A. 'Analysis for hydrogen by nuclear reaction and energy recoil detection'. *Nuclear Instruments and Methods in Physics Research Section B: Beam Interactions with Materials and Atoms*. 1992;**66**(1–2):65–82.

[175] WMA B., Linssen R.N.H., Habraken F., Van der Weg W.F., Kuiper A.E.T. 'Diffusion of hydrogen in low-pressure chemical vapor deposited silicon nitride films'. *Applied physics letters*. 1990;**56**(25):2530–2.

[176] Tavendale A.J., Alexiev D., Williams A.A. 'Field drift of the hydrogen-related, acceptor-neutralizing defect in diodes from hydrogenated silicon'. *Applied Physics Letters*. 1985;**47**(3):316–8.

[177] Johnson N.M. 'Mechanism for hydrogen compensation of shallow-acceptor impurities in single-crystal silicon'. *Physical Review B*. 1985;**31**(8):5525–8.

[178] Hallam B.J., Hamer P.G., Wenham S.R., *et al.* 'Advanced bulk defect Passivation for silicon solar cells'. *IEEE Journal of Photovoltaics*. 2014;**4**(1):88–95.

[179] Hamer P., Hallam B., Wenham S., Abbott M. 'Manipulation of hydrogen charge states for Passivation of P-type Wafers in Photovoltaics'. *IEEE Journal of Photovoltaics*. 2014;**4**(5):1252–60.

[180] Hamer P., Hallam B., Chan C., Wenham A., Sugianto A., Abbott M. Hydrogen passivation mechanisms in silicon solar cells. Proceedings of the 28th European Photovoltaic Solar Energy Conference; 2013. pp. 982–7.

[181] Sun C., Rougieux F.E., Macdonald D. 'A unified approach to modelling the charge state of monatomic hydrogen and other defects in crystalline silicon'. *Journal of Applied Physics*. 2015;**117**(4):45702.

[182] Hallam B., Abbott M., Nampalli N., Hamer P., Wenham S. 'Influence of the formation- and passivation rate of boron-oxygen defects for mitigating carrier-induced degradation in silicon within a hydrogen-based model'. *Journal of Applied Physics*. 2016;**119**(6):65701.

[183] Karzel P., Junge J., Hahn G. Mapping of hydrogen bond energies in EFG silicon samples by analysis of spatially resolved minority charge carrier lifetimes after annealing steps. 24th European Photovoltaic Solar Energy Conference; Hamburg, Germany; 2009. pp. 2023–7.

[184] Stavola M., Pearton S.J., Lopata J., Dautremont-Smith W.C. 'Vibrational spectroscopy of acceptor-hydrogen complexes in silicon: evidence for low-frequency excitations'. *Physical Review B*. 1988;**37**(14):8313–8.

[185] Hamer P., Wang S., Hallam B., *et al*. 'Laser illumination for manipulation of hydrogen charge states in silicon solar cells'. *physica status solidi (RRL) - Rapid Research Letters*. 2015;**9**(2):111–4.

[186] Chong C.M., Wenham S., Ji J., *et al*. 'LEDs for the implementation of advanced hydrogenation using hydrogen charge-state control'. *International Journal of Photoenergy*. 2018;**2018**:1–6.

[187] Wenham S.R., Hamer P.G., Hallam B.J., Sugianto A., Chan C.E., Song L. Advanced hydrogenation of silicon solar cells. US Patent 9190556; 2015.

[188] Sah C.-tang., Noyce R.N., Shockley W. 'Carrier generation and recombination in p-N junctions and P-N junction characteristics'. *Proceedings of the IRE*. 1957;**45**(9):1228–43.

[189] Hoex B., Gielis J.J.H., van de Sanden M.C.M., Kessels W.M.M. 'On the c-Si surface passivation mechanism by the negative-charge-dielectric Al2O3'. *Journal of Applied Physics*. 2008;**104**(11):113703.

Chapter 3
Hydrogen passivation of silicon surfaces

Ruy S. Bonilla[1] *and Bram Hoex*[2]

This chapter discusses the role that hydrogen takes in the passivation of silicon surfaces. The basic concepts of surface recombination and passivation are first introduced, to then examine how hydrogen can influence the dynamics of charge carriers at the surface of silicon. The different materials and synthesis techniques used to achieve optimal surface passivation are covered. In particular, the role of hydrogen in all materials and detection techniques are reviewed in detail, drawing on the latest findings in the field. Hydrogen-mediated passivation is examined in silicon dioxide (SiO_2) thin films, as well as in hydrogen-containing dielectric layers, such as plasma-enhanced chemical vapour deposition (PECVD) silicon nitride (SiN_x), PECVD amorphous silicon (a-Si), and atomic layer deposition (ALD) or PECVD aluminium oxide (AlO_x). The role of silicon oxide and nitride are examined at length, as these materials are of critical importance to the development of photovoltaic (PV) technology. SiN_x is the most widely adopted surface coating in the solar cell industry, and this has been largely due to its hydrogenation benefits and suitability as an antireflection coating. The use of thermal processes in releasing hydrogen from the dielectric are also covered, since such steps serve in the activation of surface passivation, or in the injection of hydrogen into the silicon material for bulk passivation. Last, the latest developments of hydrogenation in tunnelling oxide passivating contacts are reported, and an outlook summary is provided on the vital role that hydrogenation processes will take in the future improvement of silicon PV devices.

3.1 Recombination and passivation at silicon interfaces

3.1.1 The surface as a defect

Surfaces and interfaces in semiconductor devices create abrupt crystal discontinuities. At a bare silicon surface, for example, many atoms are only partially bonded, producing a concentration of dangling bonds with energy levels in the otherwise

[1]Department of Materials, University of Oxford, UK
[2]The School of Photovoltaic and Renewable Energy Engineering, The University of New South Wales, Australia

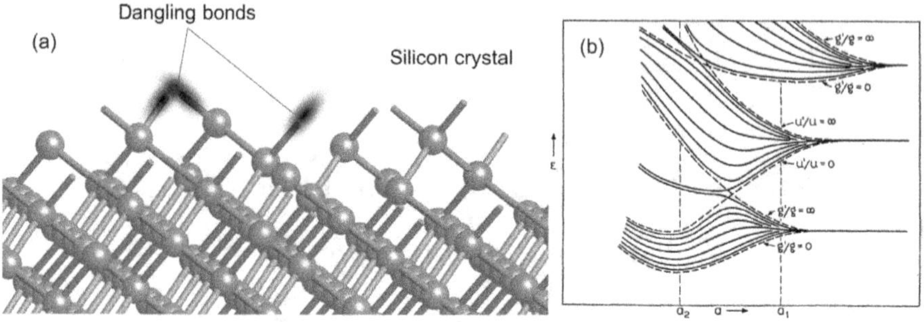

Figure 3.1 *(a) Schematic picture of a bare silicon surface with dangling bonds producing interface mid bandgap states. (b) Shockley's reconstruction of a one-dimensional lattice. Reprinted figure with permission from [2] Copyright (1939) by the American Physical Society. https://doi.org/10.1103/PhysRev.56.317.*

forbidden bandgap. Such interruptions and inhomogeneities give rise to surface energy traps or surface states [1]. A graphical depiction of dangling bonds is illustrated in Figure 3.1a. The origin of semiconductor surface states is best represented in Shockley's description of an eight atom one-dimensional array [2]. He showed how two bands of states arise from the energy levels in individual atoms, once they are arranged into a crystal. This is illustrated in Figure 3.1b. The space between atoms in the crystal is defined by the minimum total energy of the solid, and it determines the crystal lattice constant. Shockley's depiction illustrates that the increase in atomic spacing at the surface results in a deviation of the lattice constant, and thus a dispersion of the energy levels. These bandgap energy states provide effective sites for charge carrier recombination. The semiconductor surface is therefore crucial in the operation of devices, especially as the thickness of silicon wafers continues to reduce.

In general usage, the term surface refers to a solid–air interface. However, in practical solar cells, bare silicon surfaces are not present, and recombination actually takes place at interfaces between the semiconductor and other materials. Even so, it is standard practice to refer to this as surface recombination and the concept of 'surface' states is still applicable.

The effectiveness of a surface as a site for recombination is most often characterised by the surface recombination velocity (SRV), *S*, defined as:

$$S \equiv \frac{U_s}{\Delta n} \tag{3.1}$$

in units of centimetre per second. U_s indicates the surface recombination rate in $cm^{-2}s^{-1}$. The lower its value, the lower the recombination rate, and the better a surface is from the point of view of solar cell efficiency. The SRV depends on doping concentration, injection level, and doping type. The most accepted analytical description of SRV follows the Shockley-Read-Hall (SRH) recombination theory [3–5]. Using the SRH model for defect-mediated recombination, two important modifications need to be applied. First, the continuum of defect states in the bandgap due to the crystal discontinuity must be considered. Second, the physical concept of carrier lifetime becomes ambiguous since it is expressed for carriers travelling to the surface rather than in the bulk, hence the need to define the SRV. Surface recombination (U_s) is described by extending the SRH formalism to an arbitrary trap level density function [6]:

$$U_s = \int_{Ev}^{Ec} \frac{\left(n_s p_s - n_i^2\right)}{\frac{[n_s + n_1(E)]}{v_{th} D_{it}(E) \sigma_p} + \frac{[p_s + p_1(E)]}{v_{th} D_{it}(E) \sigma_n}} dE \tag{3.2}$$

where D_{it} is the energy-dependent density of interface traps, n_s and p_s are the carrier concentrations at the surface of electrons/holes, v_{th} is the thermal velocity of carriers, σ_n/σ_p are the capture cross sections for electrons/holes, and n_1/p_1 are the SRH carrier concentrations terms. The SRV is then obtained by combining (3.1) and (3.2):

$$S = \frac{1}{\Delta n} \int_{Ev}^{Ec} \frac{\left(n_s p_s - n_i^2\right)}{\frac{[n_s + n_1(E)]}{S_p} + \frac{[p_s + p_1(E)]}{S_n}} dE \tag{3.3}$$

where Δn is the excess minority carrier concentration, $S_p\left(E\right) = v_{th} D_{it}\left(E\right) \sigma_p\left(E\right)$ is the energy-dependent recombination velocity for holes and $S_n\left(E\right) = v_{th} D_{it}\left(E\right) \sigma_n\left(E\right)$ is the energy-dependent recombination velocity for electrons. Physically, S_p and S_n represent the rate at which holes and electrons are captured independently, respectively. S, on the other hand, comprises the total recombination occurring at the surface as a result of the capture of both types of carriers. An example calculation of S using (3.3) will now be discussed for the case of negligible surface charge – i.e. no band bending. In this case, the concentration of carriers in the bulk and at the surface is the same: $n_s = n_b$, $p_s = p_b$, where n_b and p_b are the bulk carrier concentrations.

Figure 3.2 illustrates the normalised SRV (S_0/S_{p0}) as a function of excess minority carrier concentration for 1.5 Ω.cm p-type silicon and different values of electron SRVs. Both S_n and S_p are considered energy independent, and will be referred to as S_0/S_{p0} and S_{p0}. Similar to the case of SRH lifetime at a single defect in the bulk, SRV at low injection levels is dominated by the capture of minority carriers, hence the normalisation by $1/S_{n0}$. This plot is useful to determine the effects that electron and hole capture rates have on total SRV, and the form SRV takes when the capture rate of electrons is different to that of holes. This difference is referred to as the asymmetry between hole and electron recombination velocities. Despite a wide variation in S_{p0}, the SRV at low injection levels is approximately constant. For high injection levels, however, both types of carriers contribute to recombination and thus the minimum S_p (better from the point of view of passivation) is controlled by whichever of S_{p0} and S_{n0} is smaller. Figure 3.2 also shows the injection-level dependence of SRV as the recombination velocity of holes becomes increasingly asymmetrical to that of electrons. This is an important characteristic since, as will be shown later, most silicon interfaces present such asymmetry. Figure 3.3 shows a plot of the normalised SRV for various doping concentrations. Surface recombination increases with doping level. This makes highly doped silicon surfaces difficult to passivate, in

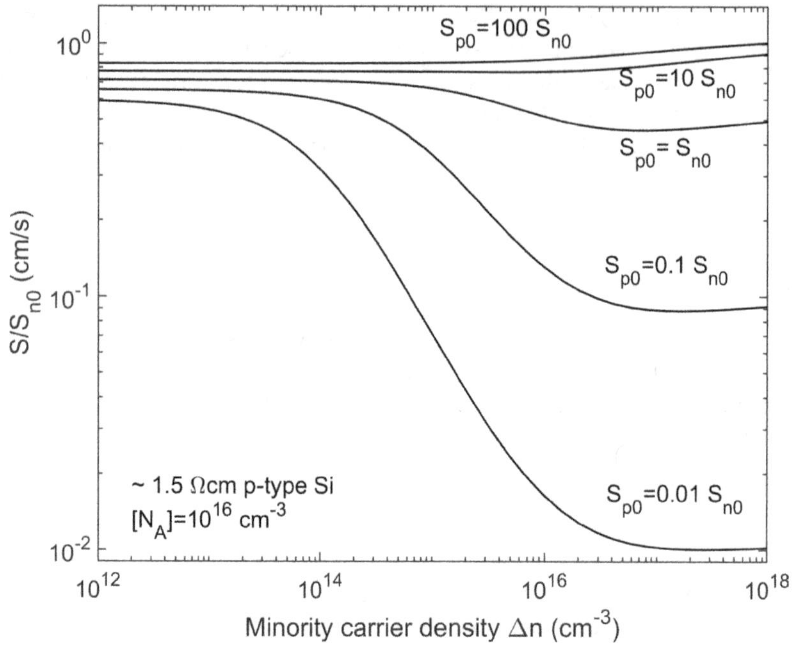

Figure 3.2 *Calculated SRV for a range of capture cross sections and excess minority carrier concentration. SRV is calculated using the extended SRH statistics. N_A represents the acceptor doping concentration.*

particular, for the injection levels solar cells normally operate at, $\Delta n \sim 10^{15}/\text{cm}^3$. In this region, S_0 is seen to increase by an order of magnitude when the surface dopant density increases by three orders of magnitude.

Examination of (3.3) and Figures 3.2 and 3.3 indicates that reductions in surface recombination can be accomplished by establishing surfaces with either fewer defect states $D_{it}(E)$, or lower capture probabilities $\sigma_p(E)$, and Q_{Si}. In addition, there is a second mechanism governing SRV. Since a recombination process requires the presence of both electrons and holes, lower recombination can be achieved by reducing the number of carriers at the surface. The minority carrier concentration at the surface normally determines the rate of recombination. This reduction can be achieved by a surface electric field that penetrates the semiconductor surface and thus modifies the surface carrier concentration. This is referred to as the field-effect passivation (FEP), charge-assisted passivation or surface-population-control mechanism. The electric field is often established by an external fixed charge density Q_f near the silicon, which creates a mirror charge inside the semiconductor, thus controlling the population of the charge carriers at the surface. Girisch *et al.* [6] and Aberle *et al.* [7] proposed a solution algorithm for (3.3) that included the contribution from photo-generated carriers and accounted for the fixed charge Q_f. It involves computing the concentrations of the surface space charge region Q_{Si}, and the charged interface states Q_{it}, where $Q_f = -(Q_{Si} + Q_{it})$, i.e. polarity of Q_{Si} and Q_{it} is opposite to Q_f, and overall the system is charge neutral. Figure 3.4

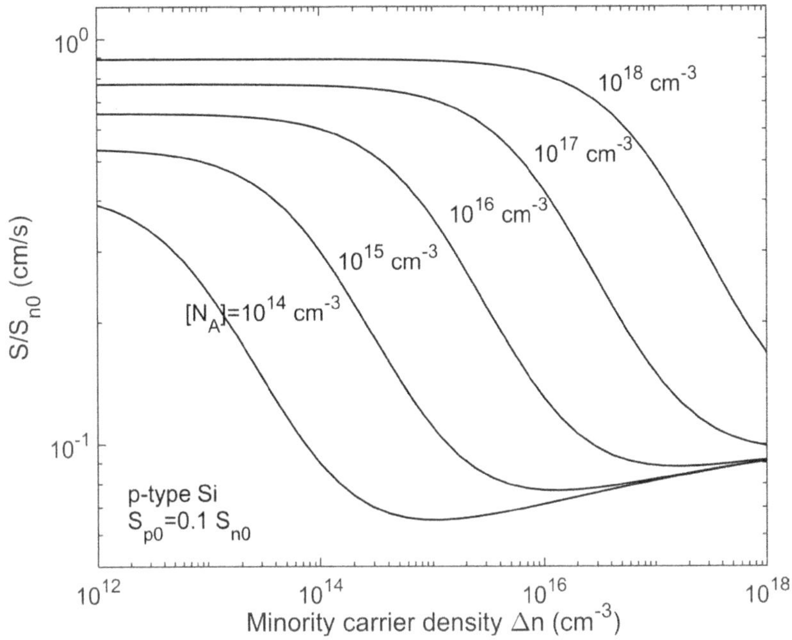

Figure 3.3 *Calculated SRV as a function of excess minority carrier concentration for different doping concentrations. SRV is calculated using the extended SRH statistics of (3.3).*

illustrates the solution to (3.3) using this algorithm, for parameters typical of a passivated *p*-type silicon surface. Here it can be evidenced that slower capture rates produce reductions in the overall SRV as previously noted. Additionally, large external charge concentrations can also reduce recombination and are therefore critical in achieving optimally performing surfaces. Positive charge normally provides better passivation than negative charge, as seen by the lower SRV values at the right of Figure 3.4. This is due to the typically higher capture velocity of electrons, compared to that of holes for most common silicon interfaces [7]. It should be noted, however, that in *p*-type silicon the positive charge drives the surface into inversion conditions and this can result in challenges such as parasitic shunting in some cases [8]. Due to this, an accumulation condition with low SRV is preferred for *p*-type silicon. Regardless of the surface type, dielectric layers with large concentrations of charge are preferred as they provide this added beneficial mechanism, while dielectrics with low charged are avoided.

3.1.2 Characterisation of surface passivation in silicon

The characterisation of surface recombination in silicon has been primarily carried out via two metrics. SRV was the first quantity to be deduced from current density measurements in *pn* junction diodes, with either a gate-controlled bias [9, 10] or photo-generation of carriers far from the junction using a bias light [6]. At present, contactless photoconductance decay (PCD) using transient or quasi-steady state

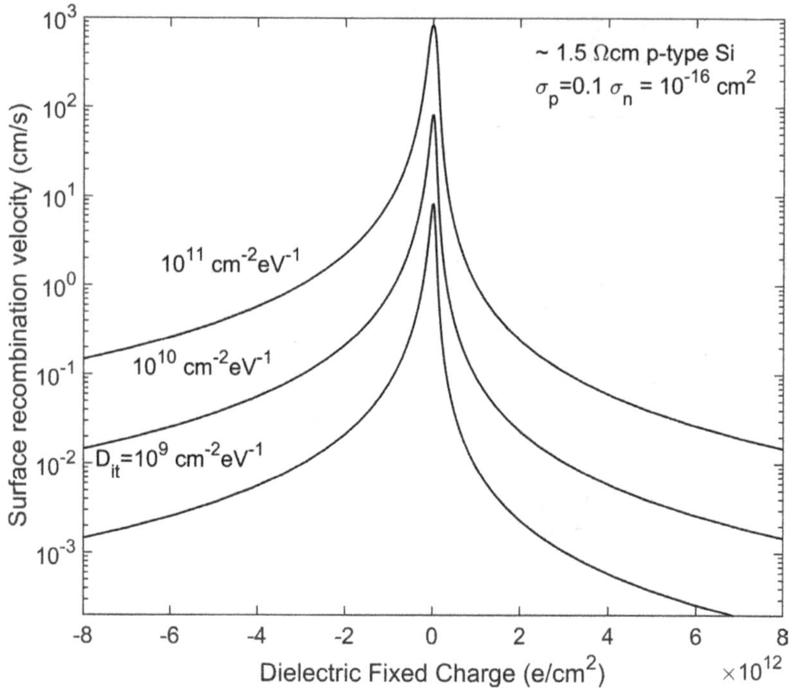

Figure 3.4 *Calculated effective SRV as a function of positive and negative dielectric charge concentration Q_f for p-type silicon with $S_{p0} = 0.1\ S_{p0}$, $\Delta n_d = 10^{15}/cm^3$, acceptor concentration $N_{acc} = 10^{16}/cm^3$, and uniform interface density D_{it}*

modes is the preferred technique to measure the effective lifetime (τ_{eff}) and extract accurate values of SRV. This technique does not require a device structure or direct contacts and can examine large areas. In PCD the change in minority carrier concentration is calculated via changes in conductance, measured using microwave [11] or inductive sensing [12, 13]. Of these two, inductive coupling PCD has become the technique of choice for minority carrier lifetime measurements in silicon in the past two decades [14, 15].

Measurements of effective lifetime contain the contributions of all recombination mechanisms in the bulk and at the surface of the specimens:

$$\frac{1}{\tau_{eff}} = \frac{1}{\tau_b} + \frac{1}{\tau_s} \tag{3.4}$$

where τ_b and τ_s are the bulk and surface lifetime components, respectively. Increasing both these lifetimes is key to improving cell performance. The bulk component, τ_b, includes the intrinsic components due to Auger and radiative recombination [1], as well as bulk SRH recombination. To date, the most accurate parametrisation of intrinsic Auger and radiative recombination was proposed by Richter *et al.* [16], with the radiative recombination term B_{rel} from [17] and B_{low} from [18]. τ_s is then found by computing

$$\frac{1}{\tau_s} = \frac{1}{\tau_{eff}} - \frac{1}{\tau_{Rad}} - \frac{1}{\tau_{Aug}} - \frac{1}{\tau_{SRH}} \tag{3.5}$$

The surface lifetime component τ_s has a complex dependence on SRV. The amount of recombination at the surface depends on the spatial variation of the injected carriers. Carriers generated far from the surface first have to diffuse to the surface before contributing to recombination. Similarly, carriers generated at the near-surface may recombine instantaneously or diffuse to the bulk. To account for this dependence, Luke and Cheng [19] developed a formalism and found a solution that indicates that the effective SRV is a function of an infinite sum of decaying exponential terms of effective lifetime, where the first mode is dominant and is written as

$$S_{eff} = \sqrt{D\left(\frac{1}{\tau_s}\right)} \tan\left(\frac{W}{2}\sqrt{\frac{1}{D}\left(\frac{1}{\tau_s}\right)}\right) \tag{3.6}$$

Here W is the specimen width and D is the ambipolar carrier diffusion coefficient. For sufficiently low SRV, $\tan(x) \approx x$, and the SRV can be approximated to the well-known relation

$$S_{eff} = \left(\frac{1}{\tau_{eff}} - \frac{1}{\tau_B}\right)\frac{W}{2} \tag{3.7}$$

This expression has been widely used to characterise SRV as its accuracy is better than 4% when recombination is the same at both surfaces and $(S_{eff}\,W)/D < 1/4$ [20]. An important distinction to draw, however, is that the *effective* SRV, S_{eff}, relates to the recombination current at a virtual surface at the edge of the space charge region, rather than the recombination occurring at the actual surface. This is because calculating and measuring the excess carrier concentration, Δn, at the actual surface is a rather involved task so that the exact S value cannot be readily obtained. This definition of S_{eff} has been widely accepted [21]. To exemplify the dependence of effective lifetime with surface recombination parameters, theoretically calculated lifetimes are shown in Figure 3.5 for a set of typical parameters of a passivated silicon surface. Lower interface state densities lead to less surface recombination and therefore higher lifetimes. Thicker wafers reduce the relative weight of SRV in the overall effective lifetime, and so lead to higher lifetime values. Figure 3.5 shows that various device and surface properties play a role in the measured lifetime and thus must be considered carefully.

The second metric used to characterise surface recombination is the surface saturation current density J_{0s} [22]. This quantity has been more generally used to assess the combined recombination from surface states and high doping near surface regions. It has become very popular as it can be used to evaluate the quality of a diffused electron or hole contact from the point of view of recombination, from the measured τ_{eff}. This metric provides a better comparison of dielectric passivation quality in specimens without an emitter (undiffused) when FEP is significant. In such undiffused specimens J_{0s} is independent of the surface dopant concentration under most practical conditions, and it is thus a useful metric of passivation quality

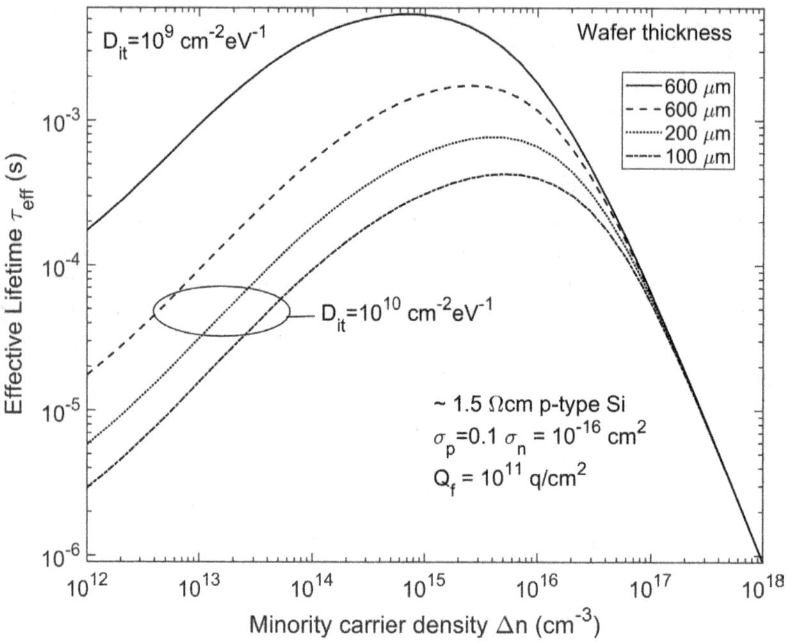

Figure 3.5 *Calculated effective lifetime as a function of wafer thickness and surface passivation quality from uniform interface density D_{it}, for* p-*type silicon, acceptor concentration N_{acc} = $10^{16}/cm^3$. The dielectric charge concentration is set to $n_{i,eff}$ =10^{11} q/cm², with$S_{n0} = 10 \times S_{p0}$.*

for dielectric films. J_{0s} can be determined from effective lifetime measurements following the iterative procedure as suggested by Kimmerle *et al.* [23], which was developed from the work of Mäckel and Verner [24], and from Kane and Swanson's well-known method [12]. In general J_{0s} is calculated from a PCD measurement of τ_{eff} as

$$J_0(\Delta n) = \frac{d}{d\Delta n}\left(n_{i,eff}^2 \cdot \sqrt{D\left(\frac{1}{\tau_s}\right)}\tan\left(\frac{W}{2}\sqrt{\frac{1}{D}\left(\frac{1}{\tau_s}\right)}\right)\right) \tag{3.8}$$

with $n_{i,eff}$ representing the effective intrinsic carrier concentration accounting for bandgap narrowing [25].

3.1.3 Dielectric coatings for surface passivation

The term 'surface passivation' covers all the physical and chemical processes carried out to a semiconductor's surface in order to reduce the surface recombination rate. From the theoretical description in Section 3.1.1, two fundamental approaches to passivate silicon surfaces arise:

1. An optimisation of the chemical interface properties by forming bonds between silicon and a different material. This reduces the number of defect states at the surface or their ability to capture charge carriers.
2. A reduction of the carrier concentration available for a recombination process. For recombination to take place, both electrons and holes must be present in the vicinity of the defect. It follows that a reduction in the surface carrier concentration of either carrier type results in a reduction of SRV. This method is normally known as FEP, or charge-assisted passivation since it uses an electric field to control the population of carriers at the surface.

The SRV of a silicon surface can often be reduced by using a dielectric or semiconducting coating. This results in a portion of the dangling bonds at the coating–silicon interface being eliminated by the formation of covalent bonds, often with oxygen, nitrogen, or silicon atoms in the coating layer. The addition of hydrogen atoms to such surface coating can further reduce SRV by saturating dangling bonds that may remain at the interface. Both these effects can be considered chemical effects.

Since charge carriers are susceptible to electric fields, two independent techniques can be used to control the population of carriers at the surface. First, the implementation of a doping profile at the silicon surface produces a built-in electric field due to the gradient of the doping concentration. When it is used on the cell's rear side, this scheme is known as the back surface field (BSF), while at the cell's front side, it is called a front surface field or floating junction. The BSF adopted its name from the process of firing aluminium into *p*-type silicon. This technique was originally proposed to melt and drive Al into a *p*-type silicon surface using a 700–900 °C heating step, and creating a p^+-p junction on the rear surface of the solar cell [26]. This junction reduced recombination since an electric field was formed at the surface so that the ratio of electrons/holes at the surface was very small. Floating junctions are n^+-p or p^+-n diffusions formed on the front of the cell, where the outer most material is not electrically connected to the cell and is therefore floating. This concept has been exploited in research and industrial solar cells that have all their electrical contacts on the back [27]. One important consequence of using BSF or floating junctions is the fact that an increase in the doping concentration produces an increase in the amount of bulk Auger recombination. A trade-off is normally achieved between the two components using doping concentrations of 10^{17}–10^{19}/cm^3, and almost always $<10^{20}$/cm^3 [28], in contrast to regions with $>10^{20}$/cm^3 for metal–silicon interface formation with screen-printed contacts.

Second, charge-assisted passivation can be implemented by installing an electric field in the dielectric coating using fixed charges associated with the dielectric. Such a field penetrates into the semiconductor and modulates the concentration of charge carriers, again reducing the ratio of minority to majority carriers. These different mechanisms are schematically presented in Figure 3.6. One aspect that must be noted is the fact that SRV scales with the surface dopant density when FEP is the dominant passivation mechanism. This was shown by McIntosh and Black in [22]. Previously, this SRV increase had been incorrectly attributed to a larger carrier

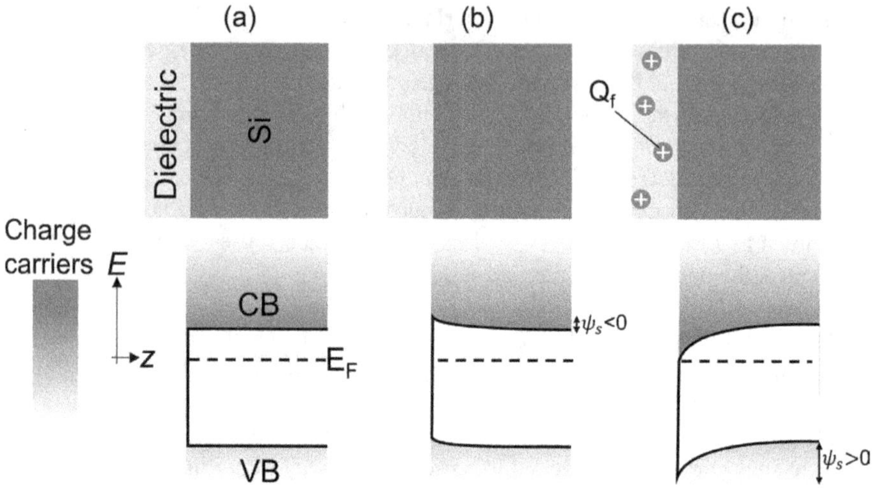

Figure 3.6 *Schematic diagram showing surface states in: (a) bare silicon; (b) chemically passivated silicon; and (c) chemically and field effect-passivated silicon*

recombination rate (S_{n0}, S_{p0}) at highly doped surfaces. It is actually found that the fundamental recombination rates (S_{n0}, S_{p0}) at a silicon surface are independent of the surface dopant density [22, 29].

3.1.4 State of the art in surface passivation of silicon

A variety of dielectric and semiconducting coatings have been used to produce extremely effective passivation of silicon surfaces. For extensive reviews on this topic, the readers are referred to [30–32]. Here, the most widely adopted dielectric coatings will be examined. Thermal oxidation of silicon has been the original and basic method to passivate silicon surfaces. Thermally grown SiO_2 can provide surface state densities as low as $10^9/(cm^{-2}\ eV^{-1})$ and it is hence one of the most effective passivation techniques [31]. However, the growth of high-quality thermal oxides requires temperatures above 900 °C, making it undesirable in terms of solar cell contamination and throughput. After thermal SiO_2, the ubiquitous industry standard for cell passivation is PECVD of hydrogenated silicon nitride (SiN_x:H) [28, 33]. The excellent surface passivation of SiN_x is due not only to a chemical reduction in the density of interface states and their capture cross-section, but also to its high concentration of fixed positive charges and hydrogen content. Thus, the passivation effect of these films is both chemical and field-effect in origin. SiN_x has two technologically important advantages over other films:

- Its refractive index makes it an excellent anti-reflective coating.
- The high hydrogen content of PECVD SiN_x layers provides a source of hydrogenation for interface and bulk passivation.

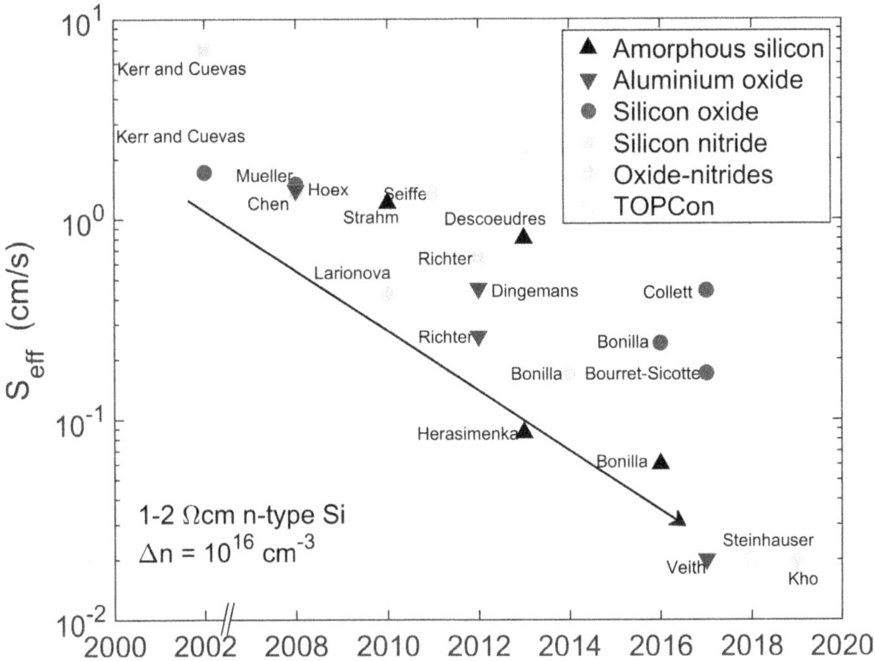

Figure 3.7 *Timeline of the best passivation technologies available at the research level. It is evident here that SRV <5 cm/s are reasonably good, and those <1 cm/s are outstanding. Please refer to [30] for full details on the data from each author.*

The only drawback in SiN_x has been the difficulty in providing effective passivation of *p*-type silicon surfaces. In *p*-type silicon, the SiN_x positive charge can lead to an inverted surface with increased recombination, or produce detrimental shunting of the *pn* junction.

A recently adopted passivation technique that circumvents this issues is a thin film of AlO_x grown via ALD [34]. Similar to SiN_x, there are multiple advantages provided by ALD AlO_x films. First, they provide a concentration of negative fixed charge, which is instrumental in the passivation of *p*-type surfaces via charge-assisted population control. Second, they also contain hydrogen that enables post-deposition hydrogenation processes in cells. Last, ALD of this film allows ultimate control of the film thickness, with layers as thin as 5 nm already providing extremely effective passivation [35]. Figure 3.7 illustrates the timeline of some of the best passivation techniques reported in the literature, in terms of their achieved SRV. Some of the most researched methods reported to date are now studied in detail, especially in their relation to hydrogen.

3.2 Hydrogen in silicon oxide thin films

3.2.1 Thermal silicon dioxide

Thermally grown SiO_2 is one of the most important dielectric films in silicon PV technology. Its importance stemmed from the extensive use of SiO_2 in the microelectronics industry as a principal component of field-effect transistors (FETs). In the context of silicon solar cells, SiO_2 came to prominence in the early works of Aberle [21], Glunz [36, 37], and Cuevas [38]. The role of hydrogen in SiO_2 passivation was, however, identified well before its adoption in solar cell technology. The first widely adopted hydrogenation process for the Si–SiO_2 interface was a high temperature anneal (500–800 °C) under a hydrogen atmosphere, which was crucial in improving the electrical properties of FETs [1]. Due to the safety issues of highly concentrated hydrogen gas, this anneal was later modified into what is now known as the forming gas anneal (FGA), consisting of a low temperature (300–500 °C) annealing step under a <5% H_2 atmosphere [39, 40]. Early evidence of the effects of hydrogenation was obtained via core photoelectron spectroscopy of the Si–SiO_2 interface [41], with complementary studies corroborating the findings via electrical techniques on metal-oxide-semiconductor devices [42, 43]. Regardless of temperature and concentration, hydrogen annealing has been widely reported to saturate free silicon dangling bonds at the Si–SiO_2 interface. Such dangling bonds are often referred to as P_b centres, and comprise either an intrinsic defect, where silicon is only back-bonded to three silicon atoms, or extrinsic defect centres, where silicon is bonded to one or two oxygen atoms [42].

A variety of electrical techniques have been used to identify the passivating effects of H at the Si–SiO_2 interface. Most prominently, capacitance-voltage (C-V), conductance voltage (G-V), and deep level transient spectroscopy have provided insights into the interface electrical properties a function of hydrogenation processes. Figure 3.8a illustrates a summary of reported measurements of D_{it} at the Si–SiO_2 interface, before and after different H passivation treatments. It is evident that

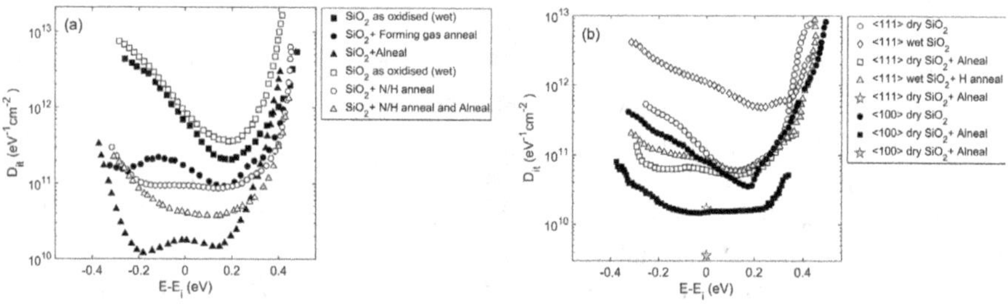

Figure 3.8 *(a) Density of interface defect states in wet oxidised <111>Si–SiO_2 as a function of hydrogenation conditions. Data adapted from Fussel [42, 44]. (b) Density of interface defect states in Si–SiO_2 for <111> and <100> surfaces, for wet and dry oxides, as a function of hydrogenation conditions. Data adapted from [7, 45, 46].*

substantial reductions in interface defect density can be attained, thanks to hydrogen saturation of bonds, reducing D_{it} by over one order of magnitude. Defect centres of the extrinsic Si^{1+} and Si^{2+} nature are primarily removed, which explains the U shape when only the intrinsic interface centres remain [42]. The states' ability to promote recombination, represented by their capture cross sections, has been reported to a lesser extent. Once the advantages of H interface passivation were discovered, most works used a form of hydrogenation without reporting data for as-grown oxides. A few works in the literature reported Si–SiO_2 capture cross sections before and after hydrogenation. Those which studied the changes showed small or no change in capture cross sections from the hydrogen processing [7, 45, 47, 48].

A second prominent method for hydrogenation is the aluminium anneal process, better known as the *alneal* after the term was coined by Deal in 1970 [49, 50]. In this process, SiO_2 is covered with a ~0.1–1 μm layer of aluminium, typically grown using physical vapour deposition, and then the dielectric is annealed in a forming gas ambient. The mechanism behind hydrogenation during the aluminium anneal was initially reported by Balk [51, 52]: at high temperature aluminium at the interface with the oxide reacts with residual water absorbed into the oxide near the surface. The reaction of aluminium and water forms an AlO_x layer and produces hydrogen atoms that migrate to the SiO_2 and passivate defects at the SiO_2–Si interface. The Al and Al_2O_3 surface films are subsequently removed to yield the final device. The improvement to the electrical properties of the Si–SiO_2 interface due to *alneal* hydrogenation has been long recognised and extensively studied [46, 52]. Figure 3.8a also illustrates typical reductions in D_{it} achieved via *alneal* processing. The benefits compare to those of a FGA, but the extent of dangling bond saturation, and thus passivation is much larger in the presence of an Al layer. Both the hydrogen and the aluminium anneals have also been reported for different silicon orientations and oxide-growth conditions, as illustrated in Figure 3.8b. For dry oxides, however, the final level of passivation obtained was better as seen by a smaller concentration of interface states. This is due to the use of higher oxidation temperature required to achieve the same oxide thickness under dry conditions. Higher temperatures allow reorganisation of bonds, leading to improved passivation.

In terms of surface orientation, <100> surfaces were more effectively passivated than <111> ones. This is due to the latter having a larger density of atoms than the former, and thus being naturally more prone to recombination. Here it is clear that the hydrogenation treatments are largely effective regardless of the surface orientation or oxide condition. In fact, the *alneal* process is one of the most successful schemes to passivate the silicon surface reported to date. In their initial work, Kerr and Cuevas [38] exploited *alnealed* SiO_2 to produce extremely effective surface passivation (~2 cm/s), and used it to characterise the Auger and radiative components of silicon bulk recombination in 2002 [53]. *Alnealed* SiO_2 has been mainly used in laboratory solar cells, most notably the record-breaking passivated emitter rear locally diffused (PERL) cell, by UNSW [54]. Using the *alneal* in combination with enhancements to the films charge, SRVs as low as 0.4 cm/s on 1 Ω·cm *n*-type silicon have been demonstrated [55]. Due to the cumbersome deposition and removal of aluminium, this method has not been adopted in the PV industry. Moreover, the refractive index of silicon oxide films is approximately 1.46, which is not suitable as an antireflection coating. Silicon oxide is normally used in combination with other dielectric layers to optimise both electronic and optical properties.

A third method to introduce hydrogen into thermal SiO_2 is the use of a plasma source to decompose a precursor hydrogen-containing gas, such that hydrogen atoms can be introduced into the SiO_2 films from the plasma. Due to the presence of a plasma directly on top of a sensitive Si–SiO_2 interface, this method requires the hydrogen ions to be generated out of line-of-sight from the silicon specimens. The technique is hence referred to as remote hydrogen plasma. Electron spin resonance and secondary ion mass spectrometry were first used to demonstrate interface defect passivation from a hydrogen plasma source [56]. It was unequivocally established that hydrogen passivation was also possible via external sources. A different implementation of this method has recently been reported and termed the shielded hydrogen plasma process [57, 58]. In this method, the hydrogen-containing plasma is isolated from the SiO_2 specimen using a palladium membrane permeable to hydrogen. Such membrane eliminates the harmful effects of ion bombardment and high plasma radiation, while keeping the benefits of a compact process.

Now that the methods to introduce hydrogen into a Si–SiO_2 interface have been covered, we turn to their impact in the recombination losses of devices. Although the amount of research on the Si–SiO_2 interface is vast, standardised measurements of SRV only became possible after the adoption of the lifetime PCD technique. This happened in the early 2000s, and therefore only SRV reports that use PCD lifetime measurements can be compared directly. Due to its effectiveness, measurements of the surface recombination activity have been primarily reported for silicon specimens that underwent a form of hydrogen passivation. Figure 3.9 illustrates the calculated SRV of Si–SiO_2 interfaces using PCD lifetime measurements. Here we include SRV for specimens with and without FGA, which show the clear benefits of hydrogenation [61, 62]. With a FGA process SRV can be reduced by more than half. In general, hydrogenated Si–SiO_2 interfaces provide extremely low values of recombination, as observed from SRV values well below 100 cm/s, and in the best cases <1 cm/s. When combined with FEP methods, S_{eff} as low as 0.24 cm/s are possible in FGA and oxidised 1 $\Omega \cdot$cm *n*-type silicon wafers [61]. For the shielded hydrogen plasma process in [57, 58] an SRV as low as 0.17 cm/s was reported in a similar thermal oxide specimen, as also included in Figure 3.9.

3.2.2 Silicon oxides grown by PECVD, ALD, or chemical anodisation

The high temperature and long processing time required to grow thermal oxides initially made them unattractive in the solar cell industry. This viewpoint has now changed since automation in large tube furnaces have enabled batch processing with throughputs in the thousands of cells per hour. The desire for faster processing methods, however, promoted substantial work into developing low-temperature synthesis of SiO_2. In this context, PECVD became a widely adopted technique to deposit thin films in PV devices. PECVD enabled low temperature, fast, and very controlled

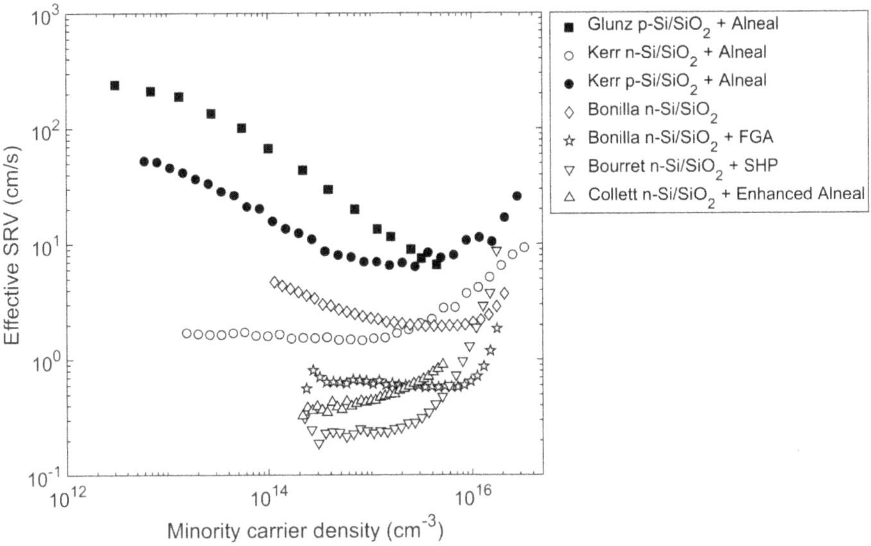

Figure 3.9 *Summary of minimum SRV achieved as a result of reduced interface recombination from hydrogenation in thermal SiO_2 and the effect of FEP. Substrates are ~1 Ω·cm <100>. Data from [36, 38, 55, 59, 60].*

synthesis of surface coatings. Additionally, with PECVD it is possible to produce silicon oxide films that already contain large amounts of hydrogen (SiO_x:H). This was achieved by devising the right chemistry in the precursor gases such that hydrogen ions are present during the film growth, and are therefore incorporated into the film. The most prevalent precursor gas used in PECVD is silane gas, which contains such beneficial hydrogen species. PECVD silicon oxides using a silane precursor have been shown to produce effective surface passivation, with SRV ~ 1.5 cm/s, in 1 Ω·cm *n*-type silicon [63]. The effect of hydrogen passivation was observed most beneficial during a post-deposition activation anneal [63]. This anneal allows the effusion of hydrogen from the dielectric coating and consequently to the SiO_x–Si interface or the Si. In this way both hydrogen passivation of the surface and the bulk is achieved, as has been described in Chapter 2. In addition, the anneal restructures the PECVD SiO_x making it similar to thermally grown SiO_2 [64]. A comparison of the SRV values reported in the literature for SiO_x films is included in Figure 3.10.

No in-depth study of the mechanisms involved in the passivation has been reported for PECVD SiO_x, yet it is known that hydrogenation plays a key role in the effectiveness of this film. Once hydrogen migrates to the interface, it can chemically passivate dangling bonds during the high-temperature treatments [68, 69], including a combination with forming gas [70] or *alneal* [71, 72] processes. Similar results have also been achieved using industrial inductively coupled plasma deposition [73, 74]. The requirement for high throughput has also driven the development of fast PECVD processing reactors. These are broadly grouped as inline or batch PECVD

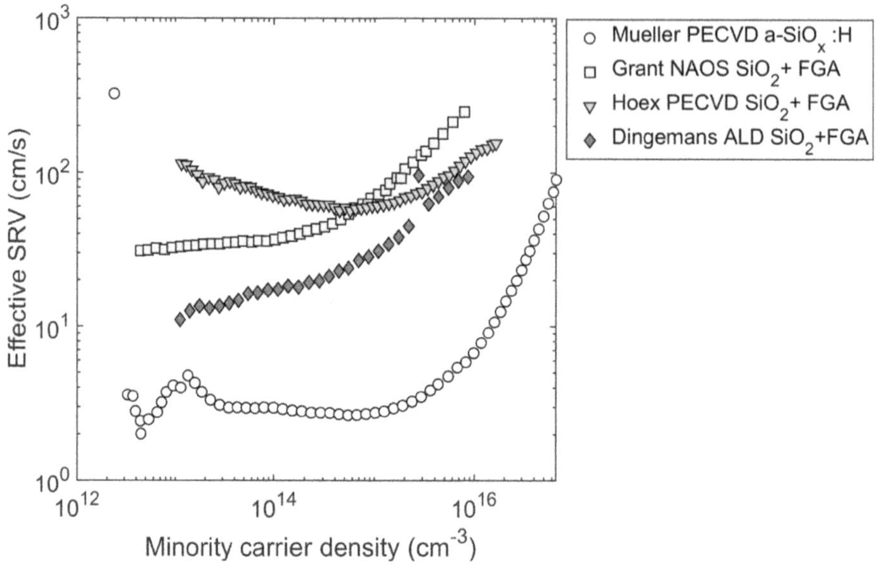

Figure 3.10 *Summary of effective SRV in PECVD and ALD synthesised SiO_2 with a strong hydrogenation effect. Data from [64–67].*

reactors. Inline production has been shown to produce only moderate passivation effectiveness in these SiO_x layers, with SRV ~100 cm/s even after temperature treatments and extrinsic hydrogen passivation via FGA [64, 65, 75, 76]. To improve the final passivation quality, these PECVD SiO_x have been combined with a second capping layer typically using SiN_x or AlO_x, as will be explored in Sections 3.3 and 3.4. In one of such double-layer schemes, a PECVD SiO_x can be deposited on a thermal SiO_2 as an advantageous strategy to exploit the hydrogen content of PECVD. Mack *et al.* [77] demonstrated SRVs <10 cm/s by first growing a thermal SiO_2 layer followed by a PECVD SiO_x film, and doing an activation anneal to transport hydrogen from the bulk of the SiO_x film towards the SiO_2–Si interface. The exact change in interface properties was not reported.

ALD has also been used as a low-temperature method to synthesise SiO_x thin films. It grew and gained substantial attention in the past decade, thanks to its versatility and the extreme control of film thickness it provides. It was introduced originally to produce AlO_x films, as will be described in Section 3.5, but can also be used to grow various other oxides. SiO_x passivation has been demonstrated using the ALD technique, yet it is only proven moderately effective [78]. The interface electrical properties were investigated with C-V and lifetime measurements and found to be poor in the as-deposited state, but largely improved after annealing [78, 79]. Annealing in the presence of a hydrogen source was also reported to promote further deactivation of interface states [78]. The best reported SRV for ALD SiO_2 is also included in Figure 3.10.

A final method covered in this section is the growth of oxide films via low-temperature chemical anodisation [80]. Very thin oxides can be grown on silicon surface using low temperature (<200 °C) chemical methods, including nitric acid [81, 82], ozone [83], and light-induced anodisation [84, 85]. It is accepted that their passivation quality originates from the processing steps that occur after growth [86, 87]. These studies have shown that such low-temperature chemical oxides produce a SiO_x–Si interface with a high defect density ~10^{12}/eV cm^2 [88]. They contain a low concentration of hydrogen, and hence require either high-temperature post-processing, or an extrinsic source of hydrogenation. When these are effectively exploited, they can produce an SRV as low as 30 cm/s, as shown in Figure 3.10. These methods have recently resurrected and studied in the context of passivated and selective contacts as covered in Section 3.6.

3.3 Hydrogen in silicon nitride thin films

The development and wide adoption of PECVD has enabled extremely effective passivation using SiN_x thin films. Much of the early work on the interface properties of the SiN_x–Si was reported by Schmidt, Cuevas, Aberle, and Hezel [28, 33, 89–100]. The understanding and techniques they developed have made SiN_x into the most ubiquitous dielectric film in solar cells [101]. The prominence of SiN_x film originated from:

1. the possibility of tuning its refractive index such that the optical properties could be optimised for antireflection
2. the large quantities of hydrogen released during PECVD deposition, which help subsequent interface and bulk passivation processes
3. the formation of positively charged states near the interface to silicon, which additionally provides FEP.

Studies of the SiN_x–Si interface have shown that the passivation originates from a combination of a slow capture rate at a moderate interface state densities (10^{11}–10^{12}/eV cm^2) [102–104], plus the added benefit of a high concentration of positive dielectric fixed charge (~10^{12} e/cm^2) providing FEP [105, 106]. A key advantage in PECVD is that the concentration of reactant precursors can be controlled so that the film's stoichiometry and hydrogen content is adjustable. Silicon-rich SiN_x films have been observed to provide passivation by the chemical reduction in recombination rates from a large concentration of Si–Si bonds, in a similar way as a-Si provides excellent interface passivation on silicon [106]. For a nitrogen-rich film, on the other hand, chemical bonding at the interface is reduced and FEP is enhanced by the formation of silicon-nitrogen charged centres. Such centres constitute a silicon atom back-bonded to three nitrogen atoms (•SiN_3) [28, 107–111]. They are typically known as the K centre, and present an amphoteric nature – i.e. they can change charge state between neutral, positive, and negative depending on the Fermi level [112]. The progression from silicon- to nitrogen-rich films also enables the useful

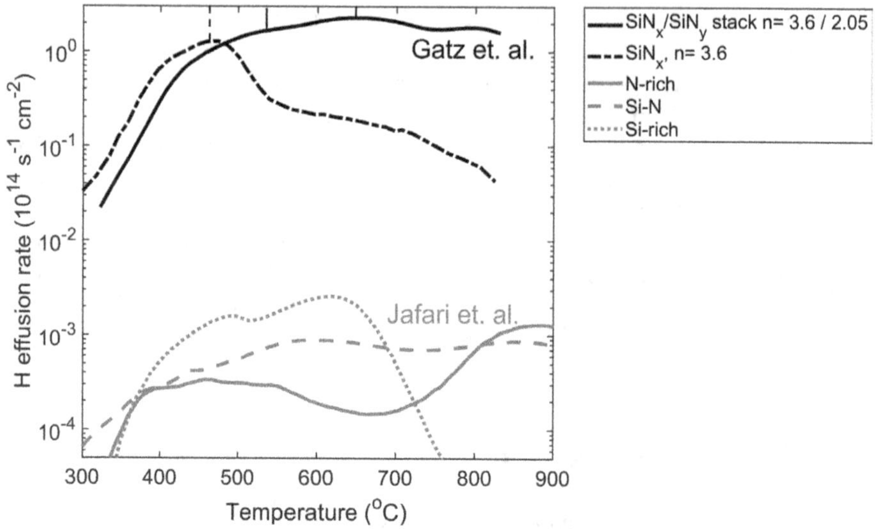

Figure 3.11 *Hydrogen effusion rate as a function of temperature. Dotted lines indicate maximal hydrogen effusion temperatures for each passivating film. Data sourced from [113, 114].*

tailoring in refractive index to achieve SiN_x films with optimal optical properties, both for antireflection and low parasitic absorption.

Besides the above, a chief advantage of PECVD SiN_x in today's PV industry is the ability to supply vast quantities of hydrogen to different regions of the cell. The processes used to redistribute hydrogen from SiN_x surface films into the bulk of the cell are covered in detail in Chapters 2 and 4. In this section, the focus lies in the supply of hydrogen for efficient interface passivation, as this has become a principal part of optimal dielectric layers. Controlling and maximising hydrogen release from dielectric thin films is a crucial subject of development. Several characterisation methods have been employed to understand hydrogen kinetics in dielectrics. Most notably, a combination of mass spectrometry and Fourier Transform Infrared Spectroscopy (FTIR) have allowed to investigate hydrogen migration and bond concentrations inside dielectric coatings. As discussed in Chapter 2, since the mass spectra of hydrogen is notoriously difficult to detect due to background hydrogen contamination, spectrometry is normally combined in effusion experiments or performed in special samples bearing deuterium rather than hydrogen. In effusion experiments, the content and thermal stability of hydrogen in the dielectrics can be recorded, while FTIR is used to identify the vibrations in chemical bonds that relate to hydrogen, including Si–H, Si–H_2, Si–N, and N–H. Using effusion experiments, Gatz *et al.* and Jafari *et al.* showed the substantial differences in hydrogen release obtained in different SiN_x stoichiometries [113, 114]. Such effusion experiments are shown in Figure 3.11. Here it is clear that changes in the precursor ratio can allow different bond configurations, which, in turn, can provide dramatically different

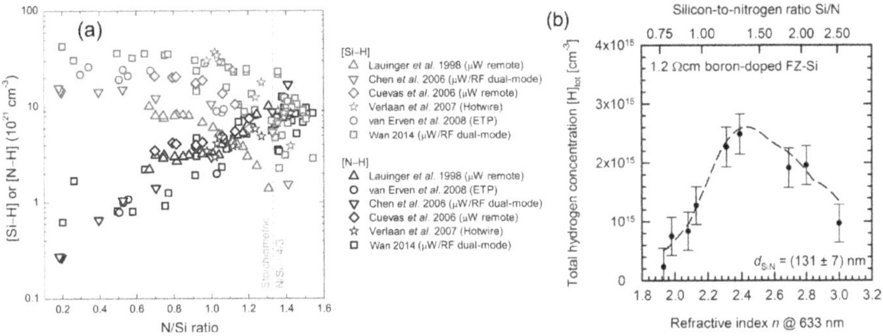

Figure 3.12 *(a) Hydrogen content inside of SiN_x films obtained from FTIR measurements. Data reproduced from [117] with permission from Y. Wan. Copyright 2014 Y. Wan. (b) Total concentration of hydrogen of a SiN_x passivation layer as a function of refractive index at a wavelength of 633 nm. Reproduced with permission from [118]. © 2019 WILEY-VCH Verlag GmbH & Co. KGaA, Weinheim.*

hydrogen release dynamics. The more silicon-rich films the lower the peak temperature for hydrogen release, and therefore the less firing-stable the film becomes. The extreme case for such films is observed in typical silicon heterojunction (SHJ) interfaces to a-Si, which can generally only tolerate processes below 250 °C to avoid excessive hydrogen release. These differences also point to the changes produced in the type and nature of hydrogen bonded to silicon and nitrogen in the dielectrics. Different bonds release hydrogen at different temperatures and can therefore be used in specific applications or cell architectures. For example, for the optimal hydrogenation of passivated emitter and rear contact (PERC) cells, maximum effusion must coincide with firing of metal contacts, and the SiN_x films must be nitrogen-rich and, thus, have a low refractive index. Similar findings have also been reported in subsequent works, e.g. [115, 116].

A large part of the hydrogen contained in SiN_x dielectrics is present in Si–H and N–H bonds. This has been characterised using FTIR of several different films as shown in Figure 3.12a. The hydrogen concentration is clearly seen to relate to the stoichiometry of the film determined by the precursor concentration that produces different N/Si ratios (*x*-axis in the figure). This data shows that silicon-rich layers exhibit more Si-H than N–H bonds as would be expected, and that the Si–H bond concentration decreases moderately as the film becomes more nitrogen-rich. On the other hand, the N–H bonds, which are initially low in the silicon-rich film, increase steeply as the film becomes more nitrogen-rich. This trend in hydrogen bonding relates directly to the chemical composition of the film, with the additional indication that increases in films' nitrogen content lead to more hydrogen concentration in the film, as seen in the rapid rise in N–H bond concentration. Once the layer becomes heavily nitrogen-rich, approaching or exceeding stoichiometry, the loss in Si–H bonds would lead to a decrease in the total concentration of H. This trend was later confirmed by Bredemeier *et al.* in [118], as illustrated in Figure 3.12 (b). Here there exists an optimal point for maximum concentration of hydrogen within the film, approximately at near or substoichimetric nitrogen concentrations with refractive index between 2.2 and 2.4. They demonstrated that such concentrations of hydrogen related direcly to defect generation in the bulk of silicon specimens exposed to elevated temperature treatements [118]. When the concentration of hydrogen is optimised and used to promote interface passivation, excellent passivation quality can be achieved.

The optimal combination of chemical hydrogenation and FEP using PECVD SiN_x were initially used to achieve SRV < 7 cm/s and 14 cm/s for 1.5 $\Omega \cdot$cm *n*- and 1 $\Omega \cdot$cm *p*-type silicon, respectively [100]. Several improvements in the passivation quality that SiN_x films have since been achieved, for example in the work by Chen *et al.* [119–121] (SRV < 1.38 cm/s in 1 $\Omega \cdot$cm *n*-type silicon), and Richter *et al.* [16] (SRV < 0.64 cm/s in 1. $\Omega \cdot$cm *n*-type silicon). High-throughput industrial tools have also been developed for the inline deposition of SiN_x for solar cells. Studies into the quality of these films indicate that the passivation has a small contribution from the chemical interface, with a large role being played by the field-effect control of carriers from the fixed charge in the film, which can be in the order of 1–5 × 10^{12} q/cm^2 [75, 105, 122]. SRV values as low as 1.82 and 2.46 cm/s in 1 $\Omega \cdot$cm *n*- and *p*-type silicon specimens have been achieved, respectively, using nitride films deposited in industrial scale inline PECVD reactors. A summary of SRVs for the most notable works on SiN_x passivation are shown in Figure 3.13. Generally, higher passivation quality has been obtained in silicon-rich high refractive index films. This has provided the challenge of trading off the passivation quality with the parasitic absorption, since silicon-rich films tend to have higher blue-light extinction coefficients [123]. A major development was achieved by Wan *et al.* [123, 125], who reported SRV < 0.67 cm/s in 0.85 $\Omega \cdot$cm *p*-type silicon, and <0.01 cm/s in 0.47 $\Omega \cdot$cm *n*-type silicon, while also achieving low-absorption PECVD SiN_x films. The star symbols in Figure 3.13 illustrate the improved films produced by an optimised dual-mode (μW/RF) PECVD reactor [102].

The effects of hydrogen has been most noticeable when SiN_x is used in a multilayer stack. For example, as a capping layer on top of a thermal or PECVD SiO_2 film. While SiO_2 provides lower interface state densities than SiN_x–Si interfaces, the H-rich nitride layers enabled a method to improve the interface electrical quality [30]. When this is added to the FEP provided by fixed charge in the SiN_x film, the double-layer oxide-nitride structures have been shown to provide some of the most outstanding passivation quality. In *n*-type silicon the field-effect reduction of holes at the surface produces better passivation than for *p*-type silicon since typically $\sigma_n > \sigma_p$ at the Si–SiO_2 interface, regardless of doping type. In *n*-type silicon, SRV between 0.2 and 0.4 cm/s have been achieved in 1–2 $\Omega \cdot$cm material [62, 126, 127]. Similar improvements have been shown possible in PECVD SiO_x–SiN_x double layers [128–131]. PECVD provided the added flexibility to tune films at the oxide–nitride boundary, obtaining variable refractive indexes from ~1.4 to 2.3, or even grading layer stacks in what are known as silicon oxynitride films. The interfacial hydrogenation is so abundant that PECVD SiN_x has even been applied to

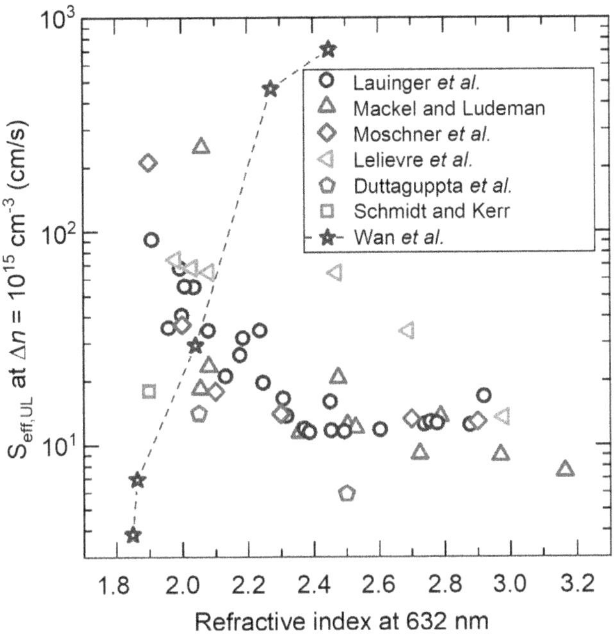

Figure 3.13 *Effective SRV as a function of refractive index at 632 nm for SiN_x PECVD passivation of* p-type *FZ silicon. Data from [90, 96, 103, 105, 106, 123, 124]. Reproduced from [102] under Creative Commons CC BY. © Author(s) 2013. http://aip.scitation.org/doi/10.1063/1.4795108.*

improve native [132], chemically grown [67], and rapid thermal oxides [133–136]. Figure 3.14 illustrates the calculated SRV for top-performing SiO_2–SiN_x double-layer stacks. The origin of such efficient passivation has been reported to be a combination of the field-effect and chemical hydrogenation of the Si–SiO_2 interface. The passivation mechanisms have been studied using second-harmonic generation spectroscopy, where it was found that the key passivation component is the reduced recombination rates due to hydrogenation of the interface states [131]. Laades *et al.* [138] and Shwab *et al.* [139] showed that hydrogen passivation played a key role in the passivation properties upon high-temperature firing. Cheng *et al.* [140] also found that the chemical passivation component played a more important role after firing the SiON/SiN_x stack, and similar stability and passivation effectiveness was shown. Due to such effective performance, a double layer of thermally grown or CVD deposited SiO_2, and PECVD SiN_x has become today's standard passivation at the front of commercial PERC silicon cells [141, 142].

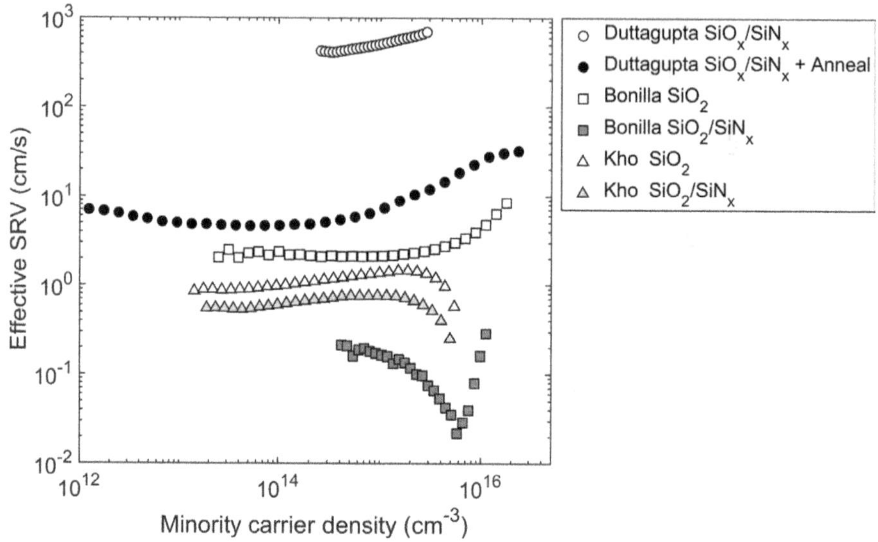

Figure 3.14 *Measurements of SRV as a function of injection density calculated from PCD lifetime for top performing SiO_2–SiN_x double-layer films on Si. Data from [61, 62, 75, 137].*

3.4 Hydrogen in amorphous silicon thin films

a-Si is typically grown using PECVD using silane (SiH_4) as the silicon precursor [143]. As the deposition process is typically done at relatively low substrate temperatures in the range of 200 °C, this results in a significant amount of hydrogen in the a-Si film and, consequently, the material is typically referred to as hydrogenated amorphous silicon (a-Si:H). Hydrogen was found to be critical for the bulk electronic properties of a-Si:H, as hydrogen lean films had a very-high bulk-defect density which limited the performance of thin-film solar cells using a-Si:H as the absorber material. a-Si:H films can also provide a very high level of surface passivation. The first report of using an a-Si:H film as a surface passivation film was by the company Sanyo, which used an intrinsic a-Si:H film to improve the V_{OC} of a poly-silicon solar cell [144].

Interestingly, they found a significant boost in the performance when an intrinsic a-Si:H film was sandwiched between the poly-silicon film and a highly doped a-Si:H film. This is due to the fact that doped a-Si:H films have a very-high bulk-defect density. These bulk defects also affect the level of surface passivation that is provided by the films as the first few nanometers of the passivation film can still act as a recombination site for charge carriers via tunnelling [145]. This also means that a-Si:H bulk degradation mechanisms such as the Staebler–Wronski effect may affect the level of surface passivation provided by a-Si:H [146]. The surface passivation mechanism of a-Si:H has been shown to predominantly be due to chemical passivation afforded by the very-low interface-defect density at the c-Si/a-Si:H

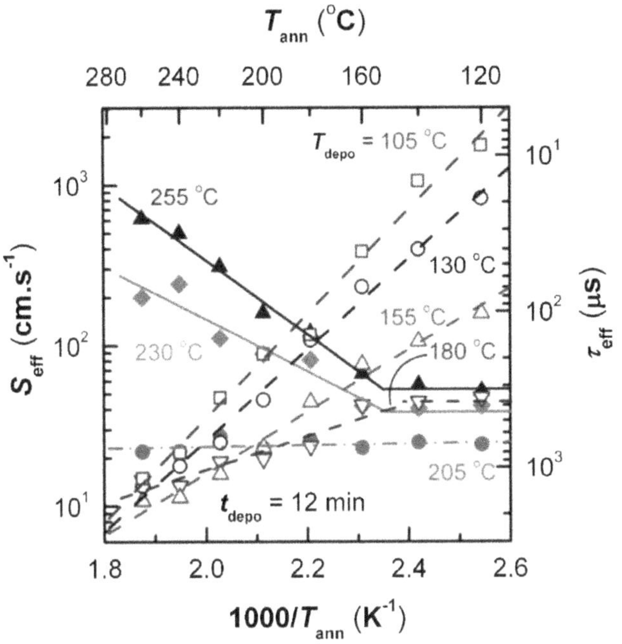

Figure 3.15 *Influence of the post-deposition annealing temperature* T_{ann} *on the effective SRV,* S_{eff} *and effective lifetime,* τ_{eff}, *of a symmetrically passivated c-Si wafer for a-Si:H films grown at various deposition temperatures* T_{depo}*. The annealing time was 30 minutes for all films while the deposition time* t_{depo} *was 12 minutes. Reprinted from [148], with the permission of AIP Publishing.*

resulting from the fact that a-Si:H is nearly identical to c-Si at the atomic scale and the high concentration of hydrogen ensures that any remaining dangling bonds are passivated. This was confirmed by real-time in-situ spectroscopic ellipsometry and second-harmonic generation measurements of a-Si:H growth on silicon complemented with exsitu minority carrier lifetime measurements, where no significant band-bending was observed in silicon. This study also revealed that epitaxial growth should be avoided at the c-Si interface as this strongly reduces the level of surface passivation [147]. Epitaxial growth has also been related to a reduction in surface passivation by post-deposition annealing [148]. It was shown that annealing can significantly improve the level of surface passivation when the interface between c-Si and a-Si:H is abrupt, while is detrimental when there is some level of epitaxial growth. This is shown in Figure 3.15, where a-Si:H films grown at temperatures below 200 °C had an abrupt interface and benefitted from annealing at temperatures up to 280 °C, while films grown at a temperature of 230 °C and 255 °C showed a strong deterioration of the level of surface passivation.

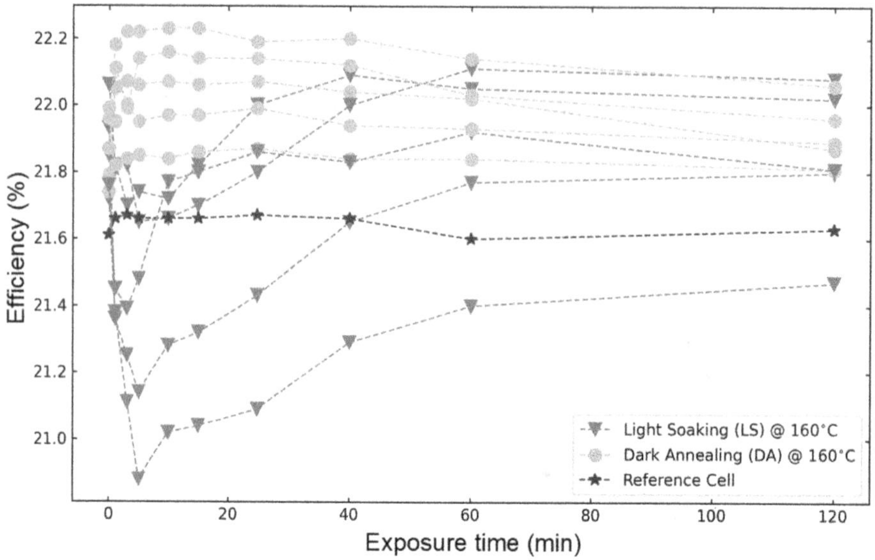

Figure 3.16 *Efficiency of industrial SHJ solar cells as a function of accelerated testing time. A clear difference in stability can be seen for solar cells that were annealed in the dark or under 1 sun illumination at 160 °C. Reprinted from [151] Copyright (2020), with permission from Elsevier.*

The main use of a-Si:H films for silicon solar cells is in SHJ solar cells. They are, for example, used in the world-record solar cell from Panasonic with an efficiency of 26.7%, which features an all-back contact design where the front surface is passivated by an intrinsic a-Si:H film, and the electron and hole collector consist of a stack of intrinsic and doped a-Si:H [149]. In the optimisation of solar cells using a-Si:H, it was also identified that the injection-level dependent lifetime is very important when optimising the solar cell performance, in particular for the fill factor of the solar cell [150]. As mentioned earlier, the level of surface passivation provided by a-Si:H correlates strongly with its bulk electronic properties, and for this reason, research groups are increasingly looking at the long-term stability of heterojunction solar cells. It was recently shown that light soaking at high temperatures can significantly reduce the efficiency of SHJ solar cells [151], with efficiency losses of more than 1% absolute as shown in Figure 3.16 and this is very likely related to changes in the level of surface passivation resulting from hydrogen migration.

Bonilla *et al.* [152] used SiO_x and SiO_x/SiN_x capping layers to determine the fundamental recombination properties for holes and electrons at the c-Si–a-Si interface. They found that the deposition of capping layers can result in the loss of some hydrogen at the c-Si–a-Si interface; however, the deposition of SiN_x:H was shown to be effective in at least partly recovering this loss. The benefit of the capping layers was that it allowed for the deposition of extrinsic corona charges onto the

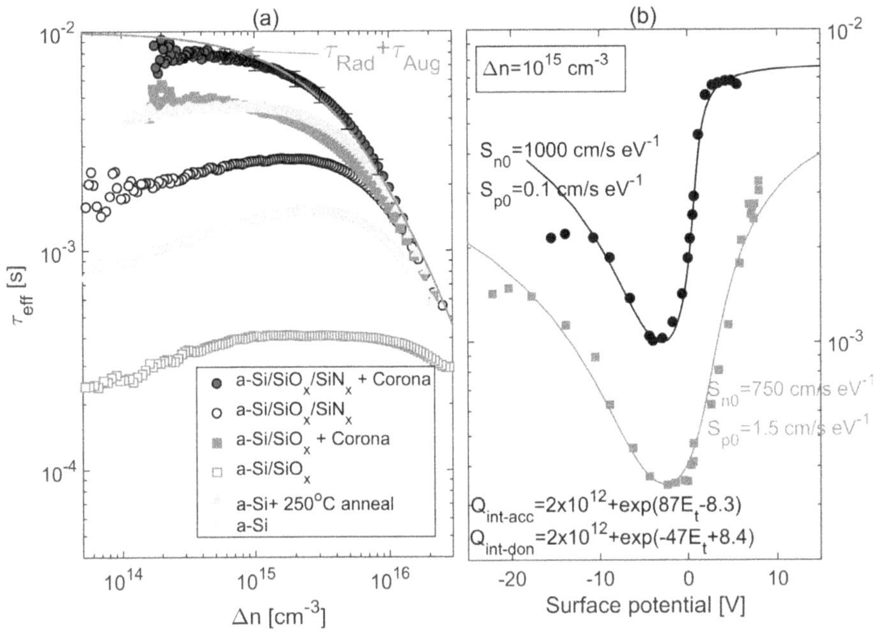

Figure 3.17 *Minority carrier lifetime as (a) function of injection level and (b) surface potential for a symmetrically passivated* n-*type c-Si wafer. The surface potential was varied by the application of corona charging. Reprinted from [152] under Creative Commons CC BY. © 2016 The Authors. Phys. Status Solidi RRL published by WILEY-VCH Verlag GmbH & Co. KGaA, Weinheim. http://doi.wiley.com/10.1002/pssr.201600307.*

surface passivation stack and thus control of the surface potential. As can be seen in Figure 3.17b, in the study by Bonilla *et al.* [152] the passivation provided by the stack could be significantly increased by the addition of corona charges to levels above the empirical intrinsic minority carrier lifetime limit published by Richter *et al.* [16]. More importantly, sweeping the surface potential also allows the extraction of S_{n0} and S_{p0} and it was found that S_{p0} was orders of magnitude lower than S_{n0}, consistent with the excellent results obtained on *n*-type c-Si surfaces. It was also found that the hydrogenation provided by the SiN_x:H deposition reduced the S_{p0} by one order of magnitude while not affecting the S_{n0} in this case. Typically, double layers of hydrogenated PECVD a-Si and SiO_2 or SiN_x have been reported in the literature with SRV as low as 0.1–1 cm/s on 1–3 Ω·cm *n*-type silicon [152–154].

3.5 Hydrogen in aluminium oxide thin films

AlO_x was first investigated for application in solar cells in 1989 by Hezel and Jaeger [98]. The result did not receive a lot of attention as the focus at that time was on

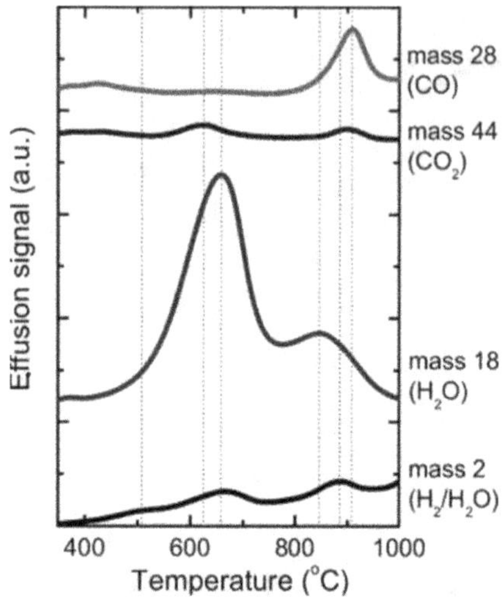

Figure 3.18 *Hydrogen effusion measurements from an AlO_x film grown using ALD. It can be seen that hydrogen is released from the film for temperatures in excess of 400 °C mainly in the form of H_2O. Reprinted from [159], with the permission of AIP Publishing. © 2009 American Institute of Physics.*

the passivation on *n*-type silicon surfaces where SiN_x:H proved to be much more effective. AlO_x typically has a moderate to very high fixed-negative-charge density, which is detrimental for the passivation of highly doped *n*-type surfaces such as phosphorus -diffused emitters [155]. The interest in AlO_x re-emerged when the PV field was looking for an effective surface passivation film for the rear of *p*-type PERC solar cells in the early 2000s. In 2006, very high levels of surface passivation were reported for AlO_x films grown using ALD, where trimethyl aluminium ($Al(CH_3)_3$) was used as the Al precursor [156]. Later, it was shown that excellent results could also be obtained using AlO_x films grown using PECVD, where again $Al(CH_3)_3$ was used as the Al-precursor [157]. Both ALD and PECVD AlO_x films contain residual hydrogen and it was identified that this played an important role in the chemical passivation provided by AlO_x films and the migration of hydrogen also partly explains the improvement in surface passivation observed after post-deposition annealing [35, 158]. The hydrogen concentration is, however, typically significantly lower than for SiN_x:H films. It was also demonstrated that a high-temperature firing-step can release hydrogen from an ALD AlO_x film, as it is exmplefied in Figure 3.18 [159].

Recently, it was also shown that ALD AlO_x can work as hydrogen diffusion barriers during the high-temperature firing-step when they are used in a stack with SiN_x:H [160]. ALD AlO_x can reduce the amount of hydrogen released to the c-Si bulk when placed in between c-Si and SiN_x, and significantly increase the amount of hydrogen released into the bulk when the ALD film is placed on top of the SiN_x

film, which blocks hydrogen effusion to the ambient . Another interesting aspect of ALD growth of AlO_x using $Al(CH_3)_3$ as the Al-precursor and H_2O as the oxygen precursor is that it can act as a source of atomic hydrogen resulting from the chemical reactions. It was shown that the deposition of an ultrathin ALD AlO_x capping layer can significantly reduce the defect density at the c-Si–SiO_2 interface similar to the *alneal* process [65].

3.6 Hydrogen in tunnelling oxide passivating contact layers

In 1983 Green and Blakers already reported the exceptional high open-circuit voltages that could be obtained using a tunnelling silicon oxide in combination with doped poly-silicon [161]. The recent interest in these contacts commenced when Fraunhofer ISE reported their first results for the so-called tunnel oxide passivated contact (TOPCon) solar cell in 2013 [162]. The TOPCon design is one of the many passivating contacts that are currently being investigated, which also includes heterojunction contacts and contacts using materials with extreme work functions [163]. All these contacts simultaneously provide a very high resistance for minority carriers (i.e., a low minority carrier recombination rate), while having a low resistance for the majority carriers. Hydrogen plays an important role in these contacts, in

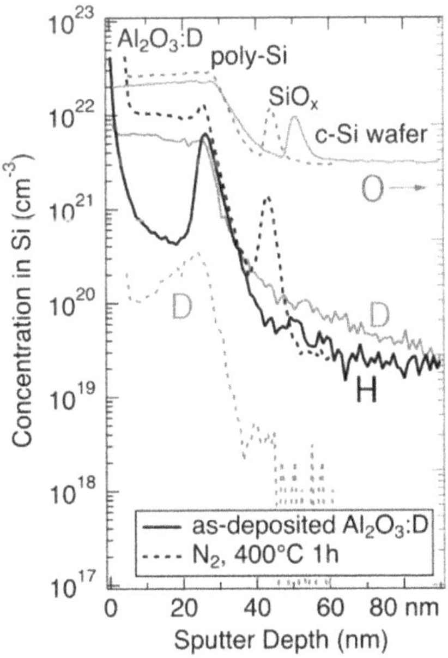

Figure 3.19 Hydrogen, deuterium, and oxyen profiles of a p-*type poly-Si/SiOx contact with a deuterated Al_2O_3 layer after a nitrogen anneal. Reprinted from [166], with the permission of AIP Publishing. © 2018 Author(s).*

particular in decreasing the minority carrier recombination rate. These structures use quantum tunnelling effects [164] and pinholes [165] to allow for charge transport in addition to interface passivation and are implemented in passivating and selective carrier contacts [162]. In the TOPCon structure, the minority carrier recombination rate is mainly determined by the interface defect density at the c-Si–SiO_2 interface. Hydrogen can passivate defect states at this interface and as such reduce the minority carrier recombination rate. It is shown that the deposition and subsequent annealing of ALD AlO_x is an effective way to hydrogenate the c-Si–SiO_2 interface in TOPCon solar cells, being significantly more effective than a FGA [166]. The presence of hydrogen was confirmed by recording the concentration of deuterium mass-charge ratio as a function of etch depth in a secondary ion mass spectroscopy experiment. The results of such experiment are illustrated in Figure 3.19, where it is clear that hydrogen provides efficient interface passivation, and it is hence crucial in the processing of novel passivating contact structures. SiN_x:H in combination with a short firing step is also shown to be an effective hydrogenation step, releasing even more hydrogen into the stack, which may even passivate defects in the bulk of the poly-silicon film [167]. It was also shown that carrier-induced redistribution of hydrogen is an effective way to reduce the interface defect density in TOPCon structures [168].

3.7 Outlook for dielectric hydrogenation processes

From this chapter, it is evident that hydrogen plays a key role in determining the performance of surface passivation films and passivating contacts. In particular, hydrogen plays a key role in reducing the density of interface defect states and is either introduced during the thin-film deposition process or in a subsequent dedicated hydrogenation process. With the projected increase in silicon solar cell efficiency, it is expected that hydrogenation will play an increasingly important role for surface passivation. This will also result in a higher sensitivity of the solar cell performance on the long-term stability of the electronic quality at the c-Si surface. Consequently, there will be an increased focus on hydrogen transport in passivation/contact films during solar cell fabrication as well as during long-term operation. The first empirical results about the instability of tunnel oxide passivating contacts have already been published, as well as instability of heterojunction solar cells under certain operating conditions. It is very likely that these instabilities are related to the migration and reactivity of hydrogen, and it will be critical to fully understand these mechanisms to guarantee long-term stable operation of next-generation solar cells featuring these kinds contacts. In Chapters 4 and 5, the use of hydrogenated passivation layers for bulk defect passivation is explored.

References

[1] Sze S.M., Li Y., K.K N. *Physics of Semiconductor Devices*. 3rd ed. Hoboken, New Jersey: John Wiley and Sons, Inc; 2007.

[2] Shockley W. 'On the surface states associated with a periodic potential'. *Physical Review*. 1939;**56**(4):317–23.

[3] Sah C.-tang., Noyce R., Shockley W. 'Carrier generation and recombination in p-n junctions and p-n junction characteristics'. *Proceedings of the IRE*. 1957;**45**(9):1228–43.

[4] Shockley W., Read W.T. 'Statistics of the recombinations of holes and electrons'. *Physical Review*. 1952;**87**(5):835–42.

[5] Hall R.N. 'Electron-hole recombination in germanium'. *Physical Review*. 1952;**87**(2):387

[6] Girisch R.B.M., Mertens R.P., De Keersmaecker R.F., De Keersmaecker R.F. 'Determination of Si-SiO$_2$ interface recombination parameters using a gate-controlled point-junction diode under illumination'. *IEEE Transactions on Electron Devices*. 1988;**35**(2):203–22.

[7] Aberle A.G., Glunz S., Warta W. 'Impact of illumination level and oxide parameters on Shockley–Read–Hall recombination at the Si-SiO$_2$ interface'. *Journal of Applied Physics*. 1992;**71**(9):4422–31.

[8] Dauwe S., Mittelstädt L., Metz A., Hezel R. 'Experimental evidence of parasitic shunting in silicon nitride rear surface passivated solar cells'. *Progress in Photovoltaics: Research and Applications*. 2002;**10**(4):271–8.

[9] Grove A.S., Fitzgerald D.J. 'Surface effects on p-n junctions: characteristics of surface space-charge regions under non-equilibrium conditions'. *Solid-State Electronics*. 1966;**9**(8):783–806.

[10] Fitzgerald D.J., Grove A.S. 'Surface recombination in semiconductors'. *Surface Science*. 1968;**9**(2):347–69.

[11] Basore P.A., Hansen B.R. *Microwave-detected Photoconductance Decay*. Proceedings of the 21st IEEE Photovoltaic Specialists Conference; 1990. pp. 374–9.

[12] Kane D.E., Swanson R.M. 'Measurement of the emitter saturation current by a contactless photoconductivity decay method'. Proceedings of the 18th IEEE Photovoltaic Specialists Conference; 1985. pp. 578–83.

[13] Yablonovitch E., Gmitter T.J. 'A contactless minority lifetime probe of heterostructures, surfaces, interfaces and bulk wafers'. *Solid-State Electronics*. 1992;**35**(3):261–7.

[14] Sinton R.A., Cuevas A. 'Contactless determination of current–voltage characteristics and minority-carrier lifetimes in semiconductors from quasi-steady-state photoconductance data'. *Applied Physics Letters*. 1996;**69**(17):2510–2.

[15] Sinton R.A., Cuevas A., Stuckings M. 'Quasi-steady-state photoconductance, a new method for solar cell material and device characterization'. Photovoltaic Specialists Conference, 1996, Conference Record of the Twenty Fifth IEEE; 1996. pp. 457–60.

[16] Richter A., Glunz S.W., Werner F., Schmidt J., Cuevas A. 'Improved quantitative description of Auger recombination in crystalline silicon'. *Physical Review B*. 2012;**86**(16):165202.

[17] Altermatt P.P., Geelhaar F., Trupke T., Dai X., Neisser A., Daub E. 'Injection dependence of spontaneous radiative recombination in c-Si: experiment,

theoretical analysis, and simulation'. NUSOD '05 Proceedings of the 5th International Conference on Numerical Simulation of Optoelectronic Devices, 2005. IEEE; 2016. pp. 47–8.

[18] Trupke T., Green M.A., Würfel P., *et al.* 'Temperature dependence of the radiative recombination coefficient of intrinsic crystalline silicon'. *Journal of Applied Physics*. 2003;**94**(8):4930.

[19] Luke K.L., Cheng Li-Jen. 'Analysis of the interaction of a laser pulse with a silicon wafer: determination of bulk lifetime and surface recombination velocity'. *Journal of Applied Physics*. 1987;**61**(6):2282–93.

[20] Sproul A.B. 'Dimensionless solution of the equation describing the effect of surface recombination on carrier decay in semiconductors'. *Journal of Applied Physics*. 1994;**76**(5):2851–4.

[21] Aberle A.G. *Crystalline Silicon Solar Cells: Advanced Surface Passivation and Analysis*. Sydney, Australia: UNSW; 2009.

[22] McIntosh K.R., Black L.E. 'On effective surface recombination parameters'. *Journal of Applied Physics*. 2014;**116**(1):014503.

[23] Kimmerle A., Greulich J., Wolf A. 'Carrier-diffusion corrected J0-analysis of charge carrier lifetime measurements for increased consistency'. *Solar Energy Materials and Solar Cells*. 2015;**142**:116–22.

[24] Mäckel H., Varner K. 'On the determination of the emitter saturation current density from lifetime measurements of silicon devices'. *Progress in Photovoltaics: Research and Applications*. 2013;**21**:850–66.

[25] Schenk A. 'Finite-temperature full random-phase approximation model of band gap narrowing for silicon device simulation'. *Journal of Applied Physics*. 1998;**84**(7):3684–95.

[26] Mandelkorn J., Lamneck J.H., Scudder L.R. 'Design, fabrication and characteristics of new types of back surface field cells'. 10th Photovoltaic Specialists Conference; Palo Alto, CA; 1973.

[27] Granek F., Hermle M., Huljić D.M., Schultz-Wittmann O., Glunz S.W. 'Enhanced lateral current transport via the front N^+ diffused layer of n-type high-efficiency back-junction back-contact silicon solar cells'. *Progress in Photovoltaics: Research and Applications*. 2009;**17**(1):47–56.

[28] Aberle A.G. 'Overview on sin surface passivation of crystalline silicon solar cells'. *Solar Energy Materials and Solar Cells*. 2001;**65**(1–4):239–48.

[29] Ma F-J D.S., Shetty K.D., Meng L., Samudra G.S., Hoex B. 'Two-dimensional numerical simulation of boron diffusion for pyramidally textured silicon'. *Journal of Applied Physics*. 2014;**116**:184103.

[30] Bonilla R.S., Hoex B., Hamer P., Wilshaw P.R. 'Dielectric surface passivation for silicon solar cells: a review'. *Physica Status Solidi*. 2017;**214**(7):1700293.

[31] Aberle A.G. 'Surface passivation of crystalline silicon solar cells: a review'. *Progress in Photovoltaics: Research and Applications*. 2000;**8**(5):473–87.

[32] Cuevas A., Wan Y., Yan D., *et al.* 'Carrier population control and surface passivation in solar cells'. *Solar Energy Materials and Solar Cells*. 2018;**184**:38–47.

[33] Schmidt J., Aberle A.G. 'Carrier recombination at silicon–silicon nitride interfaces fabricated by plasma-enhanced chemical vapor deposition'. *Journal of Applied Physics*. 1999;**85**(7):3626–33.

[34] Hossain M.A., Khoo K.T., Cui X., *et al.* 'Atomic layer deposition enabling higher efficiency solar cells: a review'. *Nano Materials Science*. 2020;**2**(3):204–26.

[35] Dingemans G., Kessels W.M.M. 'Status and prospects of Al_2O_3-based surface passivation schemes for silicon solar cells'. *Journal of Vacuum Science & Technology A: Vacuum, Surfaces, and Films*. 2012;**30**(4):040802.

[36] Glunz S.W., Biro D., Rein S., Warta W. 'Field-effect passivation of the SiO_2Si interface'. *Journal of Applied Physics*. 1999;**86**(1):683–91.

[37] Glunz S.W., Feldmann F. 'SiO_2 surface passivation layers – a key technology for silicon solar cells'. *Solar Energy Materials and Solar Cells*. 2018;**185**(8):260–9.

[38] Kerr M.J., Cuevas A. 'Very low bulk and surface recombination in oxidized silicon wafers'. *Semiconductor Science and Technology*. 2002;**17**(1):35–8.

[39] Sopori B., Deng X., Benner J., *et al.* 'Hydrogen in silicon: a discussion of diffusion and passivation mechanisms'. *Solar Energy Materials and Solar Cells*. 1996;**41-42**:159–69.

[40] Sana P., Rohatgi A., Kalejs J.P., Bell R.O. 'Gettering and hydrogen passivation of edge-defined film-fed grown multicrystalline silicon solar cells by AL diffusion and forming gas anneal'. *Applied Physics Letters*. 1994;**64**(1):97–9.

[41] Himpsel F.J., McFeely F.R., Taleb-Ibrahimi A., Yarmoff J.A., Hollinger G. 'Microscopic structure of the SiO_2/Si interface'. *Physical Review. B*. 1988;**38**(9):6084–96.

[42] Füssel W., Schmidt M., Angermann H., Mende G., Flietner H. 'Defects at the Si/SiO_2 interface: their nature and behaviour in technological processes and stress'. *Nuclear Instruments and Methods in Physics Research Section A: Accelerators, Spectrometers, Detectors and Associated Equipment*. 1996;**377**(2–3):177–83.

[43] Flietner H., Füssel W., Sinh N.D., Angermann H. 'Density of states and relaxation spectra of etched, H-terminated and naturally oxidized Si-surfaces and the accompanied defects'. *Applied Surface Science*. 1996;**104-105**:342–8.

[44] Füssel W., Schmidt M., Flietner H. 'Radiation induced degradation of Si/SiO_2 structures and the nature of defects'. *Nuclear Instruments and Methods in Physics Research Section B: Beam Interactions with Materials and Atoms*. 1992;**65**(1–4):238–42.

[45] Albohn J., Füssel W., Sinh N.D., Kliefoth K., Fuhs W. 'Capture cross sections of defect states at the Si/SiO_2 interface'. *Journal of Applied Physics*. 2000;**88**(2):842–9.

[46] Reed M.L., Plummer J.D. 'Chemistry of Si-SiO_2 interface trap annealing'. *Journal of Applied Physics*. 1988;**63**:5776.

[47] Flietner H. 'Passivity and electronic properties of the silicon/silicondioxide interface'. *Materials Science Forum*. 1995;**185-188**:73–82.

[48] Ernst M. *Deep Level Transient Spectroscopy (DLTS) Untersuchungen an Siliciumoxid/Silicium-G renzfiächen*. Hanover: Universität Hannover; 1999.

[49] Deal B.E. 'Measurement and control of dielectric film properties during semicondcutor device processing'. *Symposium on Silicon Device Processing*. 1970;**13**:36–50.

[50] Castro P.L., Deal B.E. 'Low-temperature reduction of fast surface states associated with thermally oxidized silicon'. *Journal of the Electrochemical Society*. 1971;**118**(2):280 [cited 2017 Feb 8].

[51] Balk P. *Electrochemistry Society Meeting, Buffalo*. Abstract No 111; 1965.

[52] Balk P. 'Effects of hydrogen annealing on silicon surfaces'. *Electrochemistry Society Meeting, San Francisco*. Abstract 109; 1965.

[53] Kerr M.J., Cuevas A. 'General parameterization of auger recombination in crystalline silicon'. *Journal of Applied Physics*. 2002;**91**(4):2473–80.

[54] Zhao J., Wang A., Green M.A. '24·5% efficiency silicon PERT cells on MCZ substrates and 24·7% efficiency PERL cells on FZ substrates'. *Progress in Photovoltaics: Research and Applications*. 1999;7(6):471–4.

[55] Collett K.A., Bonilla R.S., Hamer P., Bourret-Sicotte G., Lobo R., Kho T. 'An enhanced alneal process to produce SRV'. *Solar Energy Materials and Solar Cells*. 2017;**173**:50–8.

[56] Johnson N.M., Hahn S.K. 'Hydrogen passivation of the oxygen-related thermal-donor defect in silicon'. *Applied Physics Letters*. 1986;**48**(11):709–11.

[57] Hamer P., Bourret-Sicotte G., Martins G., Wenham A., Bonilla R.S., Wilshaw P. 'A novel source of atomic hydrogen for passivation of silicon solar cells'. *Physica Status Solidi RRL*. 2017;**11**(5):1600448.

[58] Bourret-Sicotte G., Hamer P., Bonilla R.S., *et al.* 'Shielded hydrogen passivation – a potential in-line passivation process'. *Physica Status Solidi*. 2017;**214**(7):1700383.

[59] Bonilla R.S., Hamer P.G., Wilshaw P.R. 'Lowest surface recombination in N-type oxidised crystalline silicon by means of extrinsic field effect passivation'. 32nd European Photovoltaic Solar Energy Conference and Exhibition; Munich, Germany; 2016. pp. 707–10.

[60] Bourret-Sicotte G., Hamer P., Bonilla R.S., Collett K., Wilshaw P.R. 'Shielded hydrogen passivation – a novel method for introducing hydrogen into silicon'. *Energy Procedia*. 2017;**124**(1-4):267–74.

[61] Bonilla R.S., Reichel C., Hermle M., Hamer P., Wilshaw P.R. 'Long term stability of c-Si surface passivation using corona charged SiO_2'. *Applied Surface Science*. 2017;**412**:657–67.

[62] Bonilla R.S., Woodcock F., Wilshaw P.R. 'Very low surface recombination velocity in N-type c-Si using extrinsic field effect passivation'. *Journal of Applied Physics*. 2014;**116**(5):054102.

[63] Mueller T., Schwertheim S., Fahrner W.R. 'Crystalline silicon surface passivation by high-frequency plasma-enhanced chemical-vapor-deposited nanocomposite silicon suboxides for solar cell applications'. *Journal of Applied Physics*. 2010;**107**(1):014504.

[64] Hoex B., Peeters F.J.J., Creatore M., Blauw M.A., Kessels W.M.M., van de Sanden M.C.M. 'High-rate plasma-deposited SiO_2 films for surface passivation of crystalline silicon'. *Journal of Vacuum Science & Technology A: Vacuum, Surfaces, and Films*. 2006;**24**(5):1823–30.

[65] Dingemans G., van de Sanden M.C.M., Kessels W.M.M. 'Excellent Si surface passivation by low temperature SiO_2 using an ultrathin Al_2O_3 capping film'. *Physica Status Solidi (RRL) – Rapid Research Letters*. 2011;**5**(1):22–4.

[66] Mueller T., Schwertheim S., Scherff M., Fahrner W.R. 'High quality passivation for heterojunction solar cells by hydrogenated amorphous silicon suboxide films'. *Applied Physics Letters*. 2008;**92**(3):033504.

[67] Grant N.E., Kho T.C., Weber K. 'High-Level silicon surface passivation by anodically grown silicon dioxide and silicon nitride stacks'. *IEEE Journal of Photovoltaics*. 2015;**5**(4):1047–52.

[68] Einsele F., Beyer W., Rau U. 'Annealing studies of substoichiometric amorphous SiO_x layers for c-Si surface passivation'. *Physica Status Solidi*. 2010;**7**:1021–4.

[69] Mueller T., Schwertheim S., Fahrner W.R. 'Application of wide-bandgap hydrogenated amorphous silicon oxide layers to heterojunction solar cells for high quality passivation'. 2008 33rd IEEE Photovolatic Specialists Conference. IEEE; 2008. pp. 1–6.

[70] Chen Z., Pang S.K., Yasutake K., Rohatgi A. 'Plasma-enhanced chemical-vapor-deposited oxide for low surface recombination velocity and high effective lifetime in silicon'. *Journal of Applied Physics*. 1993;**74**(4):2856–9.

[71] Fountain G.G., Rudder R.A., Hattangady S.V., Markunas R.J., Lindorme P.S. 'Low interface state density SiO_2 deposited at 300 °C by remote plasma-enhanced chemical vapor deposition on reconstructed Si surfaces'. *Journal of Applied Physics*. 1988;**63**(9):4744–6.

[72] Chen Z., Yasutake K., Doolittle A., Rohatgi A. 'Record low SiO_2/Si interface state density for low temperature oxides prepared by direct plasma-enhanced chemical vapor deposition'. *Applied Physics Letters*. 1993;**63**(15):2117–19.

[73] Ge J., Tang M., Wong J., *et al.* 'Investigation of wide process temperature window for amorphous silicon suboxide thin-film passivation deposited by inductively coupled PECVD'. *IEEE Journal of Photovoltaics*. 2015;**5**(3):705–10.

[74] Ge J., Tang M., Wong J., *et al.* 'Excellent silicon surface passivation achieved by industrial inductively coupled plasma deposited hydrogenated intrinsic amorphous silicon suboxide'. *International Journal of Photoenergy*. 2014;**2014**(11):1–12.

[75] Duttagupta S. *Advanced Surface Passivation of Crystalline Silicon for Solar cell Applications. PHD thesis*. Singapore: National University of Singapore; 2014.

[76] Duttagupta S., Ma F.-J., Hoex B., Aberle A.G. 'Excellent surface passivation of heavily doped P+ silicon by low-temperature plasma-deposited SiO_x/SiN_y dielectric stacks with optimised antireflective performance for solar cell application'. *Solar Energy Materials and Solar Cells*. 2014;**120**:204–8.

[77] Mack S., Wolf A., Brosinsky C., *et al.* 'Silicon surface passivation by thin thermal oxide/PECVD layer stack systems'. *IEEE Journal of Photovoltaics*. 2011;**1**(2):135–45.

[78] Dingemans G., van Helvoirt C.A.A., Pierreux D., Keuning W., Kessels W.M.M. 'Plasma-assisted ALD for the conformal deposition of SiO_2: process, material and electronic properties'. *Journal of the Electrochemical Society*. 2012;**159**(3):H277–85.

[79] Jung H., Kim W.-H., Oh I.-K., *et al.* 'Growth characteristics and electrical properties of SiO_2 thin films prepared using plasma-enhanced atomic layer deposition and chemical vapor deposition with an aminosilane precursor'. *Journal of Materials Science*. 2016;**51**(11):5082–91.

[80] Grant N.E., McIntosh K.R. 'A review on low temperature chemically formed silicon dioxide for solar cell applications'. 48th Annual Conference of the Australian Solar Energy Society; 2010.

[81] Asuha., Kobayashi T., Maida O., *et al.* 'Ultrathin silicon dioxide layers with a low leakage current density formed by chemical oxidation of Si'. *Applied Physics Letters*. 2002;**81**(18):3410–12.

[82] Mihailetchi V.D., Komatsu Y., Geerligs L.J. 'Nitric acid pretreatment for the passivation of boron emitters for N-type base silicon solar cells'. *Applied Physics Letters*. 2008;**92**(6):063510.

[83] Bakhshi S., Zin N., Ali H., *et al.* 'Simple and versatile UV-ozone oxide for silicon solar cell applications'. *Solar Energy Materials and Solar Cells*. 2018;**185**(2):505–10.

[84] Cui J., Grant N., Lennon A. 'Effective surface passivation of P-type crystalline silicon with silicon oxides formed by light-induced anodisation'. *Applied Surface Science*. 2014;**323**(3):40–4.

[85] Cui J., Wang X., Opila R., Lennon A. 'Light-Induced anodisation of silicon for solar cell passivation'. *Journal of Applied Physics*. 2013;**114**(18):184101.

[86] Grant N.E., McIntosh K.R. 'Silicon surface passivation by anodic oxidation annealed at 400°C'. *ECS Journal of Solid State Science and Technology*. 2014;**3**(2):P13–16.

[87] Grant N.E., McIntosh K.R. 'Low surface recombination velocity on (100) silicon by electrochemically grown silicon dioxide annealed at low temperature'. *IEEE Electron Device Letters*. 2010;**31**(9):1002–4.

[88] Stegemann B., Gad K.M., Balamou P., *et al.* 'Ultra-thin silicon oxide layers on crystalline silicon wafers: comparison of advanced oxidation techniques with respect to chemically abrupt SiO_2/Si interfaces with low defect densities'. *Applied Surface Science*. 2017;**395**:78–85.

[89] Schmidt J., Kerr M., Cuevas A. 'Surface passivation of silicon solar cells using plasma-enhanced chemical-vapour-deposited SiN films and thin thermal SiO_2 /plasma SiN stacks'. *Semiconductor Science and Technology*. 2001;**16**(3):164–70.

[90] Schmidt J., Kerr M. 'Highest-quality surface passivation of low-resistivity P-type silicon using stoichiometric PECVD silicon nitride'. *Solar Energy Materials and Solar Cells*. 2001;**65**(1):585–91.

[91] Schmidt J., Schuurmans F.M., Sinke W.C., Glunz S.W., Aberle A.G. 'Observation of multiple defect states at silicon–silicon nitride interfaces fabricated by low-frequency plasma-enhanced chemical vapor deposition'. *Applied Physics Letters*. 1997;**71**(2):252–4.

[92] Schmidt J., Lauinger T., Aberle A.G., Hezel R. 'Record low surface recombination velocities on low-resistivity silicon solar cell substrates'. Conference Record of the Twenty Fifth IEEE Photovoltaic Specialists Conference; 1996. pp. 413–16.

[93] Aberle A.G., Lauinger T., Schmidt J., Hezel R. 'Injection-level dependent surface recombination velocities at the silicon-plasma silicon nitride interface'. *Applied Physics Letters*. 1995;**66**(21):2828–30.

[94] Lauinger T., Schmidt J., Aberle A.G., Hezel R. 'Record low surface recombination velocities on 1 Ω cm *p* -silicon using remote plasma silicon nitride passivation'. *Applied Physics Letters*. 1996;**68**(9):1232–4.

[95] Nagel H., Berge C., Aberle A.G. 'Generalized analysis of quasi-steady-state and quasi-transient measurements of carrier lifetimes in semiconductors'. *Journal of Applied Physics*. 1999;**86**(11):6218–21 [cited 2019 Jul 19].

[96] Lauinger T., Moschner J., Aberle A.G., Hezel R. 'Optimization and characterization of remote plasma-enhanced chemical vapor deposition silicon nitride for the passivation of P-type crystalline silicon surfaces'. *Journal of Vacuum Science & Technology A: Vacuum, Surfaces, and Films*. 1998;**16**(2):530–43.

[97] Aberle A.G., Hezel R. 'Progress in low-temperature surface passivation of silicon solar cells using remote-plasma silicon nitride'. *Progress in Photovoltaics: Research and Applications*. 1997;**5**(1):29–50.

[98] Hezel R., Jaeger K. 'Low-temperature surface passivation of silicon for solar cells'. *Journal of the Electrochemical Society*. 1989;**136**(2):518–23.

[99] Kerr M.J., Schmidt J., Cuevas A., Bultman J.H. 'Surface recombination velocity of phosphorus-diffused silicon solar cell emitters passivated with plasma enhanced chemical vapor deposited silicon nitride and thermal silicon oxide'. *Journal of Applied Physics*. 2001;**89**(7):3821–6.

[100] Kerr M.J., Cuevas A. 'Recombination at the interface between silicon and stoichiometric plasma silicon nitride'. *Semiconductor Science and Technology*. 2002;**17**(2):166–72.

[101] Battaglia C., Cuevas A., De Wolf S. 'High-efficiency crystalline silicon solar cells: status and perspectives'. *Energy & Environmental Science*. 2016;**9**(5):1552–76.

[102] Wan Y., McIntosh K.R., Thomson A.F. 'Characterisation and optimisation of PECVD SiN_x as an antireflection coating and passivation layer for silicon solar cells'. *AIP Advances*. 2013;**3**(3):032113.

[103] Lelièvre J.-F., Fourmond E., Kaminski A., Palais O., Ballutaud D., Lemiti M. 'Study of the composition of hydrogenated silicon nitride SiN_x:H for efficient surface and bulk passivation of silicon'. *Solar Energy Materials and Solar Cells*. 2009;**93**(8):1281–9.

[104] De Wolf S., Agostinelli G., Beaucarne G., Vitanov P. 'Influence of stoichiometry of direct plasma-enhanced chemical vapor deposited SiN_x films

and silicon substrate surface roughness on surface passivation'. *Journal of Applied Physics*. 2005;**97**(6):063303.

[105] Duttagupta S., Lin F., Wilson M., Boreland M.B., Hoex B., Aberle A.G. 'Extremely low surface recombination velocities on low-resistivity *n* -type and *p* -type crystalline silicon using dynamically deposited remote plasma silicon nitride films'. *Progress in Photovoltaics: Research and Applications*. 2014;**22**(6):641–7.

[106] Mäckel H., Lüdemann R. 'Detailed study of the composition of hydrogenated SiN_x layers for high-quality silicon surface passivation'. *Journal of Applied Physics*. 2002;**92**(5):2602–9.

[107] Hezel R., Schörner R. 'Plasma Si nitride—A promising dielectric to achieve high-quality silicon MIS/IL solar cells'. *Journal of Applied Physics*. 1981;**52**(4):3076–9.

[108] Warren W.L., Rong F.C., Poindexter E.H., Gerardi G.J., Kanicki J. 'Structural identification of the silicon and nitrogen dangling-bond centers in amorphous silicon nitride'. *Journal of Applied Physics*. 1991;**70**(1):346–54.

[109] Bauer G.H., Fuhs W., Ley L., Kanicki J., Warren W.L., Seager C.H. 'Microscopic origin of the light-induced defects in hydrogenated nitrogen-rich amorphous silicon nitride films'. *Journal of non-crystalline solids*. 1991;**137**:291–4.

[110] Krick D.T., Lenahan P.M., Kanicki J. 'Nature of the dominant deep trap in amorphous silicon nitride'. *Physical Review B*. 1988;**38**(12):8226–9.

[111] Warren W.L., Kanicki J., Robertson J., Poindexter E.H., McWhorter P.J. 'Electron paramagnetic resonance investigation of charge trapping centers in amorphous silicon nitride films'. *Journal of Applied Physics*. 1993;**74**(6):4034–46.

[112] Walukiewicz W. 'Mechanism of Schottky barrier formation: the role of amphoteric native defects'. *Journal of Vacuum Science & Technology B: Microelectronics and Nanometer Structures*. 1987;**5**(4):1062.

[113] Gatz S., Dullweber T., Mertens V., Einsele F., Brendel R. 'Firing stability of SiN_y/SiN_x stacks for the surface passivation of crystalline silicon solar cells'. *Solar Energy Materials and Solar Cells*. 2012;**96**(1):180–5.

[114] Jafari S., Chaitanya C., Dwan J., Hirsch J., Bernhard N., Lausch D. 'Composition-dependent hydrogen effusion and diffusion mechanism within a-SiN_x:H'. *35th European Photovoltic Solar Energy Conference and Exhibition*. 2018:686–9.

[115] Wilking S., Ebert S., Herguth A., Hahn G. 'Influence of hydrogen effusion from hydrogenated silicon nitride layers on the regeneration of boron-oxygen related defects in crystalline silicon'. *Journal of Applied Physics*. 2013;**114**(19):194512.

[116] Jafari S., Hirsch J., Lausch D., John M., Bernhard N., Meyer S. 'Composition limited hydrogen effusion rate of a-SiN_x :H passivation stack'. 2019;**2147**:50004.

[117] Wan Y. *Highly Transparent and Highly Passivating Silicon Nitride for Solar Cells*. Canberra, Australia: Australian National University; 2014.

[118] Bredemeier D., Walter D.C., Heller R., Schmidt J. 'Impact of hydrogen-rich silicon nitride material properties on light-induced lifetime degradation in multicrystalline silicon'. *physica status solidi (RRL) – Rapid Research Letters*. 2019;**13**(8):1900201.

[119] Chen F., Li T., Cotter J. 'PECVD silicon nitride surface passivation for high-efficiency N-type silicon solar cells'. 2006 IEEE 4th World Conference on Photovoltaic Energy Conference. IEEE; 2006. pp. 1020–3.

[120] Chen F., Romijn I., Weeber A., Tan J., Hallam B., Cotter J. 'Relationship between PECVD silicon nitride film composition and surface and edge passivation'. 22nd Eiropean Photovoltaic Solar Energy Conference, Milan, Italy. WIP-Renewable Energies; 2007. pp. 1053–60.

[121] Chen F. *PECVD silicon nitride for N-type silicon solar cells. PHD thesis.* Kensington, Australia: UNSW; 2008.

[122] Duttagupta S., Hoex B., Aberle A.G. 'Progress with industrially-feasible excellent surface passivation of heavily-doped P-type and N-type crystalline silicon by PECVD SiO_x/SiN_x with optimised anti-reflective performance'. 28th European PVSEC WIP; 2013. pp. 993–6.

[123] Wan Y., McIntosh K.R., Thomson A.F., Cuevas A. 'Low surface recombination velocity by low-absorption silicon nitride on c-Si'. *Conference Record of the IEEE Photovoltaic Specialists Conference*. 2012;**3**(1):554–9 [cited 2016 Sep 5].

[124] Moschner J.D., Henze J., Schmidt J., Hezel R. 'High-Quality surface passivation of silicon solar cells in an industrial-type inline plasma silicon nitride deposition system'. *Progress in Photovoltaics: Research and Applications*. 2004;**12**(1):21–31.

[125] Wan Y., McIntosh K.R., Thomson A.F. 'Characterisation and optimisation of PECVD SiN_x as an antireflection coating and passivation layer for silicon solar cells'. *AIP Advances*. 2013;**3**(3):032113.

[126] Larionova Y., Mertens V., Harder N.-P., Brendel R. 'Surface passivation of N-type Czochralski silicon substrates by thermal-SiO2/plasma-enhanced chemical vapor deposition sin stacks'. *Applied Physics Letters*. 2010;**96**(3):032105.

[127] Bonilla R.S., Wilshaw P.R. 'On the c-Si/SiO_2 interface recombination parameters from photo-conductance decay measurements'. *Journal of Applied Physics*. 2017;**121**(13):135301.

[128] Hofmann M., Kambor S., Schmidt C., *et al*. 'PECVD-ONO: a new deposited firing stable rear surface passivation layer system for crystalline silicon solar cells'. *Advances in OptoElectronics*. 2008;**2008**:1–10.

[129] Hofmann M., Janz S., Schmidt C., *et al*. 'Recent developments in rear-surface passivation at Fraunhofer ISE'. *Solar Energy Materials and Solar Cells*. 2009;**93**(6–7):1074–8.

[130] Hofmann M., Kambor S., Schmidt C. 'Firing stable surface passivation using all-PECVD stacks of SiO_x:H and SiN_x:H'. 22nd European Photovoltaic Solar Energy Conference and Exhibition; 2007. p. 1030.

[131] Dingemans G., Mandoc M.M., Bordihn S., van de Sanden M.C.M., Kessels W.M.M. 'Effective passivation of Si surfaces by plasma deposited SiO_x/a-SiN_x:H stacks'. *Applied Physics Letters*. 2011;**98**(22):222102.

[132] Chowdhury Z.R., Cho K., Kherani N.P. 'High-Quality surface passivation of silicon using native oxide and silicon nitride layers'. *Applied Physics Letters*. 2012;**101**(2):021601.

[133] Rohatgi A., Doshi P., Moschner J., Lauinger T., Aberle A.G., Ruby D.S. 'Comprehensive study of rapid, low-cost silicon surface passivation technologies'. *IEEE Transactions on Electron Devices*. 2000;**47**(5):987–93.

[134] Lee J.Y., Glunz S.W. 'Investigation of various surface passivation schemes for silicon solar cells'. *Solar Energy Materials and Solar Cells*. 2006;**90**(1):82–92.

[135] Narasimha S., Rohatgi A. 'Effective passivation of the low resistivity silicon surface by a rapid thermal oxide/plasma silicon nitride stack'. *Applied Physics Letters*. 1998;**72**(15):1872–4.

[136] Agostinelli G., Choulat P., Dekkers H.F.W., Vermariën E., Beaucarne G. 'Rear surface passivation for industrial solar cells on thin substrates'. Conference Record of the 2006 IEEE 4th World Conference on Photovoltaic Energy Conversion, WCPEC-4. IEEE; 2007. pp. 1004–7.

[137] Kho T.C., Fong K., McIntosh K., *et al.* 'Exceptional silicon surface passivation by an ONO dielectric stack'. *Solar Energy Materials and Solar Cells*. 2019;**189**(Supplement C):245–53.

[138] Lawerenz A., Strutzberg H., Leschinski C., Roczen M., Maier C., Biank H.C. 'Fundamental study of silicon oxynitrides for photovoltaic applications'. 26th European Photovoltaic Solar Energy Conference and Exhibition; 2011. pp. 1653–8.

[139] Schwab C., Hofmann M., Heller R., Seiffe J., Rentsch J., Preu R. 'Effects of high-temperature treatment on the hydrogen distribution in silicon oxynitride/silicon nitride stacks for crystalline silicon surface passivation'. *Physica Status Solidi*. 2013;**210**(11):2399–403 [cited 2017 Feb 13].

[140] Cheng X., Haug H., Di S.M., Zhu J., Marstein E.S. 'Electronic Properties of a-SiOxN_y:H/SiN stacks for surface passivation of P-type crystalline silicon'. *IEEE journal of photovoltaics*. 2016;**6**(5):1103–8.

[141] Schmidt J., Peibst R., Brendel R. 'Surface passivation of crystalline silicon solar cells: present and future'. *Solar Energy Materials and Solar Cells*. 2018;**187**:39–54.

[142] Bonilla R.S., Al-Dhahir I., Yu M., Hamer P., Altermatt P.P. 'Charge fluctuations at the Si–SiO_2 interface and its effect on surface recombination in solar cells'. *Solar Energy Materials and Solar Cells*. 2020;**215**:110649.

[143] Chittick R.C., Alexander J.H., Sterling H.F. 'The preparation and properties of amorphous silicon'. *Journal of the Electrochemical Society*. 1969;**116**(1):77.

[144] Taguchi M., Tanaka M., Matsuyama T. 'Improvement of the conversion efficiency of polycrystalline silicon thin film solar cell'. Technical Digest of the Fifth International Photovoltaic Science and Engineering Conference; 1990. pp. 689–92.

[145] De Wolf S., Ballif C., Kondo M. 'Kinetics of a-Si:H bulk defect and a-Si:H/c-Si interface-state reduction'. *Physical review. B, Condensed matter and materials physics*. 2012;**85**(11):113302.

[146] Staebler D.L., Wronski C.R. 'Reversible conductivity changes in discharge-produced amorphous Si'. *Applied Physics Letters*. 1977;**31**(4):292–4.

[147] Gielis J.J.H., Van Den Oever P.J., Hoex B., Van De Sanden M.C.M., Kessels W.M.M. 'Real-time study of a-Si:H/c-Si heterointerface formation and epitaxial Si growth by spectroscopic ellipsometry, infrared spectroscopy, and second-harmonic generation'. *Physical Review B: Condensed Matter and Materials Physics*. 2008;**77**:205329.

[148] De Wolf S., Kondo M. 'Abruptness of a-Si:H/c-Si interface revealed by carrier lifetime measurements'. *Applied Physics Letters*. 2007;**90**(4):042111.

[149] Yoshikawa K., Kawasaki H., Yoshida W., *et al.* 'Silicon heterojunction solar cell with interdigitated back contacts for a photoconversion efficiency over 26%'. *Nature Energy*. 2017;**2**(5):17032.

[150] Adachi D., Hernández J.L., Yamamoto K. 'Impact of carrier recombination on fill factor for large area heterojunction crystalline silicon solar cell with 25.1% efficiency'. *Applied Physics Letters*. 2015;**107**(23):233506.

[151] Madumelu C., Wright B., Soeriyadi A., *et al.* 'Investigation of light-induced degradation in N-type silicon heterojunction solar cells during illuminated annealing at elevated temperatures'. *Solar Energy Materials and Solar Cells*. 2020;**218**(1):110752.

[152] Bonilla R.S., Reichel C., Hermle M., Wilshaw P.R. 'Extremely low surface recombination in 1 Ω cm n-type monocrystalline silicon'. *Physica Status Solidi - Rapid Research Letters*. 2016;**1**:1–5.

[153] De Wolf S., Descoeudres A., Holman Z.C., Ballif C., De Wolf S., Descoeudres A. 'High-efficiency silicon heterojunction solar cells: a review'. *Green*. 2012;**2**(1):7–24.

[154] Haschke J., Dupré O., Boccard M., Ballif C. 'Silicon heterojunction solar cells: Recent technological development and practical aspects – from lab to industry'. *Solar Energy Materials and Solar Cells*. 2018;**187**(6):140–53.

[155] Hoex B., van de Sanden M.C.M., Schmidt J., Brendel R., Kessels W.M.M. 'Surface passivation of phosphorus-diffused n+-type emitters by plasma-assisted atomic-layer deposited Al2O3'. *physica status solidi (RRL) - Rapid Research Letters*. 2012;**6**(1):4–6.

[156] Hoex B., Heil S.B.S., Langereis E., van de Sanden M.C.M., Kessels W.M.M. 'Ultralow surface recombination of c-Si substrates passivated by plasma-assisted atomic layer deposited $Al2O_3$'. *Applied Physics Letters*. 2006;**89**(4):042112.

[157] Saint-Cast P., Kania D., Hofmann M., Benick J., Rentsch J., Preu R. 'Very low surface recombination velocity on P-type c-Si by high-rate plasma-deposited aluminum oxide'. *Applied Physics Letters*. 2009;**95**(15):151502 [cited 2017 Feb 20].

[158] Hoex B., Gielis J.J.H., van de Sanden M.C.M., Kessels W.M.M. 'On the c-Si surface passivation mechanism by the negative-charge-dielectric $Al2O_3$'. *Journal of Applied Physics*. 2008;**104**(11):113703.

[159] Dingemans G., Engelhart P., Seguin R., *et al*. 'Stability of $Al2O_3$ and $Al2O_3$/a-SiN_x:H stacks for surface passivation of crystalline silicon'. *Journal of Applied Physics*. 2009;**106**(11):114907.

[160] Varshney U., Hallam B., Hamer P., *et al*. 'Controlling light- and elevated-temperature-induced degradation with thin film barrier layers'. *IEEE Journal of Photovoltaics*. 2020;**10**(1):19–27.

[161] Green M.A., Blakers A.W. 'Advantages of metal-insulator-semiconductor structures for silicon solar cells'. *Solar Cells*. 1983;**8**(1):3–16.

[162] Feldmann F., Bivour M., Reichel C., Hermle M., Glunz S.W. 'Passivated rear contacts for high-efficiency N-type Si solar cells providing high interface passivation quality and excellent transport characteristics'. *Solar Energy Materials and Solar Cells*. 2014;**120**:270–4.

[163] Allen T.G., Bullock J., Yang X., Javey A., De Wolf S., De W.S. 'Passivating contacts for crystalline silicon solar cells'. *Nature Energy*. 2019;**4**(11):914–28.

[164] Conibeer G., Green M., Corkish R., *et al*. 'Silicon nanostructures for third generation photovoltaic solar cells'. *Thin Solid Films*. 2006;**511-512**:654–62.

[165] Lancaster K., Großer S., Feldmann F., Naumann V., Hagendorf C. 'Study of pinhole conductivity at passivated carrier-selected contacts of silicon solar cells'. *Energy Procedia*. 2016;**92**(6):116–21.

[166] Schnabel M., Van De Loo B.W.H., Nemeth W., Macco B., Stradins P., Kessels W.M.M. 'Hydrogen passivation of poly-Si/SiO_x contacts for Si solar cells using $Al2O_3$ studied with deuterium'. *Applied physics letters*. 2018;**112**(20):203901.

[167] Truong T.N., Yan D., Chen W., *et al*. 'Hydrogenation mechanisms of poly-Si/SiO_x passivating contacts by different capping layers'. *Solar RRL*. 2020;**4**(3):1900476.

[168] Yang Y., Altermatt P.P., Cui Y., Hu Y., Chen D., Chen L. 'Effect of carrier-induced hydrogenation on the passivation of the poly-Si/SiO_x/c-Si interface'. *AIP Conference Proceedings*. AIP Publishing LLC; 2018. p. 040026.

Chapter 4
Hydrogen passivation of bulk defects

Brett Hallam[1], Matthew Wright[1], and Alison Ciesla[1]

The dielectric layers used in the fabrication of commercial silicon solar cells such as silicon nitride (SiN_X) antireflection coating and the rear dielectric stack of aluminium oxide (AlO_X) and SiN_X in passivated emitter and rear cell (PERC) solar cells provide an opportunity for both bulk and surface defect passivation by hydrogen without applying additional hydrogen source, such as forming gas. As discussed in Chapter 3, such layers provide excellent field-effect passivation and hydrogen passivation of interface defects at the silicon surface. However, just as importantly, hydrogen incorporated into these dielectric layers can be released during subsequent thermal processing throughout solar cell fabrication [1–5], and then diffuse into the silicon to also passivate bulk defects. The automatic incorporation of atomic hydrogen in industrial p-type screen-printed solar cells through the use of hydrogen-containing dielectric layers has been vital in the domination of p-type technologies in the photovoltaics industry, resulting in significant enhancements in the electrical quality of the silicon material [6].

In silicon solar cells fabricated using multi-crystalline, which have dominated the photovoltaics industry until 2018/2019 [7, 8], the use of plasma-enhanced chemical vapour deposition (PECVD) of hydrogenated SiN_X (SiN_X:H) has been considered vital due to the benefits of improved bulk minority carrier lifetime within the silicon [9]. The passivation of various impurities and structural defects within the highly defected multi-crystalline silicon (mc-Si) material can result in substantial improvements in effective minority carrier lifetime and, therefore, efficiency, with enhancements of up to 3–4%$_{abs}$ [10]. Ongoing development from a number of aspects including improved crystallisation techniques, hydrogen passivation processes, gettering and new cell technologies has seen the achievement of 22.8% and 24.4% efficiency for solar cells fabricated on p-type and n-type cast silicon wafers, respectively [11].

For p-type mono-crystalline silicon solar cells, which have recently taken over as the dominant wafer choice for industrial silicon solar cells, hydrogen passivation has proven beneficial for process-induced defects such as oxygen precipitates

[1]The School of Photovoltaic and Renewable Energy Engineering, The University of New South Wales, Australia

[12, 13] and the light- or carrier-induced boron–oxygen (B–O) defect [14, 15]. This defect forms under the normal operating conditions in the field and can reduce cell performance by up to $2\%_{abs}$ [16] and has plagued the photovoltaics industry for decades. This chapter will discuss the passivation of a range of bulk defects in silicon solar cells, primarily structural, impurity-related- and process-induced defects, before discussing carrier-induced defects in detail in Chapters 5 and 6, which have ultimately assisted in the fabrication of p-type Czochralski (Cz) solar cells reaching efficiencies of 24% [11, 17]. Although treated as separate categories of defects, it is noted that the defects can often fall into multiple categories. For example, impurities can lead to process-induced defects, decorate structural defects or cause carrier-induced degradation. As such, the passivation of defects should be treated with a holistic approach.

4.1 Hydrogenation of structural defects

A number of performance-limiting defects are incorporated into the silicon structure during the growth of the silicon wafer, including vacancies, grain boundaries, dislocations and slip line planes. The generation of structural defects is therefore strongly dependent on the crystallisation process used. Generally, the crystallisation technologies for industrial silicon solar cells can be broken into two broad categories*:

- cast silicon
- mono-crystalline silicon.

In some sense, structural defects incorporated into the silicon structure can be considered simple to passivate with hydrogen. These defects are typically present in the silicon prior to solar cell fabrication. As a result, the defects are available for hydrogen passivation throughout the solar cell fabrication sequence. Nonetheless, there is significant evidence within the literature related to the difficulty of passivating structural defects.

This section discusses the hydrogen passivation of a number of performance-limiting defects in silicon solar cells for a range of cast silicon and mono-crystalline silicon wafers.

4.1.1 Structural defects in cast silicon wafers

Crystalline silicon is cheaper to produce by casting methods, compared to mono-crystalline silicon produced by the Cz or Float zone (FZ) growth methods. In particular, the casting technique greatly reduces the energy usage for the crystallisation process. This has resulted in cast silicon wafers dominating the industry until

*Float zone is not considered in this textbook due to the fabrication cost being too high for commercial silicon solar cells.

2018–2019 [7, 8], where rapid reductions in the cost of Cz-grown silicon has seen a shift to a dominance of mono-crystalline wafers in the industry.

Casting techniques for silicon wafers rely on direct solidification for the crystal growth and has been used for decades [18]. For conventional mc-Si wafers, crystallisation is performed in a cast, where molten silicon is poured into the cast and is directionally solidified, with little control during the crystal growth process. This approach can result in the incorporation of extended structural defects into the crystal structure during the casting process, which can result in reduced material lifetime to less than 10 μs [19]. In particular, the crystallised silicon contains grains of different orientations, resulting in a large range of grain boundary angles, as well as a large spread in the grain size throughout the cast. These grain boundaries contain dangling bonds, and can in some sense, be considered a surface. However, given that the grain boundaries penetrate throughout the silicon wafer, they do not benefit from field-effect passivation during emitter formation. As such, chemical passivation of the grain boundaries to satisfy the dangling bonds is crucial.

In addition, dislocation clusters can form within the grain [20]. These dislocations tend to nucleate at grain boundaries with low symmetry, or due to stresses or lattice distortion around precipitates [21, 22]. This results in the rapid growth of dislocations as the silicon crystallises towards the top of the ingot [19]. Grain boundary and dislocation defects introduce defect energy levels into the band gap. However, the character, angle and the impurity contamination can cause significantly different recombination activity [23–25].

4.1.2 Passivation of grain boundaries and dislocation clusters

The transition to the use of PECVD SiN_x:H in industrial solar cells has been pivotal for improving the bulk lifetime of mc-Si through the passivation of grain boundaries and dislocation clusters. In contrast to solar cells fabricated with titanium dioxide, mc-Si solar cells fabricated using PECVD SiN_x:H show greatly reduced recombination activity [26]. The corresponding improvements in minority carrier lifetime can increase the internal quantum efficiency (IQE) and open-circuit voltage (V_{OC}) of the device [5, 27–30]. With such improvements in the electronic properties of bulk silicon, efficiency enhancements in the range of 10–15$\%_{rel}$ have been demonstrated for solar cells using hydrogen-containing dielectrics compared to those with non-hydrogen-containing dielectrics [5, 9].

4.1.2.1 Characterisation of structural defects

Commercial solar cells are large area electronic devices, and the industry standard silicon wafer size is rapidly expanding from 6-inch (approximately 156 mm × 156 mm) to larger sizes. As such, it is important to characterise spatial non-uniformities in electrical properties and cell performance. This is particularly crucial for silicon materials that contain a large concentration of structural defects, which vary spatially across the wafer such as for multi-crystalline or cast-mono wafers, and even certain structural defects in single-crystalline silicon material.

Generally, the passivation of structural defects by hydrogen has been inferred from changes in the minority carrier lifetime through a range of characterisation techniques, using thermal processes in the presence of a hydrogen source. One powerful tool to characterise the spatially dependent performance of the material, and thus the impact of hydrogenation treatments, is photoluminescence (PL) imaging [31]. This approach relies on fundamental recombination behaviour in silicon, wherein the strength of luminescence in silicon from radiative recombination is dependent on the number and strength of non-radiative recombination mechanisms present. In the PL-imaging approach, the silicon wafer is illuminated with a light source, which generates electron-hole pairs. As the carriers recombine radiatively, the silicon emits radiation with an energy equivalent to the bandgap (~1.12 eV), and this emission is captured using a charge-coupled device (CCD) camera sensitive to photons of this energy. If the material is free of non-radiative defects (e.g. grain boundaries), the amount of carriers that recombine radiatively will be high, which means the signal captured by the sensor will be large. If silicon wafers contain regions that have a large concentration of non-radiative defects, this will cause carriers to recombine through non-radiative pathways, which will reduce the strength of the emission. As such, the 'brightness' of a pixel in a PL image is indicative of the electronic quality of the material. Since the first publication on PL imaging in 2006 [31], this approach has become an integral characterisation tool for silicon solar cell manufacturers. The detection of the emission using a CCD camera and the non-contacted nature of the approach means that the measurement can be performed in high resolution and in a very short time. Additionally, the control over the illumination intensity can allow for spatial information as a function of illumination and therefore carrier injection. An example of

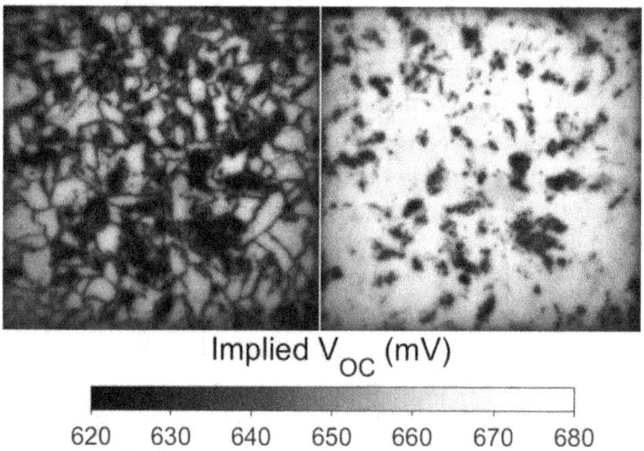

Figure 4.1 *Calibrated iV_{OC} map of a diffused conventional multi-crystalline wafer after (left) SiN_x:H deposition and (right) subsequent firing, highlighting relatively large crystal grains and large dislocation clusters. Adapted from [32].*

this approach is displayed in Figure 4.1. This clearly depicts the different material quality caused by the many grain boundaries and dislocation clusters present in mc-Si. This technique can be used to assess the changes in recombination due to a hydrogen passivation process.

Similar to PL, the electron beam-induced current (EBIC) measurement technique also uses generated carriers to measure and visualise the localised electrical quality of the silicon material. However, in this case an image is created by scanning a focused electron beam across the surface of a sample exciting electron-hole pairs using a scanning electron microscope [33]. The electrical carriers generated then either recombine or are separated and drift due to electric fields within the material and are collected by a measurement probe. Electrically active regions appear bright, particularly p-n junctions where the field is strong, while inactive regions with defects, where carriers recombine, appear dark. The EBIC contrast can be used to quantify the recombination strength of defects [34] and how it changes in response to gettering or hydrogenation processes. An example showing the use of EBIC to understand the hydrogenation of structural defects will be presented in Section 4.1.2.1.

Light beam-induced current (LBIC) is another technique that creates an image to locally measure and visualise variable quality in a silicon solar cell. Rather than using electrical excitation, a laser light beam of a certain wavelength is scanned across the cell surface and carriers are generated at a depth dependent on the incident wavelength of light. A wavelength of 404 nm will generate carriers near the illuminated surface, while 981 nm will generate carriers deeper into the silicon material and is more indicative of the bulk quality [35]. Carriers are extracted by an external circuit; the carriers collected from each point as the light beam is scanned across the cell indicate the quantum efficiency for that location and depth. If reflectance is also measured, then the localised IQE can be determined and mapped for each point [36]. An increasing number of collected carriers result in a brighter signal, representing better electrical quality, while regions of high recombination reduce the ability to collect carriers, resulting in lower or darker localised signal. See Section 4.1.5 for example.

A recent study by Tweddle *et al.* demonstrated the direct detection of hydrogen at structural defect sites using atom probe tomography and transmission Kikuchi diffraction techniques [37] (refer to Chapter 2 for details and corresponding figure).

An example highlighting the impact of hydrogenation on grain boundaries and dislocation clusters is shown in Figure 4.1 using calibrated implied open-circuit voltage (iV_{OC}) maps [38] from PL images. The first image (Figure 4.1 (left)) is shown after texturing, phosphorus emitter diffusion, which also enables gettering, and the deposition of PECVD SiN_x:H onto both surfaces of the wafer to form a symmetrical lifetime test structure. As shown, the significant recombination activity can be observed at grain boundary sites and dislocation clusters. Here in the left image, grain boundaries tend to appear as narrow lines, and dislocation clusters appear as large dark regions. In addition, grains are relatively large, but there is a large variability in grain size.

This suggests that the SiN_x:H deposition process alone, even though in the vicinity of 400 °C for several minutes,† is insufficient to passivate and that hydrogen could potentially diffuse through the wafer in this time frame. The second image (Figure 4.1 (right)) shows the corresponding image of the sample after a high-temperature firing process. The PL images highlight the significant overall improvement in effective minority carrier lifetime of the material from the firing process, and therefore, the iV_{OC}. For this sample, most grain boundaries can no longer be observed in Figure 4.1 (right) as they are readily passivated during the firing process. However, certain grain boundaries still show high levels of recombination, suggesting that conventional hydrogenation processes (CHPs) for silicon solar cells are unable to effectively passivate such defects. In addition, strong recombination activity at dislocation clusters can still be observed, although greatly reduced in extent.

4.1.2.2 Challenges of passivation

The ease of passivation for structural defects such as grain boundaries and dislocation clusters can depend on a number of factors, including characteristics of the grain boundaries and dislocations, such as grain boundary angle and impurity decoration [23–25]. A particular challenge is that structural defects can act as internal gettering sites for impurities, thereby increasing their recombination activity and also making them more challenging to passivate with hydrogen [25].

A comprehensive study by Chen *et al.* investigated the impact of grain boundary orientation using EBIC and an electron back-scattered diffraction (EBSD) technique [39]. To increase EBIC sensitivity to recombination at grain boundaries, measurements were also taken at a reduced temperature of 100 K (−173 °C). It was observed that for mc-Si with low contamination levels, the effectiveness of passivation was on dependent both the grain boundary characteristic and impurity contamination. In such material, random and small-angle grain boundaries were substantially more difficult to passivate than special Σ grain boundaries ($\Sigma3$, $\Sigma9$ and $\Sigma27$). In contrast, for mc-Si with high impurity concentrations, the effectiveness of hydrogen passivation was reduced. Rather than being dependent on the grain boundary angle, the primary impact determining the effectiveness was the impurity contamination level [40]. As a result, the effectiveness of passivation varied throughout the ingot (see Figure 4.2).

Dislocation clusters are more challenging to passivate than grain boundaries. Typically, significant recombination activity can still be observed around dislocation clusters after a firing process [41, 42], as was shown in Figure 4.1, despite substantial improvements. In particular, dislocations decorated with metallic impurities and regions with high dislocation densities have proven to be more challenging to

†This includes the time of pump-down in the PECVD chamber and in holding chambers.

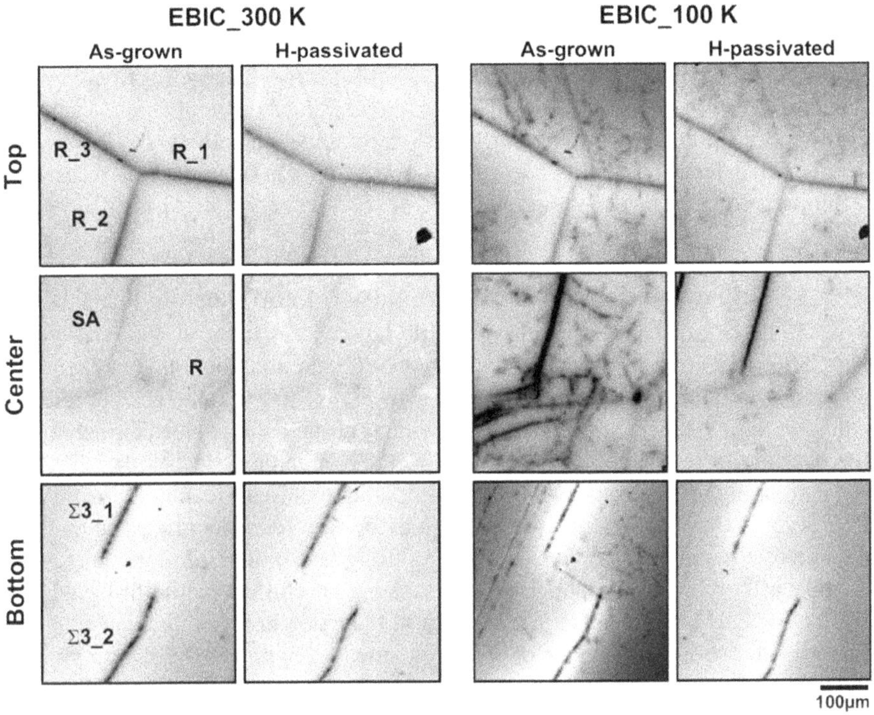

Figure 4.2 *EBIC images of grain boundaries before and after hydrogen passivation for mc-Si wafers. Reproduced from [39] with the permission of John Wiley and Sons. © 2007 WILEY-VCH Verlag GmbH & Co. KGaA, Weinheim.*

passivate [43]. It is unclear whether large regions with high dislocation densities lead to a localised depletion of hydrogen (i.e. if the concentration of defects is substantially higher than the concentration of hydrogen in the silicon), or whether it is purely due to the characteristics of the dislocation cluster that make them challenging to passivate.

The effectiveness of hydrogen passivation in mc-Si has also been reported to vary between and throughout cast silicon ingots [20, 23, 39]. This is due to the difference in characteristics of the defects and impurity concentrations throughout the ingot, with higher impurities typically occurring in the red-zone (bottom of the ingot) and the top of the ingot due to the build-up of impurity concentration as a result of segregation. This likely also has implications for the interactions of hydrogen with defects and the subsequent variations in hydrogen diffusivity [44]. Wafers near the top of the ingot are heavily impacted by recombination at dislocation clusters, while wafers from the bottom of the ingot are more heavily impacted by lower intra-grain lifetimes and larger density of recombination active grain boundaries [41].

Structural defects with deep energy levels are also reported to be more difficult to passivate than intra-grain defects within the bulk of mc-Si [26]. Hydrogen passivation can potentially shift the level of the defect (now a defect-hydrogen complex) to another level in the forbidden bandgap and still induce recombination. The passivation of deep levels can transform to low-recombination activity shallow levels at structural defect sites. Differences in the stability of hydrogen passivation has also been observed between deep level and shallow level defects [45, 46].

4.1.2.3 Impact of dielectric layer composition and thermal processing

With the deposition process of the dielectric layers insufficient to passivate grain boundaries (typically in the range of 200–400 °C), an additional thermal process is generally required to enable passivation. This is likely due to the need to release hydrogen from the dielectric layers as well as the substantially reduced diffusivity of hydrogen at low temperatures, which is impacted by trapping effects [10].

Similar to the aforementioned study comparing the use of titanium dioxide and SiN_X:H for grain boundary passivation, a study by Kim *et al.* demonstrated that the absence of a hydrogen-rich dielectric during firing had little impact on the recombination activity of grain boundaries and dislocation clusters, while firing with a PECVD SiN_X:H layer resulted in a substantial reduction in recombination activity. This provides strong indications of the involvement of hydrogen, by separating out thermal effects [47].

The hydrogenation of structural defects including grain boundaries and dislocation clusters can depend on the composition of the dielectric layer used as well as firing conditions. It is important to note that the hydrogen content alone does not necessarily dictate the effectiveness of bulk passivation that can be achieved. For example, given appropriate annealing conditions, atomic layer deposition (ALD) grown Al_2O_3, with a hydrogen content of approximately 3% [48], as can be determined using secondary ion mass spectroscopy, can provide the same passivation quality as a stack of ALD Al_2O_3 capped by a hydrogen-rich PECVD SiN_X:H layer [49], which typically has a hydrogen content in the range of 20–30% [50].

In addition, changes to the composition of the dielectric layer can significantly change the behaviour and release of hydrogen during the dielectric growth process and during subsequent thermal processes. For as-deposited SiN_X:H films, better bulk and surface passivation is typically obtained with silicon-rich films, which are more open and less dense, and more readily release hydrogen. This is believed to facilitate the diffusion of hydrogen radicals to the surface of the silicon through the dielectric layer during the PECVD process [50].

Hydrogen desorption kinetics are heavily dependent on the film composition [50], and a direct correlation between hydrogen desorption and the silicon content of the film has been demonstrated. The release of hydrogen from dielectric layers can be assessed using temperature programmed desorption (TPD). In Figure 4.3, the silicon-rich film shown with a low NH_3/SiH_4 flow ratio ($R = NH_3/SiH_4 = 2$) that results in a low ratio of nitrogen to silicon, $x = N/Si = 0.35$, and refractive index of 2.98, starts releasing hydrogen at ~400 °C, with a peak desportion at ~680 °C, and is

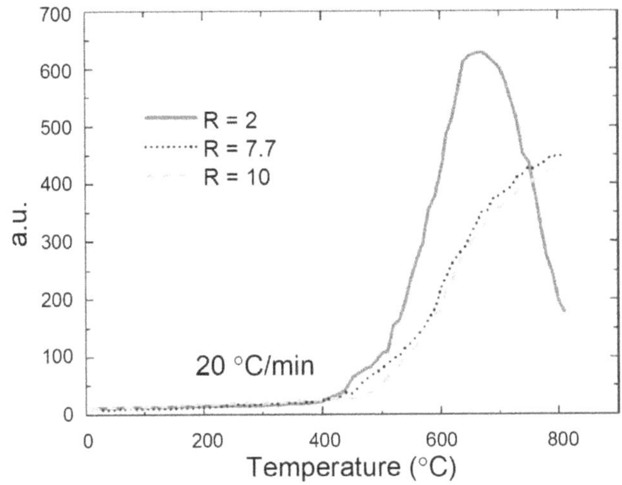

Figure 4.3 *Hydrogen desorption profiles obtained from TPD measurements for three different SiN_x:H layers. Reprinted from [50] Copyright (2009), with permission from Elsevier.*

mostly depleted from hydrogen by 800 °C. In contrast, more industrial-like SiN_x:H layers with a refractive index of 1.98 (R = 10, x = 0.81) and 2.03 (R = 7.7, x = 0.78), and hence higher NH_3/SiH_4 ratios, still show increasing hydrogen desorption at 800 °C, even though the films also started to desorb hydrogen at ~440 °C, although the release at low temperatures is much reduced compared to that of the high refractive index film.

The differences in the release of hydrogen are likely due to differences in the Si–H and N–H bond densities of the films. The Si–H and N–H bond densities, and other information on the composition can be obtained using Fourier transform infrared spectroscopy (FTIR). This can be used to highlight the impact of dielectric layer composition and changes in the dielectric layer composition induced by thermal processes on bulk defect passivation [51]. Si–H bonds have a binding energy of ~3.3 eV, which is substantially lower than the more thermally stable N–H bonds with a binding energy of 4.1 eV [52]. As a result, nitrogen-rich SiN_x:H layers,[‡] which tend to have higher N–H densities and lower refractive indexes are more thermally stable than silicon-rich SiN_x:H layers with higher Si–H bond densities, as the composition of the film approaches closer to that of amorphous silicon (a-Si:H). In addition, thermal changes including an increase in film density, which tends to create more Si–N bonds and resulting in a higher refractive index is more pronounced in silicon-rich films [51].

[‡]The ratio of nitrogen to silicon ([N]/[Si]) for SiN_x:H films can be determined using X-ray photoelectron spectroscopy (XPS) (see [50] for details).

After annealing, the Si–N bond density, as can be measured by FTIR, is generally a reasonable indicator of good bulk and surface passivation [51]. High SiN_x:H mass densities (nitrogen-rich films) enable the diffusion of atomic hydrogen during subsequent high-temperature processes, which is beneficial for passivation [53]. However, this depends on the annealing condition used. With appropriate annealing conditions, a low-density film does not necessarily provide inferior bulk defect passivation [49], although care should be taken due to the reduced thermal stability of the layers.

A recent study by Sio *et al.* investigated the impact of firing temperature and dielectric layers for the passivation of grain boundaries and dislocation clusters in p-type mc-Si [49]. It was observed that a rapid thermal process of 500 °C for 5 s was sufficient to passivate grain boundaries in the material for the dielectric layers used. In most instances, no emitter diffusions were present during firing. However, lower temperatures were only investigated for the least temperature-stable film, and hence the most likely film to release hydrogen at low temperatures. With this film, using in situ PL characterisation, it was concluded that the passivation of grain boundaries occurred during the range of 400 °C to 500 °C. However, this will be dependent on the film's ability to release sufficient quantities of hydrogen at sufficiently low temperatures, and the ability of hydrogen to diffuse through the dielectric layers, blocking layers (e.g. other dielectric layers or diffused regions) and into the silicon.

A key observation in that study was that for all films studied, a slight reduction in lifetime was observed for a process at 900 °C rather than 700 °C, with the process being less effective at passivating structural defects. In particular, the least thermally stable SiN_x:H layer resulted in significant recombination activity in grain boundaries when fired at 900 °C, and hence ineffective passivation.

In contrast, a study by Chan *et al.* determined that significant recombination activity in grain boundaries remained for a peak firing temperature of 530 °C in an industrial firing furnace (see Figure 4.4). It was observed that a temperature of 630 °C was sufficient to passivate all grain boundaries, with further improvements in the lifetime observed up until 730 °C, likely related to surface passivation. However, dielectrics were not removed after firing, and hence some changes in lifetime could also be due to changes in surface passivation. In addition, an emitter diffusion was present on all samples during firing. Similarly, an excessively high firing temperature resulted in reduced lifetime, although the highest temperature investigated was substantially lower than that in the study by Sio *et al.* In another study, grain boundary recombination was still present after a firing temperature (actual temperature ~600 °C), and a subsequent low temperature anneal led to an improvement in the passivation of the structural defects [55]. Differences in the passivation of the samples are likely impacted by the different compositions of the film, and in particular, the Si–H and N–H bond densities.

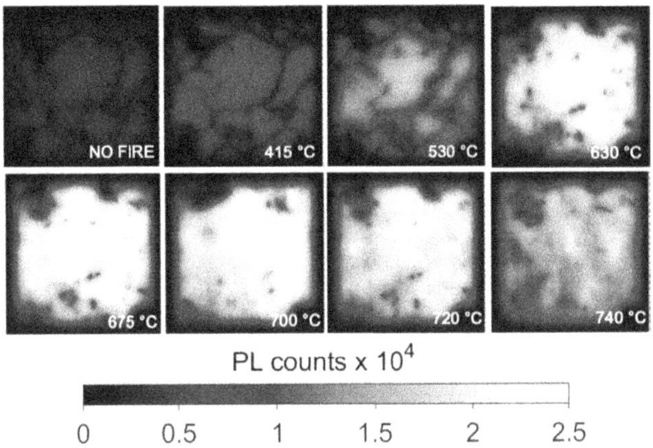

Figure 4.4 *PL images of mc-Si fired at different temperatures (unpublished) from the study of reference [54] presented at the 2017 IEEE Photovoltaic Specialists conference*

4.1.3 Dehydrogenation during extended annealing or high-temperature processes

Hydrogen passivation of bulk defects, and other defects, is reversible. This, for example, can explain the increased recombination activity after firing at 900 °C for thermally unstable SiN_X:H films [49]. Prolonged annealing at lower temperatures such as 700 °C for 3 minutes substantially reduced the effectiveness of passivation, particularly for such layers, in agreement with a study by Zhang *et al.* This is presumably due to the increased effusion of hydrogen out of the dielectric layer into the ambient, thereby reducing the concentration of hydrogen available for defect passivation in the bulk [49], and may also drive hydrogen to escape from the silicon bulk [56]. Figure 4.5 shows an example of increased recombination with extended annealing at 700 °C after an initial firing process, as indicated by the reduced iV_{OC} and hence negative ΔiV_{OC} values in the images.

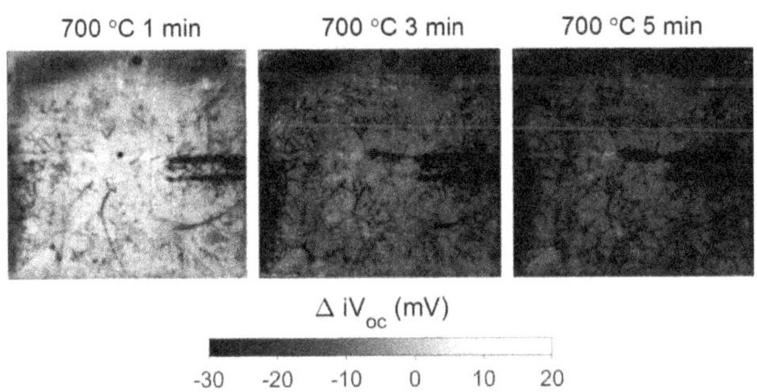

Figure 4.5 *Change in iV_{OC} (ΔiV_{OC}) induced by extended annealing at 700 °C after an initial firing process. Adapted from [57].*

The dehydrogenation of passivated defects is more clearly observed with the deliberate removal of the hydrogen source, which also reduces the impact of changes in surface passivation with additional thermal processes. For example, after firing, dielectric layers can be removed, followed by a thermal process. One study investigated the impact of the thermal process using in situ PL characterisation [49], determining that in the absence of a hydrogen source, passivated grain boundaries can become dehydrogenated, and therefore recombination active once more, for temperatures above 500 °C. In contrast, another approach to investigate the dehydrogenation is to perform ex situ PL images after the thermal process and subsequent deposition of a dielectric layer for surface passivation.

Estimated binding energies for grain boundaries are 2.5 eV [58]. Despite dislocation clusters being more difficult to passivate, they have a higher binding energy suggesting that the hydrogen passivation of dislocation clusters is more stable than that of grain boundaries with a binding energy of 3.1 eV [58]. However, studies have not investigated the impact of different grain boundary angles on kinetics of dehydrogenation in detail.

With the risk of dehydrogenation, cooling rates for solar cell processes should be considered to ensure the dielectric layers are not depleted of hydrogen (particularly for high-temperature processes) and ensure that effective bulk defect passivation is retained. Although for low annealing temperatures this may not be critical [49], for higher temperatures there is an increased risk that a slower cool down rate can lead to a further depletion of hydrogen from the dielectric layers, and therefore impact the concentration of hydrogen in the bulk of the silicon. This will be discussed in more detail in Chapters 5 and 6 for light-induced defects [59, 60].

Currently, a detailed study on the kinetics for passivation and dehydrogenation is required for a range of defects in silicon solar cells such as grain boundaries and dislocation clusters, particularly to investigate a range of dielectric and solar cell structures, along with the evolving silicon crystallisation technologies.

4.1.4 Development of cast-mono and high-performance multi-crystalline silicon material

In recent years, there have been significant efforts within the photovoltaics industry for the development of alternative crystallisation casting techniques to reduce the impact of structural defects including grain boundaries and dislocation clusters.

Modifications to the casting process, including temperature profiles in the crucible and seeding techniques, significantly alter the distribution of thermal stresses, solidification interfaces and the formation of defects [61]. The two dominant strategies for reducing harmful structural defects use opposing seeding techniques.

The first, in 2008 by Stoddard *et al.* at BP Solar, was developed using mono-crystalline seed wafers to predominately induce mono-crystalline silicon growth [62] and produce silicon material now known as 'cast-mono'. However, the cast-mono-crystalline silicon material also suffers from a large variability in crystalline quality throughout the ingot. Although the best parts of the ingot can be even better than Cz silicon material due to lower oxygen concentrations, the worst part of the

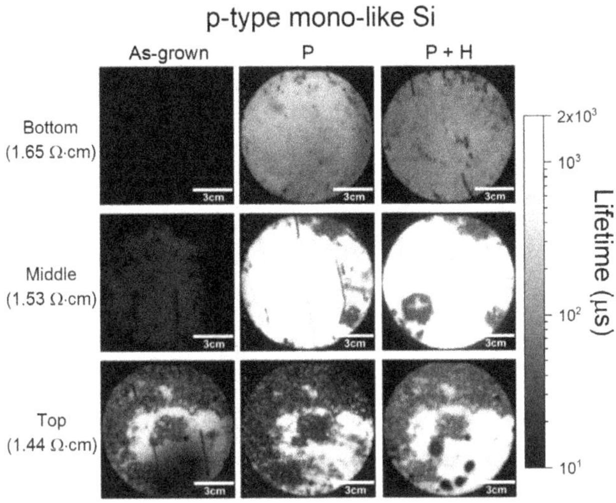

Figure 4.6 *Calibrated PL images of cast-mono silicon wafers at various heights in the ingot highlighting the impact of gettering and hydrogenation. Reproduced from [20] under Creative Commons CC-BY-NC-ND. © 2019 The Authors. Published by Elsevier B.V. https://doi.org/10.1016/j.solmat.2019.110059.*

ingot can end up being worse than mc-Si material due to high defect densities. As such, it has been challenging to achieve high yields due to the formation of multi-crystalline regions, small-angle grain boundaries and dislocations.

For example, dislocations can form due to lattice mismatch at the junctions between the seed crystals at the bottom of the ingot [62–64] or be nucleated by impurities. Due to thermal stresses, these structural defects propagate and multiply exponentially upwards and form clusters towards the surface that can be particularly electrically harmful [20, 64]. As shown in Figure 4.6, these can potentially cover large portions of a silicon wafer. The dislocations can also induce the growth of multi-crystalline material, particularly around the edge of the ingot [62, 65]. Multi-crystalline regions can also form at the ingot edges due to inhomogeneous nucleation on the crucible walls [66]. The presence of both mono- and multi-crystalline regions makes it challenging for uniformity of texturing and visual appearance, in addition to the changes in electronic quality.

Multi-crystalline regions can be minimised by intentionally introducing a functional grain-boundary near and parallel to each edge that grows vertically as the ingot crystallises, to act as a barrier to prevent the multi-crystalline regions at the edge penetrating any further into the ingot [63, 66].

Figure 4.6 shows an example of p-type cast-mono silicon material highlighting the impact of gettering and hydrogenation. Although high lifetimes can be achieved in regions of the ingot without dislocation clusters, wafers from the top of the

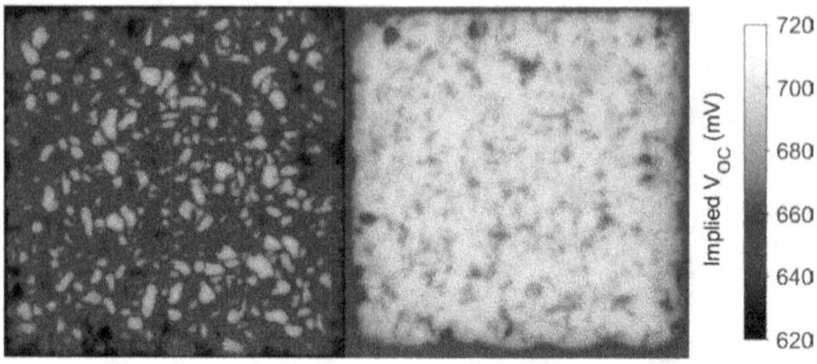

Figure 4.7 iV_{OC} *maps of p-type HPM silicon (left) before and (right) after a belt furnace firing process for hydrogenation. Adapted from [42].*

ingot with high dislocation clusters suffer from low lifetimes even after hydrogen passivation.

To address challenges with large potential dislocation densities in cast-mono material, 'high performance multi-crystalline' (HPM) material was developed by Lan *et al.* [19]. This nucleates crystal growth using a fine polycrystalline silicon seed or rough surface to enable homogeneous small grains to form, which minimise stresses, with a high density of grain boundaries and reduced dislocation density [19, 67]. In addition, seeds can be chosen to encourage grain boundaries that respond favourably to gettering and hydrogen passivation processes.

An example highlighting the small grain size, low dislocation cluster density and favourable response to hydrogen passivation is shown in Figure 4.7. This has substantially smaller grains than the conventional mc-Si wafer shown in Figure 4.1, as well as smaller and fewer dislocation clusters. The wafer responds quite favourably to the hydrogen passivation process, resulting in a relatively uniform quality of the silicon material.

The development of the HPM technology in 2012, along with challenges for cast-mono technology, greatly diminished the interest in the cast-mono technology between 2011 and 2014 [68, 69]. However, there has been renewed interest in the material, with record efficiencies of 24.4% for n-type tunnel-oxide passivated contact (TOPCon) solar cells by Jinko and p-type PERC solar cells of up to 22.8% by Canadian solar as reported in [11].

4.1.5 Benefits of an additional hydrogenation processes after firing

For defects difficult to passivate such as dislocation clusters, solar cells can potentially benefit from an additional hydrogen passivation process after metallisation firing. For example, Figure 4.8 (left) shows significant recombination activity observed in the vicinity of these structural defects on standard aluminium back-surface field (Al-BSF) solar cells directly after metallisation firing. The reduced IQE at a wavelength of 981 nm from LBIC measurements highlights the reduced collection of minority carriers throughout the bulk of the silicon due to a low lifetime in the vicinity of the dislocation clusters. However, the application of additional low-temperature (~300 °C) hydrogenation processes after the conventional co-firing

Ensuring effective passivation

To ensure effective passivation of structural defects, a comprehensive study should be performed for a range of thermal processes. When changes are made to the silicon material used, solar cell structure, fabrication process or dielectric layers, the impact on the passivation of structural defects, other bulk defects and surface should be evaluated. The optimisation of the dielectric layers for bulk defect passivation requires close attention to the optical properties of the film such as absorption, refractive index and reflection, particularly on the front surface, as well as electronic properties including surface passivation after the required thermal processing for contact formation. For solar cells with high-temperature fired contacts, films need to be firing-stable, and hence more likely to be nitrogen-rich. A sufficiently high temperature should be used to enable effective hydrogen release from the dielectric layer and rapid hydrogen diffusion and passivation. At the same time, care must be taken to avoid excessively high temperatures or long durations to avoid hydrogen depletion from the bulk and dielectric layers, and potential deterioration of surface passivation.

process can lead to substantial efficiency improvements of up to 1–2% absolute [70–72]. Figure 4.8 (right) shows the improvement in IQE around the heavily dislocated regions after an additional laser hydrogenation process. The corresponding I–V properties are shown in Table 4.1, highlighting the improvement in efficiency of 0.7% absolute from increases in the short circuit current density (J_{SC}), V_{OC} and fill factor (FF).

More recent work indicated that additional thermal hydrogenation treatments even without laser illumination can provide similar benefits [71]. The benefits have been shown to occur logarithmically over time with the most significant improvement occurring in the first second. This fast timescale indicates a passivation

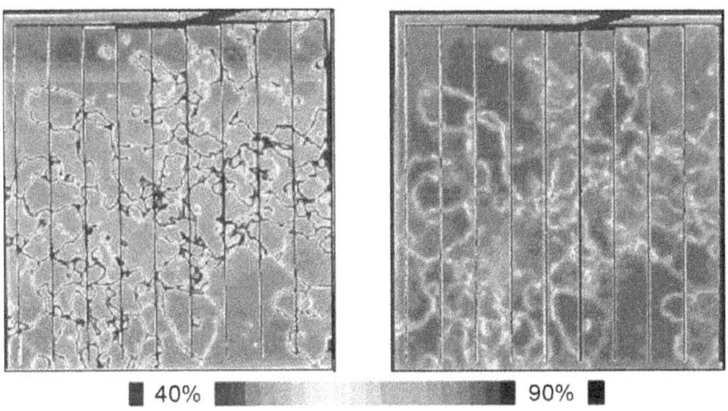

Figure 4.8 *IQE (981 nm) from a LBIC measurement of (left) a finished screen-printed solar cell, and (right) after advanced hydrogenation. Adapted from [70].*

Table 4.1 I–V properties of an Al-BSF solar cell fabricated on p-type cast-mono material before and after an advanced hydrogenation process (AHP)

Process	J_{SC} (mA/cm²)	V_{OC} (mV)	FF (%)	Eff (%)
Before advanced hydrogenation	37.0	607	76.8	17.2
After advanced hydrogenation	37.4	619	77.0	17.8

Data from [70].

reaction as a result of a change in charge state or chemical bonding arrangement of a species rather than a process dependent on diffusion of a species such as gettering [73]. The exact mechanism is not yet known, and the lower IQE still associated with the dislocation networks show that the recombination activity has not been entirely eliminated and will impact the solar cell efficiency. However, the benefits that are achieved from additional thermal hydrogenation processes highlight the weakness of conventional industrial implementations of hydrogen passivation and the importance of dedicated optimised hydrogenation processes for improving the passivation of structural defects and device performance.

4.2 Hydrogen passivation of impurity-related defects

Impurities incorporated during crystal growth can also lead to recombination active defects. For example, carbon, oxygen, nitrogen and a variety of transition metals are incorporated into the silicon lattice during crystal growth, and can result in recombination active defects [74, 75]. As a result, the type of crystallisation process used can greatly influence the concentration of impurities in the material and subsequently impact device performance. Hydrogen is believed to react with a number of impurities in silicon, allowing for a change in the electronic properties [76–80]. However, it is noted that not all hydrogen-impurity complexes are beneficial for silicon solar cells. This will be discussed in detail in Chapter 6.

Mc-Si generally has a high concentration of impurities, further compounded by structural defects such as grain boundaries and dislocation clusters, which act as efficient gettering sites for impurities [81]. For mc-Si, the 'red zone' around the edge of the cast has significantly higher impurity concentrations than in the middle of the ingot. Generally, due to the low lifetime in this material, this part of the ingot is recycled [19]. However, hydrogen passivation processes have resulted in significant changes in lifetime of such regions, as will be shown in Section 4.6.

For Cz silicon, the concentration of metallic impurities is generally much lower than that in mc-Si. This is due to the controlled cooling of the ingot as well as the absence of structural defects to act as gettering sites. As a result, carbon and oxygen are of primary concern on standard Cz silicon. However, with increasing bulk lifetimes and improved surface passivation qualities, even small concentrations of metallic impurities incorporated during crystal growth or during cell processing can significantly impact device performance [74].

Carbon generally takes on a substitutional site in the silicon lattice and is recombination inactive [82]. However, interstitial silicon can knock carbon out of the substitutional site, allowing carbon to form a number of recombination active defects (see [75] and references therein). Oxygen-related defects such as thermal donors, oxygen precipitates and boron–oxygen defects are of key concern, and have all been observed to interact with hydrogen [13, 15, 83]. A key complication for understanding the passivation of impurity-related defects such as these is that the defects can evolve throughout the solar cell process. For example, oxygen precipitates, as will be discussed in Section 4.3 for process-induced defects. Other impurity-related defects of key importance for solar cells that can be impacted by hydrogen include boron–oxygen defects, as will be discussed in detail in Chapter 5, and light- and elevated-temperature-induced degradation (LeTID) as will be discussed in Chapter 6. Impurities can also interact with vacancies in the silicon lattice [84], which can also interact with hydrogen [85].

Although Cz silicon generally has low impurity concentrations, upgraded metallurgical grade (UMG) silicon, with alternative purification processes to the conventional Siemens reactor purification process, can be subject to higher impurity concentrations. With reduced energy consumption for the purification process, this has the potential to reduce wafer fabrication cost. However, in significant concentrations, impurities can affect the ability to grow a Cz ingot, and eventually result in the nucleation of different crystal grains. The concentration of impurities depends heavily on the position in the ingot, with the concentration of most impurities with segregation coefficients larger than 1, as is the case for most impurities, being higher at the tail end of the ingot meaning that these wafers exhibit much lower lifetimes. The lower lifetimes mean that hydrogen passivation is even more critical in low-quality wafers stocks such as UMG silicon.

Figure 4.9 shows PL images of a UMG Cz wafer directly after PECVD deposition of hydrogenated silicon oxynitride ($SiON_X$:H), following a CHP involving a 400 °C anneal in a belt furnace [87]. Before hydrogenation, significant crystallographic slip line defects are observed, due to the higher impurity concentrations, resulting in a very low iV_{OC} of approximately 520 mV. The CHP treatment successfully passivates these slip line defects, leading to an enhancement in the average iV_{OC} of ~100 mV.

However, the response to hydrogenation is heavily dependent on the position in the ingot. Figure 4.10 (left) shows 1-sun iV_{OC} values of UMG Cz wafers directly after PECVD deposition of hydrogenated silicon oxynitride ($SiON_X$:H), following a 400 °C CHP, and following an AHP at 620 °C incorporating minority carrier injection. Wafers from the top of the ingot are the first to solidify during Cz crystal growth, and generally have a lower concentration of impurities. The starting iV_{OC} of the silicon material was ~610 mV and there were no visibly observable crystallographic defects. The application of the CHP does not lead to improvements in material quality, in fact, the iV_{OC} slightly reduced. This clearly highlights the care that must be taken when attempting to hydrogenate material, wherein there is a potential

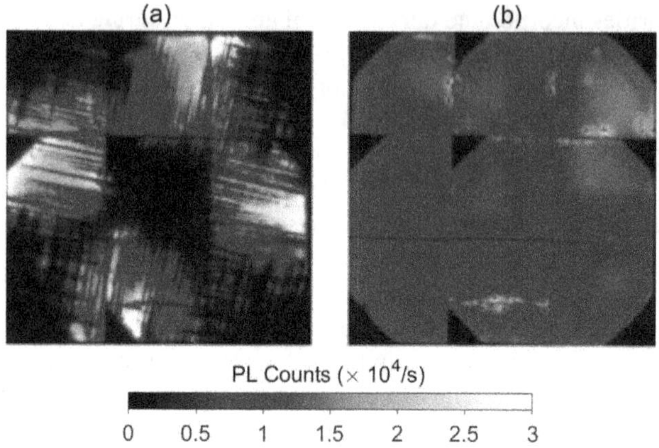

Figure 4.9 *PL images of a UMG Cz wafer with high impurity concentrations causing crystallographic defects (a) before and (b) after hydrogenation. © 2014 IEEE. Reprinted, with permission, from [86].*

to make material worse. A subsequent AHP treatment does, however, lead to significant improvements in iV_{OC} (~650 mV). This illustrates the importance of optimising hydrogen passivation processes for a given material. On the other hand, wafers at the bottom of the ingot with higher x_n values, which represents the fraction of solidified silicon, display a significantly different behaviour. Despite the very high impurity concentrations causing slip lines, a significant improvement is observed with a CHP, to result in this material being of the highest quality in the ingot. However, the material could be further enhanced through the use of AHP, leading to values of approximately 700 mV, an improvement of an astonishing 180 mV [86].

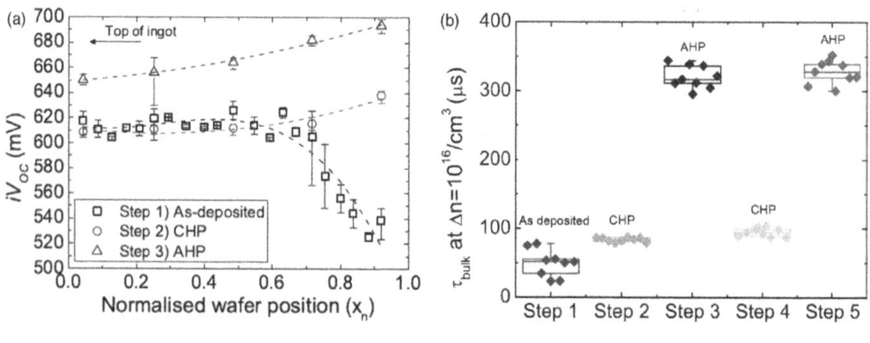

Figure 4.10 *(Left) iV_{OC} values of UMG Cz silicon wafer throughout an ingot directly after $SiON_x$:H deposition, after a CHP, and after a subsequent AHP, where x_n represents the fraction of solidified silicon. Image reproduced from [87]. (Right) Reversibility of the bulk lifetime with different hydrogenation processes. Adapted from [87].*

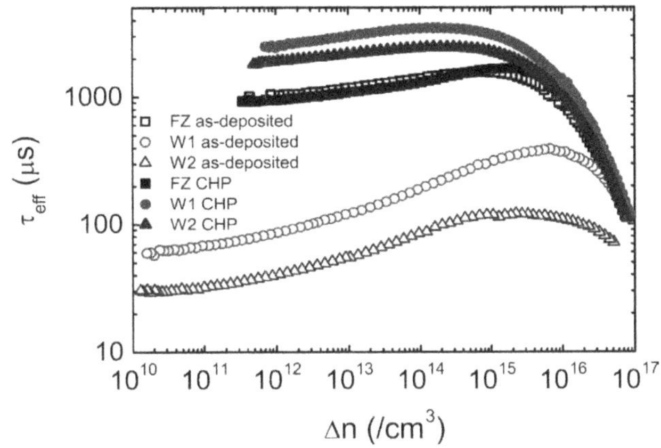

Figure 4.11 *Lifetime curves of two n-type Cz wafers (W1 and W2) with low initial lifetimes and a high-lifetime n-type FZ wafer before and after a CHP at 400 °C, resulting in multi-millisecond lifetimes for the n-type Cz wafers. Reprinted from [13], Copyright (2015), with permission from Elsevier.*

One hypothesis for this observation is that the increased concentration of n-type dopant impurities in this wafer may enhance the fractional change in hydrogen charge state. This would mean that the hydrogen atoms are more effectively able to passivate defects. Ultimately, the hydrogenation processes in the UMG wafer from the bottom of the ingot led to an increase in bulk lifetime from 8 μs to 500 μs, which would have a dramatic impact on the efficiency of a solar cell fabricated using such a wafer. The effectiveness of the hydrogen passivation on these materials with inherently very low lifetimes could have implications for ingot utilisation, as these low-quality wafers could be substantially improved by processes involved in cell manufacturing [86, 87]. A key consideration for using this material, however, is that hydrogen passivation is reversible. As shown in Figure 4.10 (right), the bulk lifetime can toggle between different states depending on which process was performed [86].

Importantly, n-type Cz wafers can also benefit from hydrogen passivation. Recent work has demonstrated that 'reject' commercial-grade n-type Cz wafers with initial bulk lifetimes of only 90 μs can benefit significantly from hydrogen passivation processes [13]. Figure 4.11 shows the lifetime of two reject n-type Cz wafers (W1 and W2) with lifetimes of less than 100 μs directly after the deposition of PECVD $SiON_X$:H, in contrast to that of a high-lifetime n-type FZ wafer with a lifetime of ~1 ms. However, after CHP at 400 °C in a belt furnace, substantial improvements in the lifetime were observed, to result in lifetime of more than 3 ms for the n-type Cz wafers, a value now suitable for solar cell manufacturing. In contrast, no change was observed for the FZ wafer [13]. However, in contrast to p-type Cz wafers, no further increases in lifetime were observed with an AHP at higher temperature processes incorporating minority carrier injection. This highlights that hydrogenation processes may need to be tailored for the material [87].

> **Never simply reject a low-lifetime wafer**
>
> Assessments of a material's suitability for solar cell manufacturing should not be purely based on the initial lifetime of the material. It should consider the impact of the natural incorporation of defect-engineering processes throughout solar cell fabrication such as gettering and hydrogenation, or what modifications can be made to the solar cell production sequence to enhance the lifetime of the material to a quality suitable for solar cell fabrication. Even 'reject' n-type Cz wafers with initial bulk lifetimes of less than 100 μs are capable of multi-millisecond bulk lifetimes.

In addition, both n-type and p-type FZ silicon wafers are susceptible to impurity-related defects [75], and the recombination activity can be modulated by hydrogen passivation processes. Nitrogen is often incorporated into FZ to improve mechanical strength in the absence of oxygen and can form recombination active defects such as vacancy-nitrogen complexes [88]. A study by Rougieux *et al.* demonstrated a significant enhancement on the lifetime of the material and a much larger improvement in lifetime for samples passivated by hydrogen-rich PECVD SiN_X:H than ALD AL_2O_3 [89]. More recent work has identified that nitrogen-vacancy defects can induce dangling bonds, which can be passivated by hydrogen [90].

A key challenge for understanding the passivation of impurity-related defects is that, often in single-crystal material, they may be approximately uniformly distributed throughout a wafer, and hence PL imaging does not highlight spatial variations with a clear passivation of a given defect, as is the case for the passivation of clearly defined structural defects such as grain boundaries of dislocation clusters. As such, a clear understanding may rely on techniques such as quasi-steady-state photo conductance (QSS-PC) to fingerprint defects of known capture cross section, or techniques such as FTIR or deel-level transient spectroscopy (DLTS). However, given the general absence of structural defects in mono-crystalline silicon material, significant changes in the bulk lifetime that occur uniformly throughout the material are likely due to impurity-related defects. Another potential challenge in using DLTS is that the fabrication process to form test structures for DLTS measurements can potentially inject hydrogen into the wafer and may unintentionally hydrogen passivate the defects to start with [91, 92].

In addition, the processes used for hydrogen passivation are performed at elevated temperatures. This creates the possibility for thermal reactions to take place, separately from hydrogen passivation. Often, hydrogen passivation may only be inferred from changes in the lifetime of a material. This, for example, has complicated the understanding of the passivation of intersitial iron (Fe_i). Mullins *et al.* recently studied the passivation of iron using non-equilibrium conditions at temperatures below 150 °C, leading to changes in the DLTS spectra. However, the passivation appeared unstable, with the DLTS signal for Fe_i-H disappearing after a 30 minute anneal at 175 °C in the absence of an applied bias [76]. With this instability, it is unlikely that the hydrogen passivation of iron plays any significant role for silicon solar cells. More recent work has demonstrated the ability to getter iron into dielectric layers such as SiN_X:H with prolonged annealing at higher temperatures [93] and suggested that this is the likely mechanisms for a reduction in the interstitial iron concentration rather than hydrogen passivation reported in previous studies [94].

4.2.1 Passivation of dopants

Dopant atoms, although essential for solar cell operation, are also essentially impurities incorporated into the silicon lattice and can interact with hydrogen. Early work by Sah *et al.* in 1983 identified the role of hydrogen in passivating boron atoms [95] and a number of studies have investigated hydrogen-dopant interactions for both donors (e.g. phosphorus) and acceptors (e.g. boron, aluminium and gallium) [95–100]. The ability for hydrogen to passivate dopants creates the possibility to change the effective conductivity of the material, through the formation of neutral hydrogen–dopant complexes. This is also the case for hydrogen passivation of oxygen-related thermal donors [83].

At low temperatures in particular, hydrogen passivation of acceptor atoms such as boron can be very effective, whereby more than 99% of acceptors can be passivated near the silicon surface provided sufficient hydrogen [101]. This is due to trapping effects, whereby hydrogen naturally assumes the H^+ charge state and is attracted to the ionised boron dopants (B^-). For example, work by Sicotte *et al.* using a shielded hydrogen plasma at 200 °C to passivate dopants in the vicinity of the silicon surface, as measured using the electro-capacitance voltage (ECV) technique for the electronically active boron dopant concentration (see Figure 4.12 (left)).

As briefly mentioned in Chapter 2, the ability of hydrogen to passivate dopants has been exploited to form selective emitters for n-type solar cells [103]. The work by Shumate *et al.* used a heavy boron diffusion over the whole surface of the solar cell. After metallisation, hydrogen was introduced by means of a hydrogen plasma for 30 minutes to locally passivate boron atoms in the non-metallised regions [103]. This resulted in increases in the sheet resistance in the non-metallised regions, as determined by four-point probe measurements (see Figure 4.12 (right)), with more than a five-fold increase in sheet resistance for a processing temperature of 150 °C. For higher temperatures, however, the passivation was less effective. In addition, subsequent annealing at temperatures higher than 150 °C led to increases in the sheet resistance, even for an annealing temperature of 175 °C. In addition, an anneal at 350 °C for 1 hour reverted the sheet resistance back to the value prior to hydrogenation, suggesting the complete splitting of the hydrogen–boron (HB) pairs. This splitting would be of significant concern for silicon solar cells if relying on the passivation of dopant atoms due to the expected changes in electrical performance.

In general, hydrogen-dopant pairs have a significantly lower activation energy for dissociation than that for structural defects passivated by hydrogen, and hence are less thermally stable [58]. As shown in Figure 4.13, work by Stavola *et al.* demonstrated that a 5 minute anneal at 175 °C could dissociate almost 50% of B–H

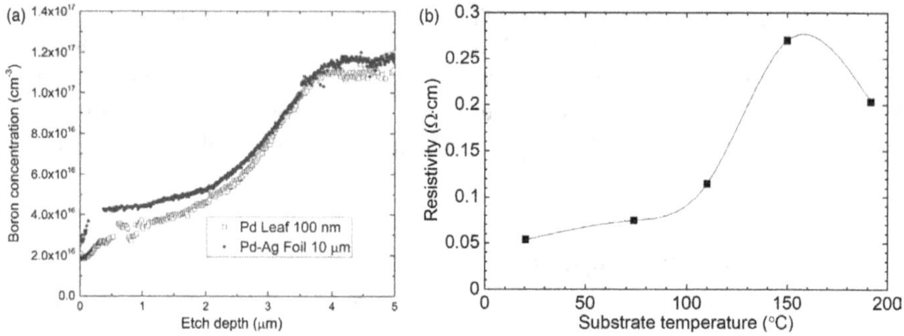

Figure 4.12 *(Left) ECV measurement highlighting the passivation of boron dopants near the silicon surface after a shielded hydrogen passivation process. Reproduced from [102] with permission from John Wiley and Sons. © 2017 WILEY-VCH Verlag GmbH & Co. KGaA, Weinheim. (Right) Change in sheet resistance with temperature induced by a hydrogen plasma for the formation of a selective emitter. © 2012 IEEE. Reprinted, with permission, from [103].*

pairs [104], which is a similar condition as that used for the module encapsulation process. However, minority carrier injection, and therefore illumination, can greatly accelerate the dissociation of HB pairs [100, 105, 106], likely due to the increased fractional concentration of H^0 and H^- to assist in the migration of hydrogen away from the dopant atoms. As a result, the operating conditions of solar cells in the field naturally tend towards dissociating HB pairs, and hence would create an unstable

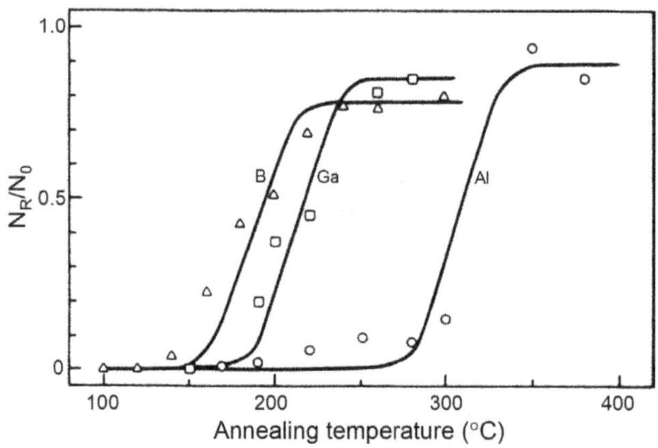

Figure 4.13 *The recovered fraction N_R/N_0 for the dissociation of hydrogen-acceptor pairs with thermal annealing for 5 minutes. Reprinted with permission from [104]. Copyright (1988) by the American Physical Society. http://dx.doi.org/10.1103/PhysRevB.37.8313.*

scenario for a solar cell relying on the passivation of dopant atoms to form a selective emitter.

4.3 Hydrogenation of process-induced defects

Process-induced defects add an extra level of complexity for hydrogen passivation in silicon solar cells compared to the structural defects that are incorporated into the silicon ingot during crystal growth. This is due to the evolution and/or formation of the defects during the cell fabrication process. It should be noted, however, that process-induced defects may in fact be caused by an evolution of structural defects or impurities incorporated into the silicon material during crystal growth, or lead to the formation of structural defects. A key example of this is oxygen-related defects such as oxygen precipitates that can be greatly impacted by thermal processing, as will be discussed in this section, and boron–oxygen-related light-induced degradation (LID) that will be discussed in detail in Chapter 5. On the other hand, the defects may be purely introduced during processing, such as laser-induced defects during processes including laser ablation as will be discussed in this section, or with laser doping as will be discussed in Chapter 8.

Although hydrogen passivation of process-induced defects is incredibly important and can increase the efficiency of silicon solar cells by more than 0.8% absolute [107], the involvement of hydrogen may often be overlooked in both industrial and laboratory solar cells. In many instances, hydrogen passivation naturally occurs after the introduction/evolution of the defects. However, it is possible that the defects have not necessarily been introduced by the process step where hydrogen passivation is typically performed such as with the use of laser ablation and low-temperature metallisation processes. As a result, the fabrication sequence may need to be modified to enable effective passivation of such defects.

4.3.1 Oxygen precipitates

A key process-induced defect that can be significantly modulated during cell fabrication is related to oxygen impurities. As such, this is of primary concern for material with high oxygen content such as Cz silicon. During the Cz silicon wafer growth process, oxygen is incorporated into the wafers from the crucible. The parameters of the Cz growth process may allow for the formation of interstitial and vacancy defects that can facilitate the formation of oxygen precipitates. Due to the rotation of the ingot during the ingot pulling process, such defects in Cz silicon are typically radially symmetric, forming ring-like structures. Such radial defects can cause significant bulk recombination that limits material quality and subsequent cell performance. With recombination activity varying across the wafer, PL imaging is a valuable technique to assess the recombination activity of such defects.

Although there are many intricacies involved in the formation of oxygen precipitates, it is well known that high-temperature processes, such as emitter formation or thermal oxidation, can cause the growth of these detrimental radial defects [12, 108, 109]. As such, wafers susceptible to oxygen ring formation may lead to significant losses on a cell level of up to 4%$_{abs}$ [110, 111]. This can significantly affect the fraction of wafers that can be utilised from an ingot with intensive high-temperature processes.

One proven approach to reduce the impact of oxygen rings is a high-temperature anneal or 'tabula rasa' process. This is conducted at temperatures of ~1 000 °C, which causes the grown-in oxygen precipitate nuclei below a critical size to dissolve [12, 112]. For this method, it is critical that the concentration of pre-existing oxygen precipitates that are sufficiently large to be stable during the tabula rasa process is low, and that the growth of oxygen precipitate nuclei during the ramp-up, is limited to avoid a further growth at elevated temperature. As such, the tabula rasa is applied before any other thermal processes such as emitter diffusion or thermal oxidation. However, the addition of a high-temperature process is typically too costly for industrial solar cell manufacturers.

Given that oxygen precipitates are typically formed prior to the deposition of hydrogenated dielectric layers during high-temperature emitter formation and thermal oxidation, when cell technologies use a subsequent firing process, hydrogen passivation can potentially be used to reduce the impact of those defects. A number of publications have indicated a reduction in the recombination activity of oxygen precipitates during processes typically used for hydrogen. For example, studies by Haunschild *et al.* demonstrated a significant reduction in the recombination activity of oxygen precipitates after a high-temperature firing process compared to that directly after the emitter diffusion process [110, 113]. Furthermore, the process was not effective at completely eliminating recombination activity, still causing a significant reduction in efficiency of almost 4%$_{abs}$ compared to cells fabricated on wafers that were not susceptible to the formation of oxygen precipitates. Although not explicitly discussed in those papers, the reduction in recombination activity was potentially due to hydrogenation [13].

Other work has demonstrated the complete passivation of oxygen precipitate defects during a high-temperature firing process [13]. Figure 4.14 displays PL images from [13] with a boron-doped p-type Cz wafer at three stages of processing: (a) directly after PECVD of SiN_x:H; (b) after the high-temperature hydrogenation process; and (c) after the hydrogenated wafer was used to fabricate a laser-doped selective emitter solar cell with a low-temperature metallisation process. In Figure 4.14a, the presence of oxygen precipitate ring defects is clearly shown. This type of circular pattern was induced during the high-temperature phosphorus emitter diffusion and thermal oxidation processes and was present in about 10% of the wafers in the studied batch. These defects, which cause significant reductions in the iV_{OC}, are attributed to oxygen precipitates, as any thermal donors present in the initial material, which can also present as ring-like structures, would have been annealed out during the high-temperature diffusion and oxidation processes [110], hence not visible after dielectric deposition. The significant variation in recombination activity between different wafers with the identical process represents a significant concern for solar cell manufacturers due to reduced yields or the mismatch it can cause for cells of different current–voltage properties when incorporated into a module.

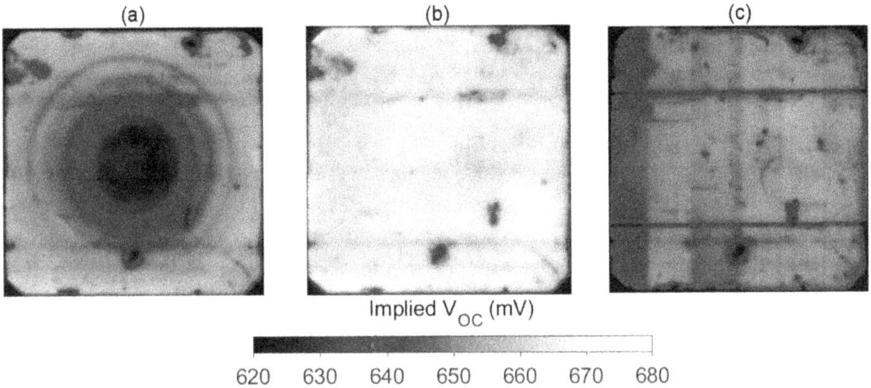

Figure 4.14 *Calibrated 1-sun implied V_{OC} mapping for p-type boron-doped silicon wafers exhibiting oxygen precipitates (a) directly after PECVD deposition of SiN_x:H, (b) after an AHP at 620 °C, (c) on a finished laser-doped selective emitter solar cell made with the same wafer. Adapted from [13].*

This wafer was then subjected to high-temperature hydrogenation process in a belt firing furnace with a peak temperature of 620 °C for 1–2 s. It is noted that this temperature is likely lower than that used for the study by Haunschild *et al.* with screen-printed contacts. This firing process used a modified power distribution to the furnace lamps by turning off the bottom lamps, thereby increasing the illumination intensity in the visible spectrum within the belt furnace, as well as enabling illumination during the cool-down process to provide minority carrier injection for hydrogen charge state control [13]. This represented an early implementation of the advanced hydrogenation technology. As shown in Figure 4.14 (b), after firing, the undesirable recombination caused by these oxygen defects was completely removed and the wafer quality was significantly improved.

The wafer was then formed into a solar cell with a laser-doped selective emitter, with an identical AHP used after laser doping to passivated laser-induced defects, as will be discussed in Chapter 8, followed by a low-temperature metallisation process with Ni/Cu/Ag plating for the n-type contact. The iV_{OC} map for this cell is displayed in Figure 4.14 (c). There is no evidence of the oxygen precipitates in the full cell, indicating that the hydrogen passivation is able to remove the oxygen precipitates and that this passivation is stable after subsequent cell processes using this laser doping and metallisation approach to cell formation.

However, no comparison was made with a more conventional firing process, or an identical thermal process in the absence of a hydrogenated dielectric layer to rule out potential thermal effects. On the other hand, given that the temperature of the hydrogenation process was significantly lower than ~1 000 °C as is used for the tabula rasa process, it is highly likely that the precipitates were passivated by hydrogen rather than being dissolved.

Subsequent work has demonstrated the passivation of oxygen rings at lower temperatures, providing further evidence for the involvement of hydrogen. For example, a study by Manshanden *et al.* used a NH_3 plasma process at 375 °C prior to dielectric deposition, demonstrating a significant reduction of recombination activity for samples passivated with low-temperature a-Si:H [114]. This impact was less pronounced for SiN_X:H passivated samples, suggesting the SiN_X:H deposition was also effective for passivating the oxygen precipitates. Annealing of the SiN_X:H passivated samples also resulted in the passivation of oxygen precipitates, particularly in the range of 350 °C to 450 °C. However, the passivation was less effective for the highest temperature of 600 °C, potentially due to a depletion of hydrogen from the dielectric layers. For samples annealed at high temperatures, passivation could also be recovered using a subsequent NH_3 plasma.

A key observation in the study of Manshanden *et al.* was that complete passivation could not be achieved in samples with a strong presence of rings [114], similar to only a partial reduction in recombination activity observed by Haunschild *et al.* [110]. This has a striking similarity for the challenges in passivating regions of mc-Si with high dislocation cluster densities. It is possible that for regions of high precipitate defect concentrations, there is insufficient hydrogen available in the vicinity to passivate all defects, essentially leading to a localised depletion of hydrogen available for defect passivation.

Another similarity to structural defects such as grain boundaries and dislocation clusters is the impact of impurity decoration on the passivation of the oxygen precipitates. Phosphorus gettering has been observed to reduce the recombination activity of precipitates by avoiding decoration by metallic impurities [115]. Using deep level transient spectroscopy, a low-temperature hydrogenation process has been observed to reduce the recombination activity of oxygen precipitates, and passivation was more effective in gettered samples [116]. However, care should be taken when applying gettering processes considering the potential growth of oxygen precipitates during high-temperature processes. This will become particularly important for n-type technologies such as n-type passivate emitter and rear, totally diffused (PERT) and TOPCon solar cells, which require multiple high-temperature processes, and the diffusion processes used may not necessarily be effective for gettering.

In order to push towards high efficiency, next-generation silicon solar cells, manufacturers are beginning to shift their focus towards n-type Cz silicon wafers. This is because unlike p-type wafers doped with boron, n-type silicon is not susceptible to LID caused by boron–oxygen complexes, as will be discussed at length in Chapter 5.

However, n-type Cz wafers are still susceptible to bulk impurities that can limit the performance, including oxygen precipitates. In fact, n-type technologies such as n-PERT and TOPCon are particularly susceptible to the formation of oxygen precipitates, as they require a boron diffusion in the process flow, which is performed at higher temperatures and for longer durations than typical phosphorous diffusions used in p-type technologies such as PERC, as well as a phosphorus diffusion process.

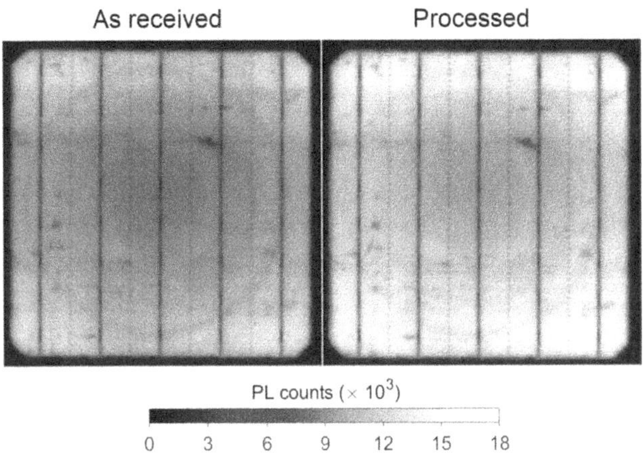

Figure 4.15 *Open circuit PL images of industrial n-PERT solar cells (left) before and (right) after the 20 s AHP treatment*

Given that both technologies use SiN_x:H and a high-temperature firing process for metallisation, a certain level of hydrogen passivation of oxygen precipitates would be expected during the firing process. Figure 4.15 (left) shows a PL image of an industrial n-PERT solar cell directly after cell fabrication, with weak recombination activity due to oxygen precipitates. As the cell had been through a high-temperature firing process, it is likely that these radial defects are related to oxygen precipitates rather than thermal donors, and that some level of passivation occurred during the firing process. Figure 4.15 (right) shows the PL image of the corresponding after an AHP at 200 °C for 20 s with a laser illumination of ~22 kW/m^2. This displays a significant increase in overall PL counts, likely due to improved surface passivation and/or a uniformly distributed impurity-related bulk defect. Lifetime samples processed in parallel suggest an improvement in both surface passivation as well as bulk lifetime. In addition, the cell shows a reduction in the presence of the radial defects. This process led to a significant increase in cell efficiency on the order of 0.3%$_{abs}$ (see Table 4.2). Another study on n-PERT solar cells demonstrated efficiency enhancements of up to 1%$_{abs}$ using an annealing process at 200 °C, although the improvement greatly depended on the ingot and position within the ingot [117].

Table 4.2 *Influence of hydrogenation process on the current–voltage characteristics of industrial n-PERT silicon solar cells*

Stage	V_{OC} (mV)	J_{SC} (mA/cm^2)	FF (%)	Efficiency (%)
Before AHP	655.9	39.74	78.84	20.55
After AHP	660.5	39.76	79.54	20.88
Change	+4.6	+0.02	+0.70	+0.33

These results indicate that hydrogen passivation, applied either on a wafer level or after cell processing, could be a potential approach to reduce the impact of oxygen precipitates on either p-type or n-type silicon wafers. However, more systematic studies are required to understand the direct impact of hydrogen treatments, either via standard high-temperature firing approaches or more advanced techniques such as high-intensity illumination or bias annealing. In particular, it is important to systematically study the impact of the type of hydrogen-containing dielectric layers on the ability to hydrogenate oxygen precipitates and clearly separate thermal effects.

4.3.2 Laser-induced defects

Another key category of process-induced defects for silicon solar cells is those introduced by laser processing. For example, laser ablation, as is used for contact definition of the p-type contact in PERC solar cells, in a process termed laser contact opening (LCO). The LCO process is performed directly after dielectric deposition to ablate the rear dielectric stack (typically AlO_X:H/SiN_X:H) to expose the underlying silicon. Subsequently, an aluminium paste can be printed over the rear surface area, covering both exposed LCO regions as well as dielectric-passivated regions. This, for example, may cover the entire rear surface of the solar cell. The paste used is non-fire through such that it does not penetrate through the rear surface dielectric stack and only makes metal–silicon interface contact at the LCO sites, and forming a local Al-BSF region through the alloying of aluminium and silicon above the eutectic temperature of 577 °C.

However, newer solar cells feature a bifacial structure. The rear contact of bifacial PERC solar cells can also feature LCO openings, where small LCO regions are defined (e.g. lines or point contacts 40-μm wide), followed by an aligned screen-printing (SP) to fully cover the LCO regions and form fingers connecting the LCO regions (e.g. 100-μm wide) also using a non-fire-through paste. Alternatively, an LCO process can be avoided if instead using a fire-through paste. However, this process leads to larger metal–silicon interface areas, defined by the printed area, and lower voltages.

To reduce the generation of laser-induced defects during LCO, ideally a picosecond UV laser (e.g. 355 nm) should be used to be primarily absorbed in the dielectric layer and have minimal absorption in the silicon, thereby causing minimal damage in the silicon, compared to using longer wavelengths (532 nm or 1064 nm) [118] and/or longer pulse lengths such as in the nanosecond range [119].

However, there are two key features of the PERC process that allow a relaxation of the laser parameters for the LCO opening on the rear surface of PERC. First, the formation of the local Al-BSF region during the high-temperature firing process melts the silicon around the LCO region as the heavily doped p^+ region is formed, which may be in the range of 5 μm to 10 μm deep, followed by the epitaxial growth of the silicon as it recrystallises. This essentially consumes the laser-induced damage. Second, during the high-temperature firing process, the release of hydrogen from dielectric layers could enable passivation of the remaining laser-induced defects [32]. As a result, cheaper lasers with relaxed processing conditions can be used for PERC without significantly impacting performance such as 532 nm lasers [17] with nanosecond pulse lengths.

On the other hand, for other metallisation schemes that do not enable a local melting/recrystallisation of damaged regions, the laser parameters are more critical, and it is preferred to use a low-damage laser process such as picosecond UV laser ablation. However, hydrogen passivation can also lead to efficiency increases through the passivation of laser-induced defects. An example of this is for solar cells with plated contacts, whereby laser ablation is used to define the contact openings on the emitter of solar cells with a homogeneous emitter. If the laser ablation is performed prior to a high-temperature firing step, there is scope for hydrogen passivation of the laser-induced defects to naturally occur during the firing step [120].

For low-temperature metallisation approaches such as those involving plated contacts, a sintering process used to form a copper diffusion barrier and improved contact resistance, can also potentially passivate laser-induced defects introduced during the laser-ablation process, leading to lower a dark saturation current density (J_0) and higher V_{OC} [121, 122]. For example, in the work by Tous *et al.* at IMEC, the low-temperature sintering process of 250 °C improved the V_{OC} by approximately 7 mV [122]. Based on the work of Hallam *et al.* [87, 107], a subsequent study by Urueña *et al.* investigated simplified hydrogenation processes based on a short belt furnace anneal (BFA) in the range of 400–500 °C for n-PERT solar cells with laser-ablated- and laser-doped n-type contacts [123]. This was compared to a more CHP based on forming gas annealing for 30 minutes. It was observed that the J_0 of the passivated phosphorus front surface field in both cases reduced after annealing, with the BFA slightly outperforming the FGA approach, despite not requiring an additional hydrogen source, and being much shorter. Of key importance, however, were the substantial reductions in the J_0 of the exposed laser-process regions, with a twofold reduction in the J_0 of samples with laser ablated contact openings from 22 800 fA/cm^2 to 12 100 fA/cm^2 (actual area), and a sixfold reduction in the J_0 of the laser-doped openings from 12 700 fA/cm^2 to 2 600 fA/cm^2. The annealing processes also slightly reduced the J_0 of the passivated boron emitter on the rear of the device, highlighting the multiple aspects of the device all improved through the addition of targeted hydrogen passivation processes. The simple BFA process was used to enable the fabrication of n-PERT solar cells with laser-ablated n-type contacts with an efficiency of almost 21%, and 22.5% for laser-doped n-type contacts [123].

The impact of processing on the passivation of laser-induced defects introduced during laser doping will be discussed in detail in Chapter 8.

4.4 Hydrogenation of light- and carrier-induced defects

Light-induced defects, or more correctly carrier-induced defects, are perhaps the most complicated category of defects to hydrogen passivate. The key challenge for such defects is that by the end of the solar cell fabrication the defects have not typically formed. Rather, the formation, or evolution into a recombination active defect occurs during operation in the field. As a result, the defects are not available for passivation during the conventional passivation process in the solar cell production sequence. This presents a significant warranty concern for solar panel manufacturers, whereby the performance of panels can degrade during operation.

Such defects are often due to impurities, and hence, can also be considered as impurity-related defects. For example, interstitial iron is a light-activated defect, due to the splitting of iron-boron pairs. Another key example is the boron–oxygen defect, further complicated in that thermal processing can alter the kinetics of the defect system, and hence, in some capacity, also process-induced. The hydrogen passivation of boron–oxygen defects will be discussed in detail in Chapter 5.

Another key carrier-induced defect caused by hydrogen and heavily impacted by processing conditions is called LeTID. This will be discussed in detail in Chapter 6.

4.5 Complementary nature of gettering and hydrogenation for silicon solar cells

Although hydrogenation is extremely important for improving the bulk lifetime of silicon through the neutralisation of recombination active defects, hydrogenation alone should not be considered as the silver-bullet solution. As discussed in Section 4.1, the decoration of structural defects with impurities can affect the response to hydrogen passivation, leaving these defects more prone to remaining recombination active in the finished device. As a result, hydrogenation should be performed in addition to other defect-engineering approaches to maximise the lifetime of silicon and even improve the effectiveness of hydrogen passivation.

A key defect engineering method that is quite complementary to hydrogen passivation is gettering. Gettering can be used to physically remove impurities from the bulk of the silicon and is particularly effective for removing metallic impurities with high diffusivities such as iron [10]. Hydrogen passivation can subsequently be used to passivate remaining defects in the material. The physical removal of impurities from the silicon also reduces the potential of forming hydrogen-defect complexes with energy levels within the bandgap, or a long-term dehydrogenation of hydrogen-defect complexes. The combined approach using both gettering and hydrogenation is especially important for cast silicon, which usually has higher concentrations of impurities and structural defects.

Where possible, defect engineering processes should be integrated into the solar cell fabrication sequence without adding processing steps to reduce fabrication cost. Therefore, the processes should ideally simultaneously enable defect engineering and the formation of the solar cell structure as typically occurs in the fabrication of industrial p-type Al-BSF and PERC solar cells (to be discussed in Section 4.5).

Predominantly, high-efficiency solar cell architectures such as silicon heterojunction (SHJ) and TOPCon solar cells rely on the use of silicon wafers with high initial bulk lifetimes. This essentially limits such devices to using high-quality n-type Cz silicon wafers, which are currently approximately 10% higher than that of p-type Cz wafers [124]. This is particularly for the case of heterojunction solar cells, where minimal processing is performed on wafers, with all processes typically performed at temperatures below 250 °C to avoid a potential 'poisoning' of the bulk silicon material throughout the fabrication [125, 126] or deterioration of surface passivation, rather than deliberately using defect engineering methods such as gettering and hydrogenation. On the other hand, TOPCon solar cells naturally benefit from gettering and hydrogenation.

Incorporating defect engineering into the fabrication of high-efficiency silicon solar cells should be seen as an opportunity to improve the electronic quality of silicon material throughout the fabrication sequence, hence reducing the impact of bulk recombination. Although this may be obvious for p-type solar cells fabricated on 'low-quality' substrates, typically with initial lifetimes below 50 μs, it should also be considered for 'high-quality' n-type silicon wafers with bulk lifetimes >1 ms. Doing so could be considered as a pathway to:

1. improve solar cell efficiencies
2. reduce cost through either an improvement in cell efficiency, relaxation of the initial silicon material quality requirements or improved ingot utilisation.

This will be discussed in further detail in Section 4.6.

4.5.1 Impact of gettering and hydrogenation on lifetime

A risk of any defect engineering process is that almost anything could happen to the lifetime of the material, ranging from a desirable improvement in lifetime, no change at all, which is undesirable considering the additional processing cost, or even more undesirably, a reduction in lifetime. In addition to gettering- and hydrogenation-related reactions resulting in changes in minority carrier lifetime, the processes are performed at elevated temperatures, raising the possibility of other thermally activated reactions and mechanisms to take place. This is particularly the case in solar-grade silicon, whereby there is not just a single defect/impurity present in the material, but a whole range of defects and impurities, which vary in composition and concentration within a given wafer, throughout ingots, and between different ingots. Some specific defects, impurities or even specific regions of a given wafer may benefit from a given defect engineering process at the detriment of increased recombination for other defects and recombination mechanisms.

A number of studies have shown negative impacts of hydrogen passivation or gettering when such processes are performed in isolation, at least in isolated regions of wafers. For example, gettering processes can activate recombination at grain boundaries [41, 49], while leading to improvements in the lifetime of intra-grain regions [41]. The response to gettering and hydrogenation can also vary throughout the ingot. In multiple studies, a significant increase in recombination activity in grain boundaries can occur in mc-Si material and dislocation clusters in cast-mono silicon material after high-temperature diffusions in the middle of the ingot [20, 127]. Although grain boundaries could be passivated during the subsequent hydrogenation process, the hydrogenation process was less effective for the passivation of dislocation clusters [20]. In addition, the background lifetime improvements in the

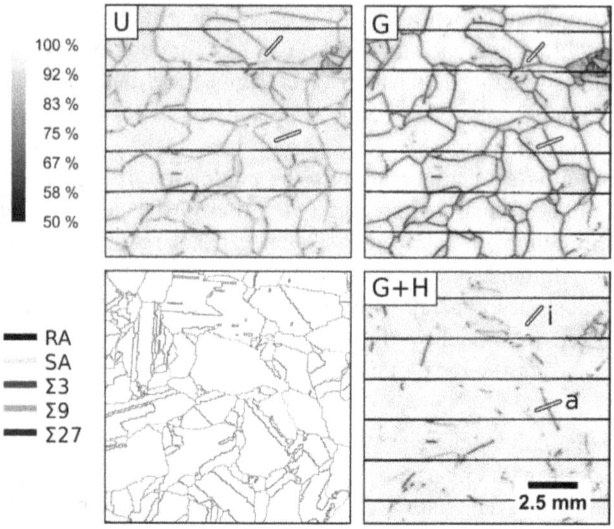

Figure 4.16 *IQE maps from spatially resolved LBIC measurements for mc-Si. Reprinted from [127], with the permission of AIP Publishing. © 2018 Author(s).*

intra-grain or dislocation-free regions after diffusion processes also depended on ingot heights. Gettering is particularly helpful for the bottom of the ingot with high impurity concentrations [20, 127].

In general, the best results of minority carrier lifetime are obtained with the application of both gettering and hydrogenation [20, 44, 47, 127]. In addition, with the impact of impurity decoration on the passivation of structural defects, hydrogen passivation is more effective on gettered grain boundaries [24], providing a further benefit of using both gettering and hydrogen passivation. Figure 4.16 shows LBIC and EBSD measurements for the response of different grain boundaries to gettering and hydrogen, highlighting an increase in recombination activity after gettering, however, a significant reduction after a subsequent hydrogen passivation process, with the passivation of most grain boundaries.

Figure 4.17 shows iV_{oc} maps that highlight the impact of gettering and hydrogenation alone as well as combined, for p-type mc-Si. It is noted that PL images were taken with the PECVD SiN_x:H still present, and hence, changes in the surface passivation are expected after the high-temperature firing process for hydrogenation. However, most importantly, the images clearly highlight the spatial variation in the changes in material quality and therefore iV_{oc}-induced by gettering and hydrogenation. The control group that did not receive any defect-engineering processes shows recombination activity in ground boundary regions, as well as significant recombination due to saw damage, indicated by horizontal marks in the PL image. The application hydrogenation alone leads to a slight improvement in the intra grain regions (however, also impacted by changes in surface passivation). However,

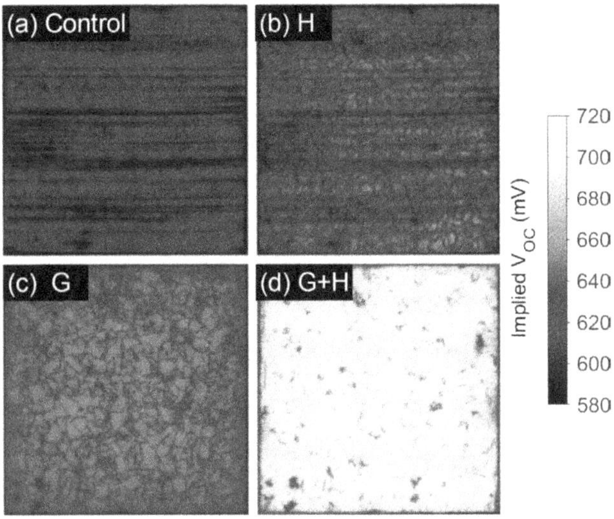

Figure 4.17 *Calibrated implied V_{OC} mapping of pre-treated p-type mc-Si wafers (a) control, no gettering or hydrogenation, (b) G, pre-gettered only, (c) H, pre-hydrogenated only, (d) G + H both pre-gettered and pre-hydrogenated. All samples are measured with symmetrical SiN_x passivation. Reproduced from [128] under Creative Commons CC BY. © 2019 The Authors. Progress in Photovoltaics: Research and Applications published by John Wiley & Sons Ltd. https://doi.org/10.1002/pip.3230.*

recombination at the grain boundary sites still severely limit the iV_{OC}, indicating that the process was not able to effectively passivate the grain boundaries. In addition, significant recombination activity associated with saw damage remained after the hydrogenation process. In contrast, gettering alone was effective at eliminating recombination activity associated with residual saw damage as well as reducing the recombination activity of grain boundaries slightly and improved intra-grain regions. The application of both gettering and hydrogenation led to the effective passivation of almost all grain boundaries, although ineffective at passivating the dislocation clusters.

Further examples highlighting the impact of gettering and hydrogenation will be shown in Section 4.6, along with the impact on cell performance.

4.5.2 Gettering and hydrogenation in Al-BSF and PERC solar cells

Until recently the conventional full area Al-BSF solar cell was the dominant industrial silicon solar cell technology [129]. This technology is a great example where critical processes used throughout the fabrication sequence to form the solar cell structure can simultaneously improve the bulk lifetime of the silicon. Since then, the industry has become completely dominated by PERC, which has relatively small

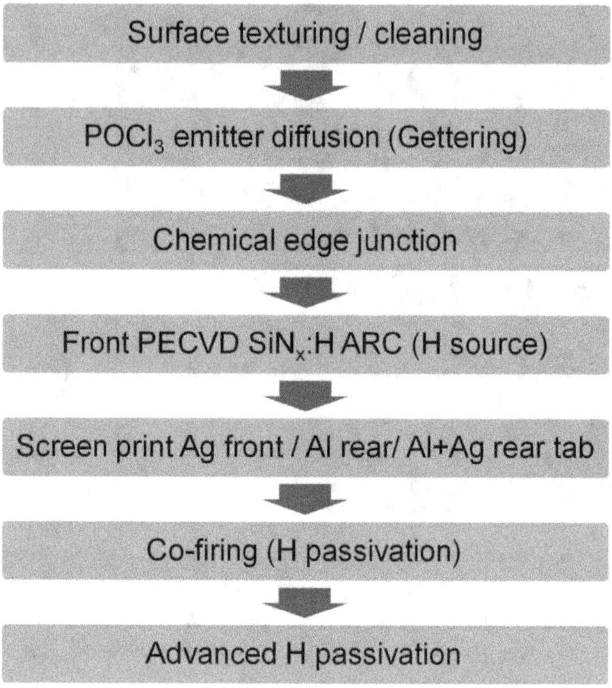

Figure 4.18 *Process flow for the fabrication of industrial screen-printed PERC solar cells and their relevance for defect engineering*

changes in the production sequence from the Al-BSF technology, compared to the changes to upgrade from Al-BSF to other technologies such as TOPCon and SHJ. As such, PERC benefits from the same basic defect engineering mechanisms as for conventional Al-BSF solar cells. The typical fabrication sequence for a standard industrial screen-printed PERC silicon solar cell is shown in Figure 4.18, with specific references to the processes that can be used for defect engineering to improve the bulk lifetime of silicon material through gettering and hydrogenation. Refer back to Chapter 2 for the changes required for Al-BSF solar cells.

Processes of key importance for gettering include the phosphorus diffusion used to form the n-type emitter of the device and subsequent chemical-edge junction isolation process. The use of $POCl_3$ diffusion to form the emitter diffusion on p-type solar cells is a relatively clean process due to the presence of chlorine in the diffusion furnace, which attracts metal impurities. In addition, the heavily diffused phosphorus-doped region formed on the surface, with a surface concentration that can easily exceed $1 \times 10^{20}/cm^3$, particularly for older generation screen-print pastes, can attract mobile impurities such as iron in a process known as gettering. The presence of electrically inactive phosphorus above the solid solubility limit can result in the formation of silicon phosphate (SiP) precipitates, which act as efficient gettering sites [130]. Although this inactive phosphorus results in a 'dead layer', which increases recombination within the emitter and impacts the blue-response of the solar cell, this region acts as an efficient gettering site for mobile impurities. This can essentially remove impurities from the bulk of the silicon, drawing the impurities to the surface of the device. This is particularly important for cast silicon materials with structural defects such as grain boundaries and dislocation clusters that can also act as efficient internal gettering sites for impurities [81]. Ideally, conditions should be used to preferentially getter impurities to the surface phosphorus-doped region rather than structural defects throughout the bulk silicon wafer, to allow the removal of impurities from structural defects to aid in hydrogen passivation. Due to the high doping concentrations and hence low minority carrier concentrations, the impurities in the heavily doped region have little impact on the lifetime of the device. Separately from gettering, during the high-temperature diffusion process, the precipitation of impurities such as iron can also occur throughout the bulk of the wafer [131].

The chemical edge junction isolation process used to remove the phosphorous-diffused layer on the rear surface of the solar cell can be used to physically remove impurities gettered to the rear diffused layer, from the device. Etching may also be performed on the front surface, although care must be taken due to the modification of the emitter sheet resistance and subsequent impact on series resistance. For example, vapours from the strong acids used for the chemical edge junction isolation can perform a slight etch-back of the front emitter, thereby also removing the most heavily phosphorus-doped region on the front surface of the device containing impurities [132]. Alternatively, a wet-chemical approach can be used [133].

Other subsequent high-temperature processes, however, can impact the effectiveness of gettering and precipitation in the material. Of particular importance are potential changes that can be induced by high-temperature hydrogenation processes such as the metallisation firing process. For example, on 1.6 Ω·cm commercial grade p-type Cz wafers with a background interstitial iron concentration[§] ($[Fe_i]$) of $2.8 \times 10^{11} \pm 1.7 \times 10^{11}/cm^3$, where a 60 Ω/□ screen-print diffusion with a surface concentration of $3.7 \times 10^{20}/cm^3$ can reduce $[Fe_i]$ almost two orders of magnitude to $5.4 \times 10^9 \pm 1.1 \times 10^9/cm^3$. However, after the high-temperature firing process, $[Fe_i]$ can undesirably increase by a factor of almost four, to $2.1 \times 10^{10} \pm 3.0 \times 10^9/cm^3$, despite the benefits from the passivation of bulk and surface defects in the device. On the other hand, for non-diffused samples, Fe_i can decrease by a factor of approximately two, to $1.4 \times 10^{11} \pm 1.1 \times 10^{11}/cm^3$. This is most likely due to a potential gettering of iron to SiN_X [93], or slight precipitation of iron in the bulk, rather than hydrogenation of iron. It is noted, that attempts to hydrogen passivate interstitial iron in the same material at lower temperatures both with and without illumination have not led to any statistically significant changes in $[Fe_i]$. This highlights the complex interactions between gettering, hydrogenation and the thermal processes used, which should be

[§]The interstitial iron concentration can be extracted using quasi-steady-state photoconductance measurements before and after a 10 s exposure to illumination to split iron–boron pairs. See [134] for details of the technique.

considered when defect engineering processes are applied for silicon solar cells. This will be discussed further in Sections 4.5 and 4.6.

Processes of key importance for hydrogenation are the dielectric layer depositions. For PERC, this is the PECVD of hydrogenated silicon nitride (SiN_X:H) on the front surface and the deposition of the rear surface dielectric stack, typically composed of AlO_X by either ALD (AL_2O_3) or PECVD (AlO_X), capped byt PECVD SiN_X:H. The deposition of the dielectric layers such as PECVD SiN_X:H themselve can act as a hydrogenation process through the generation of hydrogen radicals from silane (SiH_4) and ammonia (NH_3), which are the most commonly used precursors. However, as discussed in Section 4.1, the film composition can greatly impact the bulk and surface passivation achievable in the as-deposited state, by impacting the diffusion of hydrogen, particularly with low deposition temperatures in the vicinity of 400 °C. Although less relevant for PERC solar cells, which also feature a non-diffused p-type surface passivated by hydrogenated dielectric layers, at the front surface, the presence of a phosphorus diffusion is expected to inhibit the diffusion of hydrogen from dielectric layers, particularly at low temperatures where hydrogen naturally moves into a low-mobility charge state, although little has been done to directly investigate this experimentally.

For example, at 400 °C, in n-type silicon with a doping concentration of $1 \times 10^{20}/cm^3$, which is representative of the surface concentration of a phosphorus emitter for screen-printed PERC solar cells, almost all hydrogen is expected to be in the negative charge state (H^-), and therefore heavily impacted by electrostatic effects. The fractional concentration of the highly mobile neutral charge state (H^0) is only expected to be approximately 0.001% [6]. This concentration reduces with increasing phosphorus doping concentration. However, little has been done experimentally to investigate the impact of a phosphorus emitter on the diffusion of hydrogen, particularly at low temperatures, where it is most likely to have an impact on mobility.

A key advantage of the use of hydrogenated dielectric layers is the ability to use them as a hydrogen source for bulk defect passivation in subsequent processes. Hydrogen from such layers can be released during thermal processing, thereby avoiding the need for an external hydrogen source such as using forming gas annealing. For industrial solar cells, the metallisation co-firing process is the high-temperarure process that enables the release of hydrogen from hydrogenated dielectric layers and enables the diffusion of atomic hydrogen into and throughout the silicon wafer. The peak temperatures during the firing process, typically in the range of 700–900 °C, increases the diffusivity of each of the hydrogen charge states, as well increasing the concentration of neutrally charged hydrogen (H^0) as the silicon behaves more like intrinsic silicon [135]. At 700 °C, the fractional H^0 concentration in heavily doped n-type silicon increases by two orders of magnitude compared to that at 400 °C. Similarly, in the bulk of the silicon wafer (1 Ω·cm p-type material), the fractissivation in the device. This was aonal H^0 concentration increases by more than one order of magnitude from approximately 0.75% to 8.6% [42]. The increased diffusivity readily enables the distribution of hydrogen throughout the silicon bulk to enable the passivation of bulk defects. However, the high-temperature process can have several non-hydrogen related effects too, such as release of iron from precipitate sites.

Although a certain level of hydrogen passivation can be achieved during the conventional co-firing process for screen-printed solar cells, the process itself may not be effective for a range of defects, including carrier-induced defects. One key aspect related to hydrogen passivation that must be considered is its reversibility, leading to a reactivation of recombination activity after dehydrogenation. Although reaction rates increase with increasing temperature, meaning that defects can potentially be passivated much quicker at higher temperatures, for defect systems with a higher activation energy of the dehydrogenation reaction than the hydrogenation reaction, care must be given to ensure that a sufficiently low temperature is used to enable effective passivation. This can be further complicated by other hydrogen-related reactions taking place in the background, such as the evolution of atomic hydrogen into molecules, dimers or effusion of hydrogen [136], leading to temporal changes in the concentration of hydrogen available for defect passivation, and therefore the passivation rate. This will be discussed further in Chapter 5 for the boron–oxygen defect.

However, the release of hydrogen from dielectric layers and distribution throughout the wafer from the high-temperature co-firing process can enable a manipulation of hydrogen for defect passivation with subsequent thermal processes, as was used for improved passivation of dislocation clusters in Section 4.1 with dark or illuminated annealing. Figure 4.19 highlights the impact of a high-temperature hydrogen release from dielectric layers and subsequent activation of hydrogen passivation at low temperatures. In this example, the top half of the wafer was passivated on both sides with PECVD SiN_X:H, while the bottom half was masked during the PECVD process. Subsequently, the wafer was fired using a standard firing process with a peak temperature of 740 °C. This injected hydrogen into the bulk of the wafer in the region coated with SiN_X:H, while the lower half of the wafer was not injected with hydrogen due to the absence of the SiN_X:H layer during firing. After removal of the SiN_X:H and redeposition of PECVD SiN_X:H at ~400 °C across the whole wafer, a 2-hour AHP was then performed at 200 °C to passivate boron–oxygen defects, followed by light soaking to perform a stability test. Figure 4.19 shows the PL image after light soaking, highlighting the improved electrical quality due to the passivation of boron–oxygen defects through the deliberate injection of hydrogen in the top half of the wafer [6].

The ability to essentially store hydrogen within the device for later use to passivate defects is particularly important for eliminating LID as will be discussed in detail in Chapter 5 for the boron–oxygen defect. In additional it allows the integration of multi-stage hydrogenation processes into solar cell architectures with strict thermal budget requirements such as heterojunction solar cells as will be discussed in Section 4.6.

Figure 4.19 *Calibrated implied open-circuit voltage map of a p-type Cz wafer after an advanced hydrogen passivation process for the treatment of boron–oxygen defects and subsequent stability testing, where hydrogen was only injected into the top half of the wafer (as displayed in the figure), highlighting the effective use of the wafer as a storage for hydrogen to enable multi-stage hydrogenation processes. Reproduced with permission from [6], with the permission of John Wiley and Sons. © 2017 WILEY-VCH Verlag GmbH & Co. KGaA, Weinheim.*

4.5.3 Impact of evolving solar cell architectures and processes on gettering and hydrogen passivation

With evolving solar cell architectures and processes to constantly improve the efficiency of industrial solar cell efficiencies, it is important to understand the impact of such changes on the ability to getter impurities and hydrogenate defects. For example, the transition from the Al-BSF technology to PERC results in multiple device changes that can impact hydrogen passivation kinetics. First, the addition of hydrogen-containing dielectric layers on the rear of PERC cells is naturally expected to increase the concentration of hydrogen in the silicon bulk. Typically, the composition of the front and rear dielectric layers differ, thereby creating the opportunity for different kinetics of hydrogen release from the front and rear layers, particularly for layer with significantly different concentrations of Si–H and N–H bonds.

In addition, the improved rear surface with a rear surface passivation layer and localised rear contacts reduces rear surface recombination and, therefore, increases the effective lifetime of the solar cell structure. This translates to higher carrier concentrations in the material, which increases the fractional H^0 concentration. In addition, for hydrogen-defect systems where reaction rates, including defect formation rates, depend on the carrier concentration, this can alter the dynamics and will be discussed in detail in Chapters 5 and 6. The trend towards higher V_{OC} devices, and therefore higher carrier concentrations, will tend to relax the requirement for high-intensity illumination for AHPs. A summary of potential changes induced by changing solar cell architecture and processes are shown below.

- A change in the lifetime of the structure impacts the fractional hydrogen charge state concentrations. A higher lifetime is more favourable for the generation of H^0.
- A change in lifetime of the structure will also modulate any passivation or defect reactions with a dependence on carrier concentration (this will be discussed in Chapters 5 and 6).
- Adding a hydrogen source onto the rear of the device creates a larger hydrogen source for passivation and is expected to increase passivation rates.
- The presence of two hydrogenated dielectric layers with different compositions could lead to the release of hydrogen in different temperature regimes (e.g. a silicon-rich SiN_X:H with low-temperature release versus a nitrogen-rich SiN_X:H with high-temperature release of hydrogen).

Another key trend for solar cells is the shift towards emitters with higher sheet resistances. This is largely due to the development of improved metallisation pastes enabling effective contact to lower surface dopant concentrations. In addition, the industry is adopting the use of selective emitter structures, as will be discussed in Chapter 8, with lighter diffusions in the passivated regions to reduce the dark saturation current density (J_0) of the front surface. However, this can significantly impact the ability to getter impurities such as iron [130], and therefore have the potential to impact hydrogen passivation processes. For example, compared to the screen-print diffusion, a more lightly doped emitter (LDE) at 110 Ω/ϒ with a surface concentration of $2.4 \times 10^{20}/cm^3$, is less effective at gettering of iron, resulting in $[Fe_i]$ of $1.1 \times 10^{10} \pm 1.3 \times 10^9/cm^3$, approximately twice the concentration of that achieved for the screen-print diffusion. On the other hand, using an extremely heavy phosphorus diffusion of 35 Ω/ϒ can much more effectively getter iron, resulting in an iron concentration below the detection limit ($\sim 1 \times 10^9/cm^3$). However, such a diffusion is too heavy to use as a full-area emitter in high-efficiency solar cells due to increased Auger recombination and the formation of a dead layer.

Thermal oxidation processes may also be used to improve surface passivation such as with the use of a thin thermal oxide (5–20 nm) capped by PECVD SiN_X:H. The oxidation process also acts as a 'drive-in' process for dopants, thereby reducing the surface doping concentration and eliminating a dead layer (see Figure 4.20 (left)). However, this can also heavily impact the evolution of impurity-related defects such as iron, and the effectiveness of gettering and iron precipitation, by eliminating SiP precipitates. This is particularly the case if the oxidation temperature is higher than that used for the diffusion process [125]. For example, a short high-temperature oxidation process for 10 minutes at 890 °C can reduce the surface concentration of a screen-print phosphorus emitter by a factor of six to $5.8 \times 10^{19}/cm^3$, resulting in a substantially higher $[Fe_i]$ of $2.9 \times 10^{10}/cm^3$, thereby increasing the $[Fe_i]$ by a factor of five (Figure 4.20 (right)). Similarly, the surface dopant

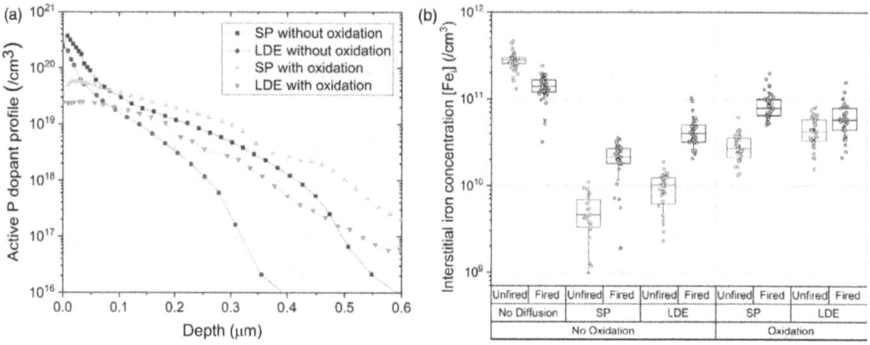

Figure 4.20 *(Left) Diffusion profiles for an emitter suitable for screen-print (SP) or a lightly-doped emitter (LDE), as well as for both diffusions after thermal oxidation. (Right) Corresponding impact on interstitial iron concentrations. Reproduced with permission from [6], with the permission of John Wiley and Sons. © 2017 WILEY-VCH Verlag GmbH & Co. KGaA, Weinheim.*

concentration of the LDE reduced an order of magnitude after thermal oxidation to $2.5 \times 10^{19}/cm^3$, reducing the effectiveness of gettering by a factor of 4.5, resulting in $[Fe_i]$ of $4.5 \times 10^{10}/cm^3$. Such iron concentrations are expected to significantly impact the performance of PERC solar cells [130]. After firing, the $[Fe_i]$ of all samples with and without thermal oxidation for the screen-print and LDE increased. However, for the samples with the heavy gettering diffusion, iron was still below the detection limit, noting that the diffusion was completely removed prior to dielectric deposition.

With the relatively high $[Fe_i]$ values returning after processes such as thermal oxidation and firing, more work could be done here to investigate different hydrogenation processes and oxidation processes to reduce the potential release of iron from getter and precipitate sites. This, for example, could include the use of lower-temperature oxidation processes, although simultaneously reducing the ability of the oxidation process to 'drive-in' dopants in the emitter.

However, the complete removal of gettering layers from the device is an effective way of limiting the potential release of impurities from gettering and precipitate sites. The use of a heavy gettering diffusion, with complete removal, can greatly reduce $[Fe_i]$ after a subsequent screen-print or LDE, to values still below the detection limit. After thermal oxidation, values of $7.8 \times 10^9 \pm 1.7 \times 10^9/cm^3$ and $1.1 \times 10^{10} \pm 2.3 \times 10^9/cm^3$ were obtained, respectively, however, substantially lower than that without the use of a deliberate gettering diffusion.

A summary of key considerations for impact of processes on the gettering of impurities is given below:

- The emitter diffusion profile and process used can greatly impact the ability to getter impurities, typically a heavy diffusion with a 'dead layer' and SiP precipitates is more effective.

- Thermal oxidation processes can release iron by a 'drive-in' of dopants to remove the dead layer, as well as potentially releasing iron from precipitate sites.
- High-temperature hydrogenation processes such as firing can release iron from precipitate sites.
- The use of a deliberate heavy diffusion for gettering with complete removal is more effective for gettering, but undesirably adds process steps and cost. Attention must also be given for the increased thermal budget and potential evolution of oxygen precipitates.

For any defect engineering process, a holistic view should be considered for other undesired impacts, for example, the release of iron from precipitate sites during firing. In addition, pure thermally related effects can occur during both processes targeted at gettering or hydrogenation.

4.5.4 Gettering and hydrogenation in high-efficiency n-type cell technologies

The use of gettering and hydrogenation of bulk defects for high-efficiency n-type solar cell technologies is not as pronounced in the literature as for p-type technologies. Work in academia often uses high-lifetime n-typer Cz or FZ material, and publications by industry often skip over key details due to commercial sensitivity. However, gettering and hydrogenation processes can also have a significant impact of n-type technologies, even for wafers starting with millisecond lifetimes. For example, recent modelling has shown that there is a potential of up to $2\%_{abs}$ efficiency loss for SHJ solar cells fabricated on n-type wafers with a bulk lifetime of 1 ms [137].

For the majority of n-type technologies, the use of hydrogen passivation is typically only discussed for surface passivation or may only be inferred from the use of hydrogen-containing dielectric layers. As mentioned in Chapter 2, publications and patents relating to the development of high-efficiency solar cells by Sanyo/Panasonic, Sunpower, Silevo and Tetrasun have eluded to the use of hydrogen-containing dielectric layers and/or subsequent processing in forming gas ambient [111, 138–142], but without any mention of bulk defect passivation in those publications.

High-efficiency n-type solar cells with conventional screen-printed metallisation techniques that require high-temperature firing processing will naturally enable some form of hydrogen passivation of bulk defects, similar to that achieved in p-type devices, provided that a hydrogen source is present such as SiN_X:H or AlO_X:H. In addition, technologies such as n-PERT, TOPCon and interdigitated back-contact (IBC) solar cells requiring thermal diffusions and high-temperature processing steps create an opportunity for gettering. For example, Generation 3 Sunpower cells use a thermal diffusion for a front surface field, which, although not mentioned in the publication, creates an opportunity for gettering/beneficial effects of gettering [111]. However, at the same time, it also creates the opportunity to kill the bulk lifetime due to the formation of oxygen precipitates as noted in the study. With appropriate processing optimisation such deterioration could be avoided. Recent work has also shown the potential to getter impurities into doped polysilicon layers [143]. The natural gettering and hydrogenation incorporated into TOPCon solar cells through the use of high-temperature processing for the formation of doped polysilicon contacts and hydrogen release from dielectric layers has led to significant improvements on the bulk lifetime of TOPCon solar cells fabricated on n-type cast-mono silicon material from 200 μs to 400 μs at $\Delta n = 1 \times 10^{14}/cm^3$, with a resultant efficiency of 23.2% [144]. However, further work is required to understand the effectiveness of gettering in full solar cell structures, including the impact of the metallisation firing process.

A clear example highlighting the benefit of gettering for high-efficiency n-type technologies was presented by Trina Solar in their fabrication of n-type IBC solar cells, resulting in an $0.2\%_{abs}$ efficiency enhancement to achieve 23.0% efficiency [145]. In this approach, a heavy diffusion was performed on the front surface followed by the complete removal. Subsequently, wafers were subjected to another diffusion process to form the required cell structure. Undesirably, this method adds process steps and therefore costs, as well as increases the thermal budget, which creates a risk of the growth of oxygen precipitates. As such, the addition of the process must be considered for each given incoming wafer material. Nonetheless, it highlights the potential benefits of incorporating gettering even for high-efficiency n-type silicon solar cells.

For solar cell technologies using metallisation scheme based on a low-temperature screen-printed paste, or approaches other than screen-printing are used, the need for high-temperature processing after dielectric deposition for the purpose of forming the device structure is often eliminated. This includes approaches such as sputtering, evaporation of aluminium and plating [121]. Using these approaches, laser-based processes such as laser ablation or laser doping may be used to open the dielectric layers in localised regions for contact definition [121]. For cases where an aluminium-based metallisation is used, care should be given to consider the impact of processes above 577 °C on melting and resolidifying the silicon, potentially with an inferior doping profile to that achievable during diffusion-based processes. Other metallisation approaches such as plated contacts may not withstand high-temperature processes.

However, the avoidance of high-temperature processing after metal deposition limits the ability to passivate bulk defects in the device. Depending on the composition of hydrogenated dielectric layers, layers may require a mid- to high- temperature firing step to effectively release hydrogen to enable improved bulk and surface passivation [51, 146]. This can be achieved by simply performing a dedicated annealing process after dielectric deposition, but before the deposition of metal layers of concern for thermal stability [87, 123]. A temperature in the range of 400–500 °C can also be used in the presence of aluminium-based metallisation layers, without significant detrimental effects [107, 123].

4.6 Gettering and hydrogenation for heterojunction silicon solar cells

An alternative device architecture to PERC that is gaining attention is the SHJ solar cell structure [147]. The world record single-junction silicon solar cell efficiency of 26.7% is held by a SHJ solar cell with IBCs, leaving the entire front surface available to receive incoming light [148]. The basic SHJ solar cell technology, with contacts on both surfaces of the solar cell, was first developed in the 1990s by Sanyo; however, research activity has recently accelerated due to the expiration of core technology patents. The excellent performance of SHJ solar cells, in particular the very high open-circuit voltage (V_{OC}), is due to the excellent surface passivation provided by thin hydrogenated amorphous silicon (a-Si:H) films. This chemical passivation on the surface due to hydrogen bonding leads to exceptionally low surface recombination velocities [149]. The use of a heterojunction (or 'passivated contact') structure also avoids Auger-related dark saturation current densities associated with diffused emitter regions [150]. A key advantage of the SHJ solar cell technology is that due to the high V_{OC} of the device, with the nominal drop in V_{OC} of 2.2 mV/K for silicon solar cells, the corresponding fractional loss of $V_{OC,}$ and therefore power at higher operating temperatures in the field, is lower.

As well as competing with other technologies in terms of single junction commercial silicon solar cells, SHJ solar cells are deemed a likely candidate to be used as the silicon bottom cell in silicon/perovskite tandem solar cells. These tandem cells, where two complementary cells are stacked together, allow for a much more efficient utilisation of the solar spectrum, which leads to higher efficiency [151, 152]. The very high voltage that SHJ cells offer means they are well suited for this application, which is predicted to capture an increasing market share when the upper efficiency threshold of single-junction silicon solar cells is reached [7].

One perceived advantage of the SHJ architecture during manufacturing is that all processes, including the PECVD for the a-Si:H deposition and the metallisation, are performed at low temperatures (<250 °C). Removing the high-temperature processes can reduce the processing cost and will avoid severe degradation or 'poisoning' of low-quality wafers that can occur due to high-temperature processing [126]. The low temperatures are also used to avoid a potential deterioration of surface passivation that can occur for annealing temperatures >250 °C after the a-Si:H deposition due to the dehydrogenation of the a-Si:H layers [153]. However, one significant drawback of the low-temperature processing is that it does not readily allow for the defect engineering processes, inherent in the process flow of conventional solar cells such as Al-BSF, PERC, n-PERT and TOPCon, which can significantly improve the quality of the bulk material during cell fabrication [6, 123, 143].

In a SHJ solar cell, the surface recombination is so low that recombination processes in the bulk become the performance-limiting recombination mechanisms. As such, all SHJ solar manufacturers use high lifetime (>1 ms) n-type Cz silicon wafers to form high-efficiency solar cells. This is also a value typically used as a cut for the manufacture of n-type ingots. The strict lifetime requirement contributes to an approximate 10% premium for the cost of an n-type Cz wafer than that of a p-type Cz wafer for the same thickness [124]. The restriction on the starting quality of the silicon wafers, along with increased silver consumption [8] and use of rare elements such as indium for the transparent conductive oxide layer, limits the cost effectiveness of this technology and makes very difficult for it to compete with PERC. If such cost issues are not overcome, this would relegate the use of SHJ cells to a premium market, which would only garner a small market share.

To partially address issues related to the increased manufacturing cost of SHJ solar cells, it would be advantageous if low-cost wafers could be used and still achieve the full potential of the excellent surface passivation provided by a-Si:H layers. However, a reduced wafer cost can typically mean the presence of performance-limiting defects. This is particularly the case for commercial-grade p-type wafers, such as p-type multi-crystalline and boron-doped p-type Cz silicon wafers.

4.6.1 Impact of gettering and hydrogenation for SHJ solar cells

To determine the potential of using low-cost p-type wafers for SHJ solar cells, a collaboration was established between the Australian National University, Arizona State University and the University of New South Wales to develop a hydrogenated heterojunction solar cell technology, in a project funded by the Australian Renewable Energy Agency.

One of the preliminary studies in 2017 investigated the impact of gettering and hydrogenation, as typically applied for p-type solar cells, as pre-treatments for wafers used to fabricate heterojunction solar cells [154]. The study contained four conditions:

- Control: no pre-fabrication gettering or hydrogenation
- Gettered only
- Hydrogenated only
- Pre-gettered and hydrogenated.

Gettering was achieved by the formation of a heavy 35 Ω/□ phosphorus emitter on both sides through a $POCl_3$ diffusion process, and a subsequent HF dip to remove the phosphosilicate glass layer. Additionally, a pre-fabrication hydrogenation step was achieved by the deposition of an industry-standard 75 nm SiN_x:H dielectric layer on both sides of the wafer by PECVD. Subsequently, the wafers underwent an industry-standard fast firing process with a peak temperature of 740 °C, as is used for p-type screen-printed solar cells. This facilitated the release of hydrogen from the SiN_x:H layer and distribution throughout the bulk of the p-type silicon wafer. The SiN_x:H layer was then removed with HF.

The defect-engineered wafers were then used to fabricate either lifetime precursor structures or full solar cells using a standard SHJ solar cell-processing sequence [155–157]. Following the deposition of intrinsic and doped a-Si:H, full cells were formed by sputtering of indium-tin-oxide (ITO) and subsequent screen-printing of low-temperature pastes and curing.

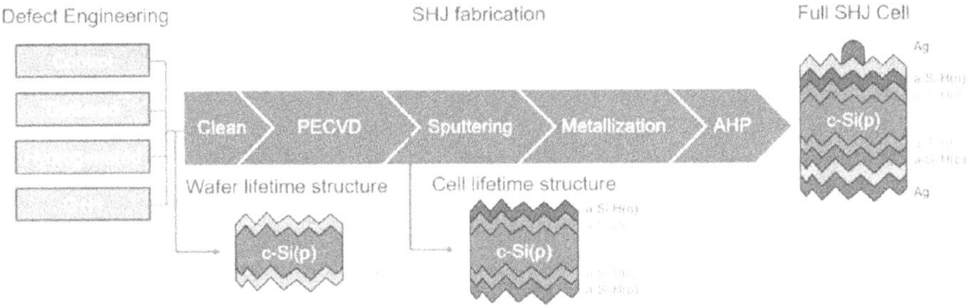

Figure 4.21 *Schematic workflow illustrating the defect engineering processes and SHJ cell and lifetime precursor fabrication steps. Image reproduced from [128] under Creative Commons CC BY. © 2019 The Authors. Progress in Photovoltaics: Research and Applications published by John Wiley & Sons Ltd. https://doi.org/10.1002/pip.3230.*

Following the conventional cell-processing sequence, an AHP incorporating minority carrier injection was used to further enhance hydrogen passivation in the device. This was achieved using a fibre-coupled 938-nm diode laser operating in quasi-continuous mode with a 0.5 ms pulse length and repetition frequency of 2 kHz.

A process time of 10 s was used with an illumination intensity >100 suns and a temperature of ~300 °C. The process sequence is shown in Figure 4.21.

The impact of pre-fabrication gettering and hydrogenation on the effective lifetime and implied V_{OC} is shown in Figure 4.22a and b. The effective lifetime of the

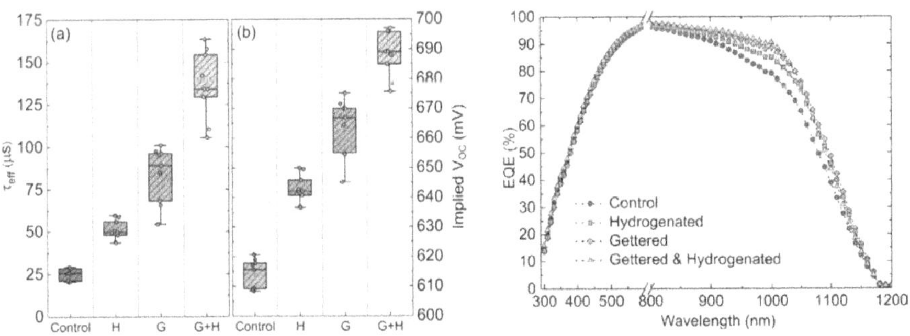

Figure 4.22 *Impact of pre-fabrication gettering (G) and hydrogenation (H) on (a) effective minority carrier lifetime extracted at a minority carrier density of $1 \times 10^{15}/cm^3$) and (b) 1-sun iV_{OC} on SHJ lifetime precursor structures fabricated using commercial grade p-type Cz silicon. (Right) Corresponding external quantum efficiency (EQE) measurements on SHJ cells. Images reproduced from [154] under Creative Commons CC-BY-NC. © 2018 The Authors. Solar RRL Published by WILEY-VCH Verlag GmbH & Co. KGaA, Weinheim. https://doi.org/10.1002/solr.201700221.*

control sample that did not receive any pre-treatment was very low, on the order of 25 μs. The application of either the individual gettering or hydrogenation processes led to an improvement of the effective lifetime to approximately 50 μs and 75 μs, respectively. However, the most significant improvement was observed in the gettered + hydrogenated group, indicating that the gettering and hydrogenation process work synergistically to eliminate the negative impact of bulk defects, to achieve lifetimes of 110–160 μs, resulting in significant improvements in the long-wavelength response [154]. Given that the J_0 for these lifetime structures with a-Si:H passivation was <5 fA/cm^2, this indicates that the changes observed are related to changes in the bulk of the material.

This improvement in effective lifetime correlated with a significant increase in the iV_{OC} (see Figure 4.22 (b)), a key measure of the maximum potential terminal voltage of a solar cell after metallisation. The iV_{OC} of the control sample was on the order of 610 mV, even lower than that of industrial Al-BSF cells, which typically achieve V_{OC} of about 640–650 mV. This is due to the absence of defect engineering, which naturally occurs during the fabrication of Al-BSF and PERC solar cells. The gettering and hydrogenation pre-treatments led to significantly improved iV_{OC}, in some cases higher than 695 mV for the gettered + hydrogenated group. Full solar cell devices were fabricated on the same batch of pre-treated commercial p-type wafers. The improvements in actual V_{OC} correlated closely with the improved iV_{OC} going from 621 mV for the control sample, to a V_{OC} for the gettered + hydrogenated group of 693 mV, an improvement of over 70 mV. This is a remarkable achievement, considering the starting lifetime of these wafers was only ~25 μs. The V_{OC} of the cells with gettering and hydrogenation is higher than for the high efficiency industrial PERC solar cells that are currently dominating the market and the record 23.83%-efficient device by LONGi [17]. The corresponding EQE measurements on the cells is shown in Figure 4.22 (right).

However, the addition of an AHP incorporating minority carrier injection after cell fabrication can lead to significant further improvements in the lifetime and subsequently device performance. For the samples with gettering, hydrogenation and an AHP, the lifetime increased to 200–300 μs [128]. The results further highlight the benefit of a comprehensive defect engineering approach and the complementary nature of gettering and hydrogenation. For example, the addition of AHP on control samples or hydrogenated, only increased the V_{OC} by 7–9 mV. Yet, on gettered samples, with a higher starting V_{OC}, the V_{OC} increased by 30 mV. For samples with gettering, hydrogenation and AHP, an increase of 16 mV was observed to 707 mV, which is higher than the record UNSW PERL solar cell fabricated with high purity p-type FZ wafers [158]. The larger increase for samples without a dedicated high-temperature process may indicate additional passivation of defects during the AHP that typically occurs during high-temperature firing. The comprehensive depiction of current–voltage characteristics is presented in Table 4.3.

Other work on p-type Cz material by Descoeudres *et al.* with starting lifetimes in the range of 4–9 ms (8–14 Ω·cm) material also demonstrated a benefit of pre-gettering, by increasing lifetimes to the range of 7–10 ms [161]. In addition, the thermal processing normalised the wafer resistivity throughout the ingot to ~6 Ω·cm

Table 4.3 *Current–voltage (I–V) results for SHJ solar cells highlighting the impact of defect engineering processes*

Reference	Material	Group	V_{OC} (mV)		J_{SC} (mA/cm^2)		FF (%)		Efficiency (%)	
			No AHP	With AHP	No AHP	With AHP	No AHP	With AHP	No AHP	With AHP
Chen *et al*. [128]	p-type Cz	C	621	630	33.3	34.6	68.9	66.3	14.2	14.5
	p-type Cz	H	628	635	34.7	33.4	66.4	70.0	14.5	14.8
	p-type Cz	G	643	673	34.4	34.2	68.9	67.9	15.2	15.6
	p-type Cz	G + H	693	707 (707*)	34.5	34.6 (34.3*)	69.6	72.1 (74.1*)	16.6	17.6 (18.0*)
Chen *et al*. [128]	p-type mc-Si	C	632	640	31.9	35.1	61.1	62.9	12.3	14.1
	p-type mc-Si	H	636	685	31.2	33.6	66.9	67.9	13.3	15.6
	p-type mc-Si	G	683	687	33.5	33.0	63.4	68.9	14.5	15.6
	p-type mc-Si	G + H	657	701 (702*)	32.1	34.9 (34.3*)	68.6	70.6 ± 0.2 (72.1*)	14.5	17.3 (9.9*)
Vicari Stefani *et al*. [159]	p-type UMG mc-Si	G + H	680.4	690.6	35.8	36.2	72.9	74.4	17.8	18.7
Basnet *et al*. [160]	n-type UMG Cz	C	678	-	36.2	-	73.4	-	18.0	-
	n-type UMG Cz	G + H	715	-	38.6	-	78.1	-	21.2	-
	n-type UMG Cz	TR + G	717	-	37.5	-	78.0	-	21.0	-
Basnet *et al*. [160]	n-type Cz	C	708	-	39.0	-	76.7	-	21.1	-
	n-type Cz	G + H	721	-	39.6	-	79.2	-	22.7	-
	n-type Cz	TR + G	720	-	39.8	-	78.4	-	22.4	-
UNSW/	n-type Cz	C	735 ± 1.4	-	38.1 ± 0.2	-	79.4 ± 1.0	-	22.3 ± 0.2	-
Hevel collaboration†	n-type Cz	G	737.5 ± 0.7	-	38.6 ± 0.1	-	80.2 ± 0.0	-	22.8 ± 0.1	-

*Independently measured at Solar Energy Research Institute of Singapore (SERIS).

†Results for a collaboration between Hevel Solar/UNSW with defect engineering performed at UNSW and cells fabricated at Hevel.

C, control (no defect engineering);G, gettering;H, hydrogenated;TR, tabula rasa.

through the annihilation of thermal donors. This resulted in significant enhancements of the resistivity-normalised lifetimes of the material from as low as ~0.3 ms/(Ω·cm), leading to efficiency enhancements in the range of 0.2%$_{abs}$. This resulted in efficiencies of large area p-type Cz SHJ solar cells of up to 23.6%, with a V_{OC} of 740 mV. Although peak efficiencies were below that of n-type Cz SHJ solar cells processed in parallel, importantly, the average efficiency of the p-type SHJ solar cells fabricated on gettered wafers was the same, if not 0.1%$_{abs}$ higher than that of n-type SHJ solar cells [161]. However, the study did not investigate hydrogenation or the potential impact of LID. The impact and treatment of LID for p-type SHJ solar cells through hydrogen passivation to enable stabilised V_{OC} values over 735 mV will be discussed in Chapter 5.

The comprehensive defect engineering incorporating gettering and hydrogenation by Chen *et al.* was extended to include p-type mc-Si wafers [128], with reduced material quality compared to that of Cz wafers [162]. The large number of lifetime-limiting crystallographic defects in mc-Si means that pre-treatment of the wafers to improve the quality is even more crucial for this material. The impact of pre-gettering and hydrogenation for this work was presented in Figure 4.17 in Section 4.5. Similarly, as for p-type Cz material, the inherent material quality of the multi-crystalline wafers was poor, resulting in the control samples only having iV_{OC} of 615 mV, gettering by itself improved the iV_{OC} to 660 mV. This was much more effective than hydrogen passivation alone, which only improved the iV_{OC} by 10 mV, resulting in an iV_{OC} of 625 mV. However, the use of both hydrogenation and gettering resulted in more than an 80 mV increase in iV_{OC} through a significant reduction in recombination activity to achieve spatially averaged iV_{OC} of more than 700 mV.

Cell performance was also greatly impacted by defect engineering. Similar to lifetime test structures, gettering alone resulted in a significant increase in the V_{OC} of cells from 632 mV to 683 mV. This was much larger than the almost negligible 4 mV increase using hydrogenation alone, to 636 mV. However, for cells, gettering alone resulted in a much higher final V_{OC} than for the group with gettering and hydrogenation (657 mV), almost 40 mV lower than the iV_{OC} of the corresponding group, suggesting a significant impact of the SHJ processing on V_{OC} of the cells, despite processing at only 200 °C. This highlights the complicated nature of the impact of gettering and hydrogenation, in this instance, with the generation of hydrogen-induced defects causing LeTID degradation, which will be discussed in Chapter 6. Furthermore, the V_{OC} of the gettered group was almost 20 mV higher than that of the lifetime test structures, suggesting a possible hydrogen passivation of the material during the SHJ process, although this may also be due to the improved surface passivation.

After AHP, substantial improvements in the V_{OC} were observed, most notably for the hydrogenated group and the group with both gettering and hydrogenation with improvements in the range of 44–49 mV. In contrast, more modest increases in the V_{OC} were observed for the control cells and cells with gettering only, in the range of 4–8 mV. The highest V_{OC} in the 2 cm × 2 cm SHJ solar cells for the gettered + hydrogenated group was 702 mV after AHP, as verified by measurements at the

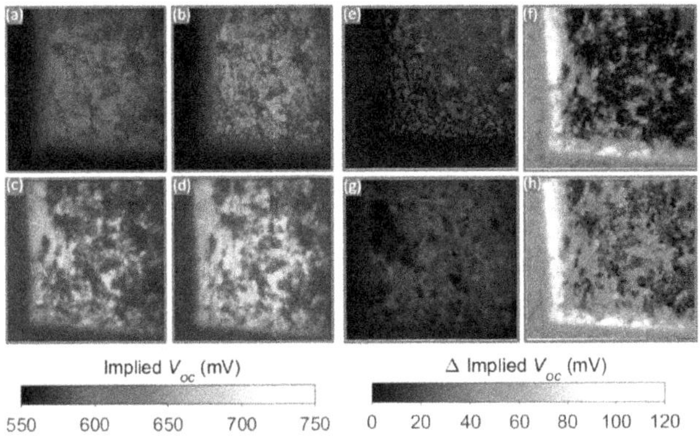

Figure 4.23 *Calibrated iV_{OC} maps (left) and ΔiV_{OC} maps (right) of UMG mc-Si wafers passivated by SiN_X:H (a) without any pre-treatment, (b) with hydrogenation, (c) gettering, and (d) both gettering and hydrogenation. The ΔiV_{OC} maps on the right represent the absolute difference between (e) hydrogenation and the control, (f) gettering and the control, (g) gettering + hydrogenation and gettering only and (h) gettering + hydrogenation and control. Image reproduced from [159] under Creative Commons CC-BY-NC-ND. © 2019 The Authors. Published by WILEY-VCH Verlag GmbH & Co. KGaA, Weinheim. https://doi.org/10.1002/pssa.201900319.*

SERIS as shown in Table 4.3. This V_{OC} is 30 mV higher than the previous record V_{OC} achieved on mc-Si solar cells [163].

Vicari Stefani *et al.* also applied this methodology to very-low-quality UMG p-type mc-Si wafers [159]. This material is much cheaper than standard Cz and high-performance mc-Si wafers, as it eliminates the energy intensive 'Siemens process' for material purification [164]. The liquid phase solidification means that UMG silicon materials contain higher concentrations of both metallic and non-metallic impurities [165]. These impurities cause recombination in the bulk of the silicon, which severely limits the carrier lifetime; thus, defect engineering approaches are crucial for this silicon material. In their study, Vicari Stefani *et al.* investigate the effectiveness of gettering, hydrogenation and advanced hydrogenation for p-type SHJ solar cells fabricated using UMG mc-Si fabricated from 'red zone' wafers, taken from the edge of the cast. The term red zone refers to UMG silicon with a high concentration of contaminants that diffused during crystallisation, which reduces the lifetime. The starting lifetime of the studied wafers was ~15 μs, indicating a very high concentration of impurities.

The response of the UMG wafers to gettering and hydrogenation was comparable to the results presented by Chen *et al.* [128] for Cz and solar grade mc-Si. In particular, gettering alone was much more effective than hydrogenation alone on lifetime samples, and the best results were obtained using both hydrogenation and gettering (see Figure 4.23). A striking feature of results from this work was large spatial variability in the response to gettering and hydrogenation. In particular, near the edge of the red-zone region, improvements in the iV_{OC} of almost 80 mV were observed after gettering, highlighting the effectiveness of the gettering process, which reduced the interstitial iron concentration by more than an order of magnitude [159]. Yet heavily dislocated regions suffered a significant reduction in iV_{OC} (see Figure 4.23 (f). The application of both gettering and hydrogenation led to effective lifetimes in the range of 70 μs. The treated wafers were subsequently used to fabricate SHJ solar cells. The highest V_{OC} on the champion cell in the group consisting of wafers undergoing gettering, hydrogenation and advanced hydrogenation treatment exhibited a V_{OC} of 690 mV, which is higher than current commercial p-type PERC solar cells. This is a remarkable achievement considering that these wafers began with a lifetime of <15 μs.

The substantial improvements demonstrated using both Cz and cast silicon materials highlight the promise of using low-cost p-type substrates for SHJ solar cells passivated using a-Si:H layers. The next scientific challenge is to address the significantly lower FF in p-type SHJ solar cells compared to that achieved by high lifetime n-type Cz solar cells, likely due to surface-related effects, which reduce the lifetime at low injection levels [128].

However, defect engineering processes are not only suitable for p-type materials for SHJ applications, but also relevant for n-type wafers. A number of studies have demonstrated improvements for n-type solar cells through various processes such as tabula rasa, gettering, rapid thermal annealing and hydrogenation [160, 166, 167]. For example, rapid thermal annealing at a temperature of 850 °C has been used to avoid a LID mechanism affecting n-type Cz material, likely related to thermal donors [167]. A combination of phosphorus diffusion gettering and hydrogenation, or a combination of tabula rasa and phosphorus diffusion gettering led to improvements in the absolute efficiency of laboratory n-type SHJ solar cells fabricated with UMG Cz wafers by $3.2\%_{abs}$ and $3.0\%_{abs}$ absolute, to reach efficiencies of 21.2% and 21.0%, respectively (see Table 4.3). The same processes for laboratory cells fabricated using electronic grade n-type Cz wafers led to efficiency increases of $1.5\%_{abs}$ and $1.2\%_{abs}$, respectively [160]. In that work, hydrogenation was achieved using the deposition of PECVD SiN_x:H and subsequent rapid thermal annealing process.

Such processes are gaining attention of industrial SHJ solar cell manufacturers. Collaborative work between the UNSW and Hevel Solar demonstrated an efficiency gain of 0.55% absolute from 22.25% to 22.80% for Hevel's standard wafers used in mass production through the use of a pre-fabrication gettering process, before the industrial n-type SHJ solar cell fabrication process. This is despite the wafers starting with multi-millisecond lifetimes, clearly highlighting that attention should also be paid to defect engineering for high-lifetime n-type wafers for opportunities to increase the lifetime and efficiency even further.

4.7 Summary

Hydrogenated dielectric layers used for the fabrication of silicon solar cells such as PECVD SiN_X:H provide an excellent source of hydrogen for bulk defect passivation. This has been critical for the success of p-type silicon solar cells technologies such as PERC. Bulk defects in silicon solar cells can be broadly placed into a number of categories such as structural, impurity-related, process-induced or carrier-induced. However, these categories can often be blurred, with the defects falling into multiple categories. This includes impurity-related defects associated with oxygen, which can evolve during solar cell processing or even during operation in the field. In addition, impurity-related defects can potentially lead to the formation of structural defects.

Typically, structural defects such as grain boundaries are the easiest to passivate, given that they are present in the material before solar cell fabrication, and are therefore available for hydrogen passivation during the fast-firing process. Nonetheless, the ease of passivation is impacted by grain boundary angle, and can be impeded by impurity decoration. Regions with high dislocation cluster densities, however, are more difficult to passivate, often with significant recombination activity remaining after the conventional firing process. It is unclear whether this is due to impurity decoration, a local depletion of hydrogen, or another reason.

Often there is a benefit of adding an extra hydrogen passivation process after firing, such as an illuminated anneal at a lower temperature of ~300 °C for challenging defects such as oxygen precipitates or dislocation clusters, which can improve efficiencies by up to 1–2% absolute. For process-induced defects, when defects are introduced before firing, such as that introduced with the LCO process for PERC or oxygen precipitates, a natural passivation may occur during solar cell fabrication. On the other hand, if defects are introduced after firing, the solar cell can benefit from the addition of another hydrogenation process.

The most challenging bulk defects are carrier-induced, as they have not formed by the time of the hydrogen passivation process. The passivation of the carrier- or light-induced boron–oxygen defect will be discussed in detail in Chapter 5.

However, hydrogenation should not necessarily be used alone. Gettering, as is used to remove metallic impurities can greatly enhance the lifetime silicon material, is quite complementary to hydrogenation and even improves the response to hydrogen passivation processes.

All materials can potentially benefit from hydrogen passivation and gettering. Rather than simply throwing out a 'reject' wafer, the impact of the solar cell fabrication process on the lifetime of the material be investigated, along with additional opportunities to improve lifetime and therefore performance. For example, the lifetime of n-type Cz wafers can be improved from <100 μs to more than 3 ms. However, care must be taken. A hydrogen passivation or gettering process beneficial for a given defect or region of the wafer, may in fact be detrimental for another.

The processes of emitter diffusion, SiN_X:H deposition and firing, naturally incorporated into fabrication sequence for PERC and TOPCon solar cells, can also be used before SHJ fabrication to result in over 80 mV improvement in the V_{OC} of p-type devices and enable p-type mc-Si solar cells with V_{OC} > 700 mV. Such processes can also benefit industrial SHJ solar cells fabricated using high-lifetime n-type Cz material, increasing efficiency by >0.5% absolute.

Key future work will involve assessing the potential use of processes such as gettering and hydrogen passivation bulk defects in solar cells with emerging passivated contact structures.

References

[1] Chen F., Romijn I., Weeber A., Tan J., Hallam B., Cotter J. 'Relationship between PECVD silicon nitride film composition and surface and edge passivation'. Proceedings of 22nd European Photovoltaic Solar Energy Conference and Exhibition, Milan; 2007. pp. 1053–60.

[2] Hallam B., Tjahjono B., Wenham S. 'Effect of PECVD silicon oxynitride film composition on the surface passivation of silicon wafers'. *Solar Energy Materials and Solar Cells*. 2012;**96**(17):173–9.

[3] Wilking S., Herguth A., Hahn G. 'Influence of hydrogenated passivation layers on the regeneration of boron-oxygen related defects'. *Energy Procedia*. 2013;**38**(5):642–8.

[4] Dubois S., Tanay F., Veirman J., Enjalbert N., Stendera J., Butté S. 'The BOLID project: suppression of the boron-oxygen related light-induced-degradation'. European Photovoltaic Solar Energy Conference; 2012. pp. 749–54.

[5] Soppe W., Rieffe H., Weeber A. 'Bulk and surface passivation of silicon solar cells accomplished by silicon nitride deposited on industrial scale by microwave PECVD'. *Progress in Photovoltaics: Research and Applications*. 2005;**13**(7):551–69.

[6] Hallam B., Chen D., Kim M., *et al*. 'The role of hydrogenation and gettering in enhancing the efficiency of next-generation Si solar cells: an industrial perspective'. *Physica Status Solidi*. 2017;**214**(7):1700305.

[7] ITRPV.I. ITRPV 2019 including maturity report. ITRPV [Internet]. 2019. Available from https://itrpv.vdma.org/en/ueber-uns [Accessed 18th January 2021].

[8] ITRPV. International technology roadmap for photovoltaic, ITRPV. 2020 [Accessed 16th February 2021].

[9] Aberle A.G. 'Overview on sin surface passivation of crystalline silicon solar cells'. *Solar Energy Materials and Solar Cells*. 2001;**65**(1):239–48.

[10] Central P., Sopori B., Central P., Sopori B., Central P., Sopori B. 'Silicon solar-cell processing for minimizing the influence of impurities and defects'. *Journal of Electronic Materials*. 2002;**31**(10):972–80.

[11] Green M., Dunlop E., Hohl-Ebinger J., Yoshita M., Kopidakis N., Hao X. 'Solar cell efficiency tables (version 57)'. *Progress in Photovoltaics: Research and Applications*. 2021;**29**(1):3–15.

[12] Falster R.J., Cornara M., Gambaro D., Olmo M., Pagani M. 'Effect of high temperature pre-anneal on oxygen precipitates nucleation kinetics in Si'. *Solid State Phenomena*. 1997;**57-58**:123–8.

[13] Hallam B., Chan C., Abbott M., Wenham S. 'Hydrogen passivation of defect-rich N-type Czochralski silicon and oxygen precipitates'. *Solar Energy Materials and Solar Cells*. 2015;**141**(4):125–31.

[14] Herguth A., Schubert G., Kaes M., Hahn G. 'A new approach to prevent the negative impact of the metastable defect in boron doped CZ silicon solar cells'. 2006 IEEE 4th World Conference on Photovoltaic Energy Conference. IEEE; 2006. pp. 940–3.

[15] Münzer K. 'Hydrogenated silicon nitride for regeneration of light induced degradation'. 24th European Photovoltaic Solar Energy Conference; 2009. pp. 1558–61.

[16] Herguth A., Schubert G., Kaes M., Hahn G. 'Investigations on the long time behavior of the metastable boron–oxygen complex in crystalline silicon'. *Progress in Photovoltaics: Research and Applications*. 2008;**16**(2):135–40.

[17] Chen R., Tong H., Zhu H., *et al.* '23.83% Efficient mono-PERC incorporating advanced hydrogenation'. *Progress in Photovoltaics: Research and Applications*. 2020;**28**(12):1239–47.

[18] Lan C.W., Hsu C., Nakajima K. 'Multicrystalline silicon crystal growth for photovoltaic applications'. *Handbook of Crystal Growth*. 2nd ed. Elsevier; 2015. pp. 373–411.

[19] Lan C.W., Yang C.F., Lan A., *et al.* 'Engineering silicon crystals for photovoltaics'. *CrystEngComm*. 2016;**18**(9):1474–85.

[20] Sio H.C., Phang S.P., Fell A., *et al.* 'The electrical properties of high performance multicrystalline silicon and mono-like silicon: material limitations and cell potential'. *Solar Energy Materials and Solar Cells*. 2019;**201**(3):110059.

[21] Takahashi I., Usami N., Kutsukake K., Stokkan G., Morishita K., Nakajima K. 'Generation mechanism of dislocations during directional solidification of multicrystalline silicon using artificially designed seed'. *Journal of Crystal Growth*. 2010;**312**(7):897–901.

[22] Ryningen B., Stokkan G., Kivambe M., Ervik T., Lohne O. 'Growth of dislocation clusters during directional solidification of multicrystalline silicon ingots'. *Acta Materialia*. 2011;**59**(20):7703–10.

[23] Gindner S., Karzel P., Herzog B., Hahn G. 'Efficacy of phosphorus gettering and hydrogenation in multicrystalline silicon'. *IEEE Journal of Photovoltaics*. 2014;**4**(4):1063–70.

[24] Sio H.C., Phang S.P., Trupke T., Macdonald D. 'Impact of phosphorous gettering and hydrogenation on the surface recombination velocity of grain boundaries in P-type multicrystalline silicon'. *IEEE Journal of Photovoltaics*. 2015;**5**(5):1357–65.

[25] Chen J., Sekiguchi T., Yang D., Yin F., Kido K., Tsurekawa S. 'Electron-beam-induced current study of grain boundaries in multicrystalline silicon'. *Journal of Applied Physics*. 2004;**96**(10):5490–5.

[26] Martinuzzi S., Périchaud I., Warchol F. 'Hydrogen passivation of defects in multicrystalline silicon solar cells'. *Solar Energy Materials and Solar Cells*. 2003;**80**(3):343–53.

[27] Lüdemann R. 'Hydrogen passivation of multicrystalline silicon solar cells'. *Materials Science and Engineering: B*. 1999;**58**(1):86–90.

[28] Chen Z., Rohatgi A., Bell R.O., Kalejs J.P. 'Defect passivation in multicrystalline-Si materials by plasma-enhanced chemical vapor deposition of SiO_2/SiN coatings'. *Applied Physics Letters*. 1994;**65**(16):2078–80.

[29] Duerinckx F., Szlufcik J. 'Defect passivation of industrial multicrystalline solar cells based on PECVD silicon nitride'. *Solar Energy Materials and Solar Cells*. 2002;**72**(1):231–46.

[30] Einhaus R., Duerinckx F., Van Kerschaver E., *et al*. 'Hydrogen passivation of newly developed EMC-multi-crystalline silicon'. *Materials Science and Engineering: B*. 1999;**58**(1):81–5.

[31] Trupke T., Bardos R.A., Schubert M.C., Warta W. 'Photoluminescence imaging of silicon wafers'. *Applied Physics Letters*. 2006;**89**(4):44107.

[32] Hallam B.J., Ciesla A.M., Chan C.C., *et al*. 'Overcoming the challenges of hydrogenation in silicon solar cells'. *Australian Journal of Chemistry*. 2018;**71**(10):743–52.

[33] Leamy H.J. 'Charge collection scanning electron microscopy'. *Journal of Applied Physics*. 1982;**53**(6):R51–80.

[34] Corkish R., Puzzer T., Sproul A.B., Luke K.L. 'Quantitative interpretation of electron-beam-induced current grain boundary contrast profiles with application to silicon'. *Journal of Applied Physics*. 1998;**84**(5473):5473–81.

[35] Paduthol A. *Contactless Specral Response Measurement of Solar Cells using Photoluminescence*. The University of New South Wales; 2019.

[36] Thantsha N.M., Macabebe E.Q.B., Vorster F.J., van Dyk E.E. 'Optoelectronic analysis of silicon solar cells by LBIC investigations and current–voltage characterization'. *Physica B: Condensed Matter*. 2009;**404**(22):4445–8.

[37] Tweddle D., Hamer P., Shen Z., Markevich V.P., Moody M.P., Wilshaw P.R. 'Direct observation of hydrogen at defects in multicrystalline silicon'. *Progress in Photovoltaics: Research and Applications*. 2019;**215**(2):1–7.

[38] Hallam B., Tjahjono B., Trupke T., Wenham S. 'Photoluminescence imaging for determining the spatially resolved implied open circuit voltage of silicon solar cells'. *Journal of Applied Physics*. 2014;**115**(4):44901.

[39] Chen J., Sekiguchi T., Yang D. 'Electron-beam-induced current study of grain boundaries in multicrystalline Si'. *Physica Status Solidi*. 2007;**4**(8):2908–17.

[40] Chen J., Yang D., Xi Z., Sekiguchi T. 'Electron-beam-induced current study of hydrogen passivation on grain boundaries in multicrystalline silicon: influence of GB character and impurity contamination'. *Physica B: Condensed Matter*. 2005;**364**(1):162–9.

[41] Sio H.C., Phang S.P., Zheng P., *et al*. 'Recombination sources in P-type high performance multicrystalline silicon'. *Japanese Journal of Applied Physics*. 2017;**56**(8):08MB16.

[42] Hallam B.J., Hamer P.G., Wenham Ciesla née., Chan C.E., Vicari Stefani B., Wenham S. 'Development of advanced hydrogenation processes for silicon solar cells via an improved understanding of the behaviour of hydrogen in silicon'. *Progress in Photovoltaics: Research and Applications*. 2018;**2020**:1–22.

[43] Sheoran M. 'Development of high-efficiency solar cells on thin silicon through design optimization and defect passivation'. *Georgia Institute of Technology*. 2009.

[44] Sheoran M., Upadhyaya A., Rohatgi A. 'Bulk lifetime and efficiency enhancement due to gettering and hydrogenation of defects during cast multicrystalline silicon solar cell fabrication'. *Solid-State Electronics*. 2008;**52**(5):612–17.

[45] Pearton S.J., Corbett J.W., Shi T.S. 'Hydrogen in crystalline semiconductors'. *Applied Physics A Solids and Surfaces*. 1987;**43**(3):153–95.

[46] Pankove J.I., Johnson N.M. *Hydrogen in Semiconductors: Hydrogen in Silicon*. **34**. Academic Press; 1991.

[47] Kim M., Hamer P., Li H., *et al*. 'Impact of thermal processes on multicrystalline silicon'. *Frontiers in Energy*. 2017;**11**(1):32–41.

[48] Dingemans G., Engelhart P., Seguin R., *et al*. 'Stability of Al_2O_3 and Al_2O_3/a-SiNx:H stacks for surface passivation of crystalline silicon'. *Journal of Applied Physics*. 2009;**106**(11):114907.

[49] Sio H.C., Phang S.P., Nguyen H.T., Hameiri Z., Macdonald D. 'Hydrogenation in multicrystalline silicon: the impact of dielectric film properties and firing conditions'. *Progress in Photovoltaics: Research and Applications*. 2020;**28**(6):493–502.

[50] Lelièvre J.-F., Fourmond E., Kaminski A., Palais O., Ballutaud D., Lemiti M. 'Study of the composition of hydrogenated silicon nitride SiNx:H for efficient surface and bulk passivation of silicon'. *Solar Energy Materials and Solar Cells*. 2009;**93**(8):1281–9.

[51] Hong J., Kessels W.M.M., Soppe W.J., Weeber A.W., Arnoldbik W.M., van de Sanden M.C.M. 'Influence of the high-temperature "firing" step on high-rate plasma deposited silicon nitride films used as bulk passivating antireflection coatings on silicon solar cells'. *Journal of Vacuum Science & Technology B: Microelectronics and Nanometer Structures*. 2003;**21**(5):2123–32.

[52] Yin Z., Smith F.W. 'Free-energy model for bonding in amorphous covalent alloys'. *Physical Review B*. 1991;**43**(5):4507–10.

[53] Dekkers H.F.W., Beaucarne G., Hiller M., Charifi H., Slaoui A. 'Molecular hydrogen formation in hydrogenated silicon nitride'. *Applied Physics Letters*. 2006;**89**(21):211914–31.

[54] Chan C.E., Payne D.N.R., Hallam B.J., *et al*. 'Rapid stabilization of high-performance multicrystalline P-type silicon PERC cells'. *IEEE Journal of Photovoltaics*. 2016;**6**(6):1473–9.

[55] Guo X. *Optimal Thermal Processes for Pre-Hydrogenation of Heterojunction Solar Cells*. University of New South Wales; 2018.

[56] Hamer P., Hallam B., Bonilla R.S., Altermatt P.P., Wilshaw P., Wenham S. 'Modelling of hydrogen transport in silicon solar cell structures under equilibrium conditions'. *Journal of Applied Physics*. 2018;**123**(4):043108.

[57] Zhang Y. 'Reactivation of defect recombination during thermal processing'. *UNSW*. 2018.

[58] Karzel P., Junge J., Hahn G. 'Mapping of hydrogen bond energies in EFG silicon samples by analysis of spatially resolved minority charge carrier lifetimes after annealing steps'. Proceedings of 24th European Photovoltaic Solar Energy Conference and Exhibition, Hamburg, Germany; 2009. pp. 2023–7.

[59] Wilking S., Ebert S., Herguth A., Hahn G. 'Influence of hydrogen effusion from hydrogenated silicon nitride layers on the regeneration of boron-oxygen related defects in crystalline silicon'. *Journal of Applied Physics*. 2013;**114**(19):194512.

[60] Eberle R., Kwapil W., Schindler F., Schubert M.C., Glunz S.W. 'Impact of the firing temperature profile on light induced degradation of multicrystalline silicon'. *Phys Status Solidi – Rapid Research Letters*. 2016;**5**:1–5.

[61] Wu B., Stoddard N., Ma R., Clark R. 'Bulk multicrystalline silicon growth for photovoltaic (PV) application'. *Journal of Crystal Growth*. 2008;**310**(7–9):2178–84.

[62] Stoddard N., Wu B., Witting I., Wagener M.C., Park Y., Rozgonyi G.A. 'Casting single crystal silicon: novel defect profiles from BP Solar's mono2 TM wafers'. *Solid State Phenomena*. 2008;**131**:1–8.

[63] Takahashi I., Joonwichien S., Iwata T., Usami N. 'Seed manipulation for artificially controlled defect technique in new growth method for quasi-monocrystalline Si ingot based on casting'. *Applied Physics Express*. 2015;**8**(10):105501–4.

[64] Hu D., Yuan S., He L., *et al.* 'Higher quality mono-like cast silicon with induced grain boundaries'. *Solar Energy Materials and Solar Cells*. 2015;**140**(2):121–5.

[65] Gu X., Yu X., Guo K., Chen L., Wang D., Yang D. 'Seed-assisted cast quasi-single crystalline silicon for photovoltaic application: towards high efficiency and low cost silicon solar cells'. *Solar Energy Materials and Solar Cells*. 2012;**101**:95–101.

[66] Kutsukake K., Usami N., Ohno Y., Tokumoto Y., Yonenaga I. 'Control of grain boundary propagation in mono-like Si: utilization of functional grain boundaries'. *Applied Physics Express*. 2013;**6**(2):025505.

[67] Buchovska I., Liaskovskiy O., Vlasenko T., Beringov S., Kiessling F.M. 'Different nucleation approaches for production of high-performance multi-crystalline silicon ingots and solar cells'. *Solar Energy Materials and Solar Cells*. 2017;**159**:128–35.

[68] ITRPV. 'International technology roadmap for photovoltaics (ITRPV.net): results 2011'. *ITRPV Ger*. 2012;**15**.

[69] ITRPV. 'International technology roadmap for photovoltaics (ITRPV.net): results 2014'. *ITRPV Ger*. 2015;**1**:Rev.1.

[70] Hallam B.J., Hamer P.G., Wang S., *et al.* 'Advanced hydrogenation of dislocation clusters and boron-oxygen defects in silicon solar cells'. *Energy Procedia.* 2015;**77**(10):799–809.
[71] Wenham A., Song L., Abbott M., *et al.* 'Defect passivation on cast-mono crystalline screen-printed cells'. *Frontiers in Energy.* 2017;**11**(1):60–6.
[72] Song L., Wenham A., Wang S., *et al.* 'Laser enhanced hydrogen passivation of silicon wafers'. *International Journal of Photoenergy.* 2015;**2015**(4):1–13.
[73] Ciesla A.M., Chan C.E., Wang S., Abbott M.D., Chong C., Wenham S.R. 'Laser hydrogenation on heavily dislocated cast-mono silicon cells'. Proceedings of 44th IEEE Photovoltaic Specialists Conference, June; 2017. pp. 2600–4.
[74] Schmidt J., Lim B., Walter D., *et al.* 'Impurity-related limitations of next-generation industrial silicon solar cells'. *IEEE Journal of Photovoltaics.* 2013;**3**(1):114–18.
[75] Grant N.E., Rougieux F.E., Macdonald D., Bullock J., Wan Y. 'Grown-in defects limiting the bulk lifetime of *p*-type float-zone silicon wafers'. *Journal of Applied Physics.* 2015;**117**(5):055711.
[76] Mullins J., Leonard S., Markevich V.P., *et al.* 'Recombination via transition metals in solar silicon: the significance of hydrogen–metal reactions and lattice sites of metal atoms'. *Physica Status Solidi.* 2017;**214**(7):1700304.
[77] Singh R., Fonash S.J., Rohatgi A. 'Interaction of low-energy implanted atomic H with slow and fast diffusing metallic impurities in Si'. *Applied Physics Letters.* 1986;**49**(13):800–2.
[78] Tavendale A.J., Pearton S.J. 'Deep level, quenched-in defects in silicon doped with gold, silver, iron, copper or nickel'. *Journal of Physics C: Solid State Physics.* 1983;**16**(9):1665–73.
[79] Yarykin N., Sachse J.-U., Lemke H., Weber J. 'Silver-hydrogen interactions in crystalline silicon'. *Physical Review B.* 1999;**59**(8):5551–60.
[80] Andersen O., Peaker A.R., Dobaczewski L., *et al.* 'Electrical activity of carbon-hydrogen centers in Si'. *Physical Review B.* 2002;**66**(23).
[81] Buonassisi T., Istratov A.A., Pickett M.D., *et al.* 'Chemical natures and distributions of metal impurities in multicrystalline silicon materials'. *Progress in Photovoltaics: Research and Applications.* 2006;**14**(6):513–31.
[82] Wang H., Chroneos A., Londos C.A., Sgourou E.N., Schwingenschlögl U. 'Carbon related defects in irradiated silicon revisited'. *Scientific Reports.* 2014;**4**:1–9.
[83] Johnson N.M., Hahn S.K. 'Hydrogen passivation of the oxygen-related thermal-donor defect in silicon'. *Applied Physics Letters.* 1986;**48**(11):709–11.
[84] Watkins G.D. 'Intrinsic defects in silicon'. *Materials Science in Semiconductor Processing.* 2000;**3**(4):227–35.
[85] Roberson M.A., Estreicher S.K. 'Vacancy and vacancy-hydrogen complexes in silicon'. *Physical Review B.* 1994;**49**(24):17040–9.
[86] Hallam B.J., Hamer P.G., Wenham S.R., *et al.* 'Advanced bulk defect passivation for silicon solar cells'. *IEEE Journal of Photovoltaics.* 2014;**4**(1):88–95.

[87] Hallam B. *High Efficiency Laser-Doped Silicon Solar Cells with Advanced Hydrogenation*. PhD thesis: University of New South Wales; 2014.

[88] Abe T., Harada H., Ozawa N., Adomi K. 'Deep level generation-annihilation in nitrogen doped FZ crystals'. *MRS Proceedings*. 1985;**59**:537–44.

[89] Rougieux F.E., Grant N.E., Barugkin C., Macdonald D., Murphy J.D. 'Influence of annealing and bulk hydrogenation on lifetime-limiting defects in nitrogen-doped floating zone silicon'. *IEEE Journal of Photovoltaics*. 2015;**5**(2):495–8.

[90] Hiller D., Markevich V.P., de Guzman J.A.T., *et al.* 'Kinetics of bulk lifetime degradation in float-zone silicon: fast activation and annihilation of grown-in defects and the role of hydrogen versus light'. *Physica Status Solidi*. 2020;**217**(17):2000436–13.

[91] Vaqueiro-Contreras M., Markevich V.P., Coutinho J., *et al.* 'Identification of the mechanism responsible for the boron oxygen light induced degradation in silicon photovoltaic cells'. *Journal of Applied Physics*. 2019;**125**(18):185704.

[92] Mchedlidze T., Weber J. 'Direct detection of carrier traps in Si solar cells after light-induced degradation'. *Physica Status Solidi (RRL) – Rapid Research Letters*. 2015;**9**(2):108–10.

[93] Liu A.Y., Sun C., Markevich V.P., Peaker A.R., Murphy J.D., Macdonald D. 'Gettering of interstitial iron in silicon by plasma-enhanced chemical vapour deposited silicon nitride films'. *Journal of Applied Physics*. 2016;**120**(19):193103.

[94] Liu A., Sun C., Macdonald D. 'Hydrogen passivation of interstitial iron in boron-doped multicrystalline silicon during annealing'. *Journal of Applied Physics*. 2014;**116**(19):194902.

[95] Sah Chih-Tang., Sun Jack Yuan-Chen., Tzou Joseph Jeng-Tao., Sah C.T., SunJ.Y.C., Tzou J.J.T. 'Deactivation of the boron acceptor in silicon by hydrogen'. *Applied physics letters*. 1983;**43**(2):204–6.

[96] Pankove J.I., Zanzucchi P.J., Magee C.W., Lucovsky G. 'Hydrogen localization near boron in silicon'. *Applied Physics Letters*. 1985;**46**(4):421–3.

[97] Johnson N.M., Herring C., Chadi D.J. 'Interstitial hydrogen and neutralization of shallow-donor impurities in single-crystal silicon'. *Physical Review Letters*. 1986;**56**(7):769–72.

[98] Pankove J.I., Wance R.O., Berkeyheiser J.E. 'Neutralization of acceptors in silicon by atomic hydrogen'. *Applied Physics Letters*. 1984;**45**(10):1100–2.

[99] Zundel T., Mesli A., Muller J.C., Siffert P., Festkrperforschung M. 'Boron neutralization and hydrogen diffusion in silicon subjected to low-energy hydrogen implantation'. *Applied Physics A Solids and Surfaces*. 1989;**48**(1):31–40.

[100] Seager C.H., Anderson R.A. 'Two-step debonding of hydrogen from boron acceptors in silicon'. *Applied Physics Letters*. 1991;**59**(5):585–7.

[101] Herrero C.P., Stutzmann M., Breitschwerdt A. 'Boron-hydrogen complexes'. *Physical Review. B*. 1991;**43**(2):1555–75.

[102] Bourret-Sicotte G., Hamer P., Bonilla R.S., *et al.* 'Shielded hydrogen passivation – a potential in-line passivation process'. *Physica Status Solidi.* 2017;**214**(7):1700383.

[103] Shumate S.D., Hutchings D.A., Mohammed H., Beilke G., Newton B.S., Young M.G. 'Self aligned hydrogenated selective emitter for n-type solar cells'. Proceedings of the 38th IEEE Photovoltaic Specialists Conference; 2012. pp. 1110–14.

[104] Stavola M., Pearton S.J., Lopata J., Dautremont-Smith W.C. 'Vibrational spectroscopy of acceptor-hydrogen complexes in silicon: evidence for low-frequency excitations'. *Physical Review B.* 1988;**37**(14):8313–18.

[105] Zundel T., Weber J. 'Boron reactivation kinetics in hydrogenated silicon after annealing in the dark or under illumination'. *Physical Review B.* 1991;**43**(5):4361–72.

[106] Yarykin N. 'Minority-carrier-enhanced dissociation of the boron–hydrogen pair in silicon'. *Physica B: Condensed Matter.* 2009;**404**(23–24):5093–5.

[107] Hallam B., Sugianto A., Mai L., *et al.* 'Hydrogen passivation of laser-induced defects for laser-doped silicon solar cells'. *IEEE Journal of Photovoltaics.* 2014;**4**(6):1413–20.

[108] Freeland P.E., Jackson K.A., Lowe C.W., Patel J.R. 'Precipitation of oxygen in silicon'. *Applied Physics Letters.* 1977;**30**(1):31–3.

[109] Voronkov V.V., Falster R. 'Grown-in microdefects, residual vacancies and oxygen precipitation bands in Czochralski silicon'. *Journal of Crystal Growth.* 1999;**204**(4):462–74.

[110] Haunschild J., Reis I.E., Geilker J., Rein S. 'Detecting efficiency-limiting defects in Czochralski-grown silicon wafers in solar cell production using photoluminescence imaging'. *Physica Status Solidi (RRL) - Rapid Research Letters.* 2011;**5**(5–6):199–201.

[111] Cousins P.J., Smith D.D., Luan H.-C., Manning J., Dennis T.D., Waldhauer A. 'Generation 3: improved performance at lower cost'. 35th IEEE Photovoltaic Specialists Conference; 2010. pp. 275–8.

[112] Walter D., Lim B., Falster R., Binns J., Schmidt J. 'Understanding lifetime degradation in czochralski-grown N-type silicon after high-temperature processing'. Proceedings of 28th European Photovoltaic Solar Energy Conference and Exhibition; 2013. pp. 699–702.

[113] Haunschild J., Glatthaar M., Demant M., *et al.* 'Quality control of as-cut multicrystalline silicon wafers using photoluminescence imaging for solar cell production'. *Solar Energy Materials and Solar Cells.* 2010;**94**(12):2007–12.

[114] Manshanden P., Bronsveld P.C.P. 'Investigation of hydrogenation in N-type wafers with ring- and disc-shaped defect zones'. *Energy Procedia.* 2016;**92**:857–66.

[115] Murphy J.D., McGuire R.E., Bothe K., Voronkov V.V., Falster R.J. 'Minority carrier lifetime in silicon photovoltaics: the effect of oxygen precipitation'. *Solar Energy Materials and Solar Cells.* 2014;**120**:402–11.

[116] Li J., Song L., Yu X., Yang D. 'Effects of iron contamination and hydrogen passivation on the electrical properties of oxygen precipitates in CZ-Si'. *Journal of Electronic Materials*. 2018;**47**(9):5039–44.

[117] Coletti G., Manshanden P., Bernardini S., *et al.* 'Removing the effect of striations in N-type silicon solar cells'. *Solar Energy Materials and Solar Cells*. 2014;**130**:647–51.

[118] Engelhart P., Grischke R., Eidelloth S., Meyer R., Schoonderbeek A., Stute U. 'Laser processing for back-contacted silicon solar cells'. *International Congress on Applications of Lasers & Electro-Optics*. 2006;**M703**.

[119] Engelhart P., Hermann S., Neubert T., *et al.* 'Laser ablation of SiO2 for locally contacted Si solar cells with ultra-short pulses'. *Progress in Photovoltaics: Research and Applications*. 2007;**15**(6):521–7.

[120] Hallam B., Urueña A., Russell R., *et al.* 'Efficiency enhancement of i-PERC solar cells by implementation of a laser doped selective emitter'. *Solar Energy Materials and Solar Cells*. 2015;**134**(4):89–98.

[121] Aleman M., de Castro A., Choulat P., Hallam B., Dang C., Russell R. 'Reducing front recombination losses to improve the efficiency of rear junction Cu-plated n-Si cells'. 29th European Photovoltaic Solar Energy Conference; Amsterdam; 2014. pp. 528–31.

[122] Tous L. 'Nickel/copper plated contacts as an alternative to silver screen printing for the front side metallization of industrial high efficiency silicon solar cells'. *KU Leuven*. 2014.

[123] Urueña A., Aleman M., Cornagliotti E., *et al.* 'Progress on large area N-type silicon solar cells with front laser doping and a rear emitter'. *Progress in Photovoltaics: Research and Applications*. 2016;**24**(March):1149–56.

[124] Wang Y. 'Mono wafer updates as substrate for high efficiency solar cell'. Oral Presentation at the 3rd International Silicon Heterojunction Workshop (Virtual); 2020.

[125] Macdonald D., Cheung A., Cuevas A. 'Gettering and poisoning of silicon wafers by phosphorus diffused layers'. Proceedings of the 3rd World Conference on Photovoltaic Energy Conversion; 2003. pp. 11–14.

[126] Taguchi M., Terakawa A., Maruyama E., Tanaka M. 'Obtaining a higher-Voc in HIT cells'. *Progress in Photovoltaics: Research and Applications*. 2005;**13**(6):481–8.

[127] Adamczyk K., Søndenå R., Stokkan G., *et al.* 'Recombination activity of grain boundaries in high-performance multicrystalline Si during solar cell processing'. *Journal of Applied Physics*. 2018;**123**(5):055705.

[128] Chen D., Kim M., Shi J., Vicari Stefani B., Yu Z., Liu S. 'Defect engineering of P-type silicon heterojunction solar cells fabricated using commercial-grade low-lifetime silicon wafers'. *Progress in Photovoltaics: Research and Applications*. 2018:1–15.

[129] Wang X., Market G.P.V. 'Oral presentation at the world solar Congress, Shanghai'. *Bloomberg New Energy Finance*. 2016.

[130] Wagner H., Dastgheib-Shirazi A., Min B., *et al.* 'Optimizing phosphorus diffusion for photovoltaic applications: peak doping, inactive

phosphorus, gettering, and contact formation'. *Journal of Applied Physics*. 2016;**119**(18):185704.

[131] Henley W.B., Ramappa D.A. 'Iron precipitation in float zone grown silicon'. *Journal of Applied Physics*. 1997;**82**(2):589–94.

[132] Hannebauer H., Dullweber T., Wyczanowski S., Weise K., Delahaye F., Doll O. 'Gas phase etch back: a new selective emitter technology for high-efficiency PERC solar cells'. 28th European Photovoltaic Solar Energy Conference; 2013. pp. 752–6.

[133] Dumbrell R., Hamer P., Lennon A., Tjahjono B. 'Emitter etch-back application for selective emitter silicon solar cells'. Proceedings of 2010 Australian Solar Energy Society Conference; Canberra, Australia; 2010.

[134] Macdonald D.H., Geerligs L.J., Azzizi A. 'Iron detection in crystalline silicon by carrier lifetime measurements for arbitrary injection and doping'. *Journal of Applied Physics*. 2004;**95**(3):1021–8.

[135] Hamer P., Hallam B., Chan C., Wenham A., Sugianto A., Abbott M. 'Hydrogen passivation mechanisms in silicon solar cells'. Proceedings of the 28th European Photovoltaic Solar Energy Conference; Paris, France; 2013. pp. 982–7.

[136] Voronkov V.V., Falster R. 'Formation, dissociation, and diffusion of various hydrogen dimers in silicon'. *Physica Status Solidi*. 2017;**254**(6):1600779.

[137] Steinkemper H., Hermle M., Glunz S.W. 'Comprehensive simulation study of industrially relevant silicon solar cell architectures for an optimal material parameter choice'. *Progress in Photovoltaics: Research and Applications*. 2016;**24**(10):1319–31.

[138] Mulligan W.P., Cudzinovic M.J., Pass T., Smith D., Kaminar N., McIntosh K. *Solar cell and method of manufacture*. US Patent 7897867; 2008.

[139] Taguchi M., Kawamoto K., Tsuge S., *et al*. 'HITTM cells—high-efficiency crystalline Si cells with novel structure'. *Progress in Photovoltaics: Research and Applications*. 2000;**8**(5):503–13.

[140] Heng J.B., Fu J., Kong B., *et al*. '>23% High-efficiency tunnel oxide junction bifacial solar cell with electroplated Cu gridlines'. *IEEE Journal of Photovoltaics*. 2015;**5**(1):82–6.

[141] Mulligan W.P., Cudzinovic M.J., Pass T., Smith D., Swanson R.M. *Metal contact structure for solar cell and method of manufacture*. US Patent 7388147; 2008.

[142] Turner A.B., Schultz-Wuttnann O., De Ceuster D., Crafts D.E. *Selective transformation in functional films, and solar cell applications thereof*. US Patent 20120186649; 2010.

[143] Liu A., Yan D., Phang S.P., Cuevas A., Macdonald D. 'Effective impurity gettering by phosphorus- and boron-diffused polysilicon passivating contacts for silicon solar cells'. *Solar Energy Materials and Solar Cells*. 2018;**179**:136–41.

[144] Liu C., Chen D., Chen Y., *et al*. 'Industrial TOPCon solar cells on n-type quasi-mono Si wafers with efficiencies above 23%'. *Solar Energy Materials and Solar Cells*. 2020;**215**:110690.

[145] Li Z., Yang Y., Zhang X., Liu W., Chen Y., Xu G. 'Pilot production of 6 Ibc solar cells yielding a median efficiency of 23% with a low-cost industrial process'. Proceedings of the 32nd European Photovoltaic Solar Energy Conference; 2016. pp. 571–4.

[146] Nampalli N., Hallam B.J., Chan C.E., Abbott M.D., Wenham S.R. 'Influence of hydrogen on the mechanism of permanent passivation of boron–oxygen defects in p-type Czochralski silicon'. *IEEE Journal of Photovoltaics*. 2015;**5**(6):1580–5.

[147] Chavali R.V.K., De Wolf S., Alam M.A. 'Device physics underlying silicon heterojunction and passivating-contact solar cells: a topical review'. *Progress in Photovoltaics: Research and Applications*. 2018;**26**(4):241–60.

[148] Yoshikawa K., Kawasaki H., Yoshida W., *et al.* 'Silicon heterojunction solar cell with interdigitated back contacts for a photoconversion efficiency over 26%'. *Nature Energy*. 2017;**2**(5):17032.

[149] Masuko K., Shigematsu M., Hashiguchi T., *et al.* 'Achievement of more than 25% conversion efficiency with crystalline silicon heterojunction solar cell'. *IEEE Journal of Photovoltaics*. 2014;**4**(6):1433–5.

[150] McIntosh K.R., Altermatt P.P. 'A freeware 1D emitter model for silicon solar cells'. Proceedings of the 35th IEEE Photovoltaic Specialists Conference; 2010. pp. 2188–93.

[151] Shen H., Walter D., Wu Y., *et al.* 'Monolithic perovskite/Si tandem solar cells: pathways to over 30% efficiency'. *Advanced Energy Materials*. 2020;**10**(13):1902840–18.

[152] Sheng R., Ho-Baillie A., Huang S., *et al.* 'Methylammonium lead bromide perovskite-based solar cells by vapor-assisted deposition'. *The Journal of Physical Chemistry C*. 2015;**119**(7):3545–9.

[153] Street R.A. *Hydrogenated Amorphous Silicon*. Cambridge, UK: Cambridge University Press; 1991.

[154] Hallam B., Chen D., Shi J., Einhaus R., Holman Z.C., Wenham S. 'Prefabrication gettering and hydrogenation treatments for silicon heterojunction solar cells: a possible path to >700 mV open-circuit voltages using low-lifetime commercial-grade p-type Czochralski silicon'. *Solar RRL*. 2018;**2**(2):1700221.

[155] De Wolf S., Descoeudres A., Holman Z.C., Ballif C. 'High-efficiency silicon heterojunction solar cells: a review'. *Green*. 2012;**2**(1):7–24.

[156] Shi J., Boccard M., Holman Z. 'Plasma-initiated rehydrogenation of amorphous silicon to increase the temperature processing window of silicon heterojunction solar cells'. *Applied Physics Letters*. 2016;**109**(3):031601.

[157] Descoeudres A., Barraud L., De Wolf S., *et al.* 'Improved amorphous/crystalline silicon interface passivation by hydrogen plasma treatment'. *Applied Physics Letters*. 2011;**99**(12):123506.

[158] Green M.A., Emery K., Hishikawa Y., Warta W., Dunlop E.D. 'Solar cell efficiency tables (version 45)'. *Progress in Photovoltaics: Research and Applications*. 2015;**23**(1):1–9.

[159] Vicari Stefani B., Weigand W., Wright M., *et al.* 'P-type upgraded metallurgical-grade multicrystalline silicon heterojunction solar cells with open-circuit voltages over 690 mV'. *Physica Status Solidi.* 2019;**216**(17):1900319.

[160] Basnet R., Weigand W., ZJ Y., Sun C., Phang S.P., Sio H.C. 'Impact of pre-fabrication treatments on n-type UMG wafers for 21% efficient silicon heterojunction solar cells'. *Solar Energy Materials and Solar Cells.* 2019;**205**.110287.

[161] Descoeudres A., Horzel J., Paviet-Salomon B., *et al.* 'The versatility of passivating carrier-selective silicon thin films for diverse high-efficiency screen-printed heterojunction-based solar cells'. *Progress in Photovoltaics: Research and Applications.* 2020;**205**(6):569–77.

[162] Michl B., Rüdiger M., Giesecke J.A., Hermle M., Warta W., Schubert M.C. 'Efficiency limiting bulk recombination in multicrystalline silicon solar cells'. *Solar Energy Materials and Solar Cells.* 2012;**98**:441–7.

[163] Schindler F., Michl B., Krenckel P., *et al.* 'Optimized multicrystalline silicon for solar cells enabling conversion efficiencies of 22%'. *Solar Energy Materials and Solar Cells.* 2017;**171**(April):180–6.

[164] Kwon J.Y., Lee D.H., Chitambar M., Maldonado S., Tuteja A., Boukai A. 'High efficiency thin upgraded metallurgical-grade silicon solar cells on flexible substrates'. *Nano Letters.* 2012;**12**(10):5143–7.

[165] Buonassisi T., Istratov A.A., Marcus M.A., *et al.* 'Engineering metal-impurity nanodefects for low-cost solar cells'. *Nature Materials.* 2005;**4**(9):676–9.

[166] Kivambe M.M., Haschke J., Horzel J., *et al.* 'Record-efficiency N-type and high-efficiency P-type monolike silicon heterojunction solar cells with a high-temperature gettering process'. *ACS Applied Energy Materials.* 2019;**2**(7):4900–6.

[167] Letty E., Veirman J., Favre W., Lemiti M. 'Bulk defect formation under light soaking in seed-end N-type Czochralski silicon wafers – effect on silicon heterojunction solar cells'. *Solar Energy Materials and Solar Cells.* 2017;**166**:147–56.

Chapter 5
The boron-oxygen defect system

Brett Hallam[1], Alison Ciesla[1], Moonyong Kim[1], and Axel Herguth[2]

In 2019, over 60% of all photovoltaic (PV) modules were fabricated using p-type Czochralski (Cz) grown mono-crystalline silicon wafers, and they are expected to remain the dominant wafer type for the foreseeable future [1]. Due to the favourable segregation coefficient of boron (~0.8), p-type Cz wafers are typically doped with boron as it enables a relatively uniform dopant density throughout the ingot. However, as first identified in 1973 [2], when boron-doped Cz silicon wafers are exposed to sunlight, the wafers experience a light-induced degradation (LID) of the minority carrier lifetime due to the formation of recombination-active boron-oxygen (B-O)-related defects [3].*

In p-type Cz silicon solar cells, B-O LID can result in the cells losing up to 10% of their output power within 48 h of exposure to illumination at room temperature as indicated by an early study by Knobloch *et al.* [4], but this very much depends on the wafer material, and solar cell structure and processing sequence used. The maximum power loss typically plateaus at 3–4%$_{\text{relative}}$ for aluminium back-surface-field (Al-BSF) cells [5] and approximately 4–6%$_{\text{relative}}$ for the passivated emitter and rear cell (PERC) [6], due to the higher efficiency structure which is more sensitive to bulk recombination [7]. Figure 5.1 highlights solar cell efficiencies for Al-BSF and PERC solar cells directly after fabrication and after subsequent degradation testing. Based on current global photovoltaic (PV) cell production with approximately 100 GW of manufacturing in 2019 and over 60 GW of this being manufactured using wafers susceptible to B-O-related LID, assuming a conservative system cost of US$1/W installed, a 5% $_{\text{relative}}$ performance loss could cost the industry approximately US$3 B per year through reduced energy yield. This would also have

*The nomenclature 'B-O defects' or 'B-O LID' in this chapter is simply is not intended for B-O to represent the specific defect composition, but rather only serve to identify the defect.

[1]The School of Photovoltaic and Renewable Energy Engineering, The University of New South Wales, Australia
[2]Department of Physics, University of Konstanz, Germany

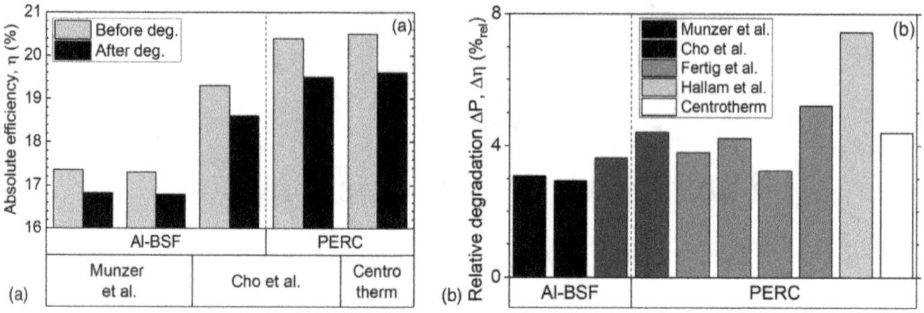

Figure 5.1 *Efficiency of p-type Cz silicon Al-BSF and PERC solar cells (a) directly after cell fabrication and after a subsequent 48 h light soak with the activation of B-O-related LID (from [5]), and (b) relative performance degradation. Data from Münzer* et al. *[8] Cho* et al. *[9], Centrotherm [10], Fertig* et al. *[11] and Hallam* et al. *[12].*

significant implications for the potential greenhouse gas emission abatement, with the degradation resulting in the inability to abate approximately 2.6 M tonnes of carbon diode equivalent (CO_2-e) emissions per year for each 60 GW of susceptible panels installed (assuming 1100 kWh/year/kW installed and an electricity emissions intensity of 0.8 kg CO_2-e/kWh).

In 2006, a key breakthrough was made by Herguth *et al.* with the demonstration that annealing at elevated temperature with either illumination or current-injection, could lead to a permanent deactivation of the B-O defect [13]. As such, the processes enabled open-circuit voltages and efficiencies to be restored to their pre-degraded values [8, 13–15] and remain stable against subsequent exposure to illumination (see Figure 5.2). This gave new hope for p-type Cz silicon solar cells to break through previously perceived efficiency limits for the stabilised performance of such cells. Although eventual recovery from B-O degradation can occur under field conditions, the timescales of degradation and recovery are highly variable depending on installation location, ranging from days to years [16, 17]. As a result, to ensure the long-term reliability of the panel performance and customer confidence, solar cells fabricated on B-O susceptible wafers should be solved during the manufacturing process, to maximise energy yield during field operation and therefore reduce the levelised cost of electricity (LCOE).

Subsequently, significant research has been undertaken by various universities and solar cell manufacturers to understand the fundamental mechanisms of the permanent recovery. In particular, hydrogen has been found to play an important role in the permanent deactivation process, made possible by the transition from hydrogen-lean dielectrics from low-pressure chemical vapour deposition (LPCVD) to hydrogen-rich dielectrics from plasma-enhanced chemical vapour deposition (PECVD) around the year 2000 [18], although its importance for solving B-O-related degradation was not fully realised until 2009 [8]. The improved understanding of B-O defects and hydrogen passivation has enabled the development of rapid

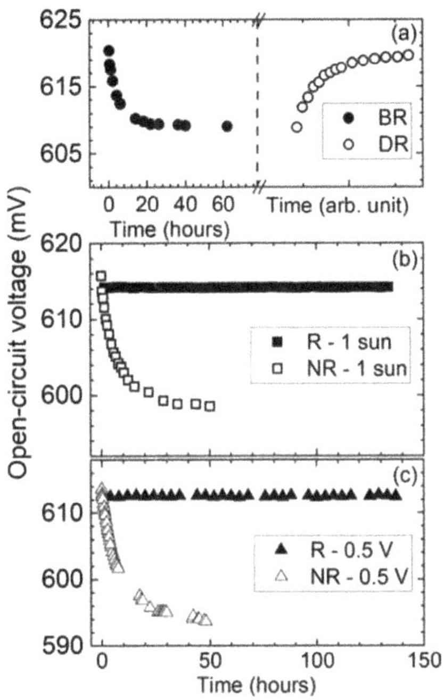

Figure 5.2 *(a) Degradation before and during recovery (BR and DR, respectively). Stability testing using either (b) 1-sun illumination, or (c) 0.5 V forward voltage at room temperature on regenerated (R) or not regenerated (NR) samples. Adapted from [13].*

solutions to eliminate B-O-related LID in timeframes suitable for mass production. This subsequently led to the widespread deployment of industrial tools for solving B-O-related LID and significant increases in the stabilised efficiency of p-type Cz silicon solar cells.

This chapter discusses the formation and elimination of B-O-related LID in p-type Cz silicon, with a particular emphasis on illuminated annealing and hydrogen passivation. Recent review papers have been published by Niewelt *et al.* [19] and Hallam *et al.* [20] on B-O defects and their mitigation, respectively [20]. The chapter also includes brand new insights into the passivation of such defects in heterojunction solar cells fabricated on boron-doped p-type Cz wafers and phosphorus/boron co-doped n-type upgraded metallurgical grade (UMG) Cz silicon. It highlights the future potential for boron-doped p-type Cz wafers to be used to fabricate high-efficiency-passivated contact solar cells with stabilised open-circuit voltages over 735 mV, contrary to the widespread belief of a need to transition to gallium-doped or n-type wafers to achieve such high voltages.

5.1 B-O-related degradation in p-type Cz silicon

B-O-related degradation in p-type Cz silicon wafers was first observed by Fischer *et al.* in 1973 [2] and is caused by the presence of both boron and interstitial oxygen (O_i). Boron is intentionally introduced to the silicon as the p-type dopant, while large amounts of O_i are unintentionally introduced to the silicon melt during the slow Cz growth process required for mono-crystalline silicon growth as the highly reactive molten silicon partially dissolves the quartz (SiO_2) crucible. The typical concentration of boron in wafers used silicon solar cells is approximately $1 \times 10^{16}/cm^3$, corresponding to a bulk resistivity of 1.5 Ω·cm. The typical concentration of O_i reported by LONGi is currently approximately 6.5×10^{17} to $9.0 \times 10^{17}/cm^3$, and this value is continuing to reduce as time goes on [21].

5.1.1 Identification of the defect and its concentration

Direct measurement of the B-O defect has typically evaded researchers. However, experimental investigations have used a number of common techniques to identify the LID associated with B-O defects, as well as estimate the defect concentration. For example, B-O LID has often been identified behaviourally by exposing samples to illumination in the range 10–1 000 mW/m², for 24–48 h at room temperature (~25 °C). Quasi-steady-state photoconductance (QSS-PC) measurements are often used to identify changes in the minority carrier lifetime induced by B-O defects. Shockley-Read-Hall (SRH) analysis through injection-dependent lifetime spectroscopy (IDLS) is often used for the QSS-PC to identify the specific presence of B-O defects. Even without precise knowledge of the capture cross sections for electrons (σ_n) and holes (σ_p), the capture cross section ratio (k) of electrons to that of holes, is known, with a value of $k = \sigma_n/\sigma_p = 10 \pm 1$ [22]. However, it should be noted that values of k in the literature after full degradation range from approximately 10 to 20, which also depends on whether a single-level or two-level defect is assumed (see [19]) and references therein for further information).†

Without exact measurements of the capture cross section values, the actual concentration of B-O defects is unknown, but estimated to be in the range of $1 \times 10^{11}/cm^3$ to $1 \times 10^{13}/cm^3$ [24, 25]. However, an apparent concentration of B-O defects can be estimated based on the increase in recombination due to LID; this estimation is known as the normalised defect density (NDD) [26], also denoted as N_t^* in some studies (e.g. [22]). The NDD of B-O defects starting from the non-recombination active dark-annealed (non-passivated) state is given by (5.1), where $\tau_{eff}(t)$ is the effective minority carrier lifetime at time t and $\tau_{eff}(t = 0)$ is the reference effective minority carrier lifetime at time $t = 0$, i.e. the lifetime before any light soaking. A key assumption for use of the NDD metric is that during testing such

†It is noted that higher capture cross section ratios have been reported in earlier studies for fast degradation. For further information see [19, 23], and references therein.

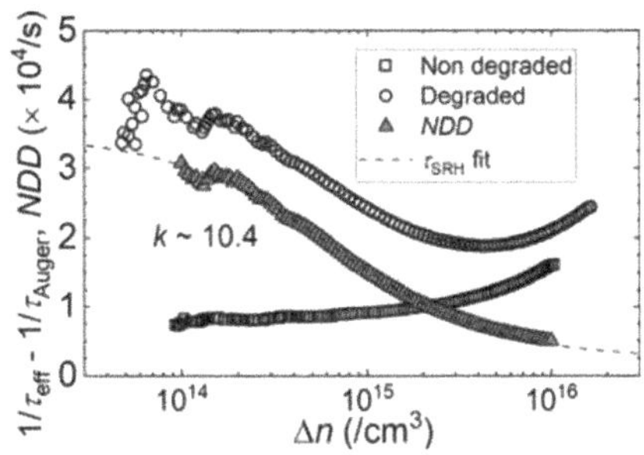

Figure 5.3 *Inverse effective minority carrier lifetime as a function of excess carrier concentration before and after degradation, and the corresponding normalised defect density and fit for SRH recombination.*

as light soaking or annealing, only a single recombination mechanism of interest, here B-O-related recombination, is modulated. With this assumption, (5.1) becomes (5.2), where τ_{BG} is the background lifetime unrelated to B-O degradation, and τ_{SRH} is B-O-related to SRH recombination [27, 28]. To account for injection-level-dependence, NDD is typically measured at a specific injection level (Δn) for a given sample, usually at 10% of the doping density, N_A, to ensure that the NDD can be compared to that determined in other studies. When measured at a given Δn for a given sample, the NDD is linearly proportional to the *true* defect concentration of B-O defects, as indicated in (5.3), where τ_{n0} is the electron lifetime related to B-O defects, N_t is the defect concentration and $v_{th,n}$ is the thermal velocity of electrons. Therefore, the NDD is a commonly used metric to quantify B-O-related LID since it has the closest physical representation to the true defect concentration. A recent study reviewing the advantages and disadvantages of using NDD is presented in [29]. Figure 5.3 shows Auger-corrected inverse lifetime curves before and after degradation, as well as the corresponding change in Auger-corrected inverse lifetime, representing the injection-dependent τ_{SRH} and therefore NDD, and the SRH fit with capture cross section ratio of approximately 10, used to identify the B-O defect.

$$\text{NDD} = \frac{1}{\tau_{eff}(t)} - \frac{1}{\tau_{eff}(t{-}0)} \tag{5.1}$$

$$\text{NDD} = \left(\frac{1}{\tau_{BG}} + \frac{1}{\tau_{SRH}}\right) - \frac{1}{\tau_{BG}} = \frac{1}{\tau_{SRH}} \tag{5.2}$$

$$\text{NDD} = \frac{1}{\tau_{SRH}} = \frac{1}{\tau_{n0}} \approx N_t \sigma_n v_{th,n} \propto N_t \tag{5.3}$$

Deep level transient spectroscopy (DLTS) has also been used to study B-O defects. However, this has proven challenging. In 2015, Mchedlidze *et al.* detected a signal related to B-O defects investigating affected cells rather than lifetime samples [25]. It was suggested that previous challenges in detecting B-O defects by many groups may be complicated by the unintentional introduction

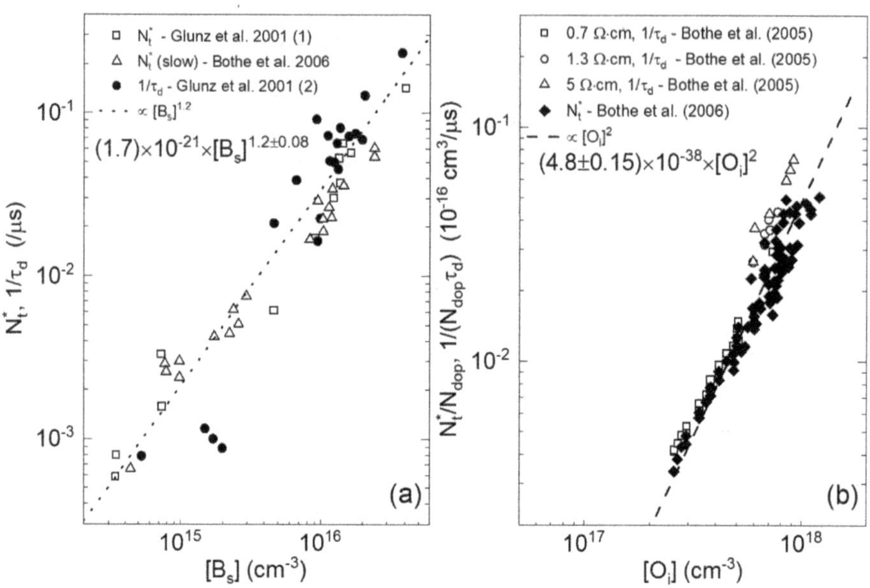

Figure 5.4 Correlation between the normalised defect density and (a) the substitutional boron (B_s)-doping concentration. Data from Glunz et al. *[36] and Bothe* et al. *[22]. (b) Corresponding dependence on the O_i concentration. Data from Bothe* et al. *[22, 32]. Dashed lines represent the fit to the data. Figures adapted from [3]*

of hydrogen during sample preparation for DLTS measurements and the known involvement of hydrogen in permanently deactivating B-O defects. More recently, Vaqueiro-Contreras *et al.* [30] identified a trap (H_{390}) using DLTS and photoluminescence that correlates with the behaviour of B-O defect precursors (the recombination inactive state prone to degradation) and disappears during light soaking. This is an evolving area and likely to shed new light on the fundamental nature of the B-O defect and associated recombination mechanisms.

5.1.2 Correlation between boron doping, the interstitial oxygen concentration and both the NDD and minority carrier lifetime

A number of studies have investigated the impact of the boron and O_i concentrations on the extent of B-O LID and observed a strong dependence on both constituents. Experimental evidence suggests that the NDD scales approximately quadratically with the concentration of O_i ([O_i]) [3, 22, 31, 32], and for wafers doped only with boron, NDD scales approximately linearly with the boron-doping concentration ([B] [26, 31–35]. Figure 5.4 shows the correlation between the doping density-corrected NDD (to account for changes in SRH recombination with differing doping densities) and the boron-doping density, N_A, and [O_i] from Schmidt *et al.* [3]. The correlations suggest that the recombination active B-O defect consists of one boron atom and an oxygen dimer (O_{2i}) [3], providing important insights into potential approaches to minimise the extent of B-O-related degradation (see Section 5.2). The linear boron dependence also applies to p-type Cz wafers co-doped with gallium [37]. B-O LID also appears in phosphorous-doped n-type Cz-Si when partially compensated with boron, albeit with different kinetics likely due to different conditions including Fermi level and carrier concentrations [19].

Bothe *et al.* [32] developed an empirical equation for B-O defect affected minority carrier lifetimes after saturated degradation ((5.4) with the exclusion of enhancement factor m). This is valid for a Δn value of 10% of the doping density, based on measured reductions in carrier lifetime and the correlations with boron and oxygen. It highlights that for 1 Ω·cm material, the expected lifetime in the degraded state for boron-doped p-type Cz silicon is in the range of approximately 10 to 700 μs. However, it should be noted that the equation presented by Bothe *et al.* is only valid for the specific sample preparation route used in that study. This included surface passivation by PECVD SiN_X:H in the mid-temperature range (400 °C) and did not include any high-temperature processes such as a phosphorus diffusion/gettering step. Thermal processes such as phosphorus diffusion can improve the degraded lifetime. In the same study Bothe *et al.* observed that their specific phosphorus diffusion step (at 860 °C) yielded an improvement in lifetime by a factor of approximately two. As such, here for (5.4), we add an enhancement factor, m, to account for improvements in lifetime. It is noted that the value of m will depend on the specific thermal process used.

$$\tau_d = 7.675 \times m \times 10^{45}[\mathrm{B_s}]^{-0.824}[\mathrm{O_i}]^{-1.748} \tag{5.4}$$

Figure 5.5 shows a contour plot of how increasing boron and oxygen concentrations are predicted to lead to reduced minority carrier lifetimes due to B-O defects (based on [32]), along with updated values for the O_i concentration in Cz silicon. Oxygen concentration has reduced significantly in the last two decades now in the range of 6.5×10^{17} to $9 \times 10^{17}/cm^3$, with further reductions in the O_i concentration expected with new and improved pulling technology of 4.5×10^{17} to $8 \times 10^{17}/cm^3$, and ongoing development of enabling ultra-low O_i in the range of 2×10^{17} to $5 \times 10^{17}/cm^3$ [21]. With the current pulling technology and an enhancement factor of two, lifetimes in the range of 36–100 μs are expected for 1 Ω·cm material.

5.1.3 Degradation characteristics

5.1.3.1 Two-stage degradation

When the degradation behaviour for B-O defects is studied with sufficiently small time steps, a two-stage degradation of the carrier lifetime, and therefore efficiency, is widely observed [3, 38, 39]. As shown in Figure 5.6, the fast degradation happens in only seconds or minutes, while the slow degradation occurs over many hours or days, with only the latter commonly observed due to its more obvious impact, accounting for approximately 80–90% of the NDD due to B-O defects. The activation energies of the two stages of defect formation in clean wafers were found to

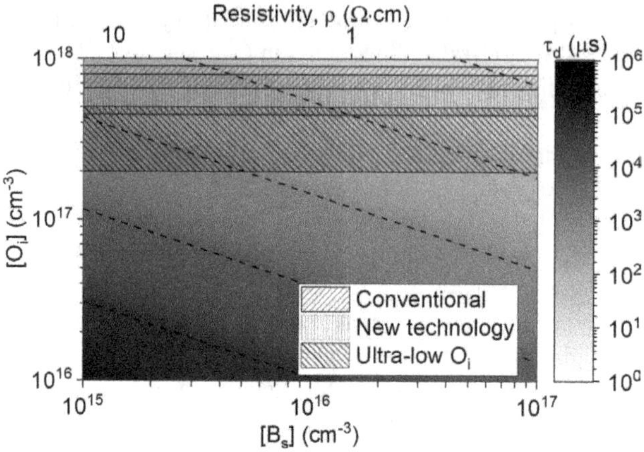

Figure 5.5 *Contour plot of the predicted minority carrier lifetime (based on [32] adding in an enhancement factor of* m = *2) due to B-O defects as a function of* O_i *concentration and resistivity for p-type silicon solely doped with boron, also highlighting expected oxygen concentrations in Cz-grown silicon using data from LONGi [21]*

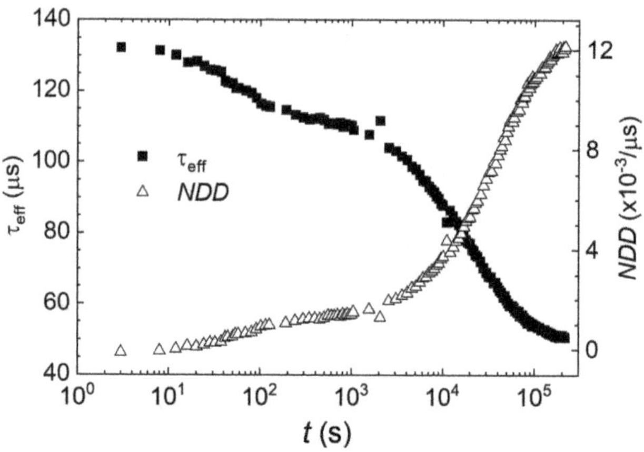

Figure 5.6 *Minority carrier lifetime,* τ_{eff} *and NDD extracted at* $\Delta n = 9.1 \times 10^{14}/cm^3$ *as a function of light soaking duration at room temperature highlighting the two-stage degradation due to B-O defects with a fast degradation and slow degradation phase*

be 0.23 ± 0.02 and 0.475 ± 0.035 eV for the fast and slow stages of degradation, respectively [22].

These different rates of degradation have been commonly assumed related to different defect centres [22]; however, it was recently proposed that this behaviour could arise from a two-stage activation of a single defect [40], sparking an on-going

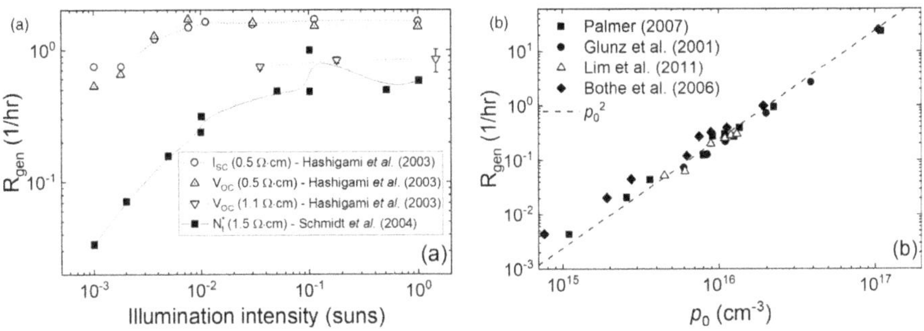

Figure 5.7 *(a) Impact of light intensity and (b) the equilibrium hole concentration on the degradation rate for B-O defect. For (b), open symbols are based on [B] and closed symbols are based on* p_0. *Data from Hashigami* et al. *[38], Schmidt* et al. *[3], Palmer* et al. *[45], Glunz* et al. *[46], Lim* et al. *[47] and Bothe* et al. *[22].*

debate [23, 41–44] including discussions on the ability to permanently deactivate defects forming in the fast and slow timescales [23, 42].

5.1.3.2 Impact of the doping concentration and illumination intensity on the degradation rates

Given the dominance of the slow timescale degradation on the overall impact of B-O defects, responsible for 80–90% of the NDD, most studies investigating B-O-related degradation purely focus on the slow timescale. In early studies it was demonstrated that at extremely low-light intensities, a linear dependence on the illumination intensity was observed. However, this saturated for intensities higher than approximately 0.1 suns (see Figure 5.7 (a)) [22, 48].‡ At this illumination intensity the excess electron concentration is on the order of $1 \times 10^{11}/cm^3$, which is in the rough vicinity of the expected concentration of B-O defects [24, 25, 49].§ This suggests a possible charging mechanism is involved in the degradation reaction. Hence, studies investigating the B-O defect have typically focused on using illumination intensities in the range of 0.1–1 sun. This understanding was also used in the early stages of developing approaches to solve B-O LID. It is important to note, however, that an incorrect definition of 1 sun can lead to a substantial variation in the estimated injection level, and therefore reaction rates for the B-O defect system. For further reading see [50].

The degradation rate is also heavily impacted by the boron-doping concentration. A number of studies have shown a quadratic dependence of the degradation

‡See Section 5.4 for modifications in the degradation rate at higher illumination intensities.

§In this context it is important to note that equilibrium electron concentration in p-type material is quite low (law of mass action), e.g. for p_0 of the order of $1 \times 10^{16}/cm^3$ at room temperature, n_0 amounts only to $1 \times 10^4/cm^3$. Hence, n_0 is almost always negligible in comparison to the injected excess electron concentration.

rate on the boron-doping concentration for silicon doped solely with boron [22, 39, 45] (see Figure 5.7 (b)).

However, studies using compensated p-type Cz silicon, where the hole concentration is not equal to the boron-doping concentration, suggested a quadratic dependence on equilibrium hole concentration p_0 rather than the boron concentration [37, 47, 51]. This implies that a sufficient supply of holes is essential for the formation of B-O defects.

On the other hand, it should not matter whether holes originate from doping or injection (generation by light). And indeed, further investigation has confirmed that degradation rate scales with total hole concentration p and not with p_0 or [B] (e.g. see [52]). Thus, the above-mentioned dependencies on [B] and p_0 are to be regarded as extreme cases where the injection level is negligible compared to equilibrium hole concentration p_0, which is equal to [B] for purely boron-doped material.

Following this, for excess electron concentrations higher than approximately $1 \times 10^{11}/cm^3$ where the defect formation rate saturates, the degradation rate is limited by the background-doping concentration of the material and hence by the supply of holes. This suggests that defect formation involves charge carrier interaction, and it opens up a possibility to accelerate degradation without the need to increase temperature. This will be discussed more in Section 5.4.

As such, solar cells fabricated on more heavily boron-doped wafers will experience faster degradation rates in addition to increased degradation extents. At room temperature, the expected timescale for full B-O-related degradation is approximately 24 h for p-type Cz silicon with a boron-doping concentration of $1 \times 10^{16}/cm^3$ [53].

For the limited studies investigating the impact of the doping concentration on the degradation rate in the fast timescale, a quadratic dependence on p_0 has also been demonstrated. It has also been observed that the $[O_i]$ does not impact the defect formation rate [22, 45].

5.1.3.3 Inducing B-O degradation with carrier injection

Although the degradation is commonly referred to as being light-induced, it is the carriers generated by the light that enable formation of the B-O defects [4, 38, 54]. Carriers generated by applying a bias in the dark or current injection can also cause such degradation of the material [4]. As such, the term carrier-induced degradation (CID) has been argued to be a more appropriate term. However, for simplicity and in alignment with the general term adopted by the industry, LID is used in this chapter, as exposure to illumination is the most relevant condition of concern for PV modules in the field to induce B-O-related degradation.

5.1.4 *Temporary recovery from B-O defects through dark annealing*

A key characteristic behaviour of B-O defects is that a short dark anneal at 200 °C for 10 min can lead to a complete recovery of the carrier lifetime, reverting the B-O defect back into a recombination inactive state. However, this recovery is only temporary, and LID can occur again with further exposure to illumination or

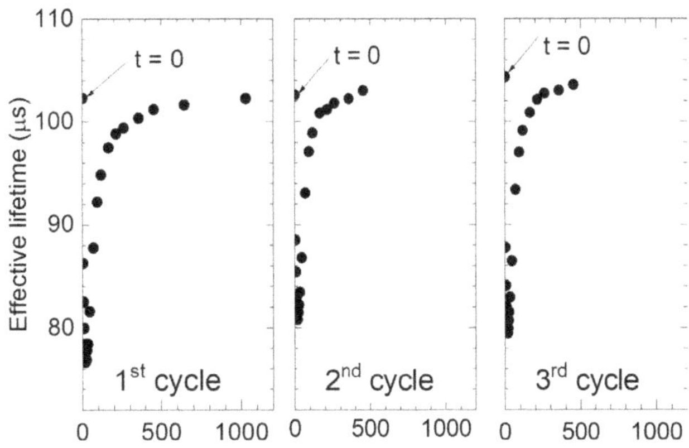

Figure 5.8 *Reversible B-O defect reactions with subsequent degradation following each dark-annealed lifetime recovery*

carrier injection. Figure 5.8 shows repeated LID cycles following dark anneals to recover the lifetime. This recovery reaction, sometimes termed defect annihilation, has an activation energy of 1.32 eV [22], which is significantly higher than that for degradation. It is presumed that this reaction is only thermally activated, although a study by Bothe *et al.* saw a slight dependence on the doping concentration [22]. As a result, during illuminated processes at different temperatures, this reaction is also active and needs to be considered. This defect annihilation reaction has important implications for eliminating B-O LID as is presented in Section 5.3. At higher temperatures, even under illumination, the system favours the recombination inactive but unstable state, meaning that with subsequent illumination at lower temperature, the silicon will again suffer from B-O LID.

5.1.5 Defect theories

The B-O defect has been investigated intensely over the past few decades. A Google Scholar search for the terms 'boron-oxygen' and 'light-induced degradation' yields over 1 100 hits. Despite this level of attention, the exact composition and nature of the B-O defect is still up for debate [40, 44, 55]. However, significant progress has been made in understanding the behaviour of B-O defects and its fairly signature LID characteristics. The general consensus is that the defect consists of a B_s atom and an interstitial oxygen dimer (O_{2i}) due to the dependence of the defect concentration [3]. Although early publications suggested the possibility of a diffusion-based mechanism of O_{2i} for defect formation [3], more recent work points towards the transformation of a latent B-O complex as first proposed by Voronkov *et al.* [56]. The recent work by Vaqueiro-Contreras *et al.* [30] suggests that the defect consists of a metastable latent defect consisting of B_sO_{2i}, in alignment with an earlier configuration estimation by Schmidt and Bothe [3]. In addition, the study suggested that the mechanism for lifetime degradation is trap-assisted Auger recombination.

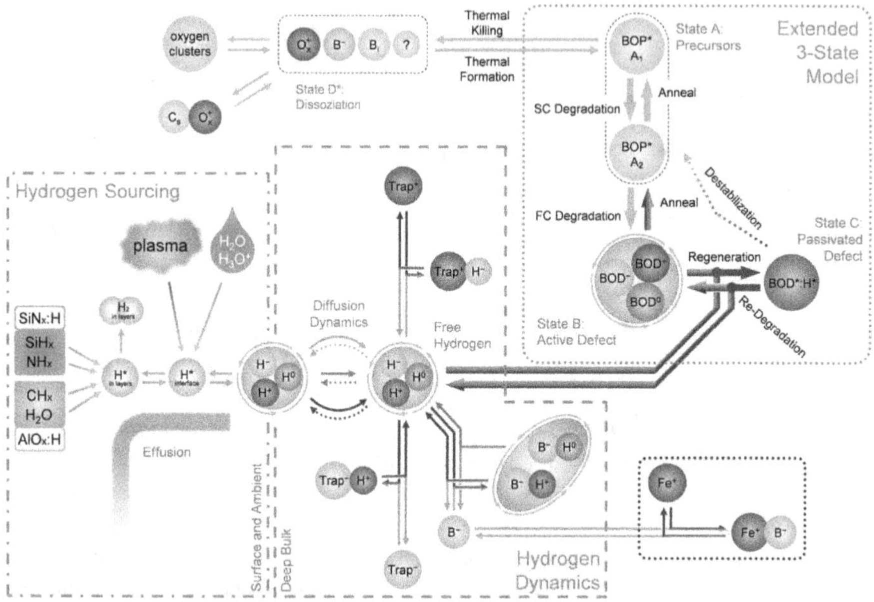

Figure 5.9 *Generalised model for the B-O defect system in Cz silicon. Reprinted from [57], with the permission of AIP Publishing. © 2018 Author(s)*

Further discussion of the defect theories is beyond the scope of this textbook and not critical to the treatment of B-O defects using hydrogenation. For further reading on the fundamental properties of the B-O defect, see the review paper published by Niewelt *et al.* [19]. Additional recent defect-related papers include [23, 30, 43, 44].

5.1.6 Generalised model for the B-O defect system

Even without knowing the exact chemical properties of the defect, enough is known about the defect behaviour to enable the highly complex B-O defect system to be summed up in a generalised model, as recently presented by Herguth and Hallam [57] (see Figure 5.9). This section serves as a brief introduction to the complex B-O defect system, various aspects of which will be discussed in detail throughout the remainder of this chapter.

Traditionally, the B-O defect system has been represented by a three-state model consisting of a recombination inactive 'precursor' state A, which is prone to subsequent LID thereby forming a recombination 'active defect' state B, and with continued carrier injection reaches a recovered stable 'passivated' state C. It is noted that the transition from state B to state C is due to hydrogen passivation. The three-state model will be discussed further in Section 5.3 and throughout this chapter due to its simplicity and benefit in understanding kinetics of the B-O defect system.

The generalised model expands on this basic three-state model to account for the many other complexities in the B-O defect system. In particular, it includes two different B-O defect precursor (BOP) A states (states A_1 and A_2) that result in the fast degradation and slow degradation. It is noted that the generalised model assumes the existence of a single B-O defect, as proposed in [40]. Here, *actual* degradation of carrier lifetime is caused by the *fast* transition from state A_2 into state B. The initial population in state A_2 is therefore responsible for *fast* degradation. The slow transition from state A_1 into state A_2 is *essentially* responsible for slow degradation, noting again that actual degradation does not occur until the subsequent and almost immediate reaction from state A_2 into state B. As such, the initial population in state A_1 is responsible for the extent of *slow* degradation. The conventional dark annealing process used to recover from degradation typically results in 80–90% of BOP being in state A_1 and 10–20% being in state A_2, therefore the majority of degradation occurring in the slow timescale. Recent work identified that the fraction of defect precursors in states A_1 and A_2 can be modulated by thermal processing, and indeed, dark annealing of a degraded sample (state B), first transitions through state A_2 before the majority is returned to state A_1 (41).

The model also highlights the three charge states of the B-O defect, with its 'negative-U' properties [58], similar to that of interstitial hydrogen. The fact that both B-O defects and hydrogen have multiple charge states means that a certain combination of charge states, is likely required for effective passivation to occur, as has been discussed in Chapter 2 and Ref [59]. It also includes the role of hydrogen sourcing from dielectric layers, and hydrogen dynamics, and how they can modulate passivation kinetics. In particular, the charge state of hydrogen, which is determined by the sample and conditions, determines how it diffuses and interacts with anything and everything in the silicon lattice such as unintentional impurities, other defects, dangling bonds etc. Hydrogen can also be lightly bound to dopant atoms such as boron to form hydrogen-boron (HB) pairs. These pairs are likely to be freed under the conditions that cause LID, to enable the passivation of B-O defects, and hence HB pairs might be a possible, but likely not the only source for passivation. This will be discussed in Section 5.4.

A key addition of the generalised model, is the presence of state D. It has been observed that, rather than illuminated processes to reach the passivated state C, B-O-related degradation can also be eliminated through certain thermal treatments, typically performed at high temperatures as mentioned in Section 5.1.2, leading to an increase in the enhancement factor in (5.4). This is believed to shift defect precursors (state A) into an additional non-recombination active state, state D, which is potentially the dissociation of latent B-O defects into their constituents and may involve a competition between carbon and boron for O_{2i}. This will be discussed further in Section 5.3.

The B-O defect system is further complicated by the interdependency with other defect systems. For example, iron-boron pairs (Fe-B) also change states under the same conditions that cause B-O LID, thereby impacting B-O studies. In particular, the dissociation of Fe-B pairs occurs on a similar timescale as the fast degradation for B-O LID and has been shown to cause apparently high capture cross section ratios during rapid degradation. Furthermore, iron is not effectively passivated by hydrogen, and as a result, iron-related recombination can be observed after permanent deactivation, which has been previously attributed to the inability to passivate fast B-O LID. Reference [23] details the impact of interstitial iron on the B-O defect system. In addition, light- and elevated temperature-induced degradation (LeTID), as will be discussed in Chapter 6, is heavily dependent on the availability of hydrogen and can occur simultaneously to B-O LID and impact the recovery/passivation of B-O defects.

5.2 Alternative silicon wafers to avoid B-O-related degradation

The known dependence of the defect concentration on the concentrations of boron and oxygen [O_i] enables potential reductions in the extent of B-O-related degradation to be achieved by reducing either the boron doping or [O_i] in the silicon material. Hence, it is interesting to consider alternative sources or modified silicon wafers for the manufacture of silicon solar cells, to ensure long-term reliability of the solar panels, which are expected to operate for more than 25 years in the field.

5.2.1 Methods to decrease the interstitial oxygen concentration

There are a number of alternative crystallisation approaches that enable reduced [O_i] in silicon wafers. The most common method used by industry, to reduce [O_i] in Cz grown wafers by up to an order of magnitude [60, 61] is to coat the walls of the crucible, with silicon nitride for example [62]. This reduces the dissolution of the crucible walls and therefore reduces the incorporation of SiO or SiO_2 into the silicon melt. Another method to reduce [O_i] in Cz silicon is by manipulating convective flows in the melt by applying magnetic fields during growth [63, 64], giving the ability to reduce LID to less than 1% $_{\text{relative}}$ [65]. However, due to the size of industrial Cz ingots, throughput and cost constraints this technique has proven challenging, particularly with the transition to larger and larger ingot diameters. Controlling the parameters used during crystal growth can also reduce [O_i], such as crucible and crystal rotation, pressure and ambient atmosphere [62]. The recently developed continuous Cz (CCz) technology can also result in lower and more uniform concentrations of O_i throughout an ingot [66]. In 2019, LONGi Solar had developed an advanced industrial scale Cz growth solution to reduce oxygen dissolution in the melt, resulting in wafers with [O_i] = $1.3 \times 10^{17}/\text{cm}^3$ (2.6 ppma) and no susceptibility to LID [67]. However, it is unclear on how representative this oxygen concentration is throughout the ingot.

An alternative low O_i mono-crystalline technology to Cz crystal growth is the float-zoning (FZ) method. FZ silicon has [O_i] typically two orders of magnitude lower than Cz silicon [68], resulting in less B-O-related LID [69]. However, FZ silicon is not practical for use in commercial cell manufacturing as the wafers are significantly more expensive than Cz wafers. FZ silicon has also been found to be affected by various other defects, for example, nitrogen-related defects or other LID effects that can significantly reduce carrier lifetimes [70–73].

5.2.2 *Casted multi-crystalline and quasi-mono-crystalline silicon*

For many years, since the identification of B-O LID, casting of silicon ingots became an industry-preferred alternative to Cz silicon, in part due to the reduced incorporation of O_i [74]. Refer back to Figure 5.5 for a contour plot of the predicted minority carrier lifetimes affected by B-O defects, highlighting the potentially large advantage of cast silicon due to the reduced oxygen concentrations. However, the casting method typically results in multi-crystalline silicon growth littered with networks of structural defects such as grain boundaries and dislocation clusters. Further, although the casting crucibles do not release as much oxygen into the melt, they are generally made from less pure materials and tend to introduce a range of other contaminants into the silicon melt. Such contamination and structural defects cause reduced carrier lifetimes [75, 76], and in turn, solar cell efficiencies. This makes these wafers less suited to high-efficiency cell structures like PERC that benefit from high carrier lifetimes.

To combine the benefits of the reduced [O_i] concentration from the casting process and overcome the performance limitation of conventional cast multi-crystalline silicon, BP (British Petroleum) Solar developed the quasi-mono technology, also known as cast-mono [77]. This was achieved by lining the bottom of a casting crucible with mono-crystalline seed ingots, to induce single crystal growth during the casting process largely without grain boundaries. An additional key benefit of this technology is the compatibility with the standard industrial alkaline texturing process for mono-crystalline silicon. However, it is challenging to maintain the mono-crystallinity throughout the cast ingot. Unintentional multi-crystalline regions tend to form at ingot edges where the melt is still in contact with the crucible, and high stresses generated during solidification, which are typically relieved across grain boundaries, can cause electrically harmful dislocations, particularly at the top of the ingot that is last to solidify [78]. Recent claims by GCL (Golden Concord Holdings Limited) are that their casted wafers are 99% mono-crystalline; however, they do not specify the yield [79]. Typically the multi-crystalline and dislocated regions reduce the usable portion of the ingot significantly, so despite initial and periodically resurging enthusiasm, such wafers have not been widely adopted [1].

In contrast, an alternative seeded casting method reduces the grain size resulting in less ingot stress and dislocations, but a higher density of less electrically harmful grain boundaries, known as 'high-performance multi' [80]. Conventional cast silicon was rapidly replaced in the 2010s as industry adopted this higher lifetime material [33]. However, during this transition and the simultaneous transition to the PERC technology, PERC fabricated from such wafers were found to suffer from new, more severe and longer term form of degradation, often referred to as LeTID (to be discussed in Chapter 6) [81–85]. Although Cz silicon was later shown to also suffer from LeTID [86], the initial discovery and severity in cast silicon led to manufacturers shifting away from using such wafers in the transition to PERC structures [33]. Generally speaking, although it is less susceptible to B-O LID than Cz, cast silicon has inherently higher bulk recombination due to the presence of crystallographic defects and other impurities as discussed in Chapter 4.

For these reasons, among others, as the industry shifts to higher efficiency cell structures, and Cz wafer costs are decreasing (the price difference is shrinking), industry demand is shifting to prefer mono-crystalline Cz wafers [87].

5.2.3 Decreasing the boron-doping concentration

Reducing the boron (p-type dopant) concentration in the wafers is another alternative to minimise B-O LID; however, this also increases the bulk resistivity. Two benefits of this approach are the reduced concentration of B-O defects and reduced sensitivity to SRH recombination [27, 28]. However, increasing the resistivity in the bulk can be detrimental for PERC solar cells, in which majority carriers flow laterally to the localised p-type contacts. In these structures, there is a trade-off between reducing recombination effects due to SRH generally and specifically B-O LID, and increasing bulk series resistance in the device. For these reasons, until recently doping around 2–5 Ω·cm has typically been used in commercial devices. However, this value has decreased towards 1 Ω·cm with the development of approaches to solve B-O-related LID.

Alternatively, the boron dopant commonly used for p-type Cz ingot growth can be replaced with other dopants. Gallium is one such dopant, and enables wafers free from B-O-related LID effects. However, similar to boron-doped wafers, they are still susceptible to other defects such as the formation of acceptor-iron pairs [88, 89]. The primary reason for widespread adoption of boron doping over gallium doping is the favourable segregation coefficient of 0.8, while that of gallium is two orders of magnitude lower at 8×10^{-3} [90, 91]. The substantially lower gallium segregation coefficient results in a significantly large variation in wafer resistivities along an ingot that reduces the usable portion of the ingot, and can lead to significant variations in cell performance. However, recent advancements in Cz ingot pulling are leading to increased interest in gallium-doped wafers, such as CCz growth, or additional co-doping with phosphorus to allow uniform resistivities throughout the ingot [1, 92]. Indium and aluminium are other possible p-type dopant substitutes. However, these have even smaller segregation coefficients of 4×10^{-4} and 2.8×10^{-3}, respectively [93, 94]. In addition, such dopants do not eliminate susceptibility to defects as aluminium also forms recombination active defects with O_i [95] and indium-doped Cz wafers have been shown to be susceptible to LID [96].

N-type wafers doped with phosphorus can be used to avoid B-O defects [34], unless the Cz silicon material is compensated with boron [47]. Additional benefits of n-type wafers are lower susceptibility to both intrinsic recombination [97] and SRH recombination due to some impurities such as iron [98], as well as a lower dark saturation current than p-type wafers with an equivalent resistivity. All of these factors mean that n-type wafers have higher lifetimes than p-type wafers. However, the minority carrier mobility is lower in n-type silicon, and n-type material is actually more susceptible to SRH recombination from metal impurities other than iron, such as chromium, cobalt, copper and nickel, which are harder to remove via gettering [98].

However, substituting the commercially dominant p-type wafers with n-type wafers is also not so straight forward, requiring significant equipment and processing modifications in the fabrication of solar cells. One option is to manufacture an entirely new cell structure suited to n-type wafers, which we are beginning to see more of commercially with increasing interest in heterojunction and back contact cells [99]. To substitute p-type wafers with n-type in a conventional homo-junction device structure, a much longer high-temperature diffusion is required to form the boron-doped p-type emitter. This is due to the much lower diffusion coefficient of boron compared to that of the phosphorus used for formation of n-type emitters on p-type wafers [90]. Modifications to screen printing and surface passivation are also required. As a result, the cell manufacturing process can be more complicated as well as more thermally and energy intensive. This increased thermal budget also created added complications as it can induce formation of oxygen precipitates [100] and other oxygen-related defects that can be more harmful than degradation due to B-O defects [101, 102], especially if B-O LID can be mitigated. Other LID effects causing up to 10% $_{\text{relative}}$ performance loss have also been observed in the seed-end of n-type Cz ingots [103]. Further, the segregation coefficient of phosphorus is only 0.35 [90, 91], leading to a larger range of bulk doping throughout an n-type ingot. N-type wafers are subsequently approximately 8–10% more expensive [104, 105].

5.2.4 Germanium and carbon co-doping

Rather than limiting the boron dopants, co-doping with germanium or carbon during the crystallisation process can also reduce the extent of degradation due to B-O defects [31, 106]. Co-doping with germanium is thought to either increase the O_i diffusivity barrier, or bond with the O_i, hence forming Ge-O complexes, thereby reducing the O_i available to form B-O defects [106]. However, approximately $1 \times 10^{20}/\text{cm}^3$ germanium doping is required to prevent LID, which increases the cost of ingot growth and tends to cause nucleation of multi-crystalline silicon making it not well suited to commercial production [106, 107]. Similarly, co-doping with carbon is thought to limit B-O defect formation as the carbon competes with boron to form complexes with oxygen [108]. Introducing other impurities such as carbon also increases the potential for the formation of other performance limiting defects. For example, recently a particularly strong recombination active carbon-oxygen-hydrogen complex was identified in n-type silicon [109]. Therefore, co-doping of boron-doped ingots with other dopants such as germanium or carbon is not a viable solution for avoiding degradation due to B-O defects, without introducing other complications and other recombination active defects to deal with.

5.2.5 Summary of wafer alternatives

The alternative wafers described in this section are all capable of reducing LID due to B-O defects. Of these, only the use of multi-crystalline silicon has been used widely in industry despite these alternatives being known for many years. The primary reason being that each of the alternatives have either technological or economic disadvantages that typically mean less cost-effective production compared to the industry dominant p-type Cz-Si (before B-O LID). On the other hand, manufacturers of p-type Cz solar cells must decide what level of degradation they, and their customers, are willing to tolerate without attempts to reduce LID, or the cost and technical challenges of adding processes to reduce LID must be considered.

With the continual increase in cell efficiencies and transition to higher efficiency structures, interest in n-type wafers has grown; however, more immediately significant is the increased demand for p-type Cz silicon, expected to makeup >50% of the growing PV production for the foreseeable future [1]. Record cell efficiencies using industrial p-type Cz wafers have increased from 20.3% in 2012 to 23.8% in 2019 [110, 111]. However, to use boron-doped p-type Cz wafers with resistivities suitable for such PERC structures with increased susceptibility to bulk degradation, B-O-related LID mitigation processes are required.

5.3 Multiple pathways of B-O LID mitigation during cell processing

This section provides a summary of two fundamentally different mechanisms that have been identified to reduce or eliminate B-O LID during cell fabrication, and the current understanding of the mechanisms by which they work. The first mechanism is purely a thermal process that is thought to modify the concentration of B-O precursors, as relatively high temperatures without light (or with minimal residual illumination emitted from infrared heating elements) trigger the reaction path from A→D in Figure 5.9. Although this mechanism is not related to hydrogen passivation, this is of particular relevance for the firing of screen-printed metal contacts, which occur at high temperatures and inadvertently releases hydrogen from dielectric layers and enables hydrogen passivation of bulk defects. This hydrogen introduced by the firing process is critical for the second B-O LID mitigation mechanism. The second method is therefore applied after firing and involves the use of annealing at low temperatures with carrier injection from illumination or an applied bias. This allows the 'regeneration' or 'passivation' of B-O defects once they have become recombination active and corresponds to the reaction pathway A→B→C in Figure 5.9. This mechanism is of particular importance in this textbook due to the role of hydrogen in the passivation process and will be discussed in detail in Section 5.4.

5.3.1 Thermal processes to modify the B-O precursor concentration – A four-state model

Thermal processes are one way to reduce the extent of B-O-related LID. Although high-temperature processes can lead to improvements in lifetime from aspects related to gettering and non B-O-related thermal effects, here we discuss thermal impacts directly related to the concentration of B-O defects. This mechanism was identified more recently and has led to the addition of State D to the conventional three-state model that was mentioned earlier in Section 5.1.6. Here, state D represents a state where the B-O defect precursors have been thermally annihilated. State

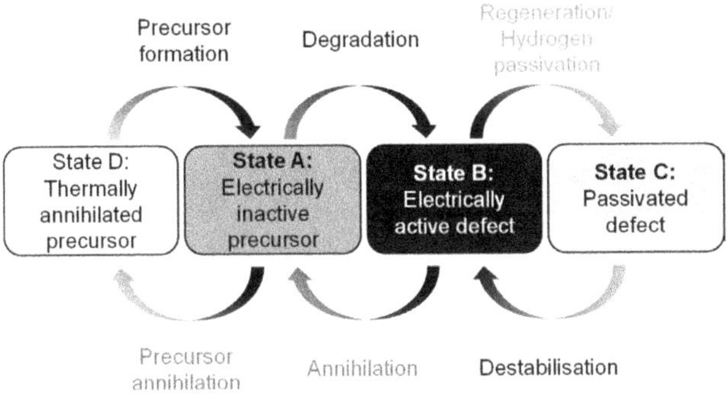

Figure 5.10 A simplified four-state diagram to represent the system of B-O defects. Adapted from [20].

A represents the recombination inactive 'precursors' that form or are transformed into a recombination 'active defect' state B under light or carrier injection, and a recovered stable 'passivated' state C. The four-state system is shown in Figure 5.10. The transition from State A →D represents the thermal annihilation of defect precursors due to high-temperature processing. Preformation of the B-O defects is not required for this annihilation reaction to occur, and is also not of concern as the same thermal processes used to annihilate the precursors will also dissociate any formed B-O defects back to their precursor state (State B →A) for precursor annihilation (State A →D). The reverse transition from State D →A is the formation of the defect precursors due to thermal processes at intermediate temperatures.

Key thermal processes that can reduce the B-O defect concentration and hence stabilise carrier lifetime include:

1. tube furnace processes such as phosphorus emitter diffusion and thermal oxidation [36, 46, 112, 113]
2. annealing at intermediate temperatures typically 400–700 °C [3, 112]
3. rapid thermal processing such as metallisation firing [114–117].

However, thermal processes tend to only reduce the extent of B-O LID rather than eliminate it completely, unlike illuminated anneals that can eliminate the impact of B-O degradation entirely. None the less, both are of significant importance for p-type silicon solar cells.

5.3.1.1 High-temperature tube-furnace processes

High-temperature emitter diffusion and thermal oxidation processes can result in a reduction in the NDD by a factor to up to approximately three [36, 46, 112, 113]. However, given the presence of oxygen in the wafers, care is required to ensure appropriate conditions are used to avoid the activation of other recombination centres, risking a reduction in the overall effective minority carrier lifetime of the material [36, 46].

5.3.1.2 Annealing at intermediate temperatures

Prolonged annealing at reduced temperatures in the range of 200–700 °C, can also modify the concentration of B-O defects [3, 112, 118]. For example, a 16 h anneal at 450 °C was observed to reduce the NDD by a factor of approximately two [112]. It is noted that an increased temperature results in a reduction in the NDD. With annealing in the range of 400–700 °C, Rougieux *et al.* observed an activation energy in the range of 0.1–0.3 eV (0.16 + 0.14/−0.04 eV), noting that this is the activation energy of the defect concentration, rather than an activation energy of the reaction rate between the two states [119]. In a lower temperature range of 200–415 °C, Walter *et al.* observed an activation energy in the same range for the reduction in defect concentration with a value of $E_a = (0.16 \pm 0.03)$ eV [118]. Although an actual reaction rate was not provided, it was determined that a 50-h anneal at 200 °C was sufficient to reach equilibrium, and similar results were obtained at 425 °C for 15 min as for a 50-h anneal. An estimation on the reaction rate at 200 °C for the reactivation of defect precursors was $k = (0.06 \pm 0.01)$/h (118). Subsequently, the reaction rate at 425 °C is $>\sim(0.045 \pm 0.005)$/h.

5.3.1.3 Rapid thermal processing such as metallisation firing

Metallisation firing, the last high-temperature process in solar cell fabrication, can also modulate the equilibrium concentration of B-O defect precursors [115–117, 120]. To our knowledge, the only report claiming complete elimination for the thermal processing is [120]. However, no LID testing was performed after the belt-furnace firing, and therefore no conclusion can be drawn on the effectiveness of the process on the long-term stability of the cells in relation to B-O LID. It is likely that the recovery observed with firing was due to a combination of defect annihilation (State B to State A), as well as a partial precursor annihilation (State A to State D).

Example data showing the reduction in the extent of B-O degradation with fast-firing is shown in Figure 5.11 from [117]. The study by Nampalli *et al.* demonstrated a reduction in the fractional defect density (FDD)[¶] with increasing peak firing temperature, although plateauing for temperatures above 625 °C. This resulted in a reduction in NDD of up to approximately 50%. In addition, Nampalli *et al.* observed the reduction was independent of whether a hydrogen-rich dielectric layer (PECVD SiN_x:H) or hydrogen-free (thermal SiO_2) was used [117]. As a result, it can be concluded that the mechanisms for the annihilation and reactivation of defect precursors are unlikely unrelated to hydrogen. It is noted, however, that this high-temperature-firing process can modulate the regeneration rate during subsequent illuminated annealing, due to the release of hydrogen from dielectric layers and distribution of hydrogen throughout the device as will be discussed in Section 5.4.

[¶]The FDD represents the normalisation of the NDD at a given time to the maximum NDD value.

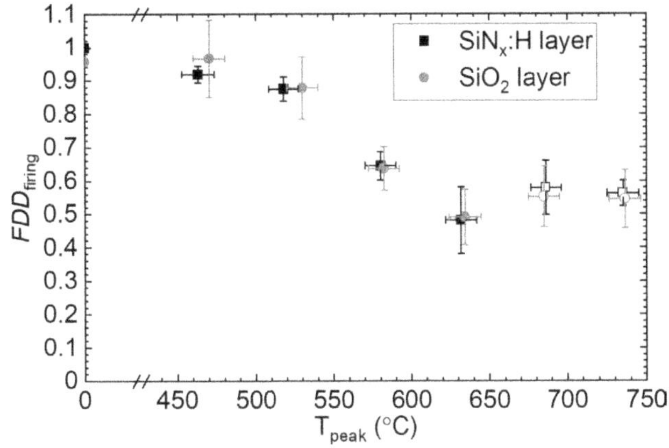

Figure 5.11 *Impact of the peak firing temperature (T_{peak}) on the FDD of B-O defects, representing the reduction in NDD occurring through thermal processing, for both SiN_x:H and SiO_2-passivated samples. Open symbols at high temperatures correspond to conditions that also activate other carrier-induced defects. Reprinted from [117] Copyright (2017) with permission from Elsevier.*

However, the response to firing is not completely clear. For example, for lower temperatures (below 675 °C), Kouhlane *et al.* demonstrated an increase of NDD (620 °C) compared to the reference sample, while the NDD decreased at higher temperatures [116]. In addition, results by Walter *et al.* suggest a substantially higher NDD for a peak temperature of 850 °C than 650 °C, contradicting the general trend of a decrease in the NDD with increasing temperature. However, the study by Walter *et al.* showed a clear trend of reducing NDD with increasing cooling rate for a given peak firing temperature [115]. This highlights the importance of rapid cooling for potentially avoiding a reactivation of defect precursors during the cool down of high-temperature processes.

With a temperature dependence of the equilibrium resulting NDD, depending on the condition of a wafer or cell, a given additional thermal process can therefore modulate the concentration of B-O defect precursors. For example, performing a prolonged anneal at 200 °C after a prior process such as activation of AlO_x passivation at 425 °C [118] or metallisation firing, can increase the defect concentration. Given that the metallisation firing process is the last high-temperature processes in silicon solar cell fabrication and subsequent processes are relatively short and performed at low temperatures, the firing process is the most likely to influence the B-O defect concentration in the final device.

5.3.1.4 Unknown nature and kinetics of precursor annihilation and reformation

The exact nature of the mechanism or mechanisms involved for the annihilation and reformation of defect precursors remains unclear. However, some possibilities include a change in the concentration of specific O_i species, likely oxygen dimers, responsible for degradation [112, 119, 121], or a competition between boron and carbon for O_{2i} [3, 57]. In addition, the dependence of the equilibrium defect precursor concentration on temperature and cooling rate is unknown, along with temperature-dependent reaction rates for transitions between States D and A, with values for precursor annihilation and reformation. Such information may provide further insights into the fundamental structure and components of the defects.

5.3.2 Annealing with illumination or carrier injection for stable passivation of B-O defects – a three-state model

It was first identified by Herguth *et al.* in 2006 that although exposure to illumination causes B-O-related LID, illuminated annealing at elevated temperatures could permanently deactivate B-O defects [13, 122]. Importantly, it was observed that the recovered lifetime after such a process remained stable when subsequently exposed to illumination at lower temperatures, typical of field operating conditions in the range of 50–75 °C. This recovery is commonly referred to as 'regeneration' or 'passivation'. The discovery of illuminated annealing as a treatment to overcome B-O LID was promising, without the trade-offs associated with the use of alternative silicon wafers that often require significant variations in the fabrication sequence.

Although in reality the B-O defect system is rather more complicated (see Figure 5.9 in Section 5.1), a three-state model, as shown in Figure 5.12, describes the high-level behaviour of the B-O defect system sufficiently [13] with an initial degradation of carrier lifetime or efficiency with exposure to illumination, followed

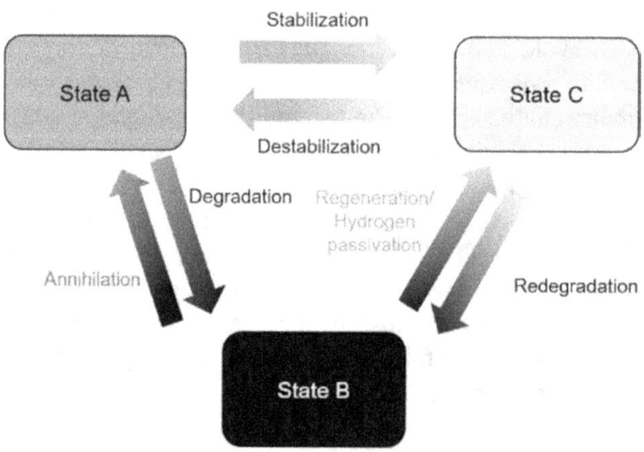

Figure 5.12 Original three-state model to describe the B-O defect system. Adapted from [13].

by recovery. Many studies have used this model as a basis for investigations into the kinetics of the B-O defect system [14, 19, 39, 44, 123–131]. Here, State A represents the recombination-inactive defect precursors, which is not stable when exposed to illumination. This is representative of the condition of wafers/cells directly after a firing process or after a typical 200 °C dark anneal for 10 min. State B represents the recombination-active B-O defect and State C is the stable 'regenerated' or 'passivated' state of the B-O defect.

The degradation of the minority carrier lifetime due to B-O-related defects is due to their transition from State A to State B with exposure to illumination or carrier injection. Defects in State B can be reverted back to State A precursors (the reverse reaction) using dark annealing processes (e.g. 200 °C for 10 min). However, the lifetime recovery is not stable, and a re-degradation of the material can occur with subsequent exposure to illumination or carrier injection.

In contrast, the 'regeneration' or 'passivation' process that occurs under illumination or current injection with somewhat elevated temperatures is a stable lifetime recovery and is represented by a transition from State B to State C. Although this original three-state model proposed by Herguth *et al.* [13] includes a possible direct transition from State A to State C for 'stabilisation', experimental results point towards an initial degradation of the carrier lifetime followed by recovery ('regeneration'). As such, unlike the thermal annihilation mechanism, eliminating B-O-related LID requires the defect to be first formed (State B) to enable regeneration or passivation to subsequently take place [132].

Importantly, defects in State C are inherently stable during subsequent exposure to illumination at operating temperatures, hence ensuring a stable performance for solar cells in the field. However, subsequent exposure to elevated temperatures in the dark for longer times can undesirably cause wafers and cells to become susceptible to B-O LID again, transitioning the system back to State A. As shown in Figure 5.13, a dark anneal at 200 °C for approximately 25 min on previously regenerated cells can result in a 15-mV reduction in V_{OC} with subsequent light soaking (~50% of the V_{OC} loss in an untreated cell).

In the initial three-state model, and many subsequent studies, this transition of the wafer from being stable in State C back to State A is represented by a 'destabilisation' reaction directly from States C and A (e.g. [125]). However, in reality, the kinetics of this reaction can be explained more accurately by a two-step process: an initial transition from State C to State B ('redegradation' in Figure 5.12), followed by an almost immediate annihilation of the defects B→A due to a much faster reaction rate under the same conditions that cause the destabilisation [132]. This rapid transition back to State A ensures that essentially no immediate degradation of the material is observed during the destabilisation of the material. The modern simplified three-state B-O defect diagram is shown in Figure 5.14.

5.3.2.1 Determination of reaction rates for kinetic modelling

To perform kinetic modelling and gain further insights on the fundamental mechanisms at play, the extraction of reaction rates for various transitions in the defect

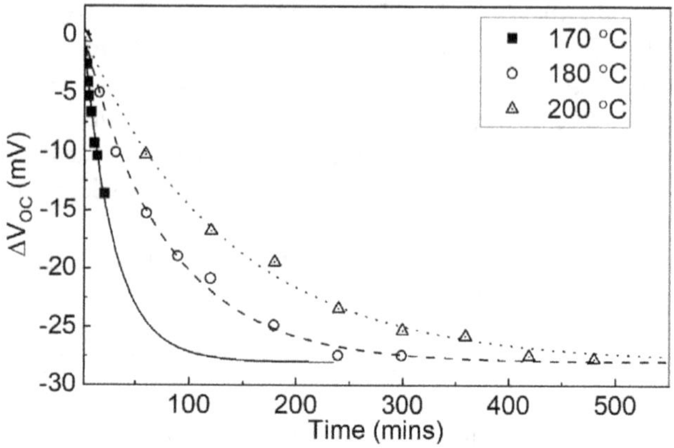

Figure 5.13 Loss of open-circuit voltage with light soaking as a function of dark annealing time after complete regeneration of cells. © 2006 IEEE. Reprinted, with permission, from [13].

system is required. This can be experimentally determined by monitoring changes in a parameter such as the effective minority carrier lifetime, NDD, V_{OC} etc. This may be performed either in situ or ex situ measurements, to determine the timescale to transition between two known states for different temperatures. For ex situ measurements, isochronal annealing may also be used.

As an example, to determine the rate of degradation, a dark annealing process at 200 °C for 10 min is typically performed to force the sample into State A. Subsequently, samples are exposed to illumination or carrier injection to cause a degradation of the minority carrier lifetime as the system transitions to State B. The curves of NDD as a function of time are then fitted against an equation to determine the reaction rate for the given condition. Typically, an exponential reaction such as (5.5) is used, where t is the light soaking time, NDD_0 is the NDD at $t = 0$, NDD_i is the amount of precursor state (i.e. the precursors available to cause LID), k_{deg} is

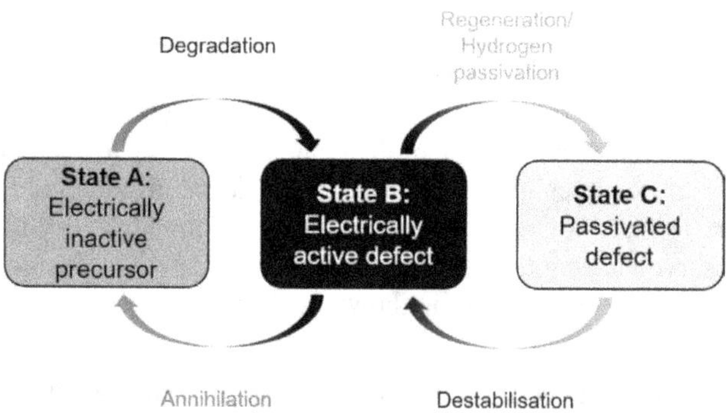

Figure 5.14 Revised three-state diagram to represent the simplified system of B-O defects. Adapted from [20].

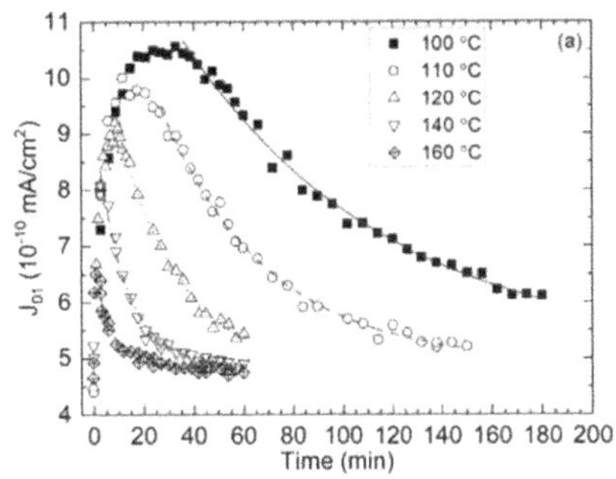

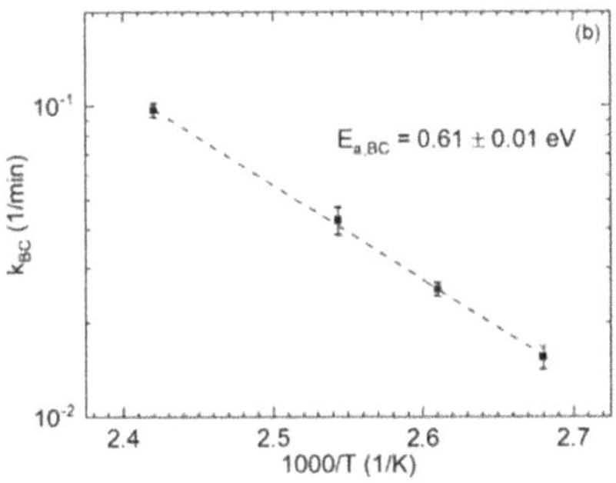

Figure 5.15 *(a) Example transition from State B into State C for illuminated annealing at different temperatures. (b) Corresponding Arrhenius plot used to extract activation energy and reaction rate. Reproduced from [133], with permission from John Wiley and Sons. Copyright © 2007 John Wiley & Sons, Ltd.*

the degradation reaction rate. Equation (5.6) shows the reaction rate k_{ij}, in terms of the attempt frequency (v_{ij}) and corresponding activation energy ($E_{a,ij}$) as determined using the Arhenius equation, where k_b is the Boltzmann constant and T is the temperature in Kelvin. Similar relationships can be obtained using other parameters such as J_{01}, as is used in Figure 5.15.

$$\mathrm{NDD}(t) = (\mathrm{NDD}_i - \mathrm{NDD}_0)\, exp\left(-k_{deg}t\right) + \mathrm{NDD}_0 \tag{5.5}$$

$$k_{ij} = v_{ij} e^{-\frac{E_{a,ij}}{k_b T}} \tag{5.6}$$

A range of different temperatures are used to investigate the temperature-dependence of the reaction rate (e.g. see Figure 5.15 (a)). An Arrhenius plot, as shown in Figure 5.15 (b), is used to extract the activation energy ($E_{A,ij}$) using (5.6), representing the temperature dependence of the reaction from state *i* to state *j*, as well as the associated characteristic frequency of the reaction (v_{ij}). It is noted that multiple reactions can take place simultaneously, and care should be taken to avoid other undesirable reactions from occurring. In particular, dark annealing can be used to suppress reactions that require carrier injection, or low temperatures can be used to essentially suppress reactions with higher activation energies that become dominant at higher temperatures. Another key note for the characteristic frequency is that it is not necessarily a constant. While this characteristic frequency is mostly temperature-independent, it can comprise other dependencies. For example, the aforementioned dependence of the degradation rate on electron and hole concentration modulates the characteristic frequency. For two materials of different doping, for example, p_0 and $p_0/2$, the characteristic frequency differs by a factor of four.

With this information, the reaction rates can then be extrapolated to other temperatures for kinetic modelling. A non-exhaustive summary of key reaction rates determined in the literature for degradation, dark annealing, regeneration and destabilisation is shown below in Table 5.1. It is noted that significant variation in the reaction rates have been reported in the literature. For example, the impact of the background-doping concentration (p_0) on the degradation rate (see Section 5.1.1.2), and a potential impact of the doping concentration on the dark annealing rate [22]. In addition, processing conditions can greatly impact the regeneration/hydrogen passivation rate, which has varied by orders of magnitude in the literature. Significant variations have also been reported for the destabilisation reaction.

5.3.3.2 Basic kinetic modelling using ordinary differential equations

With information on reaction rates for experimentally observed transitions in the B-O defect system, a kinetic model can be developed using a system of ordinary differential equations (ODEs). The system of ODEs describing Figure 5.14 is shown in (5.7). Here, N_i represents the population in state i (where i = A,B,C), k_{ij} is the reaction rate for the transition from state i to state j, (i,j = A,B,C, but $i \neq j$).

It is again noted, that this system of ODEs assumes no direct transition between States A and C. However, appropriate modifications can be made to the system to allow investigate the impact of such reactions (see [124, 132] for the modifications to the system of ODEs with all transitions assumed). Once established, an ODE solver can be used for numerical predictions as a function of time for a given initialisation condition.

$$\frac{d}{dt}\begin{pmatrix} N_A \\ N_B \\ N_C \end{pmatrix} = \begin{bmatrix} -k_{AB} & +k_{BA} & 0 \\ +k_{AB} & -\left(k_{BA}+k_{BC}\right) & +k_{CB} \\ 0 & +k_{BC} & -k_{CB} \end{bmatrix} \cdot \begin{pmatrix} N_A \\ N_B \\ N_C \end{pmatrix} \tag{5.7}$$

Assuming constant reaction rates k_{ij}, the above system can be solved analytically. The solution $N = (N_A, N_B, N_C)$ is then given by in (5.8)

$$N(t) = N_\infty + N_1 \cdot e^{-(\Omega+\omega)t} + N_2 \cdot e^{-(\Omega-\omega)t} \tag{5.8}$$

and follows a double exponential decay with effective rate constants $\Omega \pm \omega$ where the first exponential term with a higher effective rate $\Omega + \omega$ corresponds to the faster degradation phase and the second term with lower effective rate $\Omega - \omega$ corresponds to the slower regeneration/passivation phase. Effective rate constants $\Omega \pm \omega$, where Ω and ω are given by in (5.9) and (5.10), respectively,

Table 5.1 *Basic reaction rates in the B-O defect system as reported in the literature*

Reaction	$v_{0,ij}$ (/s)	$E_{a,ij}$ (eV)	k_{ij} at specified T (/s)	Reference	Note
Defect formation			(T = 25 °C)		
	-	0.3	-	[134]	
	-	0.37	-	[135]	
	-	0.38 ± 0.03	-	[136]	
	-	0.39 ± 0.05	-	[137]	
	-	0.4	-	[138]	
	-	0.44 ± 0.07	-	[13]	
	-	0.46	-	[54]	
	-	0.47	-	[45]	
	4×10^3	0.475 ± 0.035	3.5×10^{-2}	[22]	[B] = $1 \times 16/cm^3$
Defect annihilation			(T = 200 °C)		
	-	0.78 ± 0 06	-	[136]	
	-	1.24	-	[139]	
	-	1.3	-	[3]	
	1×10^{13}	1.32 ± 0.05	8.7×10^{-2}	[22]	
	1×10^{13}	1.36 ± 0.08	3.3×10^{-2}	[22]	
Defect passivation			(T = 150 °C)		
	1.7×10^8	0.61 ± 0.01	5×10	[122]	
	-	0.62 ± 0.02	-	[13]	
	-	0.64 ± 0.04	-	[122]	
	-	0.69 ± 0.06	-	[140]	Undiffused
	-	0.7	-	[133]	
	-	0.71	-	[124]	
	2×10^4	0.72	4.3×10^{-4}	[141]	
	-	0.72 ± 0.007	-	[140]	P-diffused
	-	0.9	-	[136]	
	4.6×10^9	0.98	1.7×10^{-1}	[125]	

(Continues)

Table 5.1 Continued

Reaction	$v_{0,ij}$ (/s)	$E_{a,ij}$ (eV)	k_{ij} at specified T (/s)	Reference	Note
Destabilisation			(T = 200 °C)		
	1.5×10^{7}	0.98 ± 0.01	5.5×10^{-4}	[13]	Measured using V_{OC}
	2.3×10^{7}	1 ± 0.03	5.1×10^{-4}	[13]	J_{SC}
	-	1.09 ± 0.1	-	[136]	
	5×10^{9}	1.25	2.4×10^{-4}	[125]	

> Determining reaction rates for kinetic modelling
>
> Large variations in the reaction rates are reported in the literature, which are heavily impacted by the samples and processing conditions used. When performing kinetic modelling, it is preferable to understand the reaction rates for the specific sample set and processing conditions used.

$$\Omega = \frac{1}{2} \cdot \left(k_{AB} + k_{BA} + k_{BC} + k_{CB}\right) \tag{5.9}$$

$$\omega^2 = \Omega^2 - \left(k_{AB} \cdot k_{BC} + k_{AB} \cdot k_{CB} + k_{BA} \cdot k_{CB}\right) \tag{5.10}$$

which depend on the original reaction rates k_{ij}, and correspond only to a certain rate constant if it dominates the dynamics. The pre-factors N_i are functions of the reaction rates k_{ij} and the distribution of defects in the states at $t = 0$. In particular, the long-term limit (5.11)

$$N_\infty = \begin{pmatrix} N_{A,\infty} \\ N_{B,\infty} \\ N_{C,\infty} \end{pmatrix} = \frac{1}{k_{AB} \cdot k_{BC} + k_{AB} \cdot k_{CB} + k_{BA} \cdot k_{CB}} \cdot \begin{pmatrix} k_{BA} \cdot k_{CB} \\ k_{AB} \cdot k_{CB} \\ k_{AB} \cdot k_{BC} \end{pmatrix} \tag{5.11}$$

is of interest because it shows whether a complete transition towards State C, and whether a complete mitigation of BO-LID is possible under certain treatment conditions. If, for example, degradation k_{AB} and regeneration rate k_{BC} dominate, the BO defect system will prefer the regenerated/passivated state C for sufficient long treatment times.

In general, the reaction rates k_{ij} may not be constant due to either varying temperature or varying injection. In this case, the dynamic typically still follows approximately the double exponential curve; however, an ODE solver can be used for more exact numerical predictions as a function of time for a given initialisation condition.

With the established model, the impact of variations in processing times, reaction rates or temperatures can be determined. Figure 5.16 shows reaction rates as a function of temperature assuming reaction rates from Bothe *et al.* for degradation and defect annihilation [22], and Wilking *et al.* for regeneration (assuming 2.7-suns illumination intensity) and destabilisation [125]. A key limitation for the 'regeneration' of B-O defects during illuminated annealing is that reverse reactions for 'redegradation' (State C to State B) and 'defect annihilation' (from State B to State A) occurs simultaneously as the forward transitions of 'degradation' State A to State B) and 'regeneration' (State B to State C). As shown, due to the higher activation energies of the reverse reactions rates, at higher temperatures the reverse reactions become more significant. In particular, the reaction rate for defect annihilation becomes much larger than the degradation rate. For destabilisation, however, the passivation reaction is still much higher than the regeneration/passivation rate.

For B-O mitigation processes, as indicated by both experimental results and the kinetic model (see Figure 5.17), provided that temperatures less than 230 °C

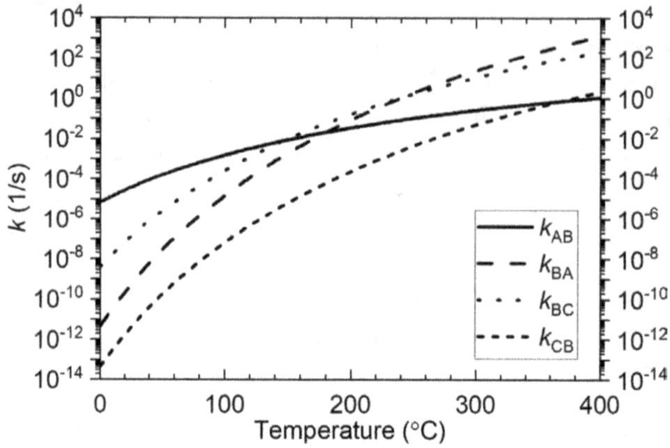

Figure 5.16 *Temperature-dependent reaction rates for the B-O defect system assuming a degradation rate and defect annihilation from Bothe et al. (v_{AB} = 10^{13}/s for [B] = 1.5×10^{16}/cm^3) [22], and regeneration rate (2.7 suns) and destabilisation rate from Wilking et al. [125]. Figure adapted from [127].*

are used, B-O-related LID can be completely eliminated by 'regeneration' with the transition the B-O defects into the stable State C [14, 124, 141]. However, as the temperature increases, the effectiveness of the mitigation is reduced [127, 142].

Another particular concern for solar cell manufacturing is processing time. In particular, mass production requires a throughput that is currently heading towards

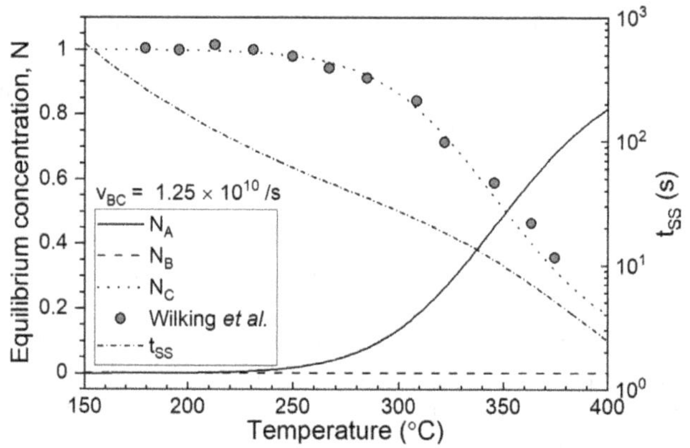

Figure 5.17 *Effectiveness of B-O LID mitigation with illuminated annealing at 2.7 suns as a function of temperature using data from Wilking et al. [125] and corresponding kinetic model of the expected effectiveness as presented in [126]. Corrected value of v_{BC} = 1.33×10^{10}/s; t_{SS} shows the time to achieve steady state (N_C = 99%) as a function of temperature with the same reaction rates.*

10 000 wafers per hour [1], or approximately 3 wafers per second. Given that the reaction rates all increase with increasing temperature, it places a minimum time for a process that can be used to achieve effective B-O LID mitigation for a given set of reaction rates.

For example, at 230 °C, a duration of approximately 85 s is required to stabilise a cell starting at State A, the natural starting state after firing. At 300 °C, only 27 s would be required (see t_{ss} in Figure 5.17); however, the effectiveness is substantially reduced to approximately 60%. Furthermore, given the exponential nature of the reaction rates, targeting the complete elimination of B-O LID may be too costly. For example, at 230 °C, the first 80% of defects can be passivated in 30 s; however, the next 15% requires an additional 26 s, a classic case of diminishing returns. This creates a trade-off between the effectiveness of B-O LID mitigation desired and the throughput and cost of the process.

Although illumination was originally thought to be required for the permanent deactivation of B-O defects, similarly to defect formation, it is actually the excess minority carriers that drive the reactions to get the B-O defects into State C [124, 128, 129, 133, 143] and therefore application of a forward bias can be used as an alternative [144]. The past decade has progressed understanding of the fundamentals of B-O LID mitigation processes based on illuminated annealing and carrier injection, which have enabled the development of rapid LID mitigation processes based on illumination or current injection [12, 144–146] as will be discussed in Section 5.4.

5.4 Fundamentals of illuminated annealing for B-O LID mitigation

This section discusses the current fundamental understanding of using illuminated anneals to mitigate B-O LID. There are still many unknowns including the exact composition of the B-O defect and its stable state (State C in Figures 5.9, 5.10, 5.12 and 5.14). However, a number of key discoveries have shed light on the role of hydrogen for defect passivation and the importance of controlling its charge state, as well as the importance of defect formation for subsequent passivation. The recognition that all of these factors are dependent on carrier injection led to development of rapid mitigation processes.

5.4.1 The role of hydrogen

In 2009, Münzer identified that hydrogen plays a critical role in the permanent deactivation of B-O LID [8]. This work found that solar cells coated with PECVD SiN_x:H degraded under illumination due to B-O LID, but then showed a recovery in V_{OC} and efficiency under further illuminated annealing. In contrast, the cells coated with a hydrogen-lean SiN_x deposited via LPCVD showed no recovery after degradation [147], despite the earlier adoption of PECVD by the industry. As such, the transition to PECVD SiN_x:H within the PV industry has been vital not only

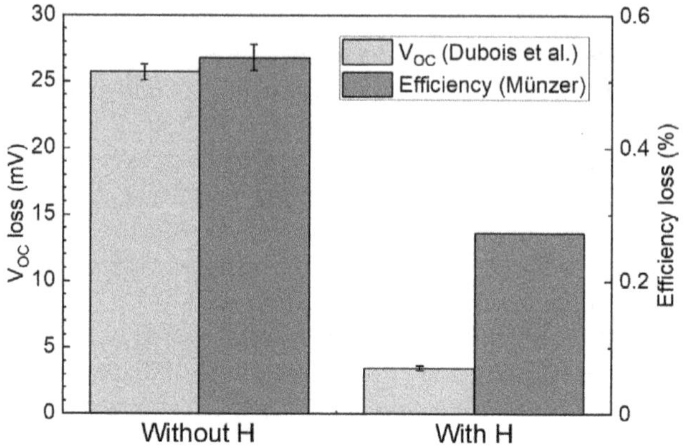

Figure 5.18 *The Voc and efficiency loss without hydrogen/using a hydrogen-lean LPCVD SiN_X compared to using PECVD SiN_X (with hydrogen). Data from [8, 15].*

for the passivation of structural defects in multi-crystalline silicon as discussed in Chapter 4, but also for the passivation of B-O defects in p-type Cz silicon solar cells.

Other studies have also identified a strong correlation between the hydrogen content in dielectric layers, such as by controlling the ammonia flux in sputtered SiN_x (without plasma pre-treatment) whereby no recovery was observed in cells without ammonia flux [148], or by fabricating solar cells with and without PECVD SiN_x:H, where V_{OC} recovery only occurred in solar cell processed with PECVD SiN_x:H [15]. Figure 5.18 shows the V_{OC} and efficiency loss from the study of Münzer *et al.* [8] and Dubois *et al.* [15] of cells with hydrogen and without hydrogen (or hydrogen-lean) cells, highlgithing the substantial reduction in loss of performance.

5.4.1.1 The importance of firing

However, the presence of hydrogen-containing dielectric layer alone does not necessarily enable regeneration. In particular, a study by Wilking *et al.* demonstrated that an identically processed sample with PECVD SiN_X:H layer in the as-deposited state did not demonstrate a recovery of lifetime during illuminated annealing. For this layer, as shown in Figure 5.19 (a), a high-temperature firing was critically required to enable the regeneration process [149], and similarly for samples with Al_2O_3 passivation layers grown by atomic layer deposition (ALD) [150].

Although firing is generally required, a key conclusion that can be drawn from the literature is that the firing process alone does not enable regeneration to occur with illuminated annealing either. For example, stripping of a deposited PECVD SiN_X:H layer prior to firing can suppress the regeneration of lifetime during subsequent illuminated annealing that occurs when samples are fired in the presence of PECVD SiN_X:H [151]. Here, the removal of the PECVD SiN_X:H prior to firing does not allow the release of hydrogen from dielectrics and distribution of hydrogen throughout the wafer, ready to activate passivation during illuminated

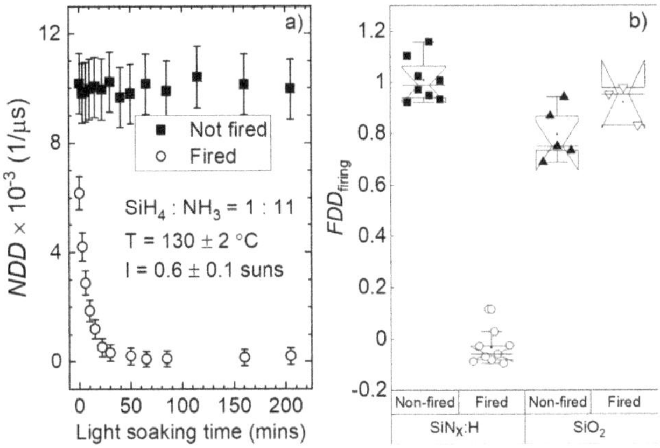

Figure 5.19 *(a) Evolution of NDD during illuminated annealing for samples coated with SiN_x:H layers with and without firing. Adapted from [149]. (b) Relative reduction in NDD after illuminated annealing (FDD) for samples with SiN_x:H and SiO_2 with and without firing, highlighting the critical role of a hydrogenated dielectric layer for regeneration. Adapted from [117].*

annealing. These results also rule out a potential role of a plasma-induced effect [152], as also concluded in another study [125]. Further evidence highlighted that firing alone does not enable a recovery during illuminated annealing was shown by Nampalli *et al.*, who compared the recovery with either hydrogen-free thermal SiO_2 passivation, or hydrogen-rich PECVD SiN_x:H [117]. In that study, no recovery was observed for SiO_2-passivated samples (see Figure 5.19 (b)).

5.4.1.2 The complicated impact of firing on B-O passivation

Part of the complication on the role of firing arises from the thermal pathway for LID mitigation by annihilating B-O defect precursors, by decreasing the LID that occurs directly after firing. In addition, the firing process simultaneously releases hydrogen from dielectric layers and allows the diffusion of hydrogen throughout the wafer, as discussed in Chapters 3 and 4. This hydrogen can then be used to passivate defects at the silicon surface and throughout the bulk, leading to changes in the background lifetime of the material, all of which can impact hydrogen passivation processes and reactions that have a dependency on the carrier injection of the material. In addition, different dielectric layers will respond differently to an identical firing process, due to the different bonding configurations within the dielectric layers.

A list of key aspects of the device that are changed by firing and can impact the passivation of B-O defects include:

- modulating the B-O defect precursor concentration, therefore changing the effective minority carrier lifetime during degradation and regeneration cycles

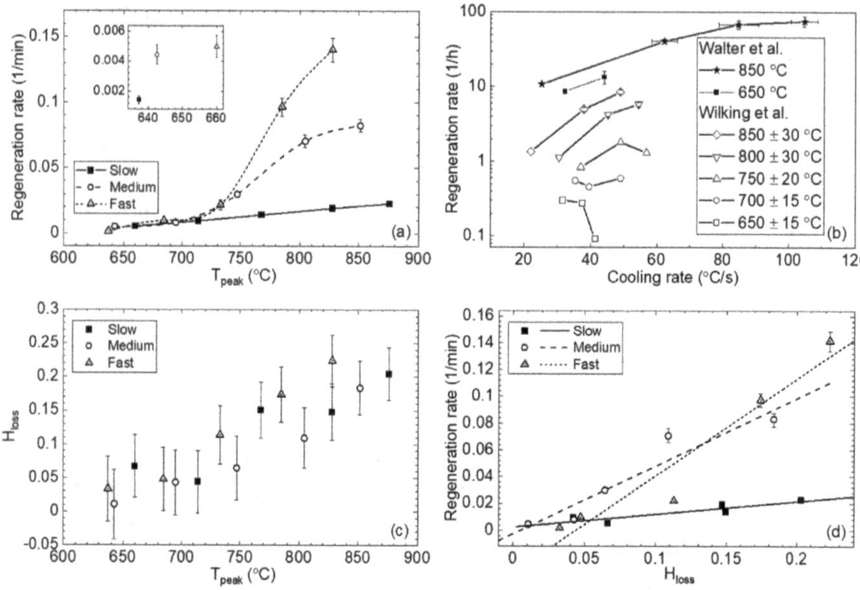

Figure 5.20 *(a) Impact of belt speed and the peak temperature (T_{peak}) and (b) cooling rate on the regeneration rate. (c) Hydrogen effusion loss of SiN_x:H layer (H_{loss}) as a function of T_{peak}. (d) The regeneration rate as a function of H_{loss}. (b) Data from [115] and [154]. Inset in (a) zooms into lower-temperature range. (a) and (c) Reprinted from [154], with the permission of AIP Publishing. © 2013 AIP Publishing LLC. (d) Adapted from [154].*

- modulating hydrogen effusion from dielectric layers, and the diffusion of hydrogen into or out of the silicon bulk, affecting the concentration of hydrogen in the silicon bulk
- a potential thermal impact and hydrogen passivation of non B-O-related recombination mechanisms changing both the bulk and surface lifetime components.

5.4.1.3 Impact of firing process parameters on passivation kinetics

In addition to the modification for the background defect concentration with firing, the regeneration rate is heavily impacted by the firing process [129, 153, 154]. For example, as shown in Figure 5.20 (a), increasing the firing temperature has been shown to result in increases in the regeneration rate during illuminated annealing and the effectiveness of LID mitigation [116, 154]. The increasing temperature correlates to an increase in the hydrogen release from dielectric layers as determined by Fourier transform infrared spectroscopy (FTIR) (Figure 5.20 (c)), therefore increasing the regeneration rate (Figure 5.20 (d)) [154]. It was also shown that

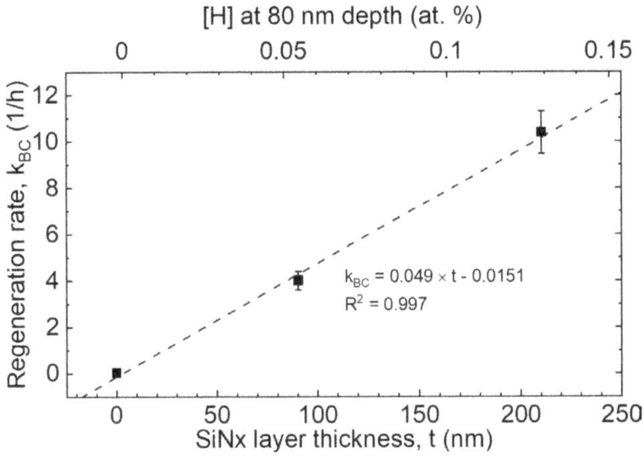

Figure 5.21 *Impact of SiN_x:H layer thickness on the regeneration rate showing a direct correlation between the hydrogen concentration at a depth of 80 nm as measured by NRRA. Adapted from [155].*

longer durations at elevated temperature accelerated regeneration rates, again with a correlation to increased hydrogen effusion from dielectric layers.

Increasing belt speed, which correlates to an increase in cooling rates during high-temperature firing (Figure 5.20 (b)) also increased the regeneration rate during illuminated annealing [115, 129], likely due to the suppression of hydrogen effusion from the bulk [154]. As such, a higher concentration of hydrogen in the bulk of the silicon is expected to lead to a higher regeneration rate [149]. Therefore, it is important to ensure a sufficiently high temperature is used for the dielectrics to allow rapid and effective mitigation of B-O LID, along with rapid cooling, while ensuring that the firing process still enables effective formation of the metal contacts and back-surface-field region. However, as will be discussed in Chapter 6, using increasing temperatures can also increase susceptibility to LeTID.

5.4.1.4 Impact of film and dielectric stack composition

With a role of hydrogen in the regeneration process, one potential method of increasing the regeneration rate is to increase the thickness of the SiN_x:H layer. Wilking *et al.* demonstrated a linear correlation between the film thickness and the regeneration rate, and further supported by nuclear resonance reaction analysis (NRRA) measurements showing a direct correlation between the [H] at a depth of 80 nm into the silicon and the regeneration rate (see Figure 5.21) [155, 156]. This is perhaps the most direct evidence of hydrogen's involvement in the B-O defect passivation process. However, for functioning silicon solar cells, the SiN_x:H layer is also used as an antireflection coating on the front surface, placing restrictions on the optical properties such as refractive index and thickness of the layers, thereby limiting the allowed film composition, without impacting on the optical performance of the device. On

Table 5.2 Impact of the composition of PECVD SiN_x:H on the passivation rate of B-O defects. Data extracted from [149].

Layer	A	B
SiH_4:NH_3 ratio	1:5.5	1:11
Thickness (nm)	85 ± 3	73 ± 3
Refractive index (630 nm)	2.15	2.05
Peak absorption (arbitrary units)	13.7 ± 0.26	15.1 ± 0.29
Passivation time constant (min)	23 ± 2	7.5 ± 5

the other hand, layers on the rear of a PERC solar cell have more flexibility for changes to the thickness and film composition.

The requirement of hydrogen for the regeneration reaction leads to a significant dependency on the composition of the layer or dielectric layers used [148, 149]. In particular, Wilking *et al.* [149] demonstrated that the recovery rate could be modulated by the film composition of SiN_x:H, even with somewhat similar refractive indexes (2.05 versus 2.15). Yet, from FTIR measurements the films displayed significantly different Si-N bond densities (integrated absorption between 700–1100/cm), with the lower refractive index film (with lower SiH_4 to NH_3 flow ratio), showing a higher Si-N bond density (see Table 5.2). It was suggested that for SiN_x:H films, a lower SiH_4:NH_3 ratio and therefore lower Si-N density results in a less dense layer. This can enhance the formation of molecular hydrogen and reduces the quantity of hydrogen into the bulk silicon [157]. As a result, a film with a higher Si-N density is expected to release more hydrogen into the bulk and, therefore, increase the regeneration rate. Another study by Bredemeier *et al.* showed a strong correlation between the refractive index and total hydrogen concentration, as was shown in Chapter 3 [158].

The presence of additional layers in between the hydrogen-rich PECVD SiN_x:H layer can also impact the regeneration rate. In particular, while a thin layer of AlO_x:H typically used on the rear surface of PERC solar cells (e.g. 8 nm thick), was observed not to modulate the regeneration rate significantly, a thicker 16 nm layer was observed to significantly reduce the regeneration rate by a factor of approximately 5 (see Table 5.3). This is despite the AlO_x layer also being a hydrogen source, and itself enabling the regeneration of B-O defects, although with a substantially reduced passivation rate (increased time constant) than for SiN_x:H. This is presuming due to the AlO_x layer acting to block hydrogen diffusion from the overlying hydrogen-rich SiN_x:H layer, such as observed by Dameron *et al.* using tritium transmission rate measurements, where a difference in ALD Al_2O_3 thickness from 5 nm to 10 nm was observed to decrease the diffusion of hydrogen by more than one order of magnitude [159]. Another recent study by Helmich *et al.* has also demonstrated the ability of Al_2O_3 to block hydrogen [160].

The impact of the AlO_x layer in modulating the regeneration rate has important implications for PERC solar cells, in which the rear surface passivation stack

Table 5.3 Impact of dielectric stack composition for ALD AL_2O_3 and PECVD SiN_x:H on the passivation rate of B-O defects suggesting the blocking of hydrogen diffusion for sufficiently thick Al_2O_3 layers. Data extracted from [149].

Dielectric layer(s)	Passivation time constant (min)
Al_2O_3 8 nm only	455 ± 56
SiN_x:H 60 nm only	44 ± 3
Al_2O_3 8 nm/ SiN_x:H 60 nm	44 ± 3
Al_2O_3 16 nm/ SiN_x:H 60 nm	225 ± 28

typically comprises AlO_x:H/SiN_x:H [154], suggesting that thin AlO_x layers should be used to ensure that sufficient hydrogen is available for the regeneration of B-O defects. The ability of AlO_x layers to contribute to the regeneration of B-O defects in PERC solar cells with AlO_x/SiN_x:H stacks might also be impacted by hydrogen effusion from the AlO_x layers during the deposition of SiN_x:H at 400–450 °C, leaving the AlO_x layer depleted of hydrogen [161], as well as the lower hydrogen concentration in AlO_x layers (~2–3% for ALD Al_2O_3 and 20% for PECVD AlO_x).

Heavily doped layers may also act to block the diffusion of hydrogen from dielectric layers into the bulk, although research into this area is limited. Wilking *et al.* demonstrated that the emitter can modulate the regeneration rate, hence highlighting that the device design can impact the ability to passivate B-O defects. However, after taking into account the change in lifetime, no change was observed [155]. It should be noted that this study was only performed with a high-temperature process for firing. A more thorough investigation comparing different emitters, particularly with firing performed at lower temperatures, is missing in the literature.

However, little work has been done with direct measurements performed, for example, with NRRA, to correlate the concentration of hydrogen in the device and regeneration rate for different dielectric layers, different emitter diffusions, as well as investigate the impact of firing at different temperatures on the concentration of hydrogen diffusing into the silicon. This is particularly important for emitter diffusions with low-temperature-firing processes where the diffused layers are more likely to have an impact on blocking hydrogen diffusion than at higher temperatures due to the silicon behaving more like intrinsic silicon. Future work could extend the work of authors such as Wilking *et al.* and Bredemeier *et al.* for different solar cell structures [155, 158].

5.4.1.5 Modulating kinetics with prior annealing

Adding a dark anneal after firing but before illuminated annealing can also modify the regeneration rate. For example, as shown in Figure 5.22, an appropriate duration such as 15 min at approximately 230 °C or 1 min at approximately 290 °C was demonstrated to enhance the regeneration rate by a factor of more than two [129].

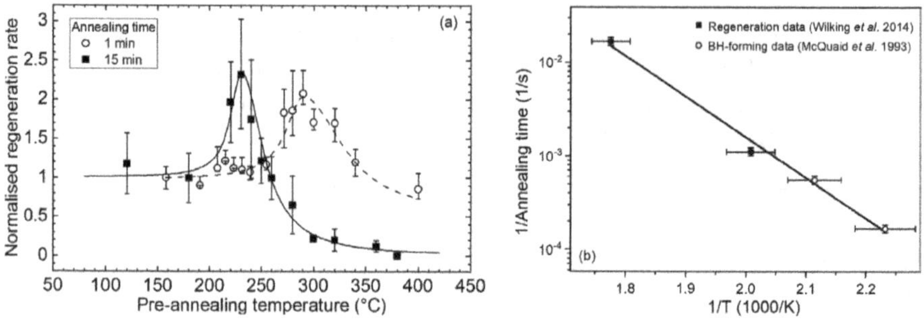

Figure 5.22 *(a) Impact of pre-annealing (dark annealing) temperature on the passivation rate of B-O defects during a subsequent illuminated anneal, and (b) close link between the passivation time constant and known boron-hydrogen (BH) forming kinetics. Reprinted from [125], Copyright (2014), with permission from Elsevier.*

However, annealing at either higher temperatures or for longer durations reduced the increase in regeneration rate, and if sufficiently high or too long, could lead to a reduction in the overall regeneration rate. From these results, Wilking *et al.* identified a correlation between the regeneration rate and the known kinetics for the forming of HB pairs, suggesting that HB pairs can act as a source of hydrogen for B-O defect passivation [125], as depicted in Figure 5.9 in Section 5.1. It is also expected that the release of hydrogen from other traps in the device, and other forms of hydrogen such as hydrogen dimers could modulate the kinetics of the B-O passivation process.

5.4.1.6 Passivation with unintentionally introduced hydrogen

It should be noted though, that there are a number of papers disputing the role of hydrogen in the regeneration process (e.g. [115, 129]). One study attempted to separate the effects of hydrogen passivation from firing by endeavouring to prepare lifetime samples and avoid the intentional introduction of hydrogen. The approach used involved firing samples without a dielectric in place and subsequently depositing hydrogen-lean ALD Al_2O_3 deposited at temperatures <200 °C or PEDOT:PSS, and avoiding subsequent high-temperature steps that could release hydrogen from dielectric layers and distribute hydrogen throughout the bulk [129]. It was observed that regeneration occurred in such samples, which led the authors to question the involvement of hydrogen. In particular, a conclusion was based on an inability of hydrogen to penetrate through a device with the processing conditions used [129] and that without intentionally adding hydrogen, the permanent deactivation occurred. The authors did note, however, that the regeneration rates for such samples were orders of magnitude lower than that of samples with a deliberate introduction of hydrogen, and that deliberately hydrogenating the material led to the highest lifetimes.

Key conclusions drawn in [129] of the inability of hydrogen to penetrate throughout the device were based on diffusivity data in [162–164] with values orders of magnitude lower than other data available in the literature, such as those from Van Wieringen *et al.* [165], as discussed in Chapter 2. By assuming higher diffusivities reported in the literature, the low-temperature processes used in the study could have concluded to enable significant hydrogen diffusion throughout the device. In particular, the duration of 10 h at 130 °C required for the permanent recovery of lifetime could enable hydrogen diffusion throughout the wafer. In addition, the AlO_x passivation activation process (260 °C for 15 min) or an illuminated annealing process at 185 °C for 20 min could also potentially enable hydrogen diffusion lengths of 160 μm and 80 μm, respectively. The results highlight the challenges faced by researchers in attempting to avoid the introduction of hydrogen into devices requiring wet chemical processing and low-temperature passivation layers, which inevitably contain hydrogen. Currently, it would appear that a safer option is using a temperature dry thermal oxide as demonstrated in [117], which avoided regeneration.

5.4.1.7 Summary of the role of hydrogen for B-O defect passivation

Hydrogen plays a key role in the regeneration of B-O defects, and the adoption of hydrogen-containing PECVD SiN_x:H has been pivotal in the development of industrial solutions to B-O LID. Any factors related to the dielectric layer used and thermal processing can influence the ability to passivate B-O defects. These include:

- composition and thickness of the dielectric layer/stack,
- thermal process used to release hydrogen from dielectric layers and distribute hydrogen throughout the wafer,
- presence of potential blocking layers reducing the concentration of hydrogen in the bulk,
- the use of additional thermal processes that can modulate the availability of interstitial hydrogen, such as interactions to form/dissociate HB pairs,
- additional impacts during illuminated annealing related to hydrogen will be discussed in the following section.

5.4.2 Role of hydrogen charge states

In addition to the degradation reaction for B-O defects requiring carrier injection, a key feature of the permanent deactivation of B-O defects with the 'regeneration' or hydrogen passivation reaction, is that it also requires carrier injection. As discussed in Chapter 2, interstitial atomic hydrogen is a negative-U impurity in silicon and has three possible charge states: positive (H^+), neutral (H^0) or negative (H^-) [166], with vastly different diffusivities and reactivities [167, 168]. Similarly, the B-O defect is also a negative-U impurity with three-charge states. Several studies have considered the influence of both B-O defect and hydrogen charge state concentrations on the passivation of B-O defects [59, 125, 149, 150, 169, 170].

With both B-O defects and hydrogen being negative-U impurities, in thermal equilibrium at low temperatures, both rest in the positive charge state, making passivation highly unlikely due to Coulombic repulsion. The dependence of

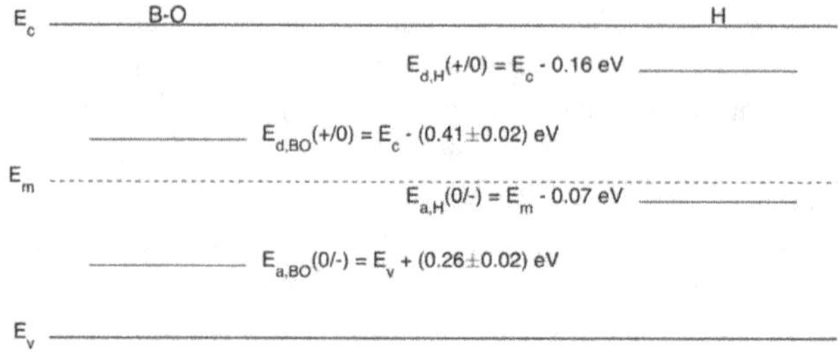

Figure 5.23 Donor (E_d) and acceptor (E_a) energy levels of B-O defects [58] and interstitial hydrogen [166], in relation to the conduction band (E_c), valence band or mid-gap (E_m)

the regeneration rate on the carrier injection is likely due to a charging of the B-O defect and hydrogen, creating a more favourable condition for defect passivation. However, it is still unclear on which charge states of the hydrogen and/or B-O defect are required for the passivation to occur, although presumably H^0 plays a key role for its enhanced diffusivity. In addition, conditions that favour a high fractional charge state concentration of H^0 result in reasonable concentrations of all three charge states. Part of the challenge in identifying the specific reactions involved arise due to the proximity of the reported B-O defect's donor and acceptor energy levels [58] with the energy level at which the hydrogen switches charge state from the majority H^+ to minority states H^- via H^0 (see Figure 5.23) [166].

One study determined a correlation between the H^0 concentrations predicted using a quasi-Fermi-based model and the B-O regeneration rate [170]. A more recent study using Sah-Shockley statistics proposed a correlation between the product of the fractional charge state of H^- and positively charged B-O defects $(B\text{-}O)^+$ [59]. With this scenario, an increasing product is realised with increasing carrier concentration up to $\Delta n = 1 \times 10^{15}/\text{cm}^3$, before it saturates.

A key factor in determinig the fractional charge state concentration of both hydrogen and B-O defects is modulating the excess carrier concentration. Two key methods to modulate the excess carrier concentration are through changes in the effective lifetime of the sample, and changes in the illumination intensity, which also follow through to change the effective lifetime of the sample during processing, and therefore excess carrier concentration.

Typically, the regeneration rate increases approximately linearly with illumination intensity [128, 129, 171–173]. However, this greatly depends on the background quality of the material and range of illumination intensity used. For high-lifetime samples, a study by Steckenreiter *et al.* observed that the regeneration rate saturated for illumination intensities higher than 2 suns [128]. On the other hand, the regeneration rate continued to increase with increasing illumination intensity (see Figure 5.24).

Higher lifetimes can also increase regeneration rate [155]. Correcting for the lifetime of the sample can also highlight apparent differences due to the solar cell structure, such as the presence of an emitter, which increases the J_0. Changes in

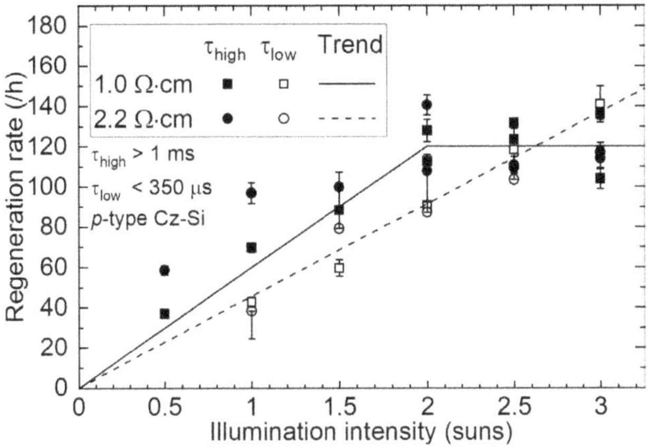

Figure 5.24 *Illumination intensity as a function of regeneration rate with high- and low-bulk-lifetime samples. Adapted from [128] under Creative Commons CC BY. https://doi.org/10.1063/1.4978266.*

lifetime can also account for at least some of the changes with reaction rates that occur with different firing conditions. For example, reanalysing the data from Figure 5.4 in [129], a strong correlation between the recovery rate time constant and the starting lifetime of the material is observed, which is essentially linked to the excess carrier concentration [174]. This is quite separate from the expected changes in the hydrogen concentration in the bulk with different firing/annealing conditions.

As such, it is vital to consider the impact of changes in the excess carrier concentration on reaction rates. A detailed study of the impact was presented by Wilking *et al.* (see [143]).

In both p-type silicon and compensated n-type silicon, an approximately linear dependence on the excess carrier density (Δn) has been reported [128, 129, 171, 172]. This accounts for the difference in material quality and illumination (see Figure 5.25). The reduction in effective lifetime with increasing carrier concentration due to J_0 and Auger effects can account for the apparent saturation of the reaction rate with increasing illumination intensity. However, the relationship is unclear at very high carrier concentrations ($>1 \times 10^{16}/cm^3$), where the change in fractional

Comparing reaction rates

When comparing reaction rates with a carrier dependence between samples, the rates should be corrected for the difference in lifetime and therefore differences in the excess carrier concentration during illuminated annealing.

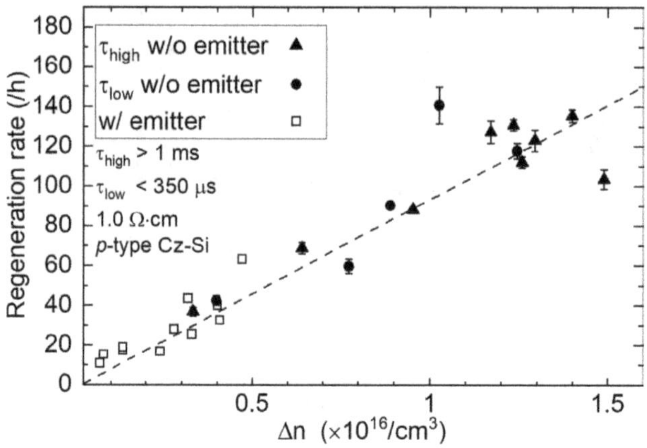

Figure 5.25 Correlation between the excess carrier concentration and the regeneration rate in high- and low-lifetime samples. Figure adapted from [128] under Creative Commons CC BY. https://doi.org/10.1063/1.4978266.

hydrogen charge states is expected to plateau. Further work is required in this area, while ensuring other reaction rates are not affected.

5.4.3 The critical role of defect formation

Illuminated annealing processes to mitigate B-O LID (and similarly for annealing processes with current injection) essentially accelerate the same degradation and regeneration mechanisms that would occur naturally for the solar panels during operation in the field [6]. Using conventional B-O LID mitigation processes at low- temperatures (50–230 °C) with low-illumination intensities <3 suns, the reaction rate of defect passivation is typically substantially faster than the degradation rate, the rate at which the defects form (refer to Figure 5.16 in Section 5.3) [127]. Given that in the hydrogen passivation pathway to eliminate B-O LID, the defects cannot be passivated/regenerated until they are first formed, the passivation process is therefore limited by the availability of the defects for passivation. This is particularly the case for screen-printed solar cells, where the last process in the conventional fabrication sequence is the high-temperature metallisation firing step, which results in B-O defects being in State A. As a result, an accelerated process for B-O LID mitigation is required to not only ensure a high reaction rate for defect passivation where possible, but also accelerate the formation of the defects/degradation to increase the availability of the defects for passivation. As discussed in Section 5.1, at low-illumination intensities, a dependence of rate of defect formation on illumination intensity was observed but saturated at approximately 0.01 suns [3, 38], where the reaction comes dependent on p_0 with a quadratic dependence, and hence the reaction rate being limited by the background boron-doping concentration in the material.

This fundamental limit imposed by the background-doping concentration and hence p_0 can be overcome by using high-intensity illumination to drive the silicon into high-injection conditions [175]. The increased reaction rate observed under high-intensity illumination has led to improved understanding that the reaction rate is dependent on the total hole concentration (p), rather than the equilibrium hole concentration, p_0 [176–178]. This discovery led to processes that enabled complete defect formation to occur within 4 s, rather than the minutes to hours required by typical defect formation processes at less than 200 °C with only approximately 1-sun equivalent illumination [179]. However, the acceleration of the reaction rate with increasing illumination intensity is not linear. In particular, the reduction of effective lifetime that occurs in high injection through J_0 and intrinsic recombination mechanisms such as Auger and radiative recombination, can limit the increase in p. For example, for sufficiently high injection levels where the lifetime is decreasing with increasing carrier concentration, an increase in illumination intensity by a factor of ten might only increase the defect formation rate by a factor of three. In contrast, larger increases in the reaction rate can be observed in regimes where the lifetime is either injection-level-independent (flat), or affected by SRH recombination, with the lifetime increasing with increasing injection level. This means, that above a certain injection level, there is a reduced benefit of adding more illumination. It is again, a case of diminishing returns.

The rate of defect formation has important implications for the extent to which B-O-related LID can be eliminated [126, 127]. Using high-intensity illumination to increase the defect formation rate for a given temperature also improves the efficacy of LID mitigation processes (the ratio of passivated to inactive/dissociated defects), thereby enabling use of higher temperatures (see Figure 5.26) [127, 131]. Increased temperatures further accelerate LID mitigation processes as the reaction rates are exponentially dependent on temperature, thereby reducing the time required for passivation (see Figure 5.26 (b)). The increase in temperature also causes increased carrier concentrations and lifetime [177, 180]. Again, there are upper limits to the temperature that can be used to prevent an incomplete passivation of B-O defects, as shown in Figure 5.26 (a) where the completeness of regeneration is reduced at high temperatures [125]. For complete BO-LID elimination the temperature should be selected to suit the defect formation rate. Increasing the defect formation rate by an order of magnitude from the typical value ($v_0 = 4 \times 10^3$/s for 1×10^{16}/cm^3 doping to $v_0 = 4 \times 10^4$/s) can increase the cut-off temperature at which the efficacy drops, from 230 °C to 300 °C [126, 142].

An alternative path to increase rate of defect passivation is to use higher doped material, with a lower resistivity in the range of 0.5–1 Ω·cm. Although more sensitive to a higher extent of degradation, kinetics are expected to be accelerated due to the higher p_0 value, and therefore a reduced necessity to use high-intensity illumination. However, to the authors knowledge, no such study has investigated this in detail. This path would have the advantage of reducing series resistance in the bulk. However, this approach would require certainty of the effectiveness of the LID mitigation to ensure a high efficiency can be maintained during operation in the field.

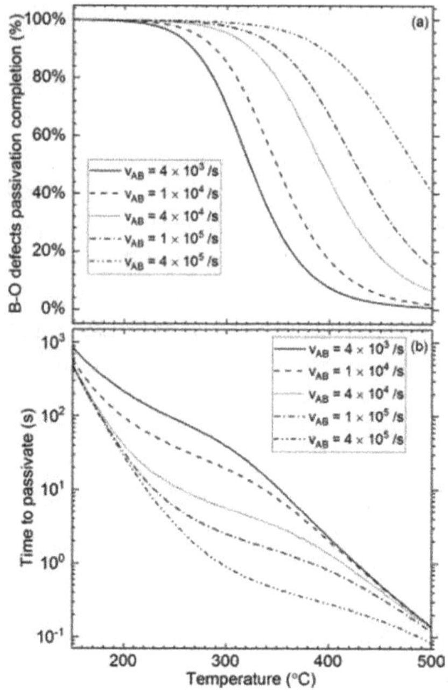

Figure 5.26 *(a) Simulated percentage of regeneration achieved for different regeneration temperatures for different defect formation rates (v_{AB}) through either high-intensity illumination or increasing the background-doping concentration. © 2016 IEEE. Reprinted with permission from [126]. (b) Corresponding impact on the time to achieve defect passivation as a function of temperature.*

5.4.4 Development of rapid laser-based processes

This section has covered recent improvements in the fundamental understanding of B-O defect formation and passivation, notably the requirement for the defect to be formed prior to passivation, and the ability for high-intensity illumination to accelerate both this and the passivation reactions. This led towards the use of laser-based processes to eliminate LID by enabling rapid defect formation and hydrogen charge state manipulation for passivation [181]. Wilking *et al.* demonstrated that localised laser-based hydrogenation processes enabled regeneration reactions with time-constants for the defect passivation reaction on the order of 0.1 s when defects were pre-formed (the samples were pre-degraded) [155]. However, after high-temperature-firing process, B-O defects are in State A, and also require defect formation.

With the improved understanding of accelerated defect formation with high-intensity illumination, Hallam *et al.* and Hamer *et al.* demonstrated the complete elimination of B-O LID in typical commercial cells using full-area laser-illumination to simultaneously form and passivate the B-O defects [182, 183]. The high-intensity

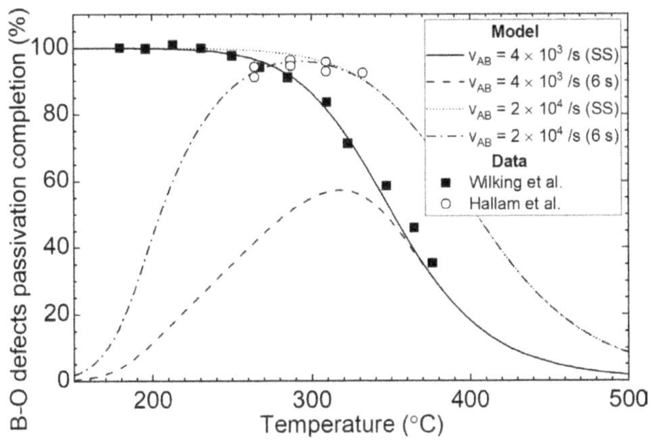

Figure 5.27 *Experimental results and simulated percentage of regeneration achieved for different regeneration temperatures for different defect formation rates through either high-intensity illumination or increasing the background-doping concentration. © 2016 IEEE. Reprinted, with permission, from [126].*

illumination allowed these rapid processes to be performed at approximately 300 °C within 6 s, while maintaining effectiveness at eliminating B-O LID, significantly higher than temperatures used in previous studies for longer durations (see Figure 5.27), and hence in a regime more suitable for mass production than using low-intensity illumination at substantially lower temperatures.

However, the cost of processing is also important to consider in mass production. With the non-linearity between the illumination intensity used and defect formation rate, there is a trade-off between the intensity used and process cost. For example, rather than using ×10 the illumination intensity for a third of the time, it may be more cost effective to use a slightly lower illumination intensity for a longer period of time, thereby reducing the overall photon flux required for B-O mitigation.

5.4.5 Long-term stability of passivation

Stability is a significant concern for solar cell manufacturers and PV consumers due to the impact it has on the LCOE cost of electricity. For B-O defects, although the destabilisation of passivated defects is of concern, in the field, conditions that cause a heating of the solar panel and hence the potential to cause a destabilisation, also provide carrier injection, and hence, enable the passivation reaction, provided that sufficient hydrogen is available in the device. This acts to reduce the impact of a potential destabilisation of the material. Even without a potential re-passivation of destabilised defects in the field such as dark annealing, an extrapolation of the destabilisation reaction rate to 60 °C suggests a destabilisation time constant of >50 years [125], and hence no concern in the field. There is uncertainty, however, on the accuracy of this, with substantial variations in the reaction rates reported. For example,

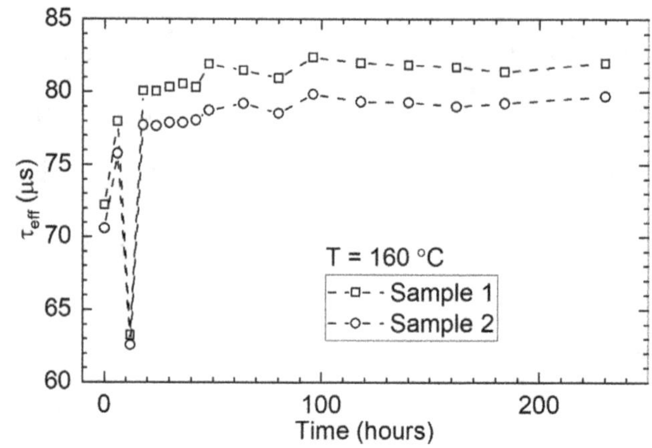

Figure 5.28 Stability of lifetime after B-O passivation with light soaking at 160 °C

a study by Hielsmair *et al.* suggested that the time constant could be only 12 000 h, equivalent to 12 years of operation in the field in California [184].

It is vital that further work is conducted to understand reaction rates in the B-O defect system while ensuring the results are not impacted by other defects such as LeTID(to be discussed in Chapter 6), and even within the B-O defect system, identifying which mechanisms are responsible, such as a destabilisation of defect passivation, or a thermal re-activation of defect precursors. For example, a study by Herguth in 2009 identified that prolonged exposure to illumination at elevated temperatures highlight a long-term instability after 30 h of processing at 120 °C. However, that paper concluded that it was insufficient to determine whether or not it was related to B-O defects or another recombination mechanism.

More recent unpublished data is shown in Figure 5.28 with light soaking at 160 °C, showing complete stability to 250 h, and the majority of the recovery still stable after 400 h. Extrapolating these results down to 60 °C and accounting for a variation in the reported activation energies for destabilisation of 0.92–1.25 eV, suggests that the passivation would be stable for 46–660 years at 60 °C, which is at least as stable as the results presented in [125].

5.5 Eliminating B-O LID in commercial solar cells

Although the majority of commercial solar cells are currently fabricated using p-type Cz wafers that are prone to B-O defects, the dominant commercial cell technologies automatically enable hydrogen passivation of bulk defects [1, 185]. Hydrogen, as required for passivation/regeneration, is introduced to the silicon bulk naturally during the conventional cell processes. This allows a relatively simple modification to be made to cell or module production processes to eliminate B-O LID.

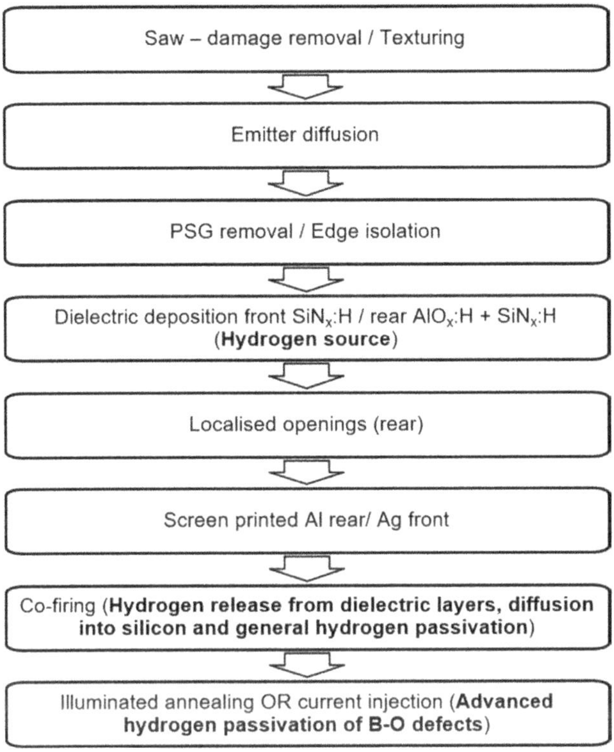

Figure 5.29 Production process for a typical commercial screen-printed PERC technology, noting steps that contribute to B-O defect passivation

5.5.1 Commercial B-O LID mitigation solutions for cells

The PERC solar cell is the dominant PV technology at present [1]. PERC cells are typically fabricated according to the processing sequence shown in Figure 5.29. The PECVD SiN_x:H dielectric applied to the front surface not only passivates the surface dangling bonds and acts as an anti-reflection coating; it is also a source of hydrogen that can be subsequently manipulated for bulk passivation. Dielectric layers such as PECVD aluminium oxide (AlO_x:H) and SiN_x:H, typically used for rear surface passivation, provide additional hydrogen sources.

The co-firing process required to form screen-printed metal contacts is the second standard but crucial process for hydrogen passivation of B-O defects. The high temperatures at the peak of the firing process can break hydrogen bonds and release it from the dielectric layers [154], and can enable the hydrogen to diffuse into and throughout the silicon bulk [186–189]. The peak temperatures in the range 700–900 °C overcome issues with hydrogen trapping by significantly increasing the fractions of neutral hydrogen (H^0) and enhancing the diffusivity [168].

Once the hydrogen is distributed throughout the silicon, the cells can go through a fast illuminated process to form the B-O defects as well as manipulate the hydrogen

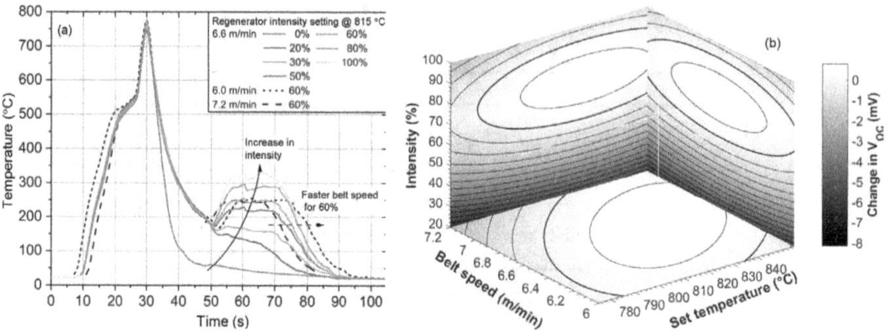

Figure 5.30 (a) Thermal profile in a Centrotherm c.FIRE REG 9.600 tool with a peak set temperature of 815 °C for various illumination intensities and belt speeds. Figure adapted from [145]. (b) Corresponding impact on open-circuit voltage loss due to LID.

charge state to passivate the B-O defects. A simple way to achieve this is to apply illumination during the natural cool-down that follows the high-temperature firing process, or alternatively it can be an additional subsequent thermal treatment. A number of commercial tools are available that can reduce the extent of B-O LID by approximately 75–80%, by companies including Despatch Industries, Schmid and Centrotherm [145, 190, 191]. The tools that integrate the LID mitigation process into a fast-firing furnace used for co-firing of contacts do not therefore require additional equipment and processes; however, changes to the firing process (such as belt speed) that may be required for future metal pastes could impact the mitigation of the B-O LID. A study by Derricks *et al.* investigated the impact of process parameters on B-O regeneration in a Centrotherm c.FIRE REG 9.600 tool where B-O mitigation is performed during the cooldown phase after fast firing [145]. It was observed that with appropriate conditions, almost complete mitigation could be achieved during conveyance in the 4 m long regeneration module (within 15 s to 30 s), in contrast to a susceptibility of the cells to LID when no illumination was provided in the regeneration module.

In particular, it investigated the illumination intensity and furnace belt speed and peak firing temperature, all of which can alter the temperature in the illuminated zone, and therefore the effectiveness of B-O mitigation. Figure 5.30 (a) highlights the impact of the illumination intensity setting and belt speed on temperature for a peak firing temperature of 815 °C. A key feature is that the variation in illumination intensity could increase the temperature from below 200 °C to above 300 °C. Figure 5.30 (b) shows the change in open-circuit voltage due to B-O for a constant belt speed of 6.6 m/s based on a design of experiments (DoE) with central composite design. It highlighted that with the chosen settings in this tool, almost complete regeneration could be achieved with an illumination intensity of 60–85%, but higher illumination intensities were unfavourable. This highlights the complex nature of the B-O mitigation process, requiring close attention to detail.

Separate tools built solely for B-O LID mitigation can therefore offer more flexibility to be optimised separately to cell processing and can be applied to any cell irrelevant of prior processing. Such standalone tools also allow PV module manufacturers who purchase cells from other cell manufacturers to perform LID mitigation processes before module encapsulation to ensure LID free modules. Several companies including Asia Neo Tech and DR laser offer such tools [12].

As illumination source technologies are also advancing, the tools for LID mitigation are incorporating increasingly higher illumination intensities to further accelerate the mitigation process and improve the passivation. Commercial tools typically use illumination intensities between 5- and 20-suns equivalent [12, 190–192]. At these illumination intensities, belt-furnace-based tools require 25–45 s processing time per wafer to be effective. To meet the requirement of one wafer processed per second for high-throughput processes (current industrial standard), multiple samples need to be processed concurrently. Laser-based tools, which have safety and cost implications, can achieve much higher illumination intensities, and enable much faster process times.

Given that the B-O passivation is driven by carrier injection, an alternative to using high-intensity illumination is to use current injection. However, due to increasing series resistance losses with higher currents, current injection does not offer the same opportunity for accelerated defect formation as with illumination (performed under open-circuit conditions) with currents used equivalent to 1–2 suns. As a result, substantially longer times are required, such as approximately 15 min at 150 °C, requiring a larger number of cells to be processed simultaneously. Commercial tools that use this method have the advantage of being able to be batch processed as cells can be coin-stacked so that current passes through an entire stack [144]. However, coin-stacking increases handling and can potentially cause significant stress on the finished cells. It can also potentially result in an inhomogeneous process and heating of the wafers that impacts the LID mitigation effectiveness. Yet, due to the much lower cost of this approach, and manufacturing solutions to address potential issues, it has become the method of choice from the industry.

One issue with performing B-O passivation during cell fabrication, is that it can be partially undone or destabilised during any subsequent thermal exposure [13], leaving the silicon prone to degrade again under illumination. LID mitigation performed on cells could therefore be compromised by temperatures experienced during module fabrication. The entire module will reach temperatures around 150 °C for up to 15 min during lamination, while localised regions of cells will be exposed to temperatures approximately 230 °C for a few seconds during the soldering of interconnects; 15 min at 150 °C should only destabilise <5% of passivated B-O defects; however, due to the large activation energy of the destabilisation reaction, slight temperature increases lead to a significantly increased portion of destabilised defects. For example, a 50 °C temperature increase to 200 °C is predicted to cause destabilisation of approximately 20% of passivated B-O defects. However, this is not likely to occur, as other aspects of the device such as the encapsulant material would also likely suffer from such an increase in temperature.

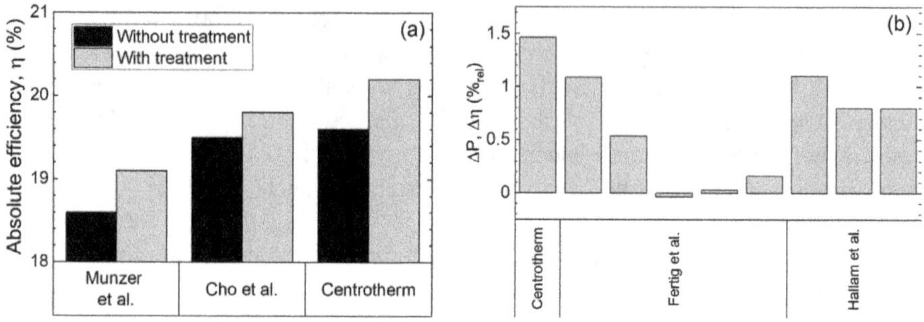

Figure 5.31 *(a) Stabilised efficiency of p-type Cz silicon Al-BSF and PERC solar cells with and without treatment to eliminate B-O-related LID. (b) Relative power (ΔP) or efficiency (Δη) change after stability test with treatment, ΔP, data from Münzer* et al. *[8], Cho* et al. *[9], Centrotherm [10], Fertig* et al. *[11] and Hallam* et al. *[12].*

Figure 5.31 highlights the difference in stabilised efficiencies with and without treatments for B-O LID along with the extent of LID after treatments, which has now been reduced to <1.5%$_{\text{relative}}$.

5.5.2 B-O LID mitigation during module lamination

To prevent potential destabilisation of passivated B-O defects during encapsulation, the LID mitigation could be performed as part of the lamination process. The advantage of this is that lamination is the final thermal process in the manufacturing of modules. This means there are no subsequent processes that could destabilise passivated defects. To achieve this, either current injection or illumination needs to be applied during the lamination process. To enable illumination, significant modifications would be required to typical lamination tools, such as transparent diaphragms. In contrast, current injection can be fairly simply applied during module encapsulation. Lee *et al.* [6] used current injection during lamination to demonstrate 80% less degradation due to B-O defects compared to a 60% reduction when B-O mitigation was performed at the cell level. The exact processing conditions were not given; however, complete regeneration/passivation of the defects requires a current injection on the order of 1 sun, with time and temperature also playing a role in the effectiveness of a B-O LID mitigation process.

5.5.3 Self-recovery in the field

Cells and modules in operation in the field are exposed to illumination and increased temperature, which have an impact on the B-O defect system and hydrogen. Modules commonly reach temperatures above 50 °C, and depending on installation location and type, can approach 100 °C [16]. Such conditions cause any B-O defect precursors (State A) to form recombination active defects (LID). With hydrogen distributed throughout the silicon from cell manufacturing processes, field operating conditions can also enable B-O defects to naturally recover [6, 16, 130]. Lee *et al.* [6] reported a recovery in open-circuit voltage corresponding to a reduction in LID of 40% in <700 h operation at maximum power-point and 50 °C. Due to the reaction dependence on excess carriers, this recovery can be accelerated by leaving modules in open-circuit conditions. Modules commonly sit in the field for at least a short time in open-circuit conditions during initial stages of installation. A full recovery can be observed in under 200 h of continuous illumination in open-circuit conditions at 60 °C [6, 82, 193]. This has led to the consideration of capitalising on the self-repairing nature of B-O LID susceptible modules by deliberately allowing regeneration to occur in the field [17]. However, the time required for recovery is also heavily dependent on the operating temperature, and hence the module mounting, geographical location and time of year of install [16]. This means that recovery times can range from days to years. Relying on a natural recovery also leads to uncertainty in module performance, potentially causing issues for system design and project financing. A stable long-term efficiency is far more desirable. Field conditions could also cause a much longer term gradual destabilisation of B-O defects, which should not be of concern as long as hydrogen exists in the cell to re-passivate the defects [17]. It is therefore preferable to ensure enough hydrogen is distributed throughout the cell during cell processing, followed by effective LID mitigation processes during cell or module fabrication.

5.6 B-O LID in p-type silicon heterojunction solar cells

With a record efficiency of 26.7% [194] and >24% full-size commercial efficiency, silicon heterojunction (SHJ) solar cells are gaining significant commercial interest [1]. Such high efficiencies can be achieved due to excellent surface passivation provided by thin intrinsic hydrogenated amorphous silicon (a-Si:H) [195]. These PECVD layers are deposited at relatively low temperatures (<250 °C) [196]. As discussed in Chapter 4, unlike PERC cells, SHJ cells do not go through any higher temperature processes such as diffusions or metal contact firing that readily enable defect engineering such as gettering and hydrogen passivation of bulk defects [185]. As such, all industrial SHJ solar cells are currently made using more expensive n-type Cz wafers, due to the higher starting lifetime and for the avoidance of B-O-related LID. It would be advantageous from a cost perspective to combine the excellent surface passivation of SHJ solar cells with cheaper p-type wafers.

Due to the strict lifetime requirement for SHJ solar cells, relatively little work exists on the fabrication of p-type SHJ solar cells, with the best efficiencies until recently being 20.7% [197]. In addition, unlike n-type SHJ cells fabricated on wafers doped solely with phosphorus, using boron-doped p-type wafers makes SHJ cells susceptible to B-O LID, and conventional SHJ solar cell sequences do not have any processes dedicated for defect engineering.

Recently, CSEM (Centre Suisse d'Electronique et de Microtechnique) demonstrated 23.6% efficient p-type Cz SHJ solar cells with a Voc of 740 mV [198] using boron-doped wafers with a moderate resistivity of 6 Ω·cm, whereby gettering was used to improve the lifetime of the material. Although slightly higher efficiencies were observed for champion cells, the average efficiency of the p-type devices in larger batches was approximately the same as that of n-type devices. A slightly higher efficiency of 23.76% was also achieved using p-type FZ silicon, the highest ever reported efficiency for a p-type SHJ solar cell. However, the work did not investigate the extent of B-O LID or treatments in the p-type Cz SHJ cells.

Collaborative work between UNSW, the Australian National University and Arizona State University, funded by the Australian Renewable Energy Agency (ARENA) and Australian Center for Advanced Photovoltaics (ACAP) has investigated approaches to integrate defect engineering into SHJ solar cells, particularly for the treatment of B-O LID. A multi-stage approach has been developed, broadly broken into three-key-processing regimes:

1. wafer-level defect engineering for gettering/thermal processing and a high-temperature introduction of hydrogen,
2. conventional SHJ solar cell processing,
3. rapid illuminated annealing for the treatment of B-O LID.

This enables a high-temperature processing for gettering and hydrogen passivation to be performed prior to the low-temperature, and hence combining processes typically used for p-type solar cells, along with the dedicated high-efficiency SHJ processes typically reserved for n-type wafers.

Chen *et al.* recently performed an analysis of light-induced degradation in p-type Cz SHJ solar cells [199]. The work tested the stability of SHJ precursor lifetime structures fabricated on boron-doped p-type Cz wafers. After 48 h light soaking at 25 °C (the typical testing conditions for B-O LID) the effective lifetime at a minority carrier density of $1 \times 10^{15}/cm^3$ reduced from approximately 220 μs to 100 μs. Injection-dependent lifetime spectroscopy allowed curve fitting to identify the defect k-value (capture cross section ratio σ_n/σ_p) of approximately 12.1, matching previous reports of B-O defects [22]. Samples that were then treated with an illuminated advanced hydrogenation process (AHP) at 300 °C for 10 s. This not only saw the lifetime increase significantly to approximately 320 μs but made the samples stable under subsequent B-O LID testing, which in fact increased the lifetime further, likely due to improvements in surface-related recombination mechanisms affecting low injection. Importantly, this highlighted the ability to achieve near millisecond lifetimes after the elimination of B-O LID when using commercial-grade boron-doped p-type Cz material with modest quality, and the lifetime after such processes appears to be limited by surface-related effects, rather than bulk effects. The rapid 300 °C process also avoided a deterioration of the surface passivation quality, which is widely believed to occur for SHJ structures for temperatures above 200 °C.

Table 5.4 presents I–V results on the impact of a high-temperature hydrogenation process and subsequent AHP treatment at approximately 295 °C for 10 s under 100-suns illumination, for large area p-type SHJ cells manufactured in an industrial environment [200]. In that study, although all wafers had a hydrogenated SiN_x:H layers deposited on both sides for lifetime analysis prior to SHJ solar cell fabrication, a comparison was given between cells that were fired and unfired, to highlight

Table 5.4 J–V characteristics of the defect-engineered p-type SHJ solar cells with and without an AHP for passivation of B-O defects before and after LID (degraded). Data from [200].

Treatment	Number of cells	J_{SC} (mA/ cm^2)	V_{OC} (mV)	FF (%)	η (%)
No AHP	8	39.6 ± 0.1	731.1 ± 1.4	73.1 ± 0.2	21.2 ± 0.1
No AHP – degraded	*	38.9	673.7	72.1	18.9
With AHP	4	39.5 ± 0	733.3 ± 0.5	74.8	21.7
With AHP – degraded	2	39.5 ± 0	729.2 ± 1.2	73.8 ± 0.1	21.3 ± 0.1
Champion stable cell*	*	39.1	736.2	76.3	22.0

*Data for the stable champion cell, externally measured at SERIS.

the impact of the firing process on improved hydrogenation. In addition, a comparison was made between cells with and without a process to treat B-O LID. Although a 735 mV V_{OC} and corresponding efficiency of 21.6% was achieved for cells that only received a gettering process, the cells were heavily susceptible to LID, with the V_{OC} reducing to 658 mV after only 48 h of light soaking at room temperature, resulting in a 3.1%$_{absolute}$ (14.3%$_{relative}$) loss in efficiency, to 18.5%. The addition of AHP, led to a slight increase in V_{OC} and efficiency, but most importantly, dramatically reduced LID extent to 0.2%$_{absolute}$ (0.9%$_{relative}$), resulting in a stabilised V_{OC} of 735 mV and efficiency of 21.6%.

Similarly, the addition of AHP on samples with a high-temperature hydrogenation process greatly reduced susceptibility to LID; however, such samples had a slightly higher extent of LID (2%$_{relative}$) than samples without a high-temperature firing process. Further work is required to understand the cause, and whether it could be due to LeTID as will be discussed in Chapter 6. However, independent I–V measurements performed at the Solar Energy Research Institute of Singapore indicated the highest stable efficiency of 22.0% with a stabilised voltage of 736 mV occurred for samples with a deliberate high-temperature hydrogenation process, although only 0.1%$_{absolute}$ higher than that of the samples without, and hence further work is required to understand if there is any statistically significant difference caused by the thermal processing and deliberate introduction of hydrogen, including the kinetics of passivating B-O defects.

Figure 5.32 shows in situ suns V_{OC} measurements for gettered samples (G) with and without AHP during light soaking for 48 h at room temperature. It highlights the significant drop in V_{OC} of approximately approximately 60 mV during light soaking and the corresponding reduction in lifetime at open-circuit and maximum power point (MPP), with an inherent bulk lifetime after degradation of <20 μs at max power point. It also highlights the stability of the V_{OC} and lifetime after the treatment of B-O defects, increasing the stabilised lifetime of the material by one order of magnitude.

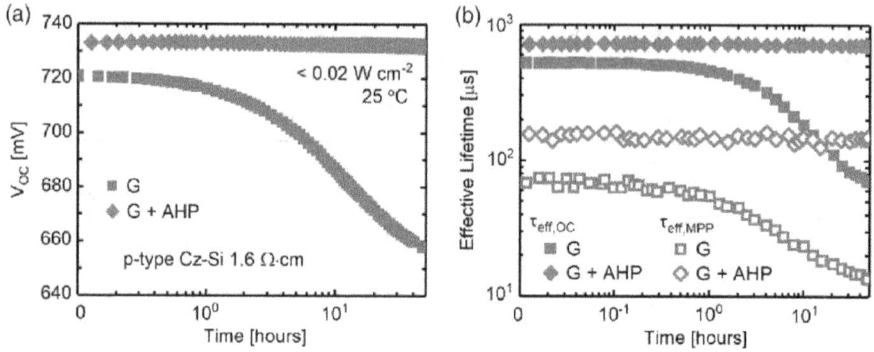

Figure 5.32 *(a) In situ open-circuit voltage measured during 48 h light soaking for a p-type SHJ cell that is G gettered only (closed square) and gettered plus AHP-treated (open square), (b) effective lifetime extracted at both open circuit (OC) and MPP as a function of light soaking. Reproduced from [200], with permission from John Wiley and Sons. © 2020 WILEY-VCH Verlag GmbH & Co. KGaA, Weinheim.*

A key result from that work was the ability to fabricate a p-type solar cell with a stabilised V_{OC} of 736 mV, using wafers heavily susceptible to B-O-related degradation. This is believed to be the highest ever stable V_{OC} for such material. Impressively, this is higher than the V_{OC} of the world-record p-type solar cell at 26.1% with a V_{OC} of 727 mV [201] and the 25% PERL cell with a V_{OC} of 706 mV [202]. This suggests that p-type Cz material may be of a sufficient quality after defect engineering to enable the fabrication of industrial solar cells >25% in the near future, substantially higher than the widely believed view in the industry. Combined with the results from CSEM, it also suggests that p-type Cz SHJ solar cells with stabilised efficiencies approaching 24% are feasible.

Another key result in that work was the ability to passivate B-O defects without a dedicated and deliberate high-temperature firing process to introduce hydrogen. However, it was unclear on whether the SiN_x:H deposition, performed on samples after the stripping of the heavy gettering diffusion layers, could have impacted the hydrogenation of the material. A study by Sun *et al.* demonstrated the ability to eliminate B-O in SHJ solar cells fabricated on compensated n-type wafers, without prior hydrogenation treatments, albeit using laboratory type tools that prolong the exposure to temperatures in the vicinity of 200 °C [203]. Importantly, the work highlighted that the deposition of PECVD a-Si:H in SHJ solar cells and subsequent annealing at 200 °C such as for the curing of the metal contacts is sufficient to distribute the hydrogenation throughout the bulk to enable the passivation of B-O defects. It was noted, however, that the deliberate hydrogenation of the material with a high-temperature process gave the best results.

Despite rapid progress on the fabrication on p-type SHJ solar cells and treatment of B-O in such devices, more work is required to understand the kinetics of B-O passivation in SHJ solar cells. In particular, whether the industrial SHJ process, with no prior exposure to hydrogen, is sufficient for the rapid passivation of B-O defects. However, based on Figure 5.9 in Chapter 2, a duration of approximately 20 min at 200 °C can potentially enable diffusion length of hydrogen of 100 μm, which can readily occur between PECVD processes and the curing of metal pastes.

5.7 Summary

Light-induced degradation due to B-O defects LID has been an issue for p-type Cz solar cells for decades. Even after decades of research there is still debate over the fundamental defect properties, however, a dependence on both the boron and oxygen concentration enables the degradation to be attributed to a form of B-O defect. In the past, strategies to deal with B-O LID involved a trade-off between accepting LID losses and economic and technological drawbacks such as more expensive alternative wafers and more complex processing. However, constant improvements in Cz pulling technology are reducing the extent of LID in p-type Cz material.

For B-O susceptible wafers, with introduction of hydrogen into commercial cells from the adoption of PECVD SiN_x:H, it was realised that B-O defects could be hydrogen passivated to form a stable recombination-inactive state under field conditions, hence eliminating LID. In the past decade, the understanding of the dependence of the defect formation and passivation reactions on the carrier injection have been greatly improved, which has led to the development of rapid treatments to passivate B-O defects. In commercial screen-printed cells, hydrogen is introduced from the hydrogen-containing dielectrics during the contact firing process. Once hydrogen is distributed throughout the silicon, subsequent thermal processing with carrier injection (by applied current or high-intensity illumination) can passivate defects to improve the bulk lifetime and reduce or even eliminate the LID. There are now a number of commercial solutions available to mitigate B-O LID implemented in mass production, enabling a reduction of the LID extent to $<1\%_{relative}$. Importantly, the methods and processes used can treat multiple different defects, not only B-O-related LID in this chapter, but also other bulk defects as discussed in the previous chapter.

The integration of such processes with p-type SHJ solar cells has enabled boron-doped p-type Cz solar cells to achieve a stabilised V_{OC} of 736 mV, with the effective lifetime affected by surface-related effects at low injection rather than bulk effects. This suggests that the standard commercial grade p-type Cz wafers will be suitable for emerging passivated contact technologies, provided that appropriate defect engineering such as gettering and hydrogenation are used, which could enable efficiencies well over 25% in the near future.

However, there are still gaps in the literature in relation to the thermal pathway for the thermal annihilation and formation of B-O defect precursors and whether this can cause a long-term instability in the field. In addition, work using advanced characterisation techniques such as NRRA to understand the role of hydrogen is limited, and in particular, gaps exist on changes in the concentration of hydrogen within the wafer for different dielectric layers, different firing conditions, and whether the presence of an emitter can impact the passivation rate for low-temperature annealing. Further work can also be conducted to better understand kinetics in the B-O defect system for passivation and destabilisation, by ensuring other reactions are 'switched off'.

References

[1] *2020 International Technology Roadmap for Photovoltaic 11th edition.* VDMA; 2020. Available from itrpv.vdma.org [Accessed 20 February 2021].

[2] Fischer H., W P. 'Investigation of photon and thermal induced change in silicon solar cells'. Proceedings of the 10th IEEE Photovoltaic Specialists Conference; 1973. pp. 404–11.

[3] Schmidt J., Bothe K. 'Structure and transformation of the metastable boron-and oxygen-related defect center in crystalline silicon'. *Physical Review B.* 2004;**69**(2):24107.

[4] Knobloch J., Glunz S.W.W., Biro D., Warta W., Schaffer E., Wettling W. 'Solar cells with efficiencies above 21% processed from Czochroalski grown silicon'. Proceedings of the 25th IEEE Photovoltaic Specialists Conference; 1996. pp. 405–8.

[5] Cho E., Lai J.-H., Ok Y., Upadhyaya A.D., Rohatgi A., Binns M.J. 'Light-induced degradation free and high efficiency p-type indium-doped PERC solar cells on Czochralski silicon'. Proceedings of the 42nd IEEE Photovoltaic Specialists Conference; 2015. pp. 1–4.

[6] Lee K., Kim M.-S., Lim J.-K., Ahn J.-H., Hwang M.-I., Cho E.-C. 'Natural recovery from lid: Regeneration under field conditions?' The 31st European Photovoltaic Specialists Conference; 2015. p. 1835.

[7] Herguth A. 'Simulation study on the impact of boron-oxygen related light-induced degradation in different silicon solar cell architectures'. AIP Conference Proceedings; 2019. p. 2147.

[8] Münzer K. 'Hydrogenated silicon nitride for regeneration of light induced degradation'. 24th European Photovoltaic Solar Energy Conference; 2009. pp. 1558–61.

[9] Cho E., Ok Y.-W., Upadhyaya A.D., *et al.* 'P-Type indium-doped passivated emitter rear solar cells (PERC) on Czochralski silicon without light-induced degradation'. *IEEE Journal of Photovoltaics.* 2016;**6**(4):795–800.

[10] Centrotherm. *Oral presentation - LID Prevention for Mono CZ-Si by Fast Regeneration.* 2015Centrotherm. Available from www.centrotherm.de [Accessed March 2016].

[11] Fertig F., Lantzsch R., Mohr A., *et al.* 'Mass production of P-type cz silicon solar cells approaching average stable conversion efficiencies of 22 %'. *Energy Procedia.* 2017;**124**(13):338–45.

[12] Hallam B.J., Chan C.E., Chen R., *et al.* 'Rapid mitigation of carrier-induced degradation in commercial silicon solar cells'. *Japanese Journal of Applied Physics*. 2017;**56**(8S2):08MB13.

[13] Herguth A., Schubert G., Kaes M., Hahn G. 'A new approach to prevent the negative impact of the metastable defect in boron doped cz silicon solar cells'. 4th IEEE World Conference on Photovoltaic Energy Conference; 2006. pp. 940–3.

[14] Lim B., Hermann S., Bothe K., Schmidt J., Brendel R. Permanent deactivation of the boron-oxygen recombination center in silicon solar cells. 23rd European Photovoltaic Solar Energy Conference; 2008. pp. 1018–22.

[15] Dubois S., Tanay F., Veirman J., Enjalbert N., Stendera J., Butté S. The BOLID project: suppression of the boron-oxygen related light-induced-degradation. 27th European Photovoltaic Solar Energy Conference; 2012. pp. 749–54.

[16] Hallam B., Bilbao J., Payne D., Chan C., Kim M., Chen D. 'Modelling the long-term behaviour of boron-oxygen defect passivation in the field using typical meteorological year data (TMY2)'. *33rd European Photovoltaic Solar Energy Conference and Exhibition*. 2016;**1**:555–9.

[17] Ciesla A.M., Bilbao J.I., Chan C.E., *et al.* 'Modeling boron–oxygen degradation and self-repairing silicon PV modules in the field'. *IEEE Journal of Photovoltaics*. 2020;**10**(1):28–40.

[18] Aberle A.G. 'Surface passivation of crystalline silicon solar cells: a review'. *Progress in Photovoltaics: Research and Applications*. 2000;**8**(5):473–87.

[19] Niewelt T., Schön J., Warta W., Glunz S.W., Schubert M.C. 'Degradation of crystalline silicon due to boron–oxygen defects'. *IEEE Journal of Photovoltaics*. 2017;**7**(1):383–98.

[20] Hallam B., Herguth A., Hamer P., *et al.* 'Eliminating light-induced degradation in commercial P-type Czochralski silicon solar cells'. *Applied Sciences*. 2017;**8**(1):10.

[21] LONGi Solar. *N-type Mono Wafer for High Efficiency Solar Cell Manufacturing*. PV Cell Tech; Kuala Lumpur, March 2019; 2019.

[22] Bothe K., Schmidt J. 'Electronically activated boron-oxygen-related recombination centers in crystalline silicon'. *Journal of Applied Physics*. 2006;**99**(1):13701.

[23] Kim M., Chen D., Abbott M., *et al.* 'Impact of interstitial iron on the study of meta-stable B-O defects in Czochralski silicon: further evidence of a single defect'. *Journal of Applied Physics*. 2018;**123**(16):161586.

[24] Voronkov V.V., Falster R., Bothe K., Lim B. 'Light-induced lifetime degradation in Boron-doped Czochralski silicon: are oxygen dimers involved?' *Energy Procedia*. 2013;**38**(1):636–41.

[25] Mchedlidze T., Weber J. 'Direct detection of carrier traps in Si solar cells after light-induced degradation'. *Physica Status Solidi (RRL) – Rapid Research Letters*. 2015;**9**(2):108–10.

[26] Glunz S.W., Rein S., Warta W., Knobloch J., Wettling W. 'On the degradation of Cz-silicon solar cells'. *Proceedings of the 2nd World Conference on Photovoltaic Solar Energy Conversion*. 1998;**1**:1343–6.

[27] Shockley W., Read W.T. 'Statistics of the recombinations of holes and electrons'. *Physical Review*. 1952;**87**(5):835–42.

[28] Hall R.N. 'Electron-hole recombination in germanium'. *Physical Review*. 1952;**87**(2):387.

[29] Herguth A. 'On the Lifetime-Equivalent defect density: properties, application, and pitfalls'. *IEEE Journal of Photovoltaics*. 2019;**9**(5):1182–94.

[30] Vaqueiro-Contreras M., Markevich V.P., Coutinho J., *et al*. 'Identification of the mechanism responsible for the boron oxygen light induced degradation in silicon photovoltaic cells'. *Journal of Applied Physics*. 2019;**125**(18):185704.

[31] Rein S., Diez S., Falster R., Glunz S.W. 'Quantitative correlation of the metastable defect in Cz-silicon with different impurities'. *Proceedings of 3rd World Conference on Photovoltaic Energy Conversion*. 2003;**2**:1048–52.

[32] Bothe K., Sinton R., Schmidt J. 'Fundamental boron-oxygen-related carrier lifetime limit in mono- and multicrystalline silicon'. *Progress in Photovoltaics: Research and Applications*. 2005;**13**(4):287–96.

[33] *2017 International Technology Roadmap for Photovoltaic*. 2017ITRPV. Available from itrpv.vdma.org [Accessed 24 June 2017].

[34] Schmidt J., Aberle A.G., Hezel R. 'Investigation of carrier lifetime instabilities in Cz-grown silicon'. The 26th IEEE Photovoltaic Specialists Conference; 1997. pp. 13–18.

[35] Lim S.Y., Rougieux F.E., Macdonald D. 'Boron-oxygen defect imaging in P-type Czochralski silicon'. *Applied Physics Letters*. 2013;**103**(9):092105.

[36] Glunz S.W., Rein S., Warta W., Knobloch J., Wettling W. 'Degradation of carrier lifetime in CZ silicon solar cells'. *Solar Energy Materials and Solar Cells*. 2001;**65**(1):219–29.

[37] Forster M., Fourmond E., Rougieux F.E., *et al*. 'Boron-oxygen defect in Czochralski-silicon co-doped with gallium and boron'. *Applied Physics Letters*. 2012;**100**(4):042110.

[38] Hashigami H., Itakura Y., Saitoh T. 'Effect of illumination conditions on Czochralski-grown silicon solar cell degradation'. *Journal of Applied Physics*. 2003;**93**(7):4240–5.

[39] Naerland T.U., Haug H., Angelskar H., Sondena R., Marstein E.S., Arnberg L. 'Studying light-induced degradation by lifetime decay analysis: excellent fit to solution of simple second-order rate equation'. *IEEE Journal of Photovoltaics*. 2013;**3**(4):1265–70.

[40] Hallam B., Abbott M., Naerland T., Wenham S. 'Fast and slow lifetime degradation in boron-doped Czochralski silicon described by a single defect'. *Physica Status Solidi (RRL) - Rapid Research Letters*. 2016;**10**(7):520–4.

[41] Kim M., Abbott M., Nampalli N., Wenham S., Stefani B., Hallam B. 'Modulating the extent of fast and slow boron-oxygen related degradation

in Czochralski silicon by thermal annealing: evidence of a single defect'. *Journal of Applied Physics*. 2017;**121**(6):053106.

[42] Voronkov V., Falster R. 'Permanent deactivation of boron-oxygen recombination centres in silicon'. *physica status solidi*. 2016;**253**(9):1721–8.

[43] Schmidt J., Bothe K., Voronkov V.V., Falster R. 'Fast and slow stages of lifetime degradation by boron–oxygen centers in crystalline silicon'. *Physica Status Solidi*. 2020;**257**(1.1900167.

[44] Hallam B., Kim M., Abbott M., *et al.* 'Recent insights into boron-oxygen related degradation: evidence of a single defect'. *Solar Energy Materials and Solar Cells*. 2017;**173**:25–32.

[45] Palmer D.W., Bothe K., Schmidt J. 'Kinetics of the electronically stimulated formation of a boron-oxygen complex in crystalline silicon'. *Physical Review B*. 2007;**76**(3):35210.

[46] Glunz S.W., Rein S., Lee J.Y., Warta W. 'Minority carrier lifetime degradation in boron-doped Czochralski silicon'. *Journal of Applied Physics*. 2001;**90**(5):2397–404.

[47] Lim B., Rougieux F., Macdonald D., Bothe K., Schmidt J. 'Generation and annihilation of boron–oxygen-related recombination centers in compensated P- and N-type silicon'. *Journal of Applied Physics*. 2010;**108**(10):103722.

[48] Hashigami H., Dhamrin M., Saitoh T. 'Characterization of the initial rapid decay on light-induced carrier lifetime and cell performance degradation of Czochralski-grown silicon'. *Japanese Journal of Applied Physics*. 2003;**42**(5A):2564–8.

[49] Uberg Naerland T., Angelskår H., Stensrud Marstein E., Nærland T.U., Marstein E.S. 'Direct monitoring of minority carrier density during light induced degradation in Czochralski silicon by photoluminescence imaging'. *Journal of Applied Physics*. 2013;**113**(19):193707.

[50] Herguth A. 'On the meaning(fullness) of the intensity unit 'suns' in light induced degradation experiments'. *Energy Procedia*. 2017;**124**(383):53–9.

[51] Macdonald D., Rougieux F., Cuevas A., *et al.* 'Light-Induced boron-oxygen defect generation in compensated P-type Czochralski silicon'. *Journal of Applied Physics*. 2009;**105**(9):93704.

[52] Hamer P., Nampalli N., Hameiri Z., *et al.* 'Boron-oxygen defect formation rates and activity at elevated temperatures'. *Energy Procedia*. 2016;**92**(17):791–800.

[53] Herguth A. 'How to properly assess boron-oxygen related degradation in crystalline silicon'. Proceedings of the 33rd European Photovoltaic Solar Energy Conference; 2017. pp. 576–9.

[54] Glunz S.W., Schaffer E., Rein S., Bothe K., Schmidt J. 'Analysis of the defect activation in Cz-silicon by temperature-dependent bias-induced degradation of solar cells'. *Proceedings of 3rd World Conference on Photovoltaic Energy Conversion*. 2003;**1**:919–22.

[55] Voronkov V., Falster R. 'The nature of boron-oxygen lifetime-degrading centres in silicon'. *Physica Status Solidi B*. 2016;**13**(10–12):712–17.

[56] Voronkov V.V., Falster R. 'Latent complexes of interstitial boron and oxygen dimers as a reason for degradation of silicon-based solar cells'. *Journal of Applied Physics*. 2010;**107**(5):53509.

[57] Herguth A., Hallam B. A generalized model for boron-oxygen related light-induced degradation in crystalline silicon a generalized model for boron-oxygen related light-induced degradation in crystalline silicon. AIP Conference Proceedings; 2018. pp. 1–5.

[58] Niewelt T., Schön J., Broisch J., Warta W., Schubert M. 'Electrical characterization of the slow boron oxygen defect component in Czochralski silicon'. *Physica Status Solidi (RRL) – Rapid Research Letters*. 2015;**9**(12):692–6.

[59] Sun C., Rougieux F.E., Macdonald D. 'A unified approach to modelling the charge state of monatomic hydrogen and other defects in crystalline silicon'. *Journal of Applied Physics*. 2015;**117**(4):45702.

[60] Nakajima K., Murai R., Morishita K., Kutsukake K., Usami N. 'Growth of multicrystalline Si ingots for solar cells using noncontact crucible method without touching the crucible wall'. *The 38th IEEE Photovoltaic Specialists Conference*. 2012;**344**(1):1830–2.

[61] Togawa S., Izunome K., Kawanishi S., Chung S.-I., Terashima K., Kimura S. 'Oxygen transport from a silica crucible in Czochralski silicon growth'. *Journal of Crystal Growth*. 1996;**165**(4):362–71.

[62] Zulehner W. 'Czochralski growth of silicon'. *Journal of Crystal Growth*. 1983;**65**(1):189–213.

[63] Hoshi K., Isawa N., Suzuki T., Ohkubo Y. 'Czochralski silicon crystals grown in a transverse magnetic field'. *Journal of The Electrochemical Society*. 1985;**132**(3):693–700.

[64] Organ A.E., Riley N. 'Oxygen transport in magnetic Czochralski growth of silicon'. *Journal of Crystal Growth*. 1987;**82**(3):465–76.

[65] Saitoh T. *'A Review of Japanese R&d for Crystalline Silicon Solar Cells. Ninth Workshop on Crystalline Silicon Solar Cell Materials and Processes'*. Breckenridge, Colorado; 1999. pp. 81–6.

[66] Xu H. 'Characterization of N-type Mono-crystalline silicon Ingots produced by continuous Czochralski (Cz) technology'. *Energy Procedia*. 2015;**77**:658–64.

[67] Lim B., Merkle A., Peibst R., Dullweber T., Wang Y., Zhou R. 'Beyond boron–oxygen deactivation: industrially feasible LID-free P-type Czochralski silicon'. *Photovoltaics International*. 2019;**42**:98–105.

[68] Sumino K., Harada H., Yonenaga I. 'The origin of the difference in the mechanical strengths of Czochralski-grown silicon and float-zone-grown silicon'. *Japanese Journal of Applied Physics*. 1980;**19**(1):L49–52.

[69] Glunz S.W., Rein S., Knobloch J., Wettling W., Abe T. 'Comparison of boron- and gallium-doped P-type Czochralski silicon for photovoltaic application'. *Progress in Photovoltaics: Research and Applications*. 1999;**7**(6):463–9.

[70] Grant N.E., Rougieux F.E., Macdonald D., Bullock J., Wan Y. 'Grown-in defects limiting the bulk lifetime of *p* -type float-zone silicon wafers'. *Journal of Applied Physics*. 2015;**117**(5):055711.

[71] Rougieux F.E., Grant N.E., Macdonald D. 'Thermal deactivation of lifetime- limiting grown-in point defects in N-type Czochralski silicon wafers'. *Physica Status Solidi (RRL) – Rapid Research Letters*. 2013;**7**(9):616–8.

[72] Niewelt T., Selinger M., Grant N.E., Kwapil W., Murphy J.D., Schubert M.C. 'Light-Induced activation and deactivation of bulk defects in boron-doped float-zone silicon'. *Journal of Applied Physics*. 2017;**121**(18):185702.

[73] Sperber D., Heilemann A., Herguth A., Hahn G. 'Temperature and light-induced changes in bulk and passivation quality of boron-doped float-zone silicon coated with SiNx:H'. *IEEE Journal of Photovoltaics*. 2017;**7**(2):463–70.

[74] Arafune K., Sasaki T., Wakabayashi F., Terada Y., Ohshita Y., Yamaguchi M. 'Study on defects and impurities in cast-grown polycrystalline silicon substrates for solar cells'. *Physica B: Condensed Matter*. 2006;**376-377**(1):236–9.

[75] Möller H.J., Funke C., Rinio M., Scholz S. 'Multicrystalline silicon for solar cells'. *Thin Solid Films*. 2005;**487**(1–2):179–87.

[76] Buonassisi T., Istratov A.A., Pickett M.D., *et al.* 'Transition metals in photovoltaic-grade ingot-cast multicrystalline silicon: assessing the role of impurities in silicon nitride crucible lining material'. *Journal of Crystal Growth*. 2006;**287**(2):402–7.

[77] Stoddard N., Wu B., Witting I., Wagener M.C., Park Y., Rozgonyi G.A. 'Casting single crystal silicon: novel defect profiles from BP Solar's mono2 TM wafers'. *Solid State Phenomena*. 2008;**131**:1–8.

[78] Guerrero I., Parra V., Carballo T., *et al.* 'About the origin of low wafer performance and crystal defect generation on seed-cast growth of industrial *mono-like* silicon ingots'. *Progress in Photovoltaics: Research and Applications*. 2014;**22**(8):923–32.

[79] Hutchins M. *The long read: change to the cast*. PV Magazine; 2019.

[80] Lan C.W., Yang C.F., Lan A., *et al.* 'Engineering silicon crystals for photovoltaics'. *CrystEngComm*. 2016;**18**(9):1474–85.

[81] Kersten F., Engelhart P., Ploigt H.-C., Stekolnikov A., Lindner T., Stenzel F. 'A new mc-Si degradation effect called LeTID'. Proceedings of the 42nd IEEE Photovoltaic Specialists Conference; 2015. pp. 1–5.

[82] Fertig F., Krauß K., Rein S. 'Light-Induced degradation of PECVD aluminium oxide passivated silicon solar cells'. *Physica Status Solidi (RRL) – Rapid Research Letters*. 2015;**9**(1):41–6.

[83] Nakayashiki K., Hofstetter J., Morishige A.E., *et al.* 'Engineering solutions and root-cause analysis for light-induced degradation in p]]>-type multicrystalline silicon PERC modules'. *IEEE Journal of Photovoltaics*. 2016;**6**(4):860–8.

[84] Zuschlag A., Skorka D., Hahn G. 'Degradation and regeneration in mc-Si after different gettering steps'. *Progress in Photovoltaics: Research and Applications*. 2017;**25**(7):545–52.

[85] Ramspeck K., Zimmermann S., Nagel H., Metz A., Gassenbauer Y., Birkmann B. 'Light induced degradation of rear passivated mc-Si solar cells'.

Proceedings of the 27th European Photovoltaic Solar Energy Conference. 2012;**1**:861–5.

[86] Chen D., Kim M., Stefani B.V., *et al.* 'Evidence of an identical firing-activated carrier-induced defect in monocrystalline and multicrystalline silicon'. *Solar Energy Materials and Solar Cells*. 2017;**172**:293–300.

[87] 2019 International Technology Roadmap for Photovoltaic 10th Edition. 2019. Available from itrpv.vdma.org [Accessed 10 February 2021].

[88] Graff K., Pieper H. 'The properties of iron in silicon'. *Journal of The Electrochemical Society*. 1981;**128**(3):669–74.

[89] Macdonald D., Roth T., Deenapanray P.N.K., Bothe K., Pohl P.,Schmidt J. 'Formation rates of iron-acceptor pairs in crystalline silicon'. *Journal of Applied Physics*. 2005;**98**(8):83509.

[90] Garandet J.P. 'New determinations of diffusion coefficients for various dopants in liquid silicon'. *International Journal of Thermophysics*. 2007;**28**(4):1285–303.

[91] Trumbore F.A. 'Solid solubilities of impurity elements in germanium and Silicon*'. *Bell System Technical Journal*. 1960;**39**(1):205–33.

[92] DeLuca J.P., Delk II F.S., Johnson B.K., Luter W.L., Middendorf N.D., Willams D.S. *Silicon Single Crystal Doped with Gallium, Indium, or Aluminium. Vol. US 9,051,6*. United States; 2015.

[93] Yu X., Zheng X., Hoshikawa K., Yang D. 'Crystal growth of indium-doped czochralski silicon for photovoltaic application'. *Japanese Journal of Applied Physics*. 2012;**51**(10):105501.

[94] Morita K., Miki T. 'Thermodynamics of solar-grade-silicon refining'. *Intermetallics*. 2003;**11**(11):1111–7.

[95] Rosenits P., Roth T., Glunz S.W., Beljakowa S. 'Determining the defect parameters of the deep aluminum-related defect center in silicon'. *Applied Physics Letters*. 2007;**91**(12):122109.

[96] Möller C., Lauer K. 'Light-Induced degradation in indium-doped silicon'. *Physica Status Solidi (RRL) – Rapid Research Letters*. 2013;**7**(7):461–4.

[97] Richter A., Glunz S.W., Werner F., Schmidt J., Cuevas A. 'Improved quantitative description of Auger recombination in crystalline silicon'. *Physical Review B*. 2012;**86**(16):165202.

[98] Schmidt J., Lim B., Walter D., *et al.* 'Impurity-related limitations of next-generation industrial silicon solar cells'. *IEEE Journal of Photovoltaics*. 2013;**3**(1):114–8.

[99] Song D., Xiong J., Hu Z., Li G., Wang H., An H. Progress in n-type Si solar cell and module technology for high efficiency and low cost. Conference Record of the IEEE Photovoltaic Specialists Conference; 2012. pp. 3004–8.

[100] Falster R.J., Cornara M., Gambaro D., Olmo M., Pagani M. 'Effect of high temperature pre-anneal on oxygen precipitates nucleation kinetics in Si'. *Solid State Phenomena*. 1997;**57-58**:123–8.

[101] Haunschild J., Reis I.E., Geilker J., Rein S. "Detecting efficiency-limiting defects in Czochralski-grown silicon wafers in solar cell production using

photoluminescence imaging''. *physica status solidi (RRL) – Rapid Research Letters*. 2011;**5**(5–6):199–201.

[102] Cousins P.J., Smith D.D., Luan H.-C., Manning J., Dennis T.D., Waldhauer A. 'Generation 3: improved performance at lower cost'. Proceedings of the 35th IEEE Photovoltaic Specialists Conference; 2010. pp. 275–8.

[103] Letty E., Veirman J., Favre W., Lemiti M. 'Bulk defect formation under light soaking in seed-end N-type Czochralski silicon wafers – effect on silicon heterojunction solar cells'. *Solar Energy Materials and Solar Cells*. 2017;**166**(January):147–56.

[104] Wang Y. 'Mono wafer requirements for advanced HE cell'. Oral Present 2018 Silicon PV Conference; 2018.

[105] Woodhouse M., Smith B., Ramdas A., Margolis R. 'Crystalline silicon photovoltaic module manufacturing costs and sustainable pricing: 1H 2018 benchmark and cost reduction roadmap'. *National Renewable Energy Laboratory*. 2019:1–46.

[106] Yu X., Wang P., Chen P., Li X., Yang D. 'Suppression of boron–oxygen defects in P-type Czochralski silicon by germanium doping'. *Applied Physics Letters*. 2010;**97**(5):051903.

[107] Taishi T., Huang X., Yonenaga I., Hoshikawa K. 'Dislocation behavior in heavily germanium-doped silicon crystal'. *Materials Science in Semiconductor Processing*. 2002;**5**(4-5 SPEC):409–12.

[108] Newman R.C. 'Defects in silicon'. *Reports on Progress in Physics*. 1982;**45**(10):1163–210.

[109] Vaqueiro-Contreras M., Markevich V.P., Halsall M.P., Peaker A.R., Santos P., Coutinho J. 'Powerful recombination centers resulting from reactions of hydrogen with carbon–oxygen defects in n-type Czochralski-grown silicon (Phys. status Solidi rrl 8/2017)'. *Rapid Research Letters*. 2017;**11**(8):1770342.

[110] Wang Z., Han P., Lu H., *et al*. 'Advanced PERC and PERL production cells with 20.3% record efficiency for standard commercial p-type silicon wafers'. *Progress in Photovoltaics: Research and Applications*. 2012;**20**(3):260–8.

[111] Chen R., Tong H., Zhu H., *et al*. '23.83% efficient mono-PERC incorporating advanced hydrogenation'. *Progress in Photovoltaics: Research and Applications*. 2020;**28**(12):1239–47.

[112] Bothe K., Schmidt J., Hezel R. 'Effective reduction of the metastable defect concentration in boron-doped Czochralski silicon for solar cells'. Proceedings of the 29th IEEE Photovoltaic Specialists Conference; 2002. pp. 194–7.

[113] Saitoh T., Wang X., Hashigami H., *et al*. 'Suppression of light degradation of carrier lifetimes in low-resistivity Cz–Si solar cells'. *Solar Energy Materials and Solar Cells*. 2001;**65**(1):277–85.

[114] Nagel H., Merkle A., Metz A., Hezel R. 'Permanent reduction of excess-carrier-induced recombination centers in solar grade Czochralski silicon by a short yet effective anneal'. Proceedings of the 16th IEEE Photovoltaic Specialists Conference; 2000. pp. 1197–200.

[115] Walter D.C., Lim B., Bothe K., Voronkov V.V., Falster R., Schmidt J. 'Effect of rapid thermal annealing on recombination centres in boron-doped Czochralski-grown silicon'. *Applied Physics Letters*. 2014;**104**(4):42111.

[116] Kouhlane Y., Bouhafs D., Khelifati N., *et al*. 'Effect of rapid thermal processing on light-induced degradation of carrier lifetime in Czochralski P-type silicon bare Wafers'. *Journal of Electronic Materials*. 2016;**45**(11):5621–5.

[117] Nampalli N., Li H., Kim M., *et al*. 'Multiple pathways for permanent deactivation of boron-oxygen defects in P-type silicon'. *Solar Energy Materials and Solar Cells*. 2017;**173**:12–17.

[118] Walter D.C., Falster R., Voronkov V.V., Schmidt J. 'On the equilibrium concentration of boron-oxygen defects in crystalline silicon'. *Solar Energy Materials and Solar Cells*. 2017;**173**(July):33–6.

[119] Rougieux F.E., Lim B., Schmidt J., Forster M., Macdonald D., Cuevas A. 'Influence of net doping, excess carrier density and annealing on the boron oxygen related defect density in compensated N-type silicon'. *Journal of Applied Physics*. 2011;**110**(6):63708.

[120] Unsur V., Hussain B., Ebong A., Unsur V., Hussain B., Ebong A. 'Complete recovery of light induced degradation of Cz silicon solar cells with rapid thermal processing'. Proceedings of the 43rd Photovoltaic Specialists Conference; 2016.

[121] Murin L.I., Hallberg T., Markevich V.P., Lindström J.L. 'Experimental evidence of the oxygen dimer in silicon'. *Physical review letters*. 1998;**80**(1):93–6.

[122] Herguth A., Schubert G., Kaes M., Hahn G. 'Avoiding boron-oxygen related degradation in highly boron doped Cz silicon'. Proceedings of the 21st European Photovoltaic Specialists Conference; 2006. pp. 530–7.

[123] Herguth A., Hahn G. 'Boron-oxygen related defects in Cz-silicon solar cells degradation, regeneration and beyond'. *Proceedings of the 24th European Photovoltaic Specialists Conference*; 2009. pp. 974–6.

[124] Herguth A., Hahn G. 'Kinetics of the boron-oxygen related defect in theory and experiment'. *Journal of Applied Physics*. 2010;**108**(11):114509.

[125] Wilking S., Beckh C., Ebert S., Herguth A., Hahn G. 'Influence of bound hydrogen states on BO-regeneration kinetics and consequences for high-speed regeneration processes'. *Solar Energy Materials and Solar Cells*. 2014;**131**(10):2–8.

[126] Hallam B.J., Abbott M.D., Nampalli N., Hamer P.G., Wenham S.R. 'Implications of accelerated recombination-active defect complex formation for mitigating carrier-induced degradation in silicon'. *IEEE Journal of Photovoltaics*. 2015;**6**(1):92–9.

[127] Hallam B., Abbott M., Nampalli N., Hamer P., Wenham S. 'Influence of the formation- and passivation rate of boron-oxygen defects for mitigating carrier-induced degradation in silicon within a hydrogen-based model'. *Journal of Applied Physics*. 2016;**119**(6):65701.

[128] Steckenreiter V., Walter D.C., Schmidt J. 'Kinetics of the permanent deactivation of the boron-oxygen complex in crystalline silicon as a function of illumination intensity'. *AIP Advances*. 2017;**7**(3):035305.

[129] Walter D.C., Schmidt J., Steckenreiter V., Walter D.C., Schmidt J. 'Two-Stage permanent deactivation of the boron-oxygen-related recombination center in crystalline silicon'. *Energy procedia*. 2017;**124**:799–805.

[130] Hallam B., Abbott M., Bilbao J., *et al.* 'Modelling kinetics of the boron-oxygen defect system'. *Energy Procedia*. 2016;**92**(5):42–51.

[131] Wilking S., Forster M., Herguth A., Hahn G. 'From simulation to experiment: understanding BO-regeneration kinetics'. *Solar Energy Materials and Solar Cells*. 2015;**142**(10):87–91.

[132] Hallam B., Hamer P., Kim M., Nampalli N., Gorman N., Chen D. 'Direct transitions between states A and C in the boron-oxygen defect system – fact or fiction?' *Proceedings of the 43rd IEEE Photovoltaic Specialists Conference;*. 2016;**1**:2430–3.

[133] Herguth A., Schubert G., Kaes M., Hahn G. 'Investigations on the long time behavior of the metastable boron–oxygen complex in crystalline silicon'. *Progress in Photovoltaics: Research and Applications*. 2008;**16**(2):135–40.

[134] M.H D., Branz H.M., Crandall R.S., Zhang S.B. 'A new mechanism for non-radiative recombination at light-induced boron-oxygen complexes in silicon'. 2005 DOE Solar Energy Technologies Program Review Meeting; Denver, 7-10 Nov 2005; 2005. p. NREL/CP-590-39021.

[135] Bothe K., Hezel R., Schmidt J. 'Understanding and reducing the Boron-Oxygen-Related performance degradation in Czochralski silicon solar cells'. *Solid State Phenomena*. 2004;**95–96**(i):223–8.

[136] Kim S.M., Chun S., Bae S., *et al.* 'Light-Induced degradation and metastable-state recovery with reaction kinetics modeling in boron-doped Czochralski silicon solar cells'. *Applied Physics Letters*. 2014;**105**(8):083509.

[137] Jourdan J., Veschetti Y., Dubois S., Desrues T., Monna R. 'Formation of Boron-doped region using spin-on dopant: investigation on the impact of metallic impurities'. *Progress in Photovoltaics: Research and Applications*. 2008;**16**(5):379–87.

[138] Schmidt J., Bothe K., Hezel R. 'Formation and annihilation of the metastable defect in boron-doped Czochralski silicon'. Proceedings of the 29th IEEE Photovoltaic Specialists Conference; 2002. pp. 178–81.

[139] Adey J., Jones R., Palmer D.W., Briddon P.R., Öberg S. 'Degradation of boron-doped Czochralski-grown silicon solar cells'. *Physical Review Letters*. 2004;**93**(5):1–4.

[140] Lim B., Bothe K., Schmidt J. 'Deactivation of the boron-oxygen recombination center in silicon by illumination at elevated temperature'. *Physica Status Solidi (RRL) – Rapid Research Letters*. 2008;**2**(3):93–5.

[141] Lim B., Bothe K., Schmidt J. 'Deactivation of the boron-oxygen recombination center in silicon by illumination at elevated temperature'. *Physica Status Solidi (RRL) – Rapid Research Letters*. 2008;**2**(3):93–5.

[142] Herguth A., Derricks C., Hahn G. 'Regeneration of boron-oxygen related degradation in Cz-Si PERC-type solar cells at high temperatures'. 33rd European Photovoltaic Solar Energy Conference and Exhibition; 2017. pp. 557–60.

[143] Wilking S., Ebert S., Beckh C., Herguth A., Hahn G. 'Of apples and oranges: why comparing bo regeneration rates requires injection level correction'. 32nd European Photovoltaic Solar Energy Conference and Exhibition; 2016. pp. 487–94.

[144] Herguth A., Hahn G. 'Towards a high throughput solution for boron-oxygen related regeneration'. 28th European Photovoltaic Solar Energy Conference and Exhibition; 2013.

[145] Derricks C., Herguth A., Hahn G., Romer O., Pernau T. 'Industrially applicable mitigation of BO-LID in Cz-Si PERC-type solar cells within a coupled fast firing and halogen lamp based belt-line regenerator – a parameter study'. *Solar Energy Materials and Solar Cells*. 2018;**2019**(195):358–66.

[146] Walter D., Pernau T., Schmidt J. 'Ultrafast lifetime regeneration in an industrial belt-line furnace applying intense illumination at elevated temperature'. 32nd European Photovoltaic Solar Energy Conference and Exhibition; 2016. pp. 469–73.

[147] Osterwald C.R., Anderberg A., Rummel S., Ottoson L. 'Degradation analysis of weathered crystalline-silicon PV modules'. Conference Record of the IEEE Photovoltaic Specialists Conference; 2002. pp. 1392–5.

[148] Krugel G., Wolke W., Geilker J., Rein S., Preu R. 'Impact of hydrogen concentration on the regeneration of light induced degradation'. *Energy Procedia*. 2011;**8**:47–51.

[149] Wilking S., Herguth A., Hahn G. 'Influence of hydrogen on the regeneration of boron-oxygen related defects in crystalline silicon'. *Journal of Applied Physics*. 2013;**113**(19):194503.

[150] Wilking S., Herguth A., Hahn G. 'Influence of hydrogenated passivation layers on the regeneration of boron-oxygen related defects'. *Energy Procedia*. 2013;**38**(5):642–8.

[151] Nampalli N., Hallam B., Chan C., Abbott M., Wenham S. 'Evidence for the role of hydrogen in the stabilization of minority carrier lifetime in boron-doped Czochralski silicon'. *Applied Physics Letters*. 2015;**106**(17):173501.

[152] Nampalli N., Hallam B.J., Chan C.E., Abbott M.D., Wenham S.R. 'Influence of hydrogen on the mechanism of permanent Passivation of Boron–Oxygen defects in P-type Czochralski silicon'. *IEEE Journal of Photovoltaics*. 2015;**5**(6):1580–5.

[153] Walter D.C., Lim B., Bothe K., Falster R., Voronkov V.V., Schmidt J. 'Lifetimes exceeding 1ms in 1-Ωcm boron-doped Cz-silicon'. *Solar Energy Materials and Solar Cells*. 2014;**131**:51–7.

[154] Wilking S., Ebert S., Herguth A., Hahn G. 'Influence of hydrogen effusion from hydrogenated silicon nitride layers on the regeneration of boron-oxygen related defects in crystalline silicon'. *Journal of Applied Physics*. 2013;**114**(19):194512.

[155] Wilking S., Engelhardt J., Ebert S., Beckh C., Herguth A., Hahn G. High speed regeneration of bo-defects: Improving long-term solar cell performance within seconds. 29th European Photovoltaic Solar Energy Conference and Exhibition; 2014. pp. 366–72.

[156] Joos S., Schiele Y., Terheiden B., Becker H.-W., Rogalla D., Hahn G. 'Fundamental studies of hydrogen at the silicon/silicon nitride interface'. *Energy Procedia*. 2014;**55**(11):786–90.

[157] Dekkers H.F.W., Beaucarne G., Hiller M., Charifi H., Slaoui A. 'Molecular hydrogen formation in hydrogenated silicon nitride'. *Applied Physics Letters*. 2006;**89**(21):211914–31.

[158] Bredemeier D., Walter D.C., Heller R., Schmidt J. 'Impact of hydrogen-rich silicon nitride material properties on light-induced lifetime degradation in multicrystalline silicon'. *Physica Status Solidi (RRL) – Rapid Research Letters*. 2019;**13**(8):1900201.

[159] Dameron A.A., Davidson S.D., Burton B.B., Carcia P.F., McLean R.S., George S.M. 'Gas diffusion barriers on polymers using multilayers fabricated by Al_2O_3 and rapid SiO_2 atomic layer deposition'. *The Journal of Physical Chemistry C*. 2008;**112**(12):4573–80.

[160] Helmich L., Walter D.C., Bredemeier D., Schmidt J. 'atomic-layer-deposited Al_2O_3 as effective barrier against the diffusion of hydrogen from SiN_x:H layers into crystalline silicon during rapid thermal annealing'. *Physica Status Solidi (RRL) – Rapid Research Letters*. 2020;**14**(12):2000367–5.

[161] Dingemans G., Beyer W., van de Sanden M.C.M., Kessels W.M.M. 'Hydrogen induced passivation of Si interfaces by Al_2O_3 films and SiO_2/$Al2O_3$ stacks'. *Applied Physics Letters*. 2010;**97**(15):152106–11.

[162] Huang Y.L., Ma Y., Job R., Ulyashin A.G. 'Hydrogen diffusion at moderate temperatures in P-type Czochralski silicon'. *Journal of Applied Physics*. 2004;**96**(12):7080–6.

[163] Capizzi M., Mittiga A. 'Hydrogen in crystalline silicon: a deep donor?' *Applied Physics Letters*. 1987;**50**(14):918–20.

[164] Newman R.C., Tucker J.H., Brown A.R., McQuaid S.A. 'Hydrogen diffusion and the catalysis of enhanced oxygen diffusion in silicon at temperatures below 500 °C'. *Journal of Applied Physics*. 1991;**70**(6):3061–70.

[165] Van Wieringen A., Warmoltz N. 'On the permeation of hydrogen and helium in single crystal silicon and germanium at elevated temperatures'. *Physica*. 1956;**22**(6):849–65.

[166] Herring C., Johnson N.M., Van de Walle C.G. 'Energy levels of isolated interstitial hydrogen in silicon'. *Physical Review B*. 2001;**64**(12):125209.

[167] Mathiot D. 'Modeling of hydroge*n* diffusio*n* i*n* *n*- and *p*-type silico*n*'. *Physical Review B*. 1989;**40**(8):5867–70.

[168] Hallam B.J., Hamer P.G., Wenham S.R., *et al*. 'Advanced bulk defect Passivation for silicon solar cells'. *IEEE Journal of Photovoltaics*. 2014;**4**(1):88–95.

[169] Hallam B.J., Wenham S.R., Hamer P.G., *et al*. 'Hydrogen passivation of B-O defects in Czochralski silicon'. *Energy Procedia*. 2013;**38**(8):561–70.

[170] Gläser M., Lausch D. 'Towards a quantitative model for bo regeneration by means of charge state control of hydrogen'. *Energy Procedia*. 2015;**77**(1):592–8.

[171] Sun C., Chen D., Rougieux F., Basnet R., Hallam B., Macdonald D. 'Kinetics and dynamics of the regeneration of boron-oxygen defects in compensated N-type silicon'. *Solar Energy Materials and Solar Cells*. 2019;**195**(10):174–81.

[172] Helmich L., Walter D.C., Bredemeier D., Falster R., Voronkov V.V., Schmidt J. 'In-situ characterization of electron-assisted regeneration of Cz-Si solar cells'. *Solar Energy Materials and Solar Cells*. 2018;**185**(May):283–6.

[173] Herguth A., Schubert G., Kaes M., Hahn G. 'Further investigations on the avoidance of boron-oxygen related degradation by means of regeneration'. Proceedings of the 22nd European Photovoltaic Solar Energy Conference and Exhibition; 2007. pp. 893–6.

[174] Hallam B.J., Hamer P.G., Ciesla née Wenham A.M., Chan C.E., Vicari Stefani B., Wenham S. 'Development of advanced hydrogenation processes for silicon solar cells via an improved understanding of the behaviour of hydrogen in silicon'. *Progress in Photovoltaics: Research and Applications*. 2020;**28**(12):1217–38.

[175] Hamer P., Hallam B., Abbott M., Wenham S. 'Accelerated formation of the boron-oxygen complex in P-type Czochralski silicon'. *Physica Status Solidi (RRL) – Rapid Research Letters*. 2015;**9**(5):297–300.

[176] Schön J., Niewelt T., Broisch J., Warta W., Schubert M.C. 'Characterization and modelling of the boron-oxygen defect activation in compensated *n*-type silicon'. *Journal of Applied Physics*. 2015;**118**(24):245702.

[177] Hamer P., Nampalli N., Hameiri Z., *et al*. 'Boron-oxygen defect formation rates and activity at elevated temperatures'. *Energy Procedia*. 2016;**92**(17):791–800.

[178] Graf A., Herguth A., Hahn G. 'Investigation on the dependence of degradation rate on hole concentration during boron-oxygen related light-induced degradation in crystalline silicon'. *AIP Advances*. 2018;**8**:085219

[179] Wilson M., Edelman P., Savtchouk A., D'Amico J., Findlay A., Lagowski J. 'Accelerated light-induced degradation (ALID) for monitoring of defects in PV silicon wafers and solar cells'. *Journal of Electronic Materials*. 2010;**39**(6):642–7.

[180] Nampalli N. *Characterisation and Passivation of Boron-Oxygen Defects in P-Type Czochralski Silicon. Phd Thesis*. University of New South Wales; 2017.

[181] Song L., Wenham A., Wang S., Hamer P., Shakil A., Hallam B. 'Laser enhanced hydrogen passivation of silicon wafers'. *International Journal of Photoenergy*. 2015;**501**:193892.

[182] Hallam B.J., Hamer P.G., Wang S., *et al*. 'Advanced hydrogenation of dislocation clusters and Boron-oxygen defects in silicon solar cells'. *Energy Procedia*. 2015;77(10):799–809.

[183] Hamer P., Hallam B., Abbott M., Chan C., Nampalli N., Wenham S. 'Investigations on accelerated processes for the boron–oxygen defect in P-type Czochralski silicon'. *Solar Energy Materials and Solar Cells*. 2016;**145**(2):440–6.

[184] Hieslmair H. Time constants of degradation, regeneration, and destabilization of the B-O related defect. 43rd IEEE Photovoltaic Specialists Conference; 2016. pp. 645–50.

[185] Hallam B., Chen D., Kim M., *et al*. 'The role of hydrogenation and gettering in enhancing the efficiency of next-generation Si solar cells: an industrial perspective'. *Physica Status Solidi*. 2017;**214**(7):1700305.

[186] Sheoran M., Upadhyaya A., Rohatgi A. 'Bulk lifetime and efficiency enhancement due to gettering and hydrogenation of defects during cast multicrystalline silicon solar cell fabrication'. *Solid-State Electronics*. 2008;**52**(5):612–17.

[187] Sheoran M., Kim D.S., Rohatgi A., *et al*. 'Hydrogen diffusion in silicon from plasma-enhanced chemical vapor deposited silicon nitride film at high temperature'. *Applied Physics Letters*. 2008;**92**(17):172107.

[188] Kleekajai S., Jiang F., Stavola M., *et al*. 'Concentration and penetration depth of H introduced into crystalline Si by hydrogenation methods used to fabricate solar cells'. *Journal of Applied Physics*. 2006;**100**(9):93517.

[189] Kleekajai S., Wen L., Peng C., *et al*. 'Infrared study of the concentration of H introduced into Si by the postdeposition annealing of a SiNx coating'. *Journal of Applied Physics*. 2009;**106**(12):123510.

[190] Hallam B., Chan C., Payne D.N.R., Lausch D., Gläser M., Abbott M. 'Techniques for mitigating light-induced degradation (LID) in commercial silicon solar cells'. *Photovoltaics International*. 2016;**33**:37–46.

[191] Pernau T., Romer O., Scheiffele W., Reichart A., Jooß W. 'Rather high speed regeneration of BO-defects: regeneration experiments with large cell batches'. *31st European Photovoltaic Solar Energy Conference*. 2015;**1**:918–20.

[192] Fischbeck G. 'Keep a lid on it'. *PV-Magazine*. 2015;**12**:44–7.

[193] Fertig F., Broisch J., Biro D., Rein S. 'Stability of the regeneration of the boron–oxygen complex in silicon solar cells during module certification'. *Solar Energy Materials and Solar Cells*. 2014;**121**:157–62.

[194] Yoshikawa K., Kawasaki H., Yoshida W., *et al*. 'Silicon heterojunction solar cell with interdigitated back contacts for a photoconversion efficiency over 26%'. *Nature Energy*. 2017;**2**(5).

[195] De Wolf S., Descoeudres A., Holman Z.C., Ballif C. 'High-efficiency silicon heterojunction solar cells: a review'. *Green*. 2012;**2**(1):7–24.

[196] Holman Z.C., Descoeudres A., Barraud L., *et al*. 'Current losses at the front of silicon heterojunction solar cells'. *IEEE Journal of Photovoltaics*. 2012;**2**(1):7–15.

[197] Bätzner D.L., Andreetta L., Frammelsberger W., Kramer R., Lachenal D., Legradic B. 'Bifacial p-type solar cells with low temperature coefficients:

heterojunction technology'. Proceedings of the 32nd European Photovoltaic Solar Energy Conference; 2016. pp. 635–8.

[198] Descoeudres A., Horzel J., Paviet-Salomon B., *et al.* 'The versatility of passivating carrier-selective silicon thin films for diverse high-efficiency screen-printed heterojunction-based solar cells'. *Progress in Photovoltaics: Research and Applications*. 2020;**28**(6):569–77.

[199] Chen D., Weigand W., Wright M., Kim M., Shi J., ZJ Y. 'Evaluating the impact of and solutions to light-induced degradation in silicon heterojunction solar cells'. Conference Record of the IEEE Photovoltaic Specialists Conference; 2019. pp. 1099–103.

[200] Vicari Stefani B., Soeriyadi A., Wright M., *et al.* 'Large-area boron-doped 1.6 ωcm P-type czochralski silicon heterojunction solar cells with a stable open-circuit voltage of 736 MV and efficiency of 22.0%'. *Solar RRL*. 2020;**4**(9):2000134–5.

[201] Haase F., Hollemann C., Schäfer S., *et al.* 'Laser contact openings for local poly-Si-metal contacts enabling 26.1%-efficient POLO-IBC solar cells'. *Solar Energy Materials and Solar Cells*. 2018;**186**(June):184–93.

[202] Green M.A. 'The path to 25% silicon solar cell efficiency: history of silicon cell evolution'. *Progress in Photovoltaics: Research and Applications*. 2009;**17**(3):183–9.

[203] Sun C., Chen D., Weigand W., *et al.* 'Complete regeneration of BO-related defects in N-type upgraded metallurgical-grade Czochralski-grown silicon heterojunction solar cells'. *Applied Physics Letters*. 2018;**113**(15):152105.

Chapter 6

Negative impacts of hydrogen in silicon

Ran Chen[1], Alison Ciesla[1], Michelle Vaqueiro-Contreras[1], Daniel Chen[1,2], Matthew Wright[1], Brett Hallam[1], and Tony Peaker[3]

As discussed in Chapter 2, hydrogen atoms can take on any of three charge states: positive (H^+), metastable neutral (H^0) or negative (H^-). As a result, hydrogen is able to form stable complexes by taking, giving or sharing a single electron and is therefore capable of bonding to basically anything in the silicon lattice. In recent decades, much research has focused on the ability of hydrogen in its various charge states to passivate and electrically neutralise surface-related defects (discussed in Chapter 3) and bulk defects (Chapter 4), particularly boron-oxygen defects responsible for light-induced degradation in p-type Czochralski (Cz) silicon (Chapter 5), to improve solar cell performance and stability. However, hydrogen can also have unintended negative effects. Hydrogen can introduce recombination active defects in silicon by forming complexes with a range of impurities or structural defects, or even form pure hydrogen-related defects. In addition to forming defects, hydrogen can harm solar cell device performance through unintentional neutralisation of dopants, passivation of contacts or counteracting field-effect surface passivation.

As the smallest atom in the periodic table, and most abundant in the universe, hydrogen is essentially unavoidable. In modern solar cells, it is naturally introduced to the silicon in relatively high concentrations during some typical fabrication processes. The bulk concentration is quite low in typical as-grown material but diffuses into the bulk from hydrogenated surface layers during the high-temperature firing of the metal contacts. Typical hydrogen concentrations have been estimated to be in the range of 10^{14}–10^{16} atoms/cm^3 [1, 2]. The impact that hydrogen has on solar cells depends on a number of factors including the composition of impurities and

[1]The School of Photovoltaic and Renewable Energy Engineering, The University of New South Wales, Sydney, NSW, Australia

[2]Sundrive Solar, Kirawee, NSW, Australia

[3]Photon Science Institute and School of Electrical and Electronic Engineering, The University of Manchester, UK

structural defects within the silicon bulk, which are determined primarily by the feedstock and growth method; the device structure including doping profiles, passivation and contacting schemes; and the cell processing conditions, in particular gases, thermal exposure and application of any intentional hydrogen control 'hydrogenation' process. Hydrogen is very reactive as an isolated ion and can readily combine with intrinsic defects, dopants and impurities.

However, coulombic repulsion prevents the reaction of mobile hydrogen ions with similarly charged defect states. It follows that reactions are dependent on bulk doping in terms of concentration and polarity (in n-type and p-type materials). This chapter will discuss many possible ways that hydrogen can negatively affect the performance of silicon solar cells. It covers various types of hydrogen-related defects that form through interaction with other species such as carbon, oxygen, metals and vacancies. More specific to silicon solar cells, it discusses a degradation mechanism attributed to hydrogen, 'light- and elevated temperature-induced degradation' (LeTID). LeTID is a newly identified form of carrier-induced degradation (CID) that has gained significant interest in academia and industry to understand its behaviours, kinetics, root cause and mitigation strategies. This chapter also discusses the ways that the behaviour of hydrogen itself can cause unintended impacts to device performance through counter doping, contact passivation, platelet formation and methods to avoid hydrogen-related defects.

6.1 Hydrogen complexes with other species

Hydrogen's high reactivity and ubiquitous presence in semiconductors have made its study of worldwide interest, for which a number of reviews and books exist since the early 1990s [3–6]. This section does not aim to make an extensive review of the existent hydrogen complex record. Instead, in this section a summary of hydrogen-containing complexes with other abundant impurities found in solar-grade silicon which can act as recombination centres, and thus act as minority carrier sinks for solar cells, are summarised. Key hydrogen-defect complexes of relevance for silicon solar cells include:

- Carbon-hydrogen
- Oxygen-hydrogen
- Carbon-oxygen-hydrogen
- Transition metal-hydrogen
- Vacancy-hydrogen.

6.1.1 Carbon-hydrogen (CH) complexes in silicon

Carbon is a very common impurity in silicon. The carbon originates primarily from the feedstock but contamination from graphite used in float zone (FZ) and Cz pullers can be significant. In Cz growth the small segregation coefficient ($k_0 = 0.058$) causes inhomogeneous incorporation of carbon along the crystal growth axis resulting in an increase of carbon content from the seed end to the tail end by a factor of 2–20. The low segregation coefficient also tends to result in higher carbon concentrations in silicon ingots grown in reusable crucibles, particularly towards the end of the crucible's life. Such is the case of the continuous Cz ingot growth method. In older electronic-grade silicon, the range of carbon concentrations was in the range of 10^{16}–10^{17}/cm^3; however, today's trends show typical electronic-grade silicon containing [C] < 2 × 10^{16}/cm^3. In cast solar silicon, the typical [C] values are much higher, sometimes as high as or even above the solubility limit at the silicon melting point (4.5 × 10^{17}/cm^3). This carbon comes from the feedstock but also from carbon monoxide in the furnace environment, which dissolves into the liquid silicon. Carbon monoxide results from silicon monoxide evaporating from the melt surface and reacting with graphite furnace components. In all cases, the last grown regions in the ingot have the highest carbon concentrations.

In FZ silicon the concentration depends primarily on the carbon concentration in the starting material, but this is reduced by the number of FZ passes. In cast silicon, the concentration in the feedstock is the key factor but due to the low segregation coefficient, the carbon tends to concentrate in the last solidified region of the ingot.

The carbon concentration in poly- and monocrystalline silicon is usually measured by room temperature absorption spectroscopy at the local mode vibrational frequency at 605/cm for which there is international standardisation. The method measures carbon on the substitutional site with a detectivity of 2×10^{15}/cm^3 in the case of a 0.5-mm-thick slice. However, the technique does not detect SiC precipitates. A more sensitive method is radio-chemical activation analysis using He^3. In this case, all carbon is measured in the silicon irrespective of its site or chemical coordination with a detectivity limit of 1 × 10^{14}/cm^3. Unfortunately, this method is complicated and expensive.

Hydrogen (H) reacts readily with substitutional carbon (C_S) in silicon to form a carbon-hydrogen (CH) complex. It has been shown using Laplace deep-level transient spectroscopy (DLTS) with uniaxial stress [7] that the most stable form of the complex is the one in which the hydrogen sits between the carbon atom and its silicon first neighbour (BC1 in Figure 6.1) [9]. Unlike isolated bond-centred

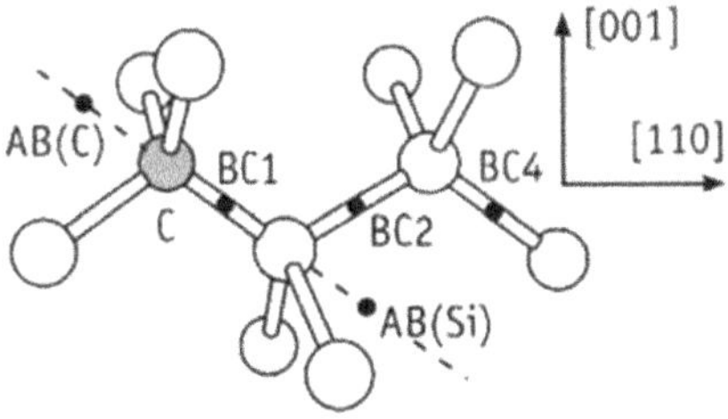

Figure 6.1 *Low-energy sites (black dots) for hydrogen near C_S on bond centre (BC) and antibonding (AB) sites. C is represented in grey, and Si in white. Reproduced from Santos* et al. *[8] under Creative Commons CC BY. https://doi.org/10.1002/pssa.201700309 © 2017 The Authors.*

hydrogen, which has a negative-U behaviour [10], the C-H-Si defect shows a positive-U ordering with its donor and acceptor states located at E_v+0.33 eV and E_c–0.16 eV, respectively [9, 11].

6.1.2 Oxygen-hydrogen (OH) complexes in silicon

Like carbon, oxygen is a ubiquitous impurity in silicon with a concentration dependent on the method of growth. The solubility of oxygen in silicon at the melting point of silicon is $2.6 \times 10^{18}/\text{cm}^3$. In FZ material grown in vacuum or argon, often with the addition of some nitrogen, the oxygen concentration is rarely above $1 \times 10^{16}/\text{cm}^3$. In Cz silicon, the oxygen concentration is much higher. The oxygen concentration results from the dissolution of the quartz crucible counteracted by the evaporation of silicon monoxides from the crucible surface. The actual concentration depends on many factors including furnace design and rotation speed but is usually in the range of $6 \times 10^{17}/\text{cm}^3$ to $1 \times 10^{18}/\text{cm}^3$ [12].

The source of oxygen contamination in cast silicon is mostly dissolution of the quartz crucible into the silicon melt; therefore, the oxygen concentration at the bottom of the ingot is higher than any other region. Here the concentration can be as high as $6 \times 10^{17}/\text{cm}^3$ whereas at the top half of the ingot it is often at the $2 \times 10^{17}/\text{cm}^3$ level [13].

Oxygen occupies an interstitial site in silicon and is not electrically active. Its concentration in silicon is usually measured using local mode, room temperature, optical absorption at 1107/cm. Several internationally agreed (but different) calibration standards exist, so it is usual to cite which standard is used. Using optical absorption, the detection limit of oxygen at room temperature is approximately $5\times10^{15}/\text{cm}^3$ in a 0.5-mm-thick slice; however, an order of magnitude improvement can be achieved by undertaking the measurement at liquid helium temperatures. As in the carbon case the optical absorption only measures the oxygen on a specific site, chemical methods have to be used if the measurement is to include molecules, complexes and precipitates.

Hydrogen reacts quite readily with interstitial oxygen (O_i) to form an oxygen-hydrogen (OH) complex. The low-energy sites for hydrogen to exist in an OH complex in silicon are shown in Figure 6.2. We know from DLTS measurements that the electrical levels of OH are similar to those of isolated H again with negative-U behaviour (9). Consequently the typical donor and acceptor charge states are identified as being at E_c – 0.17 eV and E_c – 0.68 eV, respectively [14].

6.1.3 Carbon-oxygen-hydrogen (COH) complexes in silicon

Similar to the readily achieved interactions of hydrogen with single carbon and single oxygen atoms that can often be found in various silicon materials, hydrogen has been shown to combine with substitutional carbon (C_s) and O_i complexes bonded to a shared silicon atom, to form carbon-oxygen-hydrogen (COH) complexes. The carbon-oxygen (CO) species have been identified to produce an infrared (IR) absorption band at 1104/cm in silicon [15–17]. Thermal stability studies of the CO complexes demonstrate that at O_i-diffusing temperatures (>450 °C) oxygen can be

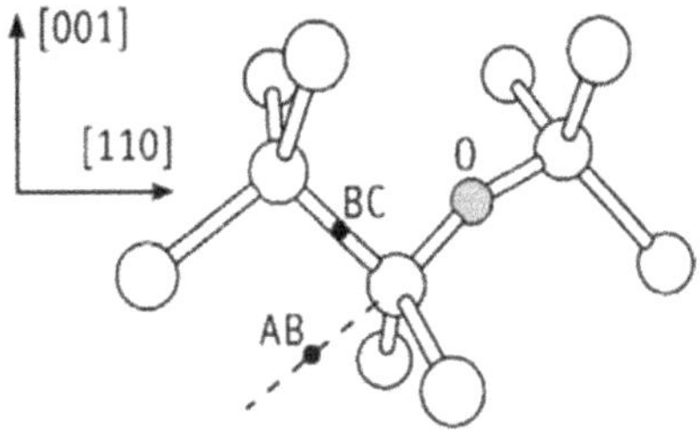

Figure 6.2 *Low-energy sites (shown in black) for hydrogen near interstitial oxygen (O_i) (shown in grey). Reproduced from Santos* et al. *[8] under Creative Commons CC BY. © 2017 The Authors. https://doi.org/10.1002/pssa.201700309.*

effectively trapped by the C_s-Si complex resulting in increased CO concentrations [16, 17]. Even though these complexes are not expected to be electrically active, a rather strong electrical activity detected by DLTS and minority carrier transient spectroscopy (MCTS) has been associated with the presence of these complexes in hydrogenated silicon crystals [18–21].

According to theoretical calculations, three possible ground-state structures of the COH complexes have been proposed by Santos *et al.* [8] and are shown in Figure 6.3, which correspond to its different charge states: negative, neutral and

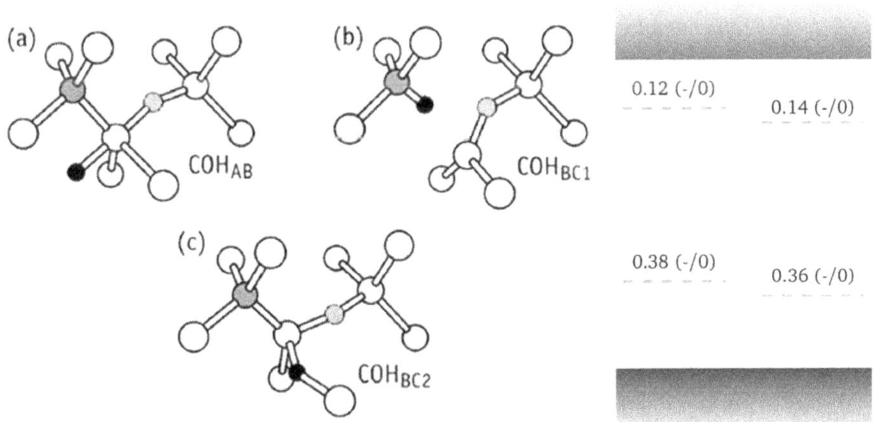

Figure 6.3 *(left) Ground-state structures for COH in the negative (a), neutral (b) and positive (c) charge states. Si, C, O and H atoms are shown in white, grey, light grey and black, respectively. Reproduced from Santos* et al. *[8] under Creative Commons CC BY. © 2017 The Authors. https://doi.org/10.1002/pssa.201700309; (right) Energy band diagram showing COH charge state energy levels located at E_C−0.12 eV, E_c−0.14 eV, E_V+0.38 eV and E_V+0.36 eV.*

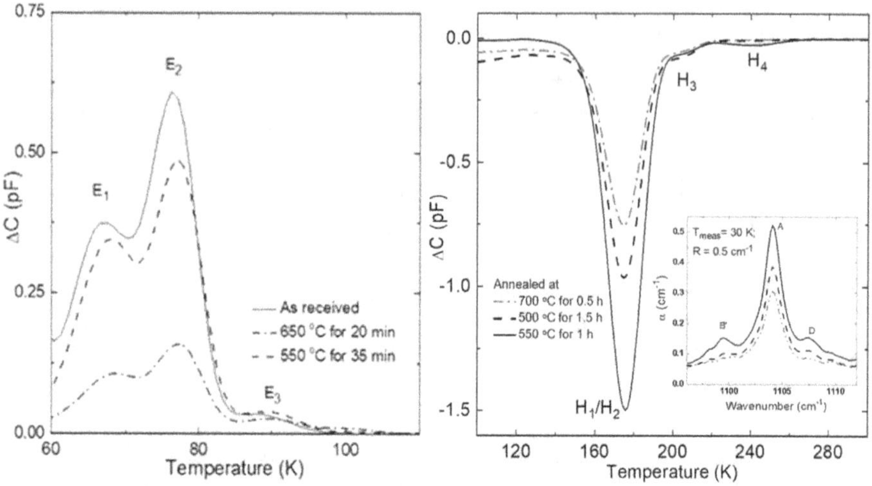

Figure 6.4 (left) DLTS spectra measured from samples of a Cz n-type silicon wafer as it was received, after 20 min anneal at 650 °C and after 35 min at 550 °C. (right) MCTS spectra of samples from another C-rich Cz n-type silicon wafer after annealed at 700 °C, 500 °C and 550 °C for 30, 90 and 60 mins, respectively. The inset shows the 1104/cm absorption band measured from sister samples of the same wafer used for MCTS measurements after the same heat treatments. Both DLTS and MCTS spectra are recorded after 1HF:7HNO$_3$ wet etch of the samples and with measurement parameters: rate window = 50/s; U_b = −2.0 V; t_p = 10 ms. Reproduced from Vaqueiro et al. *[20].*

positive. It is shown that in the negative charge state of the complex, there exists a C_s atom in the vicinity of a H^- occupying an antibonding position in the silicon lattice resulting in a H^--Si-O-Si linear defect that introduces an acceptor level of energy close to that of the measured H_1/H_2 traps.

Vaqueiro *et al.* [22] showed that the presence of the predominant H^- in phosphorous-doped Si:O+C crystals give rise to a family of charge carrier traps in both halves of the silicon band gap. Four of these traps, namely E_1, E_2, H_1 and H_2, increase in concentration depending on the [C] and [O] present in the crystal and show a depth distribution similar to the expected distribution of hydrogen in the subsurface of the studied crystals.

Figure 6.4 shows the observed DLTS/MCTS peak formation of two carbon-rich silicon samples and their relation to the CO species formation measured by IR absorption at low temperature. Full electrical parameterisation of the traps' properties from transient spectroscopy has been conducted to show that two of these are majority carrier traps (E_1/E_2) and are located at $E_C - 0.12$ eV and $E_C - 0.14$ eV [22], and two more are minority carrier traps (H_1/H_2) and are located in the bottom half of the gap at $E_V + 0.38$ eV and $E_V + 0.36$ eV [20]. Results presented in [22] strongly suggest that the electron/hole trap pairs H_1/E_1 and H_2/E_2 are related to two energy levels of the same COH complex in the material. High concentrations of these traps in the range of $1 \times 10^{13}/cm^3$ to $5 \times 10^{13}/cm^3$ have been found in a range of commercial Cz and continuous Cz silicon crystals; however, it is the direct measurement of the capture cross section of the hole traps that reveals the significance of such traps in terms of minority carrier lifetimes. The acceptor nature of the traps, their energy levels, capture cross sections in the order of 10^{-16} cm^2, and calculated concentrations result in a rather strong recombination pathway associated with the centres H_1 and H_2 which will limit the carrier lifetime to only a few tens of microseconds. It can then be expected that cells made from these crystals will suffer from low open-circuit voltages and in turn limited power conversion efficiencies to only a few percent.

6.1.4 Transition metal-hydrogen complexes in silicon

The tendency for atomic hydrogen to bind and interact with various transition metals (TMs) in silicon has been extensively documented in [23] and references therein. One of these interactions, discussed in detail throughout this book, is the electrical deactivation of previously active metallic impurities through hydrogen passivation. However, the reaction of a TM with hydrogen does not necessarily result in passivation, indeed in some cases, it can result in an increase in recombination activity.

The configuration of TM-hydrogen complexes can often involve more than one hydrogen atom with the strength of the recombination activity dependent upon the position of the associated trap energy level within the bandgap. It is known that the act of hydrogen passivation can result from hydrogen satisfying the four dangling bonds associated with the TM (chemical passivation), causing either an elimination or a reduction in recombination activity, which involves a shift of the energy level towards the conduction or valence band edges. In some cases, the energy level can be shifted into the bands resulting in no recombination or trapping activity. However, in the formation of certain TM-hydrogen complexes, increases in recombination can be realised if the defect level, initially residing close to the band edge is then relocated closer to the centre of the bandgap.

There are several examples of interactions whereby the formation of a TM-hydrogen complex can lead to a shift in energy level and, in a few cases, this has been documented for successive additions of hydrogen together with changes in the capture cross section. Some examples of the energy change of TM defects upon hydrogenation are given in the following references: cobalt [24, 25], titanium [26], vanadium [27], chromium [28], iron [29], gold [30], copper [31] and silver [32]. It is worth noting that the vast majority of these complexes are not stable above 473 K, often dissociating into their constituent parts with reformation only possible at lower temperatures.

6.1.5 Vacancy-hydrogen complexes in silicon

Vacancies can be grown into crystalline silicon. For example, in Cz-grown silicon the concentration of grown-in vacancies is determined by the V/G ratio, where V is the growth rate and G is the axial temperature gradient [33]. This means that the concentration of vacancies incorporated can be minimised; however, it is a trade-off against both higher throughput and increased incorporation of interstitials [33]. In addition, many processes such as plasma treatments, dielectric deposition, contact sputtering or implantation can dislodge a silicon atom from its lattice position causing a vacancy with dangling bonds, and an interstitial (*V–I* Frenkel pair) [34]. Both *V* and *I* are highly mobile, even at low temperatures, and can annihilate each other or bond with other defects/impurities to form complexes. *V–I* pairs are less likely to recombine in p-type where self-interstitials become trapped at dopants, leaving such material more prone to vacancies [34]. Similar to hydrogen, a vacancy is a negative-U system [10] with stable charge states V^{++}, V^0 and V^-, with the dominant charge state determined by the Fermi level and temperature [34]. Hydrogen is well known for tying up or passivating dangling bonds, including those of vacancies. A single vacancy can stably bond with up to four hydrogen atoms (VH_n with $n = 1, 2, 3, 4$). VH_n complexes where $n < 4$ have electrically active levels in the gap, while VH_4 is electrically inactive [35]. VH_n complexes where $n > 1$ are not mobile at temperatures below which they break up, while VH_1 is very mobile at low temperatures, perhaps even more so than hydrogen alone [34]. The hydrogen and vacancy interaction leads to covalent bonds that face towards the centre of the vacancy along the [111] direction (approx. trigonal axis) [34, 36]. Figure 6.5 shows the calculated neutral configurations of VH_n with $n = 1, 2, 3, 4$ including the symmetry, internuclear distances and degrees of bonding [34]. The dissociation energy for VH_n complexes is in the order of 2.2–2.6 eV [34].

Infrared absorption is commonly used to study VH_n configurations. The local vibrational modes (LVM) detected an increase from approximately 2000/cm to approximately 2260/cm as the number of hydrogen atoms involved increases; this is due to a shortening of Si–H bonds as the hydrogens repel each other [36]. The many possible configurations lead to similar IR bands or spectrums, so many VH_n complexes have not been identified with a high degree of certainty [34]. Certain modes of the single-vacancy-hydrogen complexes that have been identified include the neutral charge state of VH_1 (VH^0) which has been attributed to LVM 2038/cm; the B_1 and A_1 modes of VH_2 attributed to LVMs 2121/cm and 2144/cm, respectively; LVMs 2155/cm and 2185/cm have been suggested to belong to VH_3; the T_2 and A_1 modes of VH_4 are attributed to LVMs at 2223/cm and 2257/cm, respectively [36]. LVMs 2165/cm and 2190/cm previously attributed to VH_3 are now attributed to the E and A modes of a V_2H_6 complex, respectively, highlighting the similarities and understandable confusion. The ability for both vacancies and hydrogen to bond to any number of other defects and impurities, and form aggregates/complexes consisting of several vacancies and several hydrogen atoms complicates (semi) empirical models, and makes it difficult to resolve specific complexes IR signals and energy levels. For example, the neutral charge states of vacancy hydrogen complexes VH^0 and $V2H^0$, as well as the phosphorus-vacancy complex PV^0, have very similar electron paramagnetic resonance spectral parameters [7].

6.2 Light- and elevated temperature-induced degradation

LeTID is a relatively newly identified form of CID in silicon that has gained significant interest in academia and industry. At the time of writing, there have been

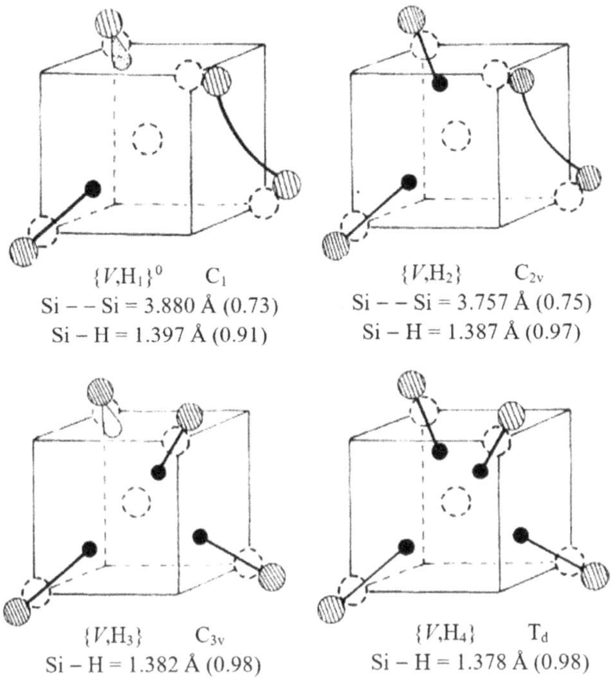

Figure 6.5 *Calculated neutral configurations of single-vacancy-hydrogen complexes,* VH_n, *with* n = *1, 2, 3, 4 including the symmetry, internuclear distances and degrees of bonding. Reprinted with permission from [34] Copyright (1994) by the American Physical Society. https://doi.org/10.1103/PhysRevB.49.17040.*

over 170 publications from more than 40 research institutes globally, focused on understanding the behaviours, kinetics, root cause and mitigation strategies of the defect. A detailed review has been written recently, which can be referred to for further information [37]. Although it is still not entirely understood, there is a general consensus that hydrogen plays a role in the formation of this defect.

The first observation of this unique degradation behaviour was reported at *Schott Solar* by Ramspeck *et al.* in 2012 [38]. It was not until 2015 when the degradation phenomenon was coined 'LeTID' by Kersten *et al.* upon disclosing the accelerated degradation behaviour observed under illumination at elevated temperatures [39, 40]. However, LeTID can also be induced using applied bias, or thermal anneals in the absence of light, which reveals that the degradation is in fact induced by carriers and accelerated via excess carrier injection [41–43], making this mechanism more accurately ascribed as a form of CID. The term CID, however, has led to a somewhat confusing position as the term is commonly used to describe a range of other carrier-induced defects in silicon including boron-oxygen-related light-induced degradation (B-O LID), copper-induced light-induced degradation (Cu-LID) [44, 45], and surface defects among others. In an attempt to distinguish LeTID from the other reported CID mechanisms, some groups later began to describe the phenomenon as specific to mc-Si (e.g. mc-CID and mc-LID) [46–50]; however, it was later found that the defect appears in all types of silicon wafers [51]. A perhaps more accurate name was later proposed, 'hydrogen-induced degradation', due to the strong evidence for hydrogen being involved in the degradation process (either as a precursor for defect formation or as the defect itself) [52].

Despite being a misnomer due to the lack of a need for illumination, the name 'LeTID' has largely been adopted by the industry as the most standard and recognisable terminology. For this reason, and the fact that the degradation in normal operation conditions will mostly occur via illumination at elevated temperatures, throughout this chapter, the phenomenon is referred to as LeTID and the recombination active defect as the 'LeTID-related defect'.

6.2.1 The impact of LeTID – cells, modules and systems

Since 2012, numerous reports of the impact of LeTID on a wide range of wafers, cells, modules and even some installed systems have been published. However, LeTID degradation time frames under normal operating conditions in the field are in the range of several years. In Table 6.1, a list of studies on 'untreated' p-type mc-Si passivated emitter and rear cells (PERC) and PERC modules reported for LeTID throughout the literature is compiled. The extent of apparent degradation varies significantly between different studies. Petter *et al.* presented findings detailing degradation up to $16\%_{rel}$ on high-performance mc-Si PERC cells [53] while Kersten *et al.* presented field data of mc-Si PERC modules installed in both Germany and Cyprus [40]. These modules were observed to have up to $7\%_{rel}$ power loss after three years of field operation. Using this data, and corresponding laboratory tests, a value of 15 years out in the field for degradation was predicted for the specified climate with regeneration assumed to occur on a much longer timescale. Given the timescale for recovery out in the field, and increased extent of degradation, LeTID would be of greater concern than B-O LID. Bearing in mind the standard 25- to 30-year manufacturer's performance warranty of PV modules, this would have a disastrous impact on consumers and energy suppliers if left untreated.

One complication in assessing the degree of degradation and comparison between different studies, however, is the conditions used for testing. As seen in Table 6.1, the range of testing conditions shows large inconsistencies in the temperature, time and carrier injection technique used. This raises the need for a standard and ideally accelerated LeTID test for effective production quality checks and comparable research studies, which currently does not exist [64]. The International Electrotechnical Commission (IEC) has since attempted to draft and implement a standard of testing in both cells and modules, although there is still debate over the testing conditions. Currently, IEC 61215-2 ED2 ('Terrestrial Photovoltaic (PV) Modules – Design Qualification and Type Approval – Part 2: Test Procedures') provides the specifications of LeTID testing on modules: 75 ± 3 °C with current-injection set to a value equal to the difference between the short-circuit current of the module and the current at MPP (i.e. $I_{SC} - I_{MPP}$) for a duration of 162 hours. This

Table 6.1 Reported relative power and efficiency losses on p-type cells and modules

Source	Type	Deg. (%$_{rel}$)	Conditions	
Petter *et al.* [53]	Cell	Up to 16	η	75 °C, CI (1-sun equiv.) 200 h
Chan *et al.* [49]	Cell	12.7	η	70 °C, 0.46 kW/m^2, 480 h
Krauß *et al.* [54]	Cell	11.2	η	80 °C, 0.8 kW/m^2, 325 h
Luka *et al.* [55]	Cell	≈10	η	75 °C 1-sun, 48 h
Sen *et al.* [56]	Cell	6	η	75 °C, 1 kW/m^2, 250 h
Ramspeck *et al.* [38]	Cell	5-6	η	75 °C 0.4 kW/m^2, 400 h
Deniz *et al.* [57]	Cell	4.4	η	75 °C, CI (J_{SC} equiv.), 45 min
Sio *et al.* [58]	Cell	4.3	η	65 °C, 1-sun, 5 h
Padmanabhan *et al.* [59]	Cell	4.3	η	90 °C, 1-sun, 21 h
Chunlan *et al.* [60]	Cell	3.2	η	70 °C, 0.8 sun, 30 h
Fertig *et al.* [61]	Module	11	W	85 °C, CI (1 kW/m^2 equiv.), 425 h
Kersten *et al.* [39]	Module	11	W	85 °C, CI MPP, 400 h
Nakayashiki *et al.* [62]	Module	10	W	Outdoors – Singapore, 56 days
Fokuhl *et al.* [63]	Module	9.2	W	85 °C, CI (I_{MPP}), 400 min
	Module	7.2	W	85 °C, CI ($I_{SC} - I_{MPP}$), 400 min
Kersten *et al.* [40]	Module	7.5	W	Outdoors – Germany
	Module	≈7	W	Outdoors – Nicosia, Cyprus, 3 years

CI, current injection; MPP, maximum power point (current equivalent); W, relative power loss (Watts).

is repeated until the difference in module power under standard test conditions is less than 1% between two consecutive 162-hour stress tests. On the other hand, an LeTID testing standard of individual solar cells remains in the drafting stages, with IEC 63202-1 ('Measurement of Light-Induced Degradation of Crystalline Silicon Photovoltaic Cells') used as a guideline for testing: 60 ± 5 °C with steady-state illumination (irradiance of 1000 ± 50 W/m^2) until a difference in power of less than 0.5% is observed between consecutive measurements. Unfortunately, these prescribed conditions cannot effectively isolate LeTID from other thermal, light- and carrier-induced defects such as B-O LID.

6.2.2 The key behaviours of LeTID

6.2.2.1 Dependence on firing on degradation extent

It is now well known that LeTID degradation extent is heavily dependent on the peak temperature of the firing process used for solar cell metallisation. This dependence was identified early on by Bredemeier *et al.*, Chan *et al.*, Nakayashiki *et al.*

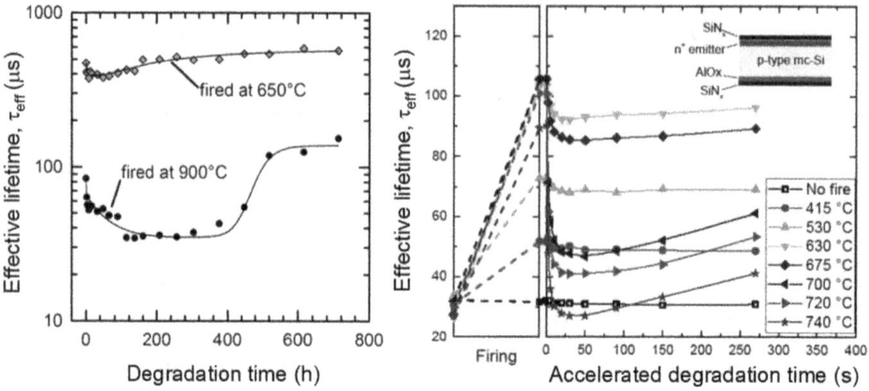

Figure 6.6 LeTID testing of mc-Si lifetime test structures (left) τ_{eff} ($\Delta n = 1 \times 10^{15}$ cm^{-3}) of samples fired at 650 °C and 900 °C degraded under 1-sun at 75 °C. Figure adapted from Bredemeier et al. *[65] under Creative Commons CC BY. https://doi.org/10.1063/1.4944839 © Author(s) 2016. (right) τ_{eff} ($\Delta n = 9.1 \times 10^{14}$ cm^{-3}) of samples after PECVD, after firing, and degraded under 44.8 kW/m² at 140 °C under 44.8; © [2016] IEEE. Reprinted, with permission, from [49].*

and Eberle *et al.*, finding that firing triggers LeTID and higher temperatures lead to more severe degradation [49, 62, 65, 66]. More recent publications have used this firing-temperature dependence as an indicator for identifying LeTID in other silicon wafers such as p-type Cz [67] and n-type Cz [68]. Figure 6.6 (left) compares mc-Si wafers fired with a set peak temperature of 650 °C and 900 °C, demonstrating that after firing at lower temperatures there is no degradation [65]. Similarly, Chan *et al.* fired mc-Si wafers using a range of peak temperatures from 415 °C to 740 °C (actual wafer temperature) with the subsequent degradation under light soaking plotted in Figure 6.6 (right). Here, LeTID does not appear to occur if wafers were not fired or were fired below approximately 650 °C. However, it was later demonstrated by Eberle *et al.* that LeTID after a high 800 °C peak firing temperature could be suppressed using slow cooling rates [66]. It, therefore, appears as though LeTID requires both a high peak firing temperature and a relatively fast cooling rate. Due to limitations modifying belt furnace thermal profiles, it was unclear from the work of Eberle *et al.* whether it was the rate of cooling or the total thermal budget that was suppressing the degradation.

The work of Chan *et al.* not only confirmed increasing severity of LeTID with increased firing temperature but also observed that firing with temperatures above 630 °C caused a reducing improvement in the absolute τ_{eff} measured immediately after firing [49]. Chen *et al.* [43] later attributed this reduced lifetime directly after firing to a pre-formation of LeTID that occurs during the firing process, and which varies in extent depending on the cooling rate. Kim *et al.* [69] found that a critical temperature of 765 °C or higher combined with a cooling rate >50 °C/s was required for pre-formation of LeTID defects. This effect can be seen in other studies whereby samples have undergone firing at high temperature before degradation and recovery and the final lifetime is much higher than the lifetime directly after firing, with at least some of this increase due to the sample starting with a partially degraded lifetime after firing. The critical dependence on firing conditions led to the first hypotheses regarding the root cause of LeTID: a possible involvement of hydrogen released from surface dielectric films [70] or a dissociation of metallic precipitates in the bulk [71].

6.2.2.2 A universal defect in silicon

It was originally thought that LeTID only affected p-type mc-Si wafers and mc-Si PERC solar cells. An early study by Ramspeck *et al.* identified LeTID in p-type Ga-doped mc-Si PERC solar cells, which differentiated the defect from both B-O LID and interstitial iron activation effects [38]. In 2015, cases of unknown light-induced degradation (LID) began appearing on Cz structures with aluminium oxide (AlO_x) dielectric layers on the rear [72]. However, the first confirmations that LeTID also affected monocrystalline silicon came from studies in 2017 by Fertig *et al.* [73] and Chen *et al.* [43]. Fertig *et al.* demonstrated that PERC modules stabilised for B-O LID would continue to experience degradation at 75 °C in MPP conditions [73]; however, in modules the defect recombination properties could not be studied. Chen *et al.* performed a systematic study on p-type Cz-Si and mc-Si symmetrically passivated test structures to compare the defect recombination parameters in each material and the degradation behaviour, as depicted in Figure 6.7 (top). A dark anneal was used to form the defect and prevent the formation of B-O LID, thereby isolating the LeTID [74, 75]. To verify the existence of LeTID, injection-dependent lifetime spectroscopy (IDLS) analysis of the Shockley-Read-Hall (SRH) statistics was used to correlate with the known defect parameters (refer to Table 6.2) as measured on mc-Si [43]. Many prior observations of degradation in monocrystalline silicon wafers have since been attributed to LeTID including several studies that observed CID on FZ silicon wafers [47, 76, 81, 86–88]. Niewelt *et al.* observed similar degradation when light soaking p-type FZ-Si samples at 75 ± 5 °C; the samples were passivated with Al_2O_3/SiN_x and fired above 650 °C (up to 900 °C) (Figure 6.7 bottom) [47]. The defect recombination properties observed in FZ, Cz and mc-Si are compared in Table 6.2 in Section 6.2.2.4. The discovery that FZ is also affected by LeTID provides opportunities to study the defect on wafers that have significantly lower concentrations of crystallographic defects and metallic impurities than mc-Si [89].

N-type silicon wafers have been less susceptible to previously identified forms of CID; however, recent work has demonstrated that LeTID also affects n-type silicon [58, 68, 83, 90]. In 2014, Renevier *et al.* observed bulk degradation of n-type Cz wafers with SiO_2/SiN_x passivation after rapid thermal annealing [91]. However, only wafers with at least one side boron-diffused degraded. No degradation was observed in samples passivated with Si-rich SiN_x or stoichiometric aluminium oxide (Al_2O_3) or in samples with double-sided phosphorus diffusions. The authors speculated that the degradation may come from an interaction between the boron-diffusion and the

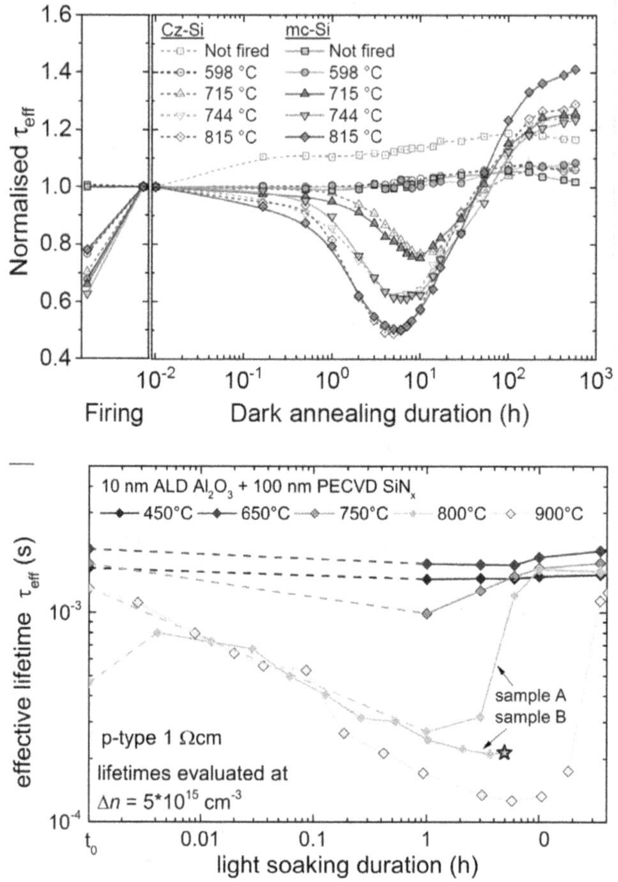

Figure 6.7 *LeTID in various wafer types: (top) Cz and mc-Si τ_{eff} (normalised to the value directly after firing) fired at various temperatures and degraded using 175 °C dark anneal. Reprinted from [43] Copyright (2017), with permission from Elsevier. (bottom) FZ τ_{eff} (at $\Delta n = 5 \times 10^{15}$ cm^{-3}) with various thermal treatments after deposition degraded at 75 °C and 1-sun illumination. Reprinted from [47], with the permission of AIP Publishing. © 2017 Author(s).*

SiO_2 layers; however, no specific root cause was discussed. Chen *et al.* had similar findings in 2018 when studying n-type Cz wafers with boron-, phosphorus- or no diffusion [83]. It was found that a diffusion (of either polarity) was required during contact firing for bulk degradation to occur. The boron-diffused wafers had identical degradation timescales and firing dependence to LeTID in p-type, thereby indicating that it is, in fact, LeTID occurring in n-type silicon. In contrast with Renevier *et al.*, the n-type wafers with phosphorus-diffusions also degraded; however, the timescale was several orders of magnitude faster. This could explain why previous studies

Table 6.2 LeTID capture cross section ratios (k-values) and apparent defect energy levels in various silicon materials as reported in the literature in chronological order

Source/Ref.	Type	Method	k-value	Energy level
Grant *et al.* [76]	(p)FZ-Si	IDLS	26	-0.31 eV $< E_t - E_i < 0.19$ eV
Bredemeier *et al.* [77]	(p)mc-Si	IDLS	20 ± 7	None stated
Nakayashiki *et al.* [62]	(p)mc-Si	IDLS	28.5	-0.26 eV $< E_t - E_i < 0.24$ eV
Morishige *et al.* [78]	(p)mc-Si	IDLS	$26 < k < 36$	-0.27 eV $< E_t - E_i < 0.13$ eV
Vargas *et al.* [48]	(p)mc-Si	TIDLS	Upper bandgap: 49 ± 21 Lower bandgap: 56 ± 23	Upper bandgap: $E_t - E_i = 0.21 \pm 0.05$ eV Lower bandgap: $E_t - E_i = -0.32 \pm 0.05$ eV
Jensen *et al.* [79]	(p)mc-Si	TIDLS	Upper bandgap: 23.5 ± 5.6 Lower bandgap: 23.9 ± 5.5	Upper bandgap: $E_t - E_i = 0.1 \pm 0.07$ eV Lower bandgap: $E_t - E_i = -0.21 \pm 0.06$ eV
Chen *et al.* [43]	(p) mc-Si (p) Cz-Si	IDLS	33.4 ± 1.5 39.4 ± 4.9	Assumed midgap
Niewelt *et al.* [80]	(p)FZ-Si	IDLS	35	Assumed midgap
Sperber *et al.* [81]	(p)FZ-Si	IDLS	20 ± 2	None stated
Niewelt *et al.* [47]	(p)FZ-Si	IDLS	35* ± 10	None stated
Fung *et al.* [82]	(p)mc-Si	IDLS	$27 < k < 35$	None stated
Kim *et al.* [69]	(p)Cz-Si	IDLS	36 ± 2	None stated
Chen *et al.* [83]	(n)Cz-Si	IDLS	0.028 ± 0.003	Assumed midgap
Sen *et al.* [84]	(p)mc-Si	IDLS	29 ± 4	Assumed midgap
Winter *et al.* [85]	(p)mc-Si	IDLS	29* ± 1	None stated

*Q-factor = *k*-value × 1.21

looking for LeTID in n-type FZ wafers did not see degradation. Of course, it is possible the defect did not form or was not recombination active, however, now appears possible it could have occurred but on a faster timescale [47]. Interestingly, both Sio *et al.* and Vargas *et al.* studied LeTID in n-type mc-Si wafers and found that under illumination the rate of degradation in n-type mc-Si was slower than in p-type [58, 90]. Chen *et al.* carried out a systematic study to compare the kinetics of LeTID in n-type and p-type both under illumination and in the dark [68]. While there remains the possibility that it could be different defects causing the degradation in the different wafer types, the evidence suggests that it is the same recombination active defect. This has important implications for understanding the nature and properties of the LeTID defect as will be discussed in Section 6.2.5.

6.2.2.3 LeTID dependence on dielectrics

In addition to the strong relationship between LeTID extent and peak firing temperature, studies have repeatedly shown that the degradation is heavily dependent on the dielectric passivation layers. Initially, an increased degradation was shown on mc-Si PERC compared to Al-BSF solar cells [38], which led to the suggestion that aluminium gettering may be the prevention mechanism [61, 72]. However, the time spent above the melting point of aluminium during co-firing is typically short (<5 s), which limits the gettering efficacy only to defects close to the rear-surface or very fast-diffusing impurities, and therefore unlikely to be the reason [92]. Similarly, Fertig *et al.* suggested that the LeTID was related to the AlO_x rear surface passivation layer, rather than a bulk defect [61]. Indeed, it is clear now that the dielectric plays a major role in providing a source of defect precursors as shown in mc-Si by Padmanabhan *et al.* [59] and Vargas *et al.* [93], and in Cz-Si and FZ-Si by Sperber *et al.* in [94] and [95], respectively.

Further studies on the role of the passivation layers revealed that only samples deposited with hydrogen-containing dielectrics such as hydrogenated silicon nitride (SiN_x:H) or hydrogenated aluminium oxide (AlO_x:H) degraded. Kersten *et al.* investigated various combinations of Al_2O_3 and SiN_x:H dielectrics on p-type mc-Si wafers both with and without n-type emitters and using different tools [70]. It was shown that varied dielectrics can lead to different degradation rates, and the authors concluded that LeTID only occurs on samples that have dielectric layers present during firing. Vargas *et al.* [96] and Varshney *et al.* [97] observed that the LeTID extent in mc-Si is dependent on the SiN_x:H passivation layer. Their results showed that increased degradation occurred as a result of higher deposition temperatures and thicker films, respectively. This will be further discussed in consideration of the root cause in Section 6.2.3.

6.2.2.4 LeTID characterisation

It can be difficult to identify a specific defect in silicon out of thousands of different potential defects. However, some defects can be identified due to unique 'fingerprints' using certain characterisation techniques. Lifetime spectroscopy is one of these commonly used techniques, either IDLS or the more advanced, temperature- and injection-dependent lifetime spectroscopy (TIDLS) [55]. Specifically, fitting the SRH-related lifetime and J_0, to isolate from the intrinsic- and surface-related components of the lifetime, can determine the capture cross section ratio (*k*-value) of the active defect. The *k*-value is the ratio of the defect's electron and hole capture cross sections, and is given by [56]

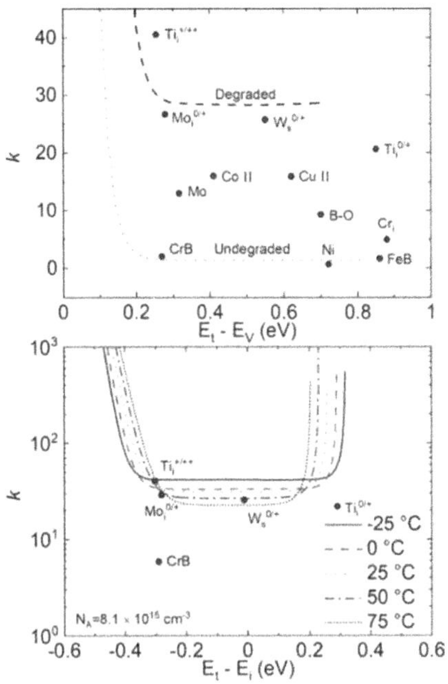

Figure 6.8 *(top)* k-*value in the un-degraded and degraded states plotted against the trap level relative to the valence band, showing* k-*values of approximately 1.4 and 28.5, respectively, between 0.3 eV and 0.8 eV above* E_V*. Adapted from [78]. (bottom)* k-*value plotted against the trap level relative to the intrinsic energy level for LeTID in a p-type mc-Si wafer with lifetimes measured at temperature (T) from −25 °C to 75 °C with 25 °C increments. Dots are defect parameters from the literature. Adapted from [79].*

$$k = \frac{\sigma_n}{\sigma_p} = \frac{\tau_{n0}}{\tau_{p0}} \times \frac{v_{th,h}}{v_{th,e}} \tag{6.1}$$

where σ_n and σ_p are the electron and hole capture cross sections, $v_{th,h}$ and $v_{th,e}$ are the thermal velocities of holes and electrons, and τ_{n0} and τ_{p0} are the capture time constants for the electrons and holes. A number of studies have used TIDLS to study LeTID; Nakayashiki *et al.* [78] (Figure 6.8 (top)), Morishige *et al.* [78], Jensen *et al.* [79] and Vargas *et al.* [48] (Figure 6.8 (bottom)) all note a possible involvement of a tungsten substitutional donor ($W_s^{0/+}$), a titanium interstitial double donor ($Ti_i^{+/++}$) and a molybdenum interstitial donor ($Mo_i^{0/+}$). However, the susceptibility of FZ-Si wafers to LeTID, which contain low concentrations of such metals means these metals are not likely to be the root cause [80, 98]. It is possible that the defect responsible for LeTID has not previously been characterised, and as such correlating even the most accurate SRH data to existing known defects may not reveal the specific defect.

From Figure 6.8, it can be seen that the *k*-value for the deep-level defect plateaus over a range of energy levels incorporating midgap. Assuming that a large

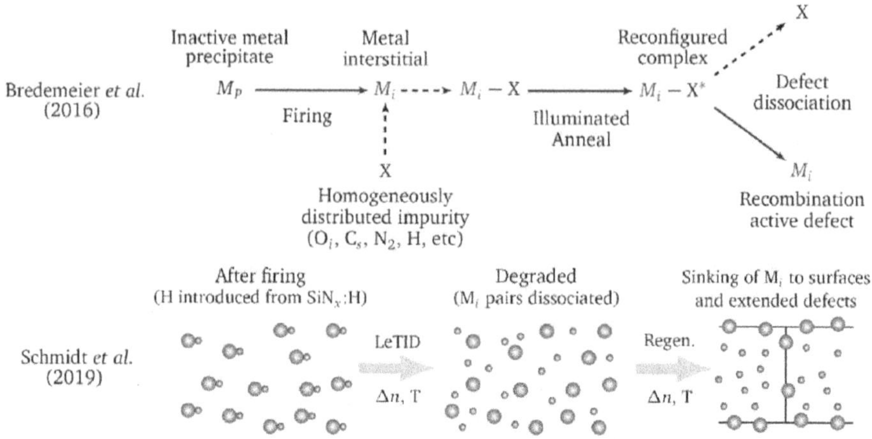

Figure 6.9 A depiction of a proposed model for LeTID involving metallic impurities by Bredemeier et al. *[77]. and a more recent adaption that includes dissociation of a M_i-H complex as in [100]. Reproduced from [101].*

range of defect energy levels located far enough from the band edges behaves like a midgap defect, then a defect with an unknown energy level may reasonably be assumed to be at midgap for determining the *k*-value. Based on this, a simple quality of fit can be used, assuming a defect with a mid-gap energy level, to assess whether the samples being measured are likely to contain LeTID or not [99]. Table 6.2 collates defect *k*-values and apparent defect energy levels (extracted or assumed) for various LeTID-affected materials as reported in the literature.

6.2.3 *The search for the root cause of LeTID*

Over the years, many studies have attempted to pinpoint the defect that causes LeTID, and many candidates have been proposed and investigated. This section looks at the evidence for and against each, with a particular focus on the highly plausible role of metallic impurities and/or hydrogen.

6.2.3.1 Metallic impurities

One of the first proposed causes was metallic impurities. Figure 6.9 shows a proposed model by Bredemeier *et al.* that consists of latent metallic precipitates (M_p) grown into the ingot [65]. During high-temperature firing (>650 °C), the metal precipitates dissolve to become mobile, probably interstitial, metal atoms (M_i), and do not (or only partly) dissolve at lower firing temperatures. The rapid cooling after firing causes the M_i atoms to weakly bond with homogeneously distributed impurities such as O_i, C_S, N_2 or H to form an M_i-X complex that is recombination inactive. More recently, Schmidt *et al.* narrowed the proposed latent complex to be a metal-hydrogen (M_i-H) complex (Figure 6.9) [100]. To explain observations that there is an initial fast-forming recombination centre, it is proposed that under illuminated annealing, the complex changes configuration to form a slightly more recombination active complex M_i-X* [77]. Sustained illuminated annealing causes the M_i-X* complex to dissociate into its constituents thereby releasing the recombination active metallic impurity M_i which causes the degradation. Over time, the mobile interstitial diffuses to the wafer surface or other sinks such as crystallographic defects to enable the LeTID recovery.

In a subsequent experiment, Bredemeier *et al.* studied LeTID in wafers of different thickness and concluded that the regeneration can be explained by diffusion of the recombination active defect towards the surface [102]. Based on this, the range of diffusion coefficient for the defect would be $D = (5 \pm 2) \times 10^{-11}$ cm²/s, which matches closely with Co and nickel (Ni). However, it also closely matches that of hydrogen, and so it was also proposed that, alternatively, the recovery could be a result of hydrogen in-diffusing from the surface and a passivating the defect. Ni-rich precipitates were observed in a study by Jensen *et al.* [98], and again by Deniz *et al.,* specifically located in dislocations or stacking faults in the latter [57]. However, these localised observations of Ni cannot explain the mostly homogeneous nature of degradation observed within mc-Si grains [80] and in fact, denuded (or defect-free) zones of approximately 100 μm have been observed around mc-Si grain boundaries [46, 80, 103]. This rules out any slow diffusing species (such as $W_s^{0/+}$, $Mo_i^{0/+}$ and $Ti_i^{+/++}$) since they could not move far enough during the thermal processes of solar cell manufacture [98]. In 2018, Wagner *et al.* conducted a systematic study along the length of a Cz ingot to correlate the severity of LeTID with various detected impurities including chromium, iron, copper, aluminium, nickel, titanium, substitutional carbon and interstitial oxygen (Cr, Fe, Cu, Al, Ni, Ti, C_S and O_i) [12]. A linear correlation was found with the concentration of aluminium and a quadratic correlation with the O_i concentration above a certain amount ($O_i \cong 1 \times 10^{18}$/cm³).

A range of techniques has been used with reported observations of fast-diffusing metallic impurities in LeTID affected samples. Luka *et al.* and Jensen *et al.* used elemental micro-analysis techniques (energy-dispersive X-ray spectroscopy and micro-X-ray fluorescence, respectively) and found copper precipitates in the grain boundaries of samples susceptible to LeTID [98, 104]. It is well known that copper can cause light-induced degradation (Cu-LID) in contaminated p-type silicon [45, 105, 106], primarily due to clustering and precipitation of copper silicide (Cu_3Si) [107]. Yli-koski *et al.* expanded a copper precipitation model as a way of explaining the effects of a dark annealing process applied before LeTID light soaking. Consistent with a potential defect related to metal precipitation and dissolution, the study found that anneals at higher temperatures (300 °C) for several hours reduced the extent of LeTID, while cooler (<250 °C) or shorter (0.5 hours) anneals caused much stronger LeTID [108].

However, confirmation that LeTID can also affect n-type silicon may rule out the possibility that copper is the cause of LeTID as discussed in [83] by Chen *et al.* Even though Cu-LID has been seen in n-type silicon, only very lightly doped wafers (>5 kΩ) seem to be affected [107]. In n-type silicon, lower concentrations are required for initiation of precipitate nucleation due to the Fermi-level being closer to the precipitate's electron neutrality level (~E_C − 0.2 eV) [105]. As a result, not only would the copper contamination have more severe effects on the carrier lifetime for identical concentrations in n-type, but the effects of precipitation would be seen immediately after in-diffusion of the copper rather than after illumination or excess carrier injection [105, 109].

One final behaviour that indicates the potential of a metallic defect is that impurity gettering procedures can impact the severity of LeTID. This was first identified by Zuschlag *et al.* [110] and again later by others [111, 112]. Zuschlag *et al.* found that samples that underwent phosphorus diffusion gettering (PDG) and emitter removal, prior to passivation and firing showed reduced LeTID, thereby indicating the removal of a defect constituent or precursor during gettering and thermal processes. A study by Chakraborty *et al.* claimed to also observe reduced LeTID on wafers and cells after gettering [112]. However, the lifetime test structures were not fired, which is a now known requirement for observing LeTID, and the differences in the cells' emitter profiles with and without gettering make it challenging to draw conclusions about the impact of gettering on the LeTID extent [112]. In addition, studies by Sperber *et al.* and Chen *et al.* have contradicted this role of gettering [83, 113]. Sperber *et al.* found very similar degradation in both non-gettered and gettered samples using a 840 °C, approximately 45 min $POCl_3$ diffusion on p-type FZ-Si wafers, thereby concluding that phosphorus gettering does not significantly affect LeTID defects in FZ-Si [113]. Chen *et al.* found that LeTID only appeared in n-type Cz-Si wafers if diffused layers were present on the surfaces during firing, irrespective of whether the diffusion was removed after firing, prior to degradation. LeTID did not occur on samples without a diffusion present during firing. This work concluded that during firing the diffused layer acted as a catalyst and was required to introduce LeTID precursors into the wafer [83]. This directly contradicts the other findings that gettering helps suppress LeTID degradation.

Despite the conflicting evidence on the role the metallic impurities might play in LeTID, there is little doubt that these impurities can act as a storage for hydrogen precursors as expanded upon in Section 6.2.4. Although not directly involved as a recombination active complex, these impurities can certainly supply hydrogen to the subsequent formation of LeTID upon dissociation. Further work is required to decipher whether any of these M_i-H complexes are in fact recombination active and responsible for LeTID.

6.2.3.2 Hydrogen: a growing consensus

Many findings have led to speculation of the involvement of hydrogen, from the firing temperature dependence to the requirement for hydrogenated dielectric layers to be present during firing. However, there have been many different theories as to the exact role hydrogen plays in LeTID. Both Eberle *et al.* and Chan *et al.* hypothesised that hydrogen could activate the defect or even be part of the defect itself [66, 74]. The increased degradation after higher temperature firing would then be due to a higher defect concentration as a result of more hydrogen being released into the silicon from the hydrogenated dielectric. Chan *et al.* proposed that hydrogen may both cause the degradation and then also enable regeneration whereby one hydrogen atom activates the defect, and a second hydrogen atom subsequently passivates the defect [74]. Nakayashiki *et al.* proposed an alternative theory whereby illumination causes a charge state change and dissociation of a previously passivated defect, freeing the defect to a recombination active form. In this theory, dark annealing shifts the quasi-Fermi levels back towards the equilibrium values so that charge states revert back and the defect is re-passivated by the hydrogen [62]. This theory is similar to that discussed earlier by Bredemeier *et al.* whereby a metallic impurity is initially passivated by hydrogen after firing [65]. However, this model cannot explain that the defect has also been observed to both form and regenerate under the same dark conditions [43]. Alternatively, a model of slow hydrogen out-diffusion from the silicon bulk is one that can explain the defect formation followed by a slow recovery after prolonged durations of annealing.

6.2.4 The role of hydrogen in LeTID

6.2.4.1 The impact of hydrogenated dielectric films

A number of studies have investigated the correlation between hydrogen within or diffusing from hydrogen-containing dielectrics and the LeTID severity [71, 96, 97, 114, 115]. Vargas *et al.* used Fourier-transform infrared spectroscopy (FTIR) to measure the fractional bond densities of nitrogen–hydrogen (N–H) and silicon–hydrogen (Si–H) in various SiN_x:H layers before and after firing with different peak temperatures. The diminished bond densities after firing indicate the hydrogen fraction released, which was compared against the extent of LeTID (maximum normalised defect density (NDD)) and found a correlation, as shown in Figure 6.10 (top) [96]. Varshney *et al.* used a similar idea, assuming larger amounts of hydrogen would be released from thicker layers (i.e. larger H source volume), this work compared varied SiN_x:H layer thicknesses and also found a correlation of increased LeTID defect with increased layer thicknesses, as shown in Figure 6.10 (bottom) [116]. Bredemeier *et al.* confirmed the correlation between SiN_x:H thickness and LeTID extent, however, unlike the previous work, in this work the LeTID extent was found to saturate for layers >105 nm thick [114].

The assumptions in these works were subsequently disputed by Walter *et al.* [117] and Bredemeier *et al.* [114]. Walter *et al.* calculated the total hydrogen introduced to the silicon during firing based on bulk boron–hydrogen (B–H) pair densities and noticed that higher hydrogen content layers did not necessarily lead to greater hydrogen in-diffusion to the bulk [117]. This suggested that it may be other properties of the SiN_x:H layers that determine how much hydrogen diffuses into the silicon [117]. Bredemeier *et al.* verified this suggestion, finding that in addition to the hydrogen concentration, the atomic density of the SiN_x:H layer has a large impact on the hydrogen incorporated into the silicon and the extent of LeTID [114]. Specifically, nitrogen-rich films which have a lower refractive index (RI) (<2.2 at 633 nm) are denser and limit the atomic hydrogen diffusing into the silicon. Conversely, silicon-rich dielectrics have a more open structure that promotes

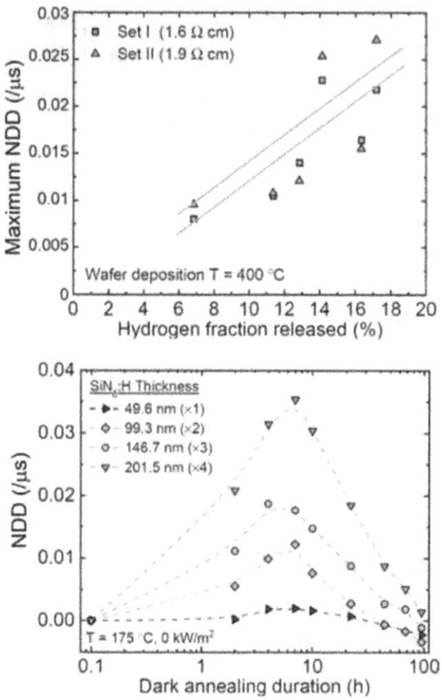

Figure 6.10 *(top) LeTID degradation extent (maximum NDD) plotted against the hydrogen fraction released from SiN_x:H layers on two sets of mc-Si wafers. © 2019 IEEE. Reprinted, with permission, from [96]. (bottom) NDD with degradation time for mc-Si samples with varied SiN_x:H thicknesses. © 2019 IEEE. Reprinted, with permission, from [116].*

movement and release of hydrogen. However, in layers with lower atomic densities (RI > 2.4), the hydrogen atoms tend to form dimers that restrict the in-diffusion into the silicon, again causing reduced hydrogen concentrations in the silicon bulk [114].

6.2.4.2 Direct correlation between hydrogen and LeTID

Most studies that have investigated a correlation between hydrogen and the extent of LeTID have done so in one of two ways: (1) manipulate the hydrogenated SiN_x:H dielectric layers or (2) modify the firing profiles or thermal treatments which affect hydrogen diffusion. More recently, more direct methods have demonstrated hydrogen's role in causing LeTID [95, 100, 118]. Jensen *et al.* used microwave-induced remote hydrogen plasma to introduce hydrogen directly into some p-type mc-Si wafers and heated others for identical durations without the hydrogen plasma; only the samples exposed to the hydrogen showed LeTID [118]. This confirms that it is the hydrogen rather than the thermal process that activates the defect, and the thermal process (typically firing) is only required for the introduction of hydrogen into the silicon, rather than direct defect activation. Bourret-Sicotte *et al.* observed an LeTID related defect on p-type FZ-Si wafers after a shielded hydrogen plasma treatment [119]. These results help to explain LeTID observations in wafers that had dielectric layers removed after firing, as the hydrogen would have already infiltrated the silicon [120].

Interestingly, Sperber *et al.* demonstrated a unique case of LeTID forming without firing in p-type FZ-Si [95]. Degradation occurred directly after dielectric deposition with similar severity to wafers that were fired at 850 °C. It was suggested that the plasma-enhanced chemical vapour deposition (PECVD) step at 400 °C may have allowed enough hydrogen to enter the silicon and cause LeTID. However, it is worth noting that deposition temperatures of approximately 400 °C are common for PECVD, and other studies have not seen the same LeTID formation without firing. It is possible the difference in [95] could be due to the use of direct PECVD (Oxford Instruments Plasmalab 100) rather than remote PECVD. Layers deposited in this way can contain up to 10 times more hydrogen due to a chemical difference in how silane is used that results in larger concentrations of smaller hydrogen-containing molecules (e.g. SiH_2, SiH_3) [121]. In addition, direct exposure to the hydrogen-containing plasma is well known for its ability to introduce hydrogen into silicon, typically for defect passivation [122].

It is challenging to directly quantify low concentrations of hydrogen in silicon due to the detection limits of characterisation tools. Walter *et al.* [117] proposed quantifying the hydrogen concentration by the change in silicon resistivity since hydrogen neutralises dopant atoms (e.g. forming B–H pairs in p-type) [123, 124]. Using this method, work by Schmidt *et al.* was one of the first to be able to demonstrate a direct correlation between the hydrogen concentration in the silicon and LeTID defect concentration [100]. In this work, FZ-Si wafers for hydrogen content measurements were processed in parallel with mc-Si wafers for defect concentration. Samples were coated with SiN_x:H layers with varied Si:N ratios, which led to different hydrogen concentrations in the silicon bulk after firing. The maximum defect concentration on mc-Si wafers was compared with the bulk hydrogen concentration in the FZ-Si wafers (measured after H_2 dimer dissociation) and can be seen in Figure 6.11.

6.2.4.3 Analysis using DLTS

To date, there is not enough conclusive evidence to rule out any involvement of metallic impurities, other non-metallic impurities or a theory of hydrogen alone. In fact, very little is understood about the exact defect composition or configuration. Some insight into the defect and its characteristics have been provided by DLTS measurements by Chunlan *et al.* and Mchedlidze *et al.* [60, 125]. Chunlan *et al.* measured degraded p-type mc-Si PERC cells and identified two deep-level traps with energy levels $E_V + 0.485$ eV and $E_V + 0.295$ eV, associated with interstitial iron and iron–hydrogen (Fe–H) pairs, respectively. Due to the metastability of Fe_i in silicon, and activation energies extracted from DLTS spectra overlapping, the inclusion of other TMs could not be excluded. Similarly, Mchedlidze *et al.* also measured p-type mc-Si PERC solar cells and identified two minority carrier traps at

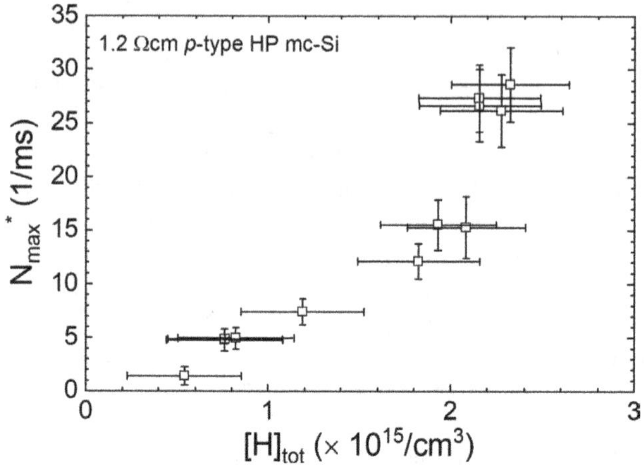

Figure 6.11 *Maximum effective defect concentrations, N_{max}^* on mc-Si sister wafers, plotted against total the hydrogen content $[H]_{tot}$ in corresponding boron-doped FZ-Si wafers processed in parallel; both surfaces were coated with SiN_x:H films with different Si:N ratios before firing at 900 °C. © 2019 IEEE. Reprinted, with permission, from [100].*

energy levels $E_C - 0.19$ eV and $E_C - 0.34$ eV. Although the exact species could not be determined, this work concluded that the defects that caused the degradation likely formed in the bulk of the silicon near structural defects and had an intrinsic origin.

6.2.5 LeTID mitigation

Although the defect is not yet fully understood, and the constituents not yet confirmed (aside from a growing consensus that hydrogen plays a critical role), enough is known about LeTID behaviour to be able to manipulate and treat it. As the PV industry is rapidly shifting to PERC, finding commercial solutions to suppress or mitigate LeTID has been of critical importance. Many techniques have been suggested and studied in the literature, through changing steps in the solar cell manufacturing process or through an additional post-processing treatment on finished solar cells. Here, they will be described in the order that they could be applied as part of the cell manufacturing process.

Before processing even begins, characteristics and choice of the silicon wafer can have an impact on LeTID [62]. The degradation extent has been shown to reduce with wafer thickness [102] and as a result, use of thinner wafers is one possible way to reduce LeTID. However, reducing the wafer thickness could lead to yield and handling issues, particularly for screen-printed cells, which would need to be considered. Wafers from different ingots, and even within the same ingot [41, 53] have shown different LeTID behaviour, and so it could be possible to select wafers with lower susceptibility to LeTID. However, in a study of various p-type mc-Si wafers, Hanwha Q Cells noted that there were 'no LeTID free wafers found on the market'

[53]. Only using parts of ingots is likely to be prohibitively expensive in terms of silicon wastage as well as requirements for wafer-level screening of LeTID susceptibility (if such a test was even available) and is therefore not an ideal solution.

Since it has been reported that a phosphorus gettering process can reduce LeTID [110, 111], it is possible that an optimised diffusion gettering could suppress LeTID. Similarly, the properties of the dielectric layer could be modified as the composition [93, 96, 100], thickness [102, 126] and stoichiometry [114] have been shown to significantly impact LeTID, most likely due to changes in the hydrogen that diffuses into the silicon. For example, SiN_x:H with RI < 1.9 or > 3 has been shown to reduce LeTID [114]. However, changing the SiN_x:H properties will have other impacts on the optical and electrical performance of the solar cell. Silicon-rich (high RI) SiN_x:H layers can have better surface passivation but tend to absorb some of the short and medium wavelength light causing parasitic losses [127]. In addition, without also modifying the co-firing process, changing the dielectric layer hardness would affect the penetration depth of the fired Ag fingers, causing poor ohmic contact or shunting. Changing the dielectric thickness would have a similar effect on the contacts, and also impact the anti-reflection properties.

The most commonly used method for studying and modifying the behaviour of LeTID is by modification of the contact firing profile, also likely due to the way it changes the hydrogen diffusion into the silicon. Nakayashiki *et al.*, Bredemeier *et al.* and Chan *et al.* all showed that degradation was reduced by firing the cells or wafers at lower peak firing temperatures [49, 62, 65]. Nakayashiki *et al.* tested contact firing solar cells at 960 °C and 860 °C and assembled a module of each type; the higher temperature firing lost 8% of its power output after 50 days outdoors while the lower temperature firing lost only 3%. Reducing the firing temperature beyond this is not likely to be compatible with existing screen-print pastes. Fortunately, Eberle *et al.* have shown that LeTID can be suppressed in lifetime samples fired at high peak temperatures if a slow cooling rate was used. However, this method would again require careful optimisation in coordination with metallisation to ensure ohmic contact formation.

Several annealing treatments have been developed that can be applied post-processing to suppress LeTID, using illumination [49, 50, 54, 56], current-injection [39, 50] or in the darkness [74, 108, 128, 129]. The various treatments indicate that there are two likely paths to LeTID mitigation: injection of carriers at elevated temperatures, or a purely thermal effect similar to that as for B-O LID as discussed in Chapter 5. In a similar way to how laser treatments have been used to accelerate and solve B-O LID (see Chapter 5), both Payne *et al.* and Krauß *et al.* demonstrated laser-illuminated anneals to accelerate the degradation and regeneration of LeTID [130–132]. Also similar to B-O LID, LeTID is a carrier-induced defect, so an equivalent process in the dark with current-injection could be applied to the cell. However, such a process would be much slower due to the lower carrier densities compared with what can be achieved through high-intensity illumination.

'Re-firing' the cells in the belt furnace but at a lower temperature than contact firing has been shown to suppress LeTID, although without careful optimisation can affect the contacts and device fill factor (FF) [49, 56]. LeTID has also been able to be

Table 6.3 Summary of various suppression or mitigation strategies at a wafer or cell level as highlighted throughout the literature

Source/Ref.	Treatment
Zuschlag *et al.* [110]	Use of PDG to suppress LeTID
Skorka *et al.* [111, 134]	Use of PDG to suppress LeTID
Chakraborty *et al.* [112]	Use of PDG to suppress LeTID
Eberle *et al.* [66]	Reduction of the cooling rate during firing using a rapid thermal processing (RTP) furnace
Chan *et al.* [49]	Secondary post-metallisation firing process (480–660 °C) + laser processing (200 °C, >40 suns)
Sen *et al.* [84]	Reduction of the cooling rate during firing using an RTP furnace
Bredemeier *et al.* [135]	Post-firing anneals in a commercial in-line furnace-based tool (Centrotherm c.REG)
Sen *et al.* [133]	Low-temperature annealing (~650 °C) prior to metallisation firing to suppress the issues associated with increased FF
Sen *et al.* [56]	Application of a two-step moderate temperature annealing process post-metallisation firing in a belt furnace
Sharma *et al.* [128]	Optimisation of firing conditions with reductions in peak firing temperature and modification of belt speed. Use of post-metallisation dark annealing at moderate temperatures (300–550 °C)
Sharma *et al.* [136]	Incorporation of post-firing forming gas annealing
Yli-koski *et al.* [137]	Application of a low-temperature (200–300 °C) long duration (>18.5 h) anneal to suppress subsequent LeTID formation
Varshney *et al.* [116]	Reductions in the thickness of SiN_x films to reduce in-diffused hydrogen content
Varshney *et al.* [138]	Implementation of ALD deposited Al_2O_3 films as a blocking layer for hydrogen in-diffusion during firing
Bredemeier *et al.* [114]	Tuning of SiN_x films to high (towards 3) or low (towards 1.9) RIs to reduce hydrogen mobility or density, respectively
Bredemeier *et al.* [102]	Reducing the thickness of the silicon wafer resulting a supposed faster diffusion of impurities to the surface
Payne *et al.* [42, 50]	High-intensity laser (44.8 kW/m^2) and high-temperature process (140 °C) to accelerate defect formation and recovery rates
Wang *et al.* [139]	Use of single- or double-current injection annealing processes (260 °C, 14.5 A) on finished solar cells

suppressed through annealing in the dark [74, 108, 128]. However, similar to the 're-firing' approach, longer duration or higher temperature dark annealing can harm the FF [129]. To prevent this issue, it could be possible to perform a dark anneal prior to contact firing as demonstrated by Sen *et al.* on lifetime samples [133]; however, any changes in SiN_x density would require re-optimisation of the co-firing process. Alternatively, Hamer *et al.* found that dark annealing the cell with a reverse bias (RB) voltage applied would prevent the FF degradation [129]. It is hypothesised that hydrogen at the metal/silicon contact interface causes the FF drop, and the reverse biased electric field repels the hydrogen from the interface. A summary of the various existing treatments for LeTID is presented in Table 6.3.

Most of the above methods aim to reduce the amount of hydrogen that diffuses into and remains in the bulk of the silicon. However, it is important to note that this could have other detrimental effects on device performance. As discussed throughout Chapters 2 to 5, hydrogen is essential for passivating surface and bulk defects, especially structural defects such as grain boundaries in mc-Si wafers and B-O defects responsible for LID in p-type Cz wafers. The B-O defect regeneration rate and effectiveness of LID mitigation are dependent on the hydrogen concentration released from the dielectric (SiN_x:H) layers [140, 141]. Varshney *et al.* showed that a reduced SiN_x:H thickness, or the inclusion of an atomic layer deposited (ALD) Al_2O_3 barrier layer under the SiN_x:H, suppressed LeTID in mc-Si wafers but also reduced the B-O regeneration rate in Cz-Si wafers [97]. However, this complication may be avoided through the transition from boron-doped p-type wafers towards gallium doping. Reducing the bulk hydrogen concentration is therefore likely to impact the effectiveness of commercial hydrogenation processes used for preventing other types of LID or passivating other defects. When using such hydrogen control techniques, it is crucial to optimise and find a balance to ensure high enough hydrogen concentrations exist for passivation while removing any excess that may contribute to LeTID.

There are now several commercial LeTID treatment tools in the market. These include both dark annealing with current-injection (CI) tools, and illuminated annealing tools based on lasers light-emitting diodes, or modified belt furnaces [126]. The former has been widely adopted by industry due to the ability to batch coin stack, thereby enabling high throughput. Since 2017, there have been several examples demonstrating LeTID stable modules in commercial environments. One of the earlier developments by Hanwha Q Cells, Q.ANTUM technology, has demonstrated both LeTID and LID stable Cz-Si and mc-Si modules with degradation reduced to less than $1\%_{rel}$ (see Figure 6.12). Additionally, several large players within the industry have also reported LeTID reduction to below 1% relative to others receiving 'anti-LeTID' certification.

6.2.6 *Models for LeTID*

Early on, Sperber *et al.* proposed a three-state defect model with transitions to explain the observed degradation and recovery behaviour (see Figure 6.13) [81]. This preliminary model resembles that commonly used to represent the B-O defect system (discussed in Chapter 5), as initial observations of LeTID showed similar behaviour. In this model, under continued exposure to LeTID inducing conditions, the defect transitions from recombination inactive precursor state 'A', into a degraded active defect state 'B' before transitioning to a stable recovered state 'C'. The experiments in this study showed that dark anneals cause an unstable lifetime recovery, possibly indicating a reversible reaction from defect state B back to inactive state A, similar to B-O dissociation. Also similar to the B-O defect system, later works observed a fast and slow two-step degradation [142]. However, the simplified single transition from A to B in this model is widely used since the fast-forming component is often negligible or not observable under typically accelerated testing conditions [143].

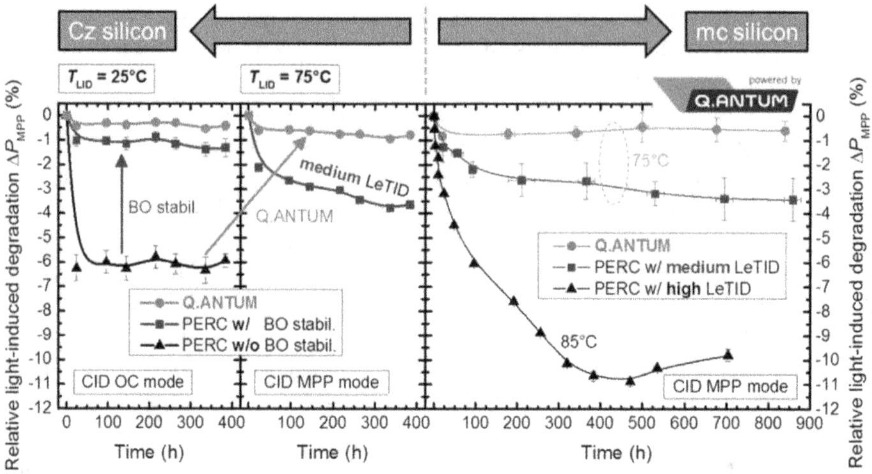

Figure 6.12 *Relative degradation in maximum power output of Cz-Si and mc-Si undergoing LeTID and B-O LID stability testing. Reprinted from [73] Copyright (2017), with permission from the Authors.*

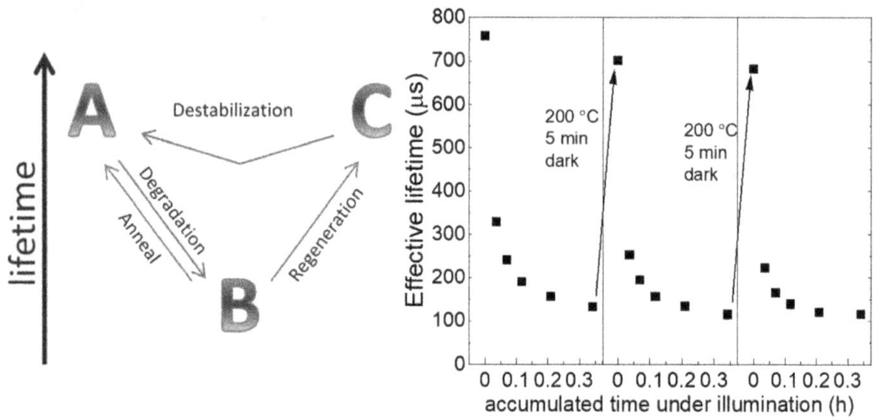

Figure 6.13 *(left) Proposed three-state LeTID defect model based on observations in boron-doped FZ-Si [81]; (right) Effective lifetime during LeTID testing at 80 °C under approximately 1-sun illumination (a) on a fresh sample, (b) after LeTID testing and 5 min, 200 °C dark anneal, (c) after further LeTID testing and another 5 min, 200 °C dark anneal. Reproduced from [81] with permission from John Wiley and Sons. © 2016 WILEY-VCH Verlag GmbH &Co. KGaA, Weinheim.*

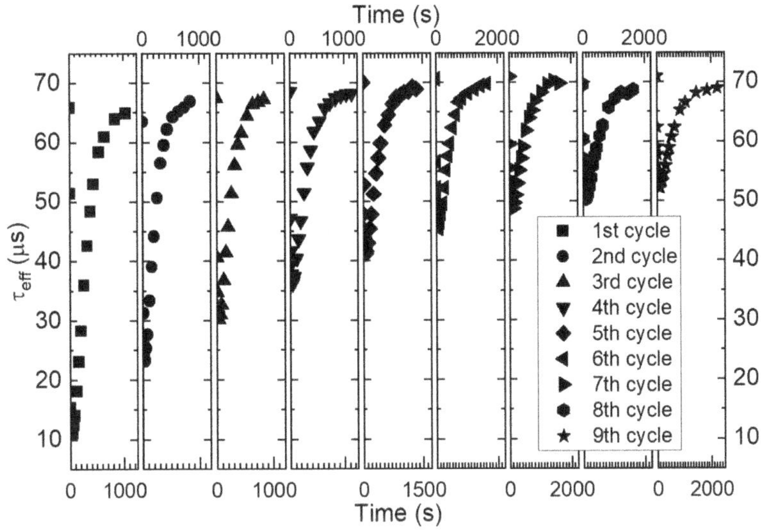

Figure 6.14 *Effective lifetime measured during successive LeTID degradation and regeneration cycles at 135 °C under 25.7 kW/m² illumination (squares) and after 8 min 232 °C dark anneals (triangles) preceding each degradation cycle. Reprinted from [82] Copyright (2018) with permission from Elsevier.*

Degradation and recovery of LeTID is now known to differ from B-O in multiple ways. One, in particular, is that the degradation and recovery processes occur simultaneously unlike B-O defect formation which can occur independently [144]. Further experiments by Sperber *et al.* demonstrated that dark anneals could also destabilise the defect in the stable recovered state C [81]. The carrier lifetime did not change due to dark annealing so it was postulated that a direct transition from state C to A occurs; however, a transition through state B could not be ruled out. If the latter, this would be similar to the B-O defect system; however, in this case the reduced subsequent degradation indicates only a partial destabilisation.

The three-state model was later expanded by Fung *et al.* who found that the impact of dark annealing was not repeatable and could therefore not be explained by a destabilisation [82]. In this work, a sample was LeTID tested and subjected to an 8 min, 232 °C dark anneal each time it recovered, initiating another degradation and recovery cycle. However, as shown in Figure 6.14, each successive cycle showed a reduced degradation extent despite identical dark annealing conditions. The authors concluded that rather than destabilising state C back to state A, the dark anneals introduce new defect precursors into state A. To account for this, they introduced to the model a recombination inactive fourth 'reservoir' state R to feed into state A (see Figure 6.15).

The reducing extent of degradation with each successive cycle could be explained by the gradual depletion of a reservoir with a fixed volume after firing. However, as

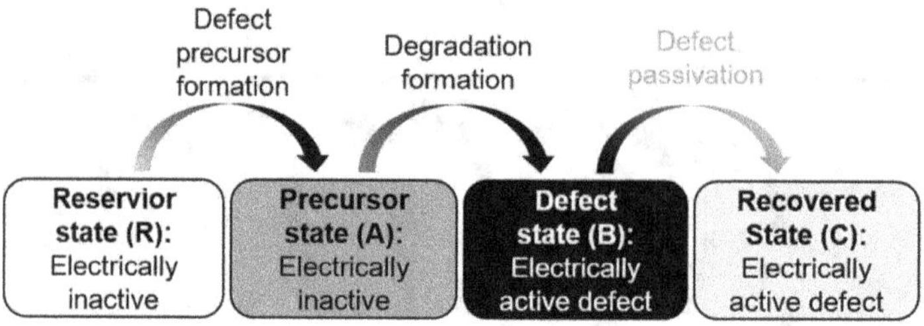

Figure 6.15 A four-state model for LeTID. This model was based on the three-state B-O LID model by Herguth et al. *[145]. State R represents a reservoir of precursors that feed into state A. Adapted from [82].*

the reduction in degradation is less with each cycle, it appears that the amount of degradation after successive dark anneals converges at a non-zero amount. It is possible that this limit is due to some reversal from state B or C back to A. Fung *et al.* conducted further investigation to understand the proposed reservoir state [143]. Fung *et al.* used varied temperature dark anneals to show a correlation between LeTID and both B-O regeneration and boron–hydrogen (B–H) pairs (see Figure 6.16 (left)). In the case of the former, Fung *et al.* annealed sister sets of p-type mc-Si and Cz-Si

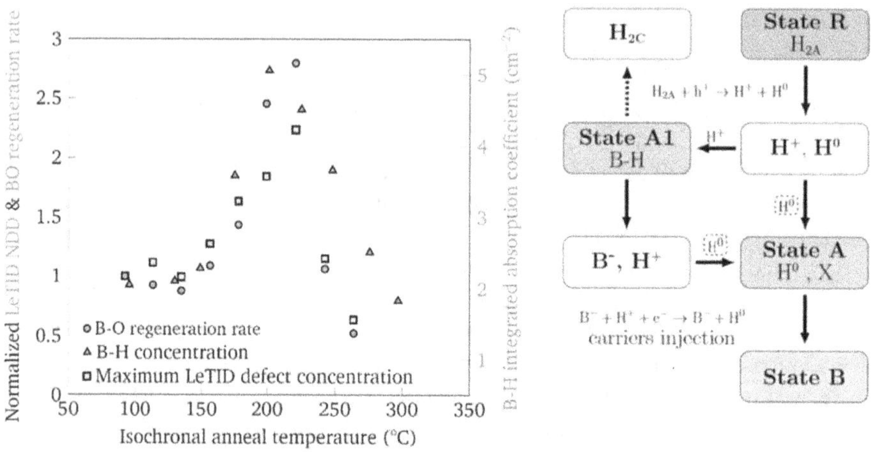

Figure 6.16 (left) NDD of LeTID and B-O regeneration rate following different temperature dark anneals, with the B-H-integrated absorption coefficient during dark anneal extracted from McQuaid et al. *[146]. (right) A proposed LeTID defect model including a reservoir state R (H_2A molecules), two precursor states A1 (B–H pairs) and unspecified A, defect state B and a dark-annealed state of H_2C molecules, with hydrogen transitioning through different interstitial charge states. Reprinted from [143] with the permission of AIP Publishing. © 2018 Author(s).*

wafers and found the same trends and maxima for the subsequent LeTID extent in mc-Si and rate of B-O passivation in Cz-Si, both of which are dependent on hydrogen. The latter is based on the work by Voronkov *et al.* [147] and Wilking *et al.* [148] that showed –H pair formation after annealing processes and correlated with work by McQuaid *et al.* [146], who used FTIR, to show B–H pair concentrations after different annealing temperatures, again showing the same trends. Fung *et al.* suggested that B–H pairs could therefore play a role in both regeneration of B-O defects and formation of LeTID.

In further developing a model to understand LeTID, Fung *et al.* also considered the hydrogen behaviour expected due to thermal processes according to literature to develop a model shown in Figure 6.16 (right) [143]. Voronkov *et al.* note that rapid cooling after firing could freeze in hydrogen dimers (H_{2A}) in concentrations above the equilibrium concentration [147]. These molecules can interact with holes and dissociate during annealing [149]), freeing up interstitial hydrogen that is attracted to the boron dopants to form B–H pairs. At lower temperatures, an increase in annealing temperature increases the transition to and completeness of B–H pair formation; however, as temperatures continue to increase, the reverse debonding processes increase causing a net decrease in B–H pairs [150]. Carrier injection can also break B-H bonds, having a similar effect at lower temperatures. This behaviour has been observed in Cz, as Wilking *et al.* found the B–H pair concentration and interstitial hydrogen concentration that forms under illumination correlate directly with the passivation rate of B-O defects, indicating that the hydrogen released from B–H pairs is freed to then passivate the B-O defects [140]. In mc-Si, the interstitial H released may cause the formation of LeTID.

The hypothesis of Voronkov *et al.* that hydrogen molecules (H_{2A}) are dominant after firing could explain the reservoir state of LeTID defect precursors that deplete over time [147]. In the context of LeTID, Fung *et al.* proposed that the reservoir of H_{2A} molecules interacts with holes and dissociates into interstitial H^0 and H^+ atoms [143]. The H^+ atoms are free to neutralise and bond to nearby boron atoms forming B–H pairs (precursor state A1), while the H^0 can form part of another inactive precursor state A. The B–H pair concentration that forms precursor state A1 is altered by the dark annealing processes thereby affecting the subsequent LeTID behaviour, as observed in works such as [74]. If true, hydrogen released from the LeTID defect could re-bond with boron, reverting back to the precursor state (B-H pairs) possibly explaining the reversibility observed in Figures 6.14 and 6.16. In this case, the level of reversibility (repeatable degradation after convergence of successive cycles) should correlate with the doping concentration as yet to be investigated. Increasing the temperature and illumination for LeTID testing will not only transition state A to active defect state B but also free additional B–H pairs to add more defect precursors to state A. In this model, state C was omitted, and an H_{2C} dimer state that is more stable than the H_{2A} reservoir state was included which forms above a certain concentration of B–H pair due to additional annealing. In any case, it is highly likely that both molecular hydrogen and that bound to impurities such as boron dopants contribute to LeTID and may vary between wafers.

6.2.7 LeTID in p-type silicon heterojunction solar cells

Silicon heterojunction (SHJ) solar cells, as discussed in Chapter 3, have gained interest in the photovoltaics community, due to the very high efficiency that this cell architecture can achieve [151]. The primary advantage of this approach, compared to the dominant PERC solar cell design, is the higher V_{OC}, enabled by the superior passivation of thin hydrogenated amorphous silicon (a-Si:H) layers [152]. Due to the excellent surface passivation, SHJ solar cells are extremely sensitive to recombination in the bulk. As such, SHJ solar cells are typically fabricated using high lifetime n-type Cz silicon wafers. These high lifetime n-type wafers are more expensive than p-type wafers used for PERC or Al-BSF solar cells. This high wafer cost could severely limit the commercial prospect of this technology. As such, it would be advantageous to utilise lower-cost p-type substrates to help unlock the potential of the excellent passivation scheme in SHJ solar cells. Multiple reports have shown the potential of p-type Cz [153] and multi-crystalline wafers [120, 154]. These reports suggest that these wafers require a pre-treatment hydrogenation process to passivate unwanted bulk defects in these low-quality materials before the SHJ solar cell is made. In light of recent observations showing that hydrogen can induce bulk degradation, it is important to understand the balance between hydrogen passivation and hydrogen-induced recombination in p-type SHJ solar cells. This idea is supported by Chen *et al.*, who showed that p-type multi-crystalline silicon wafers that have undergone a pre-hydrogenation treatment have lower than expected V_{OC} when fabricated into SHJ solar cells [120].

To investigate this phenomenon and to assess the impact of SHJ fabrication on the lifetime of p-type mc-Si wafers, Chen *et al.* [120] designed a comprehensive experiment that analysed the lifetime at each stage of the fabrication process. The results of this investigation are displayed in Figure 6.17 (top). First, a p-type mc-si wafer that had undergone a gettering and hydrogenation pre-treatment was measured with the SiN_x:H passivation after firing. This is shown as stage I in the figure. Subsequently, the SiN_x layer was removed with HF and the wafer surface was cleaned. Then, an intrinsic a-Si:H layer was deposited on both sides of the wafer at 250 °C using PECVD, shown as stage II. This led to a 40% reduction in the effective lifetime. Following this, the wafer was subjected to a thermal treatment at 200 °C to simulate the deposition of the a-Si:H(n) doped layer (stage III), the a-Si:H(p) doped layer (stage IV) and a hotplate anneal to simulate the metallisation paste curing (stage V). Following these thermal treatments, the average effective lifetime degraded by a maximum of 80%. This would have significant implications for SHJ solar cells fabricated using this material and can explain the significantly lower than expected V_{OC} for pre-hydrogenated p-type SHJ solar cells in Chen *et al.* [120]. To understand the nature of these changes, Figure 6.17 (bottom), displays the injection dependent lifetime for a single sample that underwent this process flow. In transitioning from the fired SiN_x film to the deposition of intrinsic a-Si:H (stage I and stage II), an increase in the lifetime at high injection levels is observed. This is representative of the better surface passivation quality in the a-Si:H films. This simultaneous improvement in surface passivation as the effective lifetime reduces

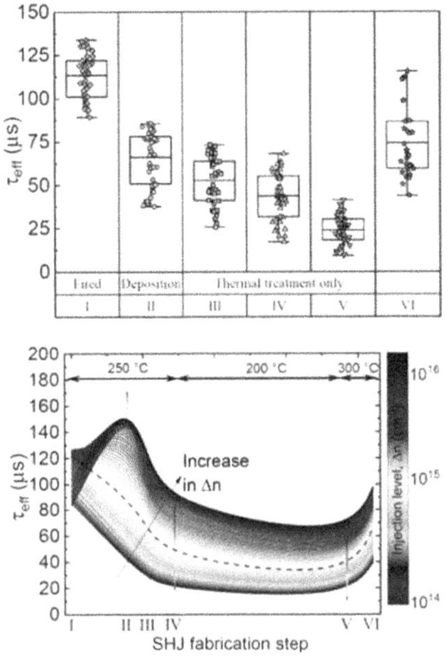

Figure 6.17 *(top) Effective lifetime (extracted at a minority carrier density of 1×10^{15} cm^{-3}) of a p-type mc-si wafer at various stages of the SHJ fabrication process. Stage I refers to the fired wafer with SiN_x surface passivation, stage II is after the intrinsic a-Si:H deposition and stages III–VI represent a simulated thermal treatment representing subsequent process in the SHJ fabrication sequence. Adapted from [120]. (bottom) Injection resolved lifetime of a single sample undergoing the same treatment.*

by 40% implies that a significant problem with bulk recombination has been induced. The reduction in lifetime at low injection levels, shown in Figure 6.17 (bottom) provides further evidence of this. PL images indicated that this reduction in lifetime is spatially uniform.

To investigate the nature of this process-induced bulk defect, IDLS was used to extract the electron-hole capture cross section ratio, k_{Defect}. This was achieved by analysing the change in the lifetime curve for one sample between the initial fired SiN_x state (stage I) and the point of maximum degradation (stage V). From this analysis (not shown), the value of the k_{Defect} for the unidentified bulk defect causing degradation was found to be 33 ± 4. This value closely matches the *k*-value of LeTID, as shown in previous studies [43, 155]. This provides conclusive evidence that LeTID can be induced during the SHJ fabrication procedure, and that p-type mc-si silicon wafers that have been treated with a pre-hydrogenation step are particularly susceptible to this. This observation also provides insights into previous failed attempts to produce high-efficiency SHJ solar cells using lower quality p-type wafers.

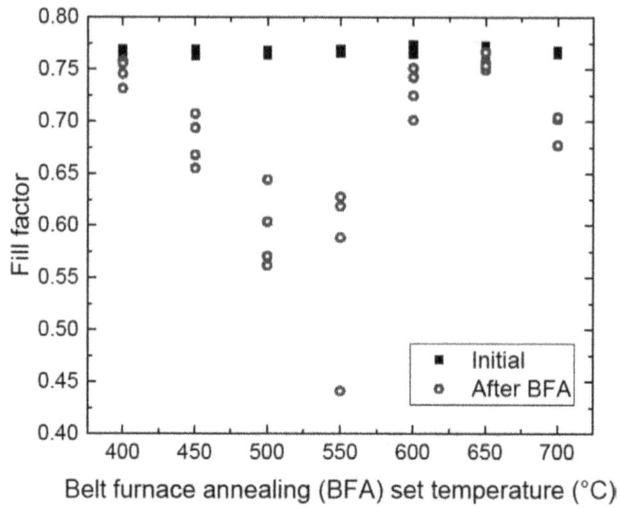

Figure 6.18 *FF of industrial silicon solar cells before and after a post-processing belt furnace anneal for set temperatures ranging from 400 °C to 700 °C. Reproduced from [157] with permission of John Wiley and Sons. © 2017 WILEY-VCH Verlag GmbH & Co. KGaA, Weinheim.*

6.3 Negative effects due to hydrogen behaviour

6.3.1 Hydrogen-induced contact resistance

As discussed in the previous section, one method to reduce the impact of LeTID in industrial silicon solar cells is the application of a thermal treatment after the metallisation firing step. The application of a belt furnace anneal (BFA), performed at a lower temperature than the co-firing metallisation step, has been shown to suppress the impact of CID [49]. However, CID is known to predominantly impact the V_{OC}. Although the V_{OC} has been stabilised, using this low temperature annealing process can result in an undesirable reduction in the cell FF. This effect is thought to be caused by an increase in the contact resistance on the front silver contacts. An initial hypothesis by Peral *et al.* suggested that the increased resistance was caused by a thickening of the glass layer at the silver/silicon interface. However, a large part of this change in contact resistance was shown to be unstable at room temperature [156], indicating that another mechanism may also be present.

Chan *et al.* studied the cause of this increased contact resistance in p-type mc-Si PERC solar cells [157]. In their study, they varied the temperature of a post-production BFA from 400 °C to 700 °C. Figure 6.18 shows the impact of the BFA on the FF of industrially sourced mc-Si PERC solar cells. The average FF before the treatment was 76.7% with a very tight distribution. The BFA had a large impact on the FF, and this impact was heavily temperature-dependent. The application of a

BFA with an intermediate temperature (450 – 550 °C) caused a significant reduction in FF. For example, the cells that underwent a BFA at 550 °C dropped to 56.9%, with a standard deviation of 8.7%. Comparison with suns-V_{OC} measurements indicated that the reduction in FF was related to an increase in series resistance (R_s), as no significant change in the pFF was seen after the BFA at intermediate temperatures. Interestingly, this change in the R_s characteristics induced by the BFA was not stable in the presence of an electric field. When a forward bias (FB) was applied to the cell in the dark, the R_s reduced. After the bias was removed, the R_s increased back to a very similar value. This instability suggests the changes may be related to mobile, negatively charged particles moving away from the surface during the bias, and slowly migrating back to the surface when the electric field was removed. In this case, the movement of hydrogen is a likely candidate for the changed contact resistance, as it is highly mobile and would be predominantly in the negatively charged state in the emitter. Transmission line measurements (TLMs) confirmed that the reduced FF was related to an increase in R_S and specifically identified as an increase in contact resistance at the screen-printed metal fingers. After higher temperature BFA, a less reversible increase would occur. It was hypothesised that this stable component was related to changes in the structure of the Ag–Si interface, as suggested by Peral *et al.* [156].

Hamer *et al.* extended this analysis by studying the increased contact resistance under different bias conditions [129]. In their study, they used a modified experimental setup to allow for in situ monitoring of the contact resistance under various bias conditions. Instead of performing a post-cell anneal in a firing furnace, a heated stage in the dark was used to treat the cells at 350 °C. All samples investigated in this report were p-type mc-Si PERC cells. Figure 6.19 (top) shows the TLM data for samples subjected to a RB current of 1 mA and FB of 1 mA, 5 mA, 25 mA and 50 mA. Figure 6.19 (bottom) shows the evolution of the measured resistance of a sample as the applied current was changed from 1 mA to 50 mA over time. This clearly displays the impact of the applied current on the contact resistance. Increasing the value leads to a significant drop in resistance, followed by a stabilisation. These results neatly match the observation of Chan *et al.* [157]. By systematically annealing the samples at either a reverse or FB of 0.2 V, the authors showed that this change in contact resistance was almost completely reversible by applying subsequent biased annealing. This process reversibility is a trademark of a hydrogen-induced change. To provide a clearer picture of the possible involvement of hydrogen movement for the increased resistance, the authors provide a model of the re-distribution of hydrogen according to the elevated temperature and applied bias explored in the experimental studies. This model provides a clear picture that the electric field would result in a redistribution of hydrogen at the surface and suggests that the change in contact resistance correlates with the redistribution of hydrogen in various charge states at the surface.

Hamer *et al.* performed further analysis to study the phenomenon in different materials and cell architectures [158]. First, to understand the influence of the bulk material, multi- and mono-PERC solar cells subjected to a FB of 0.2 V were compared. The rate of change of R_s in multi-crystalline solar cells was

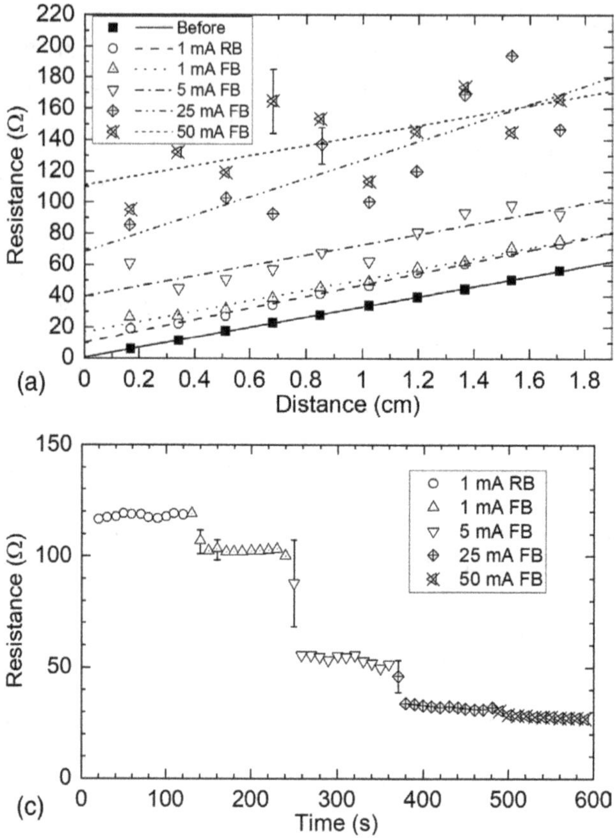

Figure 6.19 *(top) TLMs after firing (before) and after post-fire annealing at 353 °C for 2 hours after subsequent application of a variety of bias conditions. (bottom) Resistance measurements during the application of various biases. FB and RB stand for FB and RB, respectively. Reprinted from [129] Copyright (2018) with permission from Elsevier.*

significantly lower than for mono-PERC cells. It is expected that the change in R_s is related to the release of hydrogen from bound forms. The increased number of impurities and crystallographic defects in multi silicon leads to higher binding energy, thus it is more difficult to release hydrogen in this material. This idea aligns with the observation of slower induced changes in multi-PERC cells. Following this, PERC cells were compared with Al-BSF solar cells. Using an Arrhenius plot for annealing temperatures in the range from 350 °C to 450 °C, it was demonstrated that the rate constants for changes in resistance values were two orders of magnitude higher for PERC samples. This result aligns with the expectation that PERC cells will have a larger concentration of hydrogen in the bulk, as a result of replacing a hydrogen sink (Al-Si interface)

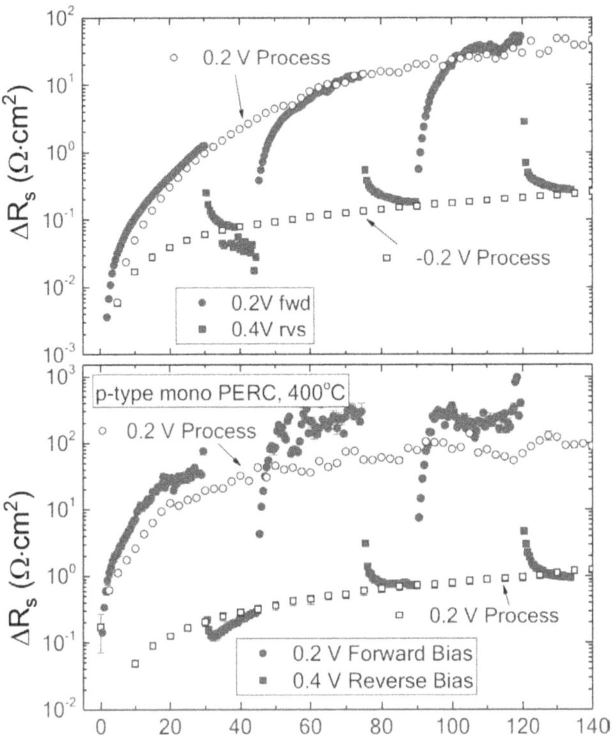

Figure 6.20 *Change in area normalised series resistance as the applied bias is switched from +0.2 V to –0.4 V during annealing at 400 °C for (A) multi PERC and (B) mono PERC solar cells. The open symbols represent samples that were annealed at a fixed bias of either +0.2 V (circle) or –0.2 V (square). © 2018 IEEE. Reprinted, with permission, from [158].*

with a second hydrogen source. Finally, the reversibility of the process for multi- and mono-PERC cells was compared. Figure 6.20 shows the evolution of change in R_s over time, performed at 400 °C, for both multi- (A) and mono- (B) PERC solar cells. The applied bias was switched between +0.2 V and –0.4 V. The open circles indicated the changes for a sample that was kept at either +0.2 V or –0.2 V. It is apparent that the process is highly reversible and that changing the bias can rapidly switch the state of the sample.

6.3.2 *Neutralisation of charge and dopants*

As discussed in Chapter 2, interstitial atomic hydrogen can act as a dopant. Whether it acts as an acceptor or donor is determined by the charge state, which is dependent on the position of the Fermi level [159]. In p-type silicon, the Fermi level is closer to the valence band and there is an excess of holes (deficiency of electrons),

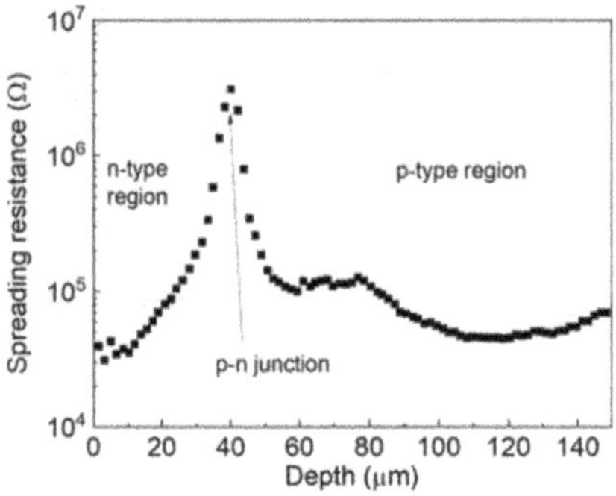

Figure 6.21 *A p-n junction formed on p-type Cz wafers using a 30-minute hydrogen plasma treatment at 450 °C. Reprinted from [160] Copyright (2000) with permission from Elsevier.*

so hydrogen readily loses its electron to become H^+, thereby acting as a donor and counter doping the p-type dopants. The H^+ can then bond to and directly neutralise the negatively charged p-type dopants which are drawn together by coulombic attraction. Conversely, in n-type silicon, when the Fermi level is closer to the conduction band (above the H(+/-) level), there is an excess of electrons, hydrogen readily takes an electron to become H^- thereby acting as an acceptor and counter doping the n-type dopants. The H^- can then also bond to and directly neutralise positively charged n-type dopants. Due to the positioning of the H(+/−) level above midgap (where hydrogen has an equal probability of being H+ or H−), when the Fermi-level is at midgap in intrinsic silicon, atomic hydrogen will still become almost entirely H^+. This means that in high enough concentrations, hydrogen can actually entirely neutralise the p-type dopants and continue acting as donors to turn the silicon n-type, thereby creating p-n junctions using hydrogen as an n-type dopant. At room temperature, hydrogen prefers to take on bound forms; however, once all dopants are neutralised, it only takes a small concentration of interstitial atomic (H^+) hydrogen to turn the silicon n-type. Job *et al.* created p-n junctions on p-type Cz wafers using a 30-minute hydrogen plasma treatment at 450 °C, as shown in Figure 6.21, although this was attributed to hydrogen enhanced formation of oxygen thermal donors [160]. Barakel and Martinuzzi, later created p-n junctions with a PV response using hydrogen ion implantation in p-type FZ silicon (i.e. low oxygen) due to the formation of hydrogen-related shallow donors [161]. The counter doping and neutralising nature of hydrogen is commonly capitalised upon for studying the movements of hydrogen in silicon; for example, Tavendale *et al.* studied hydrogen movement in phosphorus-doped n-type where over half of the respective background dopants were deactivated [162].

The ability for hydrogen to neutralise dopants has been suggested as a mechanism for easily forming selective emitters [163]. However, in general dopants in solar cells are specifically designed to maximise device performance and are required for electrical conductivity, so any counter doping can lead to negative unintended effects such as increased series resistance. In the same way that hydrogen can counter dope, by neutralising the charged dopants, hydrogen can also neutralise other charges. This could be particularly harmful for interfaces that rely on field effect or surface charge for passivation. Dopants or charge near the surface or metal contacts are intended to shield the active device from regions of high recombination; high concentrations of hydrogen could neutralise and reduce this barrier effect increasing the recombination of minority carriers at such high recombination interfaces. Such effects could potentially have an increasing impact on solar cell devices as hydrogen concentrations have been increasing and solar cells are being pushed to their performance limits.

6.3.3 Formation of hydrogen-induced platelets

When hydrogen is introduced to silicon in large concentrations the hydrogen can bond to silicon atoms and precipitate, forming ordered planar defects called platelets [164, 165]. There are two known fundamentally different types of platelets that have been commonly observed and studied. The first form primarily in n-type silicon when a high enough concentration of hydrogen exists in the absence of defects, such platelets are oriented along the {111} crystallographic plane [166, 167]. The second type form in either polarity silicon following ion implantation, when high concentrations of hydrogen coincide with the damaged regions of highest strain, and as a result form in planes perpendicular to the wafer surface , for example, in the [100] and [111] direction [168]; this type of platelet formation forms the basis for the Smart-cut wafer lift-off technique [169]. After it was identified in the 1980s that hydrogen could induce defects and was not always beneficial as previously thought, investigations into the mechanisms really took off [170, 171]. The fundamentals underlying the two different platelet types are now generally understood.

The first type, {111} platelets, is generally found within 0.1 μm of the surface and the density is proportional to the near-surface hydrogen concentration, which is much higher in n-type [171]. Such platelets are typically formed using a two-step process:

1. Nucleation – which occurs at lower temperatures and is suppressed above 250 °C [172]
2. Growth – occurs rapidly at higher temperatures (above 250 °C), but the platelets anneal out at temperatures higher than 400 °C [173].

They can be formed by any means of hydrogenation e.g. direct and remote plasma exposure, ion implantation or reactive ion etching. First-principles pseudopotential total energy calculations predict that in the perfect (undamaged) silicon lattice, the lowest energy configuration during the early stages of platelet formation when no lattice dilation exists arises when H_2^* complexes aggregate to form a double-layer $H_2^*[H_2^*]_N^D$, as depicted in Figure 6.22 (left) [165, 174]. The $W_s^{0/+}$ complex consists of a Si–H bond with the H sitting at the BC site, and another Si–H

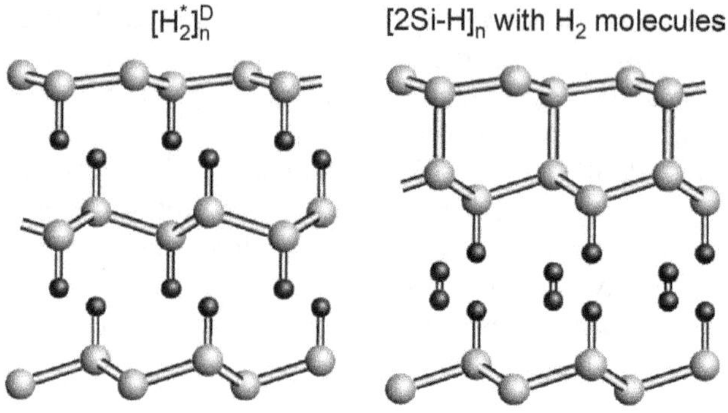

Figure 6.22 *(left)* $[H_2^*]^D n$ *preferred when no lattice dilation, early stages. Transforms into (right) 2Si-H with* H_2 *molecules. Reprinted with permission from [174], Copyright 2001 by the American Physical Society. https://doi.org/10.1103/PhysRevLett.86.1773.*

bond sitting at the antibonding interstitial site [175]. As the temperature increases and the $\left[H_2^*\right]_n^D$ structure grows laterally after nucleation, the inner surfaces repel each other thereby increasing the lattice dilation. As this happens, the $[2Si\text{-}H]_n$ structure with H_2 molecules as depicted in Figure 6.22 (right) becomes more energetically favourable [174]. In this latter configuration, hydrogen passivates the silicon dangling bonds; these bonds align along the {111} platelet axis [166] thereby passivating the adjacent planes of silicon. Si–H bonds have been detected commonly in platelets using Raman spectroscopy [166, 167, 171]. H_2 gas molecules, also detected by Raman spectroscopy, sit interstitially within and fill the platelet void [176]. Such platelets' characteristics are typically tens of nm in length, with a lattice dilation of 0.3 nm or 20–30% and contain a high density of hydrogen with pressures of the order of 1 GPa [177]. Correlating the platelet characteristics and density with secondary ion mass spectrometry (SIMS) reveals that platelet formation corresponds with the peak hydrogen concentration, most of the hydrogen in this region is included within the platelets [166], and the platelets trap hydrogen in concentrations far exceeding the level expected due to in-diffusion [178]. Interestingly, Nickel *et al.* observed that platelets do not form within 40 nm of grain boundaries [178]. Subsequently, the same team observed that platelet generation was highly Fermi-level dependent, requiring the background E_F to be within 0.3 eV of E_C, and as E_F approached E_C the platelet density increased monotonically [167]. Their observations led to the realisation that platelets nucleate (H_2^* complexes begin to form) at a specific E_F, where there are equal concentrations of H^+ and H^-. This occurs in heavily doped n-type silicon as E_F bends from close to E_C in the bulk towards midgap where it is pinned at the surface, crossing over the H(+/−) level, approximately E_C–E_F = −0.38 eV. Formation of the H_2^* complexes is dependent on the availability of H^- atoms, which are only abundant at $E_C - E_F \leq -0.38$ eV. This work also identified a minimum concentration of $10^{17}/cm^3$ hydrogen is required for platelet generation [167].

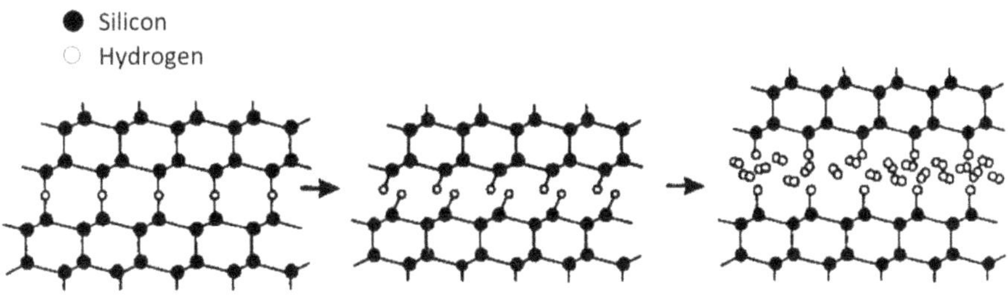

Figure 6.23 *{100} platelet formation and progression into H_2 gas bubbles. Reprinted from [179] with permission from AIP Publishing. https://doi.org/10.1063/1.1494844.*

The second platelet type is also formed through a two-stage process in this case:

1. Nucleation (typically ion implantation) – hydrogen atoms get trapped at and passivate dangling bonds (implantation-induced lattice defects), as identified by infrared spectroscopy
2. Growth – thermal annealing causes hydrogen atoms to rearrange within the damaged region forming H_2 gas bubbles and high pressure, observable as blisters [179].

These platelets have been shown to nucleate and grow, thereby also causing the wafer cleavage to occur at the depth of the maximum damage and strain, rather than the maximum hydrogen concentration [168, 180]. Figure 6.23 shows the progression of {100} platelet formation. Strain associated with the implantation damage reduces the energy needed for H located at the BC site (typical in p-type, as discussed in Chapter 2), to form Si–H–Si bond. During the anneal, single hydrogen atoms are replaced by two hydrogen atoms, resulting in two {100} hydrogen passivated silicon surfaces. This already reduces the fracture strength of the silicon, but when the platelet subsequently fills with H_2 gas bubbles it causes the blistering and wafer cleavage.

More recently, and more relevant to silicon solar cells, SiN_x:H deposited by PECVD have been shown to introduce a near-surface damage layer [181, 182]. Transmission electron microscopy images have identified platelets and bubbles, likely nucleated during the initial stages of the plasma deposition process when the silicon is bare [181]. These defects can trap and store hydrogen that can be subsequently released during solar cell firing processes for bulk passivation [181]. Such near-surface damage can cause poor carrier lifetimes at low carrier injection thereby harming solar cell efficiencies [182].

6.4 Summary

Previous chapters have discussed the power of hydrogen to passivate both the surfaces and bulk of silicon solar cells, however, the presence of hydrogen can also have unintended effects. This chapter highlighted the potential negative impacts that hydrogen can have on silicon and specifically on industrial solar cells. The behaviour of hydrogen and the multitude of defects that it can form are challenging to specifically characterise for a number of reasons: hydrogen is the smallest atom and difficult to detect directly; it is highly mobile in many forms, even at room temperature, and can be impacted by measurement techniques; it can take on multiple charge states and configurations; it is difficult to eliminate in 'control' samples; hydrogen can bond with almost anything to form complexes with such similar IR signals or energy levels they can be near impossible to differentiate. Some of the common hydrogen-containing complexes that can form in silicon were discussed including carbon complexes C-H, O-H and C-O-H, TMs and vacancy hydrogen complexes.

Of great importance to industrial solar cells is the relatively recently discovered degradation, LeTID, which can cause a degradation in performance of up to 16_{rel}, even more severe than B-O LID as discussed in Chapter 5. A key concern for LeTID is that it can affect not only p-type mc-Si but also Cz and FZ grown materials, including n-type wafers. There is mounting evidence to suggest that hydrogen is involved in the formation of LeTID in silicon solar cells. The defect could have a significant impact on cell performance for decades in the field. Although it is not entirely understood, commercial solutions exist to reduce or eliminate the impact of LeTID to below $1.5\%_{rel}$. Potential changes in the fabrication include modifications to the dielectric layers, changes in firing conditions and cooling profiles, or the addition of a process at the end of cell fabrication incorporating illuminated annealing or current injection.

Hydrogen can also have other negative impacts on cell performance, such as the observed increase in contact resistance during thermal treatments used to treat LeTID, neutralising dopants or charge required for conductivity or passivation, and the formation of hydrogen-induced platelets. There has been a general push for more hydrogen in industrial solar cells for improved passivation; however, this chapter highlights that more hydrogen is not always a good thing and must be managed accordingly. As silicon solar cells continue to increase in performance and be pushed to efficiency limits, they become more sensitive and prone to defects, managing the balance and control of hydrogen for passivation while avoiding such negative effects will be particularly important for the long-term success and stability of silicon solar cells.

Acknowledgement

ARP would like to thank EPSRC (UK) for funding via grant EP/TO25131/1.

References

[1] Hallam B.J., Hamer P.G., Ciesla née Wenham A.M., Chan C.E., Vicari Stefani B., Wenham S. 'Development of advanced hydrogenation processes for silicon solar cells via an improved understanding of the behaviour of hydrogen in silicon'. *Progress in Photovoltaics: Research and Applications*. 2020;**28**(12):1217–38.

[2] Kleekajai S., Jiang F., Stavola M., *et al.* 'Concentration and penetration depth of H introduced into crystalline Si by hydrogenation methods used to fabricate solar cells'. *Journal of Applied Physics*. 2006;**100**(9):93517.

[3] Jacques P., Noble J. *Hydrogen in Semiconductors*. **34**; 1991. pp. 1–629.

[4] Pearton R.J., Corbett J.W., Stavola M. *Hydrogen in Cystalline Semiconductors. Springer Series in Materials Science 16*. Springer-Verlag Berlin Heidelberg; 1992.

[5] Norbert N. *Hydrogen in Semiconductors II*. **61**. San Diego, USA: Academic Press; 1999. pp. 1–526.

[6] Chevallier J., Pajot B. *Solid State Phenomena, interactions of hydrogen with impurities and defects in semiconductors*; 2002. pp. 85–6.

[7] Dobaczewski L., Peaker A.R., Bonde Nielsen K., Nielsen K.B. 'Laplace-transform deep-level spectroscopy: the technique and its applications to the study of point defects in semiconductors'. *Journal of Applied Physics*. 2004;**96**(9):4689–728.

[8] Santos P., Coutinho J., Öberg S., *et al.* 'Theory of a carbon-oxygen-hydrogen recombination center in n-type Si'. *Physica Status Solidi*. 2017;**214**(7):1700309.

[9] Andersen O., Peaker A.R., Dobaczewski L., *et al.* 'Electrical activity of carbon-hydrogen centers in Si'. *Physical Review B*. 2002;**66**(23).

[10] Coutinho J., Markevich V.P., Peaker A.R. 'Characterisation of negative-U defects in semiconductors'. *Journal of Physics: Condensed Matter*. 2020;**32**(32):323001.

[11] Stübner R., Scheffler L., Kolkovsky V., Weber J. 'Single and double acceptor-levels of a carbon-hydrogen defect in N-type silicon'. *Journal of Applied Physics*. 2016;**119**(20):205709.

[12] Wagner M., Wolny F., Hentsche M., *et al.* 'Correlation of the LeTID amplitude to the aluminium bulk concentration and oxygen precipitation in PERC solar cells'. *Solar Energy Materials and Solar Cells*. 2018;**187**(June):176–88.

[13] Kerkar F., Kheloufi A., Dokhan N., *et al.* 'Oxygen and carbon distribution in 80 kg multicrystalline silicon ingot'. *Silicon*. 2020;**12**(3):473–8.

[14] Nielsen K.B., Dobaczewski L., Søgård S., Nielsen B.B. 'Acceptor state of monoatomic hydrogen in silicon and the role of oxygen'. *Physical Review B*. 2002;**65**(7):75205.

[15] Newman R.C., Smith R.S. 'Vibrational absorption of carbon and carbon-oxygen complexes in silicon'. *Journal of Physics and Chemistry of Solids*. 1969;**30**(6):1493–505.

[16] Shirakawa Y., Yamada-Kaneta H., Mori H. 'Annealing behavior of carbon-oxygen complexes in silicon crystals observed by low-temperature infrared absorption'. *Journal of Applied Physics*. 1995;**77**(1):41–6.

[17] Shirakawa Y., Yamada-Kaneta H. 'Annealing behavior and atomic composition of substitutional carbon–oxygen complexes in silicon crystals'. *Journal of Applied Physics*. 1996;**80**(7):4199–201.

[18] Yoneta M., Kamiura Y., Hashimoto F. 'Chemical etching-induced defects in phosphorus-doped silicon'. *Journal of Applied Physics*. 1991;**70**(3):1295–308.

[19] Tokuda Y., Katoh I., Ohshima H., Hattori T. 'Observation of hydrogen in commercial Czochralski-grown silicon wafers'. *Semiconductor Science and Technology*. 1994;**9**(9):1733–5.

[20] Vaqueiro-Contreras M., Markevich V.P., Halsall M.P., Peaker A.R., Santos P., Coutinho J. 'Powerful recombination centers resulting from reactions of hydrogen with carbon–oxygen defects in n-type Czochralski-grown silicon (Phys. status Solidi rrl 8/2017)'. *Rapid Research Letters*. 2017;**11**(8):1770342.

[21] Gwozdz K., Stübner R., Kolkovsky V., Weber J. 'Carbon-hydrogen defects with a neighboring oxygen atom in N-type Si'. *Applied Physics Letters*. 2017;**111**(3):032102.

[22] Vaqueiro-Contreras M., Markevich V.P., Mullins J., *et al.* 'Lifetime degradation of N-type Czochralski silicon after hydrogenation'. *Journal of Applied Physics*. 2018;**123**(16):161415.

[23] Mullins J., Leonard S., Markevich V.P., *et al.* 'Recombination via transition metals in solar silicon: the significance of hydrogen–metal reactions and lattice sites of metal atoms'. *Physica Status Solidi*. 2017;**214**(7):1700304.

[24] Jost W., Weber J., Lemke H. 'Hydrogen–cobalt complexes in p-type silicon'. *Semiconductor Science and Technology*. 1996;**11**(4):525–30.

[25] Jost W., Weber J., Lemke H. 'Hydrogen-induced defects in cobalt-doped N-type silicon'. *Semiconductor Science and Technology*. 1996;**11**(1):22–6.

[26] Leonard S., Markevich V.P., Peaker A.R., Hamilton B. 'Passivation of titanium by hydrogen in silicon'. *Applied Physics Letters*. 2013;**103**(13):132103.

[27] Mullins J., Markevich V.P., Halsall M.P., Peaker A.R. 'Vanadium in silicon: lattice positions and electronic properties'. *Applied Physics Letters*. 2017;**110**(14):142105.

[28] Sadoh T., Watanabe M., Nakashima H., Tsurushima T. 'Deep levels of chromium-hydrogen complexes in silicon'. *Journal of Applied Physics*. 1994;**75**(8):3978–81.

[29] Sadoh T., Tsukamoto K., Baba A., *et al.* 'Deep level of iron-hydrogen complex in silicon'. *Journal of Applied Physics*. 1997;**82**(8):3828–31.

[30] Deixler P., Terry J., Hawkins I.D., *et al.* 'Laplace-transform deep-level transient spectroscopy studies of the G4 gold–hydrogen complex in silicon'. *Applied Physics Letters*. 1998;**73**(21):3126–8.

[31] Yarykin N., Weber J. 'Deep levels of copper-hydrogen complexes in silicon'. *Physical Review B*. 2013;**88**(8):1–7.

[32] Yarykin N., Sachse J.-U., Lemke H., Weber J. 'Silver-hydrogen interactions in crystalline silicon'. *Physical Review B*. 1999;**59**(8):5551–60.

[33] Falster R. 'Intrinsic point defects in silicon: a unified view from crystal growth, wafer processing and metal diffusion'. *Solid State Phenom.* 2005;**108–109**:1–10.

[34] Roberson M.A., Estreicher S.K. 'Vacancy and vacancy-hydrogen complexes in silicon'. *Physical Review B*. 1994;**49**(24):17040–9.

[35] Myers S.M., Baskes M.I., Birnbaum H.K., *et al.* 'Hydrogen interactions with defects in crystalline solids'. *Reviews of Modern Physics*. 1992;**64**(2):559–617.

[36] Lavrov E.V., Weber J., Huang L., Nielsen B.B. 'Vacancy-hydrogen defects in silicon studied by Raman spectroscopy'. *Physical Review B*. 2001;**64**(3):1–5.

[37] Chen D., Vaqueiro Contreras M., Ciesla A., *et al.* 'Progress in the understanding of light- and elevated temperature-induced degradation in silicon solar cells: a review'. *Progress in Photovoltaics: Research and Applications*. 2020;**2**.

[38] Ramspeck K., Zimmermann S., Nagel H., Metz A., Gassenbauer Y., Brikmann B. Light induced degradation of Rear Passivated mc Si solar cells. Proceedings of the 27th European Photovoltaic Solar Energy Conference; 2012. pp. 861–5.

[39] Kersten F., Engelhart P., Ploigt H.-C., *et al.* 'Degradation of multicrystalline silicon solar cells and modules after illumination at elevated temperature'. *Solar Energy Materials and Solar Cells*. 2015;**142**:83–6.

[40] Kersten F., Fertig F., Petter K., *et al.* 'System performance loss due to LeTID'. *Energy Procedia*. 2017;**124**:540–6.

[41] Kersten F., Engelhart P., Ploigt H.-C., Stekolnikov A., Lindner T., Stenzel F. 'A new mc-Si degradation effect called LeTID'. Proceedings of the 42nd IEEE Photovoltaic Specialists Conference; 2015. pp. 1–5.

[42] Payne D.N.R., Chan C.E., Hallam B.J., *et al.* 'Acceleration and mitigation of carrier-induced degradation in P-type multi-crystalline silicon'. *Physica Status Solidi (RRL) – Rapid Research Letters*. 2016;**10**(3):237–41.

[43] Chen D., Kim M., Stefani B.V., *et al.* 'Evidence of an identical firing-activated carrier-induced defect in monocrystalline and multicrystalline silicon'. *Solar Energy Materials and Solar Cells*. 2017;**172**:293–300.

[44] Lindroos J., Savin H. 'Review of light-induced degradation in crystalline silicon solar cells'. *Solar Energy Materials and Solar Cells*. 2016;**147**:115–26.

[45] Inglese A., Lindroos J., Vahlman H., Savin H. 'Recombination activity of light-activated copper defects in *p* type silicon studied by injection- and temperature-dependent lifetime spectroscopy'. *Journal of Applied Physics*. 2016;**120**(12):125703.

[46] Luka T., Großer S., Hagendorf C., Ramspeck K., Turek M. 'Intra-grain versus grain boundary degradation due to illumination and annealing behavior of multi-crystalline solar cells'. *Solar Energy Materials and Solar Cells*. 2016;**158**(6):43–9.

[47] Niewelt T., Selinger M., Grant N.E., Kwapil W., Murphy J.D., Schubert M.C. 'Light-Induced activation and deactivation of bulk defects in boron-doped float-zone silicon'. *Journal of Applied Physics*. 2017;**121**(18):185702.

[48] Vargas C., Zhu Y., Coletti G., *et al.* 'Recombination parameters of lifetime-limiting carrier-induced defects in multicrystalline silicon for solar cells'. *Applied Physics Letters*. 2017;**110**(9):092106.

[49] Chan C.E., Payne D.N.R., Hallam B.J., *et al.* 'Rapid stabilization of high-performance multicrystalline P-type silicon PERC cells'. *IEEE Journal of Photovoltaics*. 2016;**6**(6):1473–9.

[50] Payne D.N.R., Chan C.E., Hallam B.J., *et al.* 'Rapid passivation of carrier-induced defects in P-type multi-crystalline silicon'. *Solar Energy Materials and Solar Cells*. 2016;**158**(1):102–6.

[51] Ciesla A., Wenham S., Chen R., Chan C., Chen D., Hallam B. Hydrogen-Induced Degradation. Proceedings of the 7th IEEE World Conference Photovoltaic Energy Conversion; 2018. pp. 1–8.

[52] Chen R., Tong H., Zhu H., *et al.* '23.83% efficient mono-PERC incorporating advanced hydrogenation'. *Progress in Photovoltaics: Research and Applications*. 2020;**28**(12):1239–47.

[53] Petter K., Hubener K., Kersten F., Bartzsch M., Fertig F., Kloter B. 'Dependence of LeTID on brick height for different wafer suppliers with several resistivities and dopants'. 9th International Work Crystalline Silicon Solar Cells; 2016.

[54] Krauß K., Brand A.A., Fertig F., Rein S., Nekarda J. 'Fast regeneration processes to avoid light-induced degradation in multicrystalline silicon solar cells'. *2017 IEEE Photovoltaic Specialists Conference*. 2017;**6**(6):1–3.

[55] Luka T., Eiternick S., Frigge S., Hagendorf C., Mehlich H., Turek M. 'Investigation of Light Induced Degradation of Multi-crystalline PERC cells'. Proceedings of the 31st European Photovoltaic Solar Energy Conference and Exhibition; 2015. pp. 826–8.

[56] Sen C., Chan C., Hamer P., *et al.* 'Eliminating light- and elevated temperature-induced degradation in P-type PERC solar cells by a two-step thermal process'. *Solar Energy Materials and Solar Cells*. 2020;**209**(February):110470.

[57] Deniz H., Bauer J., Breitenstein O. 'Nickel precipitation in light and elevated temperature degraded multicrystalline silicon solar cells'. *Solar RRL*. 2018;**2**(9):1800170.

[58] Cheong H., Wang H., Wang Q., Sun C., Chen W., Jin H. 'Light and elevated temperature induced degradation in P-type and N-type cast-grown multicrystalline and mono-like silicon'. *Solar Energy Materials and Solar Cells*. 2018;**182**:98–104.

[59] Padmanabhan M., Jhaveri K., Sharma R., *et al.* 'Light-induced degradation and regeneration of multicrystalline silicon Al-BSF and PERC solar cells'. *Physica Status Solidi (RRL) – Rapid Research Letters*. 2016;**10**(12):874–81.

[60] Zhou C., Zhou S., Ji F., Wang W. 'Deep level transient spectroscopic investigation of carrier trap defects in p-type mc-Si PERC solar cells

after elevated temperature light soaking'. 36th European Photovoltaic Solar Energy Conference and Exhibition; 2019. pp. 147–50.

[61] Fertig F., Krauß K., Rein S. 'Light-Induced degradation of PECVD aluminium oxide passivated silicon solar cells'. *Physica Status Solidi (RRL) – Rapid Research Letters*. 2015;**9**(1):41–6.

[62] Nakayashiki K., Hofstetter J., Morishige A.E., *et al*. 'Engineering solutions and root-cause analysis for light-induced degradation in p-type multicrystalline silicon PERC modules'. *IEEE Journal of Photovoltaics*. 2016;**6**(4):860–8.

[63] Fokuhl E., NEEDSFIXING G. 'Letid – a comparison of test methods on module level'. *36th European Photovoltaic Solar Energy Conference and Exhibition*. 2019;**4**(1):75–84.

[64] Pander M., Luka T., Jäckel B., Bauer J., Daßler D., Ebert M. 'Prediction of potential power/ yield loss From letid susceptible'. 36th European Photovoltaic Solar Energy Conference and Exhibition; 2019. pp. 810–5.

[65] Bredemeier D., Walter D., Herlufsen S., Schmidt J. 'Lifetime degradation and regeneration in multicrystalline silicon under illumination at elevated temperature'. *AIP Advances*. 2016;**6**(3):35119.

[66] Eberle R., Kwapil W., Schindler F., Glunz S.W., Schubert M.C. 'Firing temperature profile impact on light induced degradation in multicrystalline silicon'. *Energy Procedia*. 2017;**124**(1):712–7.

[67] Hallam B., Hamer P., Kim M., Nampalli N., Gorman N., Chen D. 'Direct transitions between states A and C in the Boron-Oxygen defect system – fact or fiction?' Proceedings of the 43rd IEEE Photovoltaic Specialists Conference; 2016. pp. 2430–3.

[68] Chen D., Hamer P., Kim M., *et al*. 'Hydrogen-induced degradation: explaining the mechanism behind light- and elevated temperature-induced degradation in N- and P-type silicon'. *Solar Energy Materials and Solar Cells*. 2020;**207**(207):110353.

[69] Kim M., Chen D., Abbott M., Wenham S., Hallam B. 'Role of hydrogen: formation and passivation of meta-stable defects due to hydrogen in silicon'. AIP Conference Proceedings; 2018. p. 130010.

[70] Kersten F., Heitmann J., Müller J.W. 'Influence of Al_2O_3 and SiN_x Passivation layers on LeTID'. *Energy Procedia*. 2016;**92**(0):828–32.

[71] Bredemeier D., Walter D.C., Schmidt J. 'Lifetime degradation in multicrystalline silicon under illumination at elevated temperature: indications for the involvement of hydrogen'. AIP Conference Proceedings; 2018.

[72] Krauss K., Fertig F., Menzel D., Rein S. 'Light-Induced degradation of silicon solar cells with aluminium oxide passivated rear side'. *Energy Procedia*. 2015;**77**:599–606.

[73] Fertig F., Lantzsch R., Mohr A., *et al*. 'Mass production of P-type Cz silicon solar cells approaching average stable conversion efficiencies of 22 %'. *Energy Procedia*. 2017;**124**(13):338–45.

[74] Chan C., Fung T.H., Abbott M., *et al.* 'Modulation of carrier-induced defect kinetics in multi-crystalline silicon PERC cells through dark annealing'. *Solar RRL.* 2017;**1**(2):1600028.

[75] Chan C.E., Fung T.H., Payne D.N.R., Chen D., Abbott M.D., Ciesla A.M. 'Investigation of carrier-induced defect behavior in p-type multicrystalline silicon'. Proceedings of the 44th IEEE Photovoltaic Specialists Conference; 2017. pp. 2576–81.

[76] Grant N.E., Rougieux F.E., Macdonald D., Bullock J., Wan Y. 'Grown-in defects limiting the bulk lifetime of *p* -type float-zone silicon wafers'. *Journal of Applied Physics.* 2015;**117**(5):055711.

[77] Bredemeier D., Walter D., Herlufsen S., Schmidt J. 'Understanding the light-induced lifetime degradation and regeneration in multicrystalline silicon'. *Energy Procedia.* 2016;**92**(0):773–8.

[78] Morishige A.E., Jensen M.A., Needleman D.B., *et al.* 'Lifetime spectroscopy investigation of light-induced degradation in P-type multicrystalline silicon PERC'. *IEEE Journal of Photovoltaics.* 2016;**6**(6):1466–72.

[79] Jensen M.A., Zhu Y., Looney E.E., Morishige A.E., Vargas C., Hameiri Z. 'Assessing the defect responsible for LeTID temperature- and injection- dependent lifetime spectroscopy'. Proceedings of the 44th IEEE Photovoltaic Specialists Conference; 2017.

[80] Niewelt T., Schindler F., Kwapil W., Eberle R., Schön J., Schubert M.C. 'Understanding the light-induced degradation at elevated temperatures: similarities between multicrystalline and floatzone P-type silicon'. *Progress in Photovoltaics: Research and Applications.* 2018;**26**(8):533–42.

[81] Sperber D., Herguth A., Hahn G. 'A 3-state defect model for light-induced degradation in boron-doped float-zone silicon'. *Physica Status Solidi (RRL) – Rapid Research Letters.* 2017;**11**(3):1600408.

[82] Fung T.H., Kim M., Chen D., *et al.* 'A four-state kinetic model for the carrier-induced degradation in multicrystalline silicon: introducing the reservoir state'. *Solar Energy Materials and Solar Cells.* 2018;**184**:48–56.

[83] Chen D., Hamer P.G., Kim M., *et al.* 'Hydrogen induced degradation: a possible mechanism for light- and elevated temperature induced degradation in N-type silicon'. *Solar Energy Materials and Solar Cells.* 2018;**185**(March):174–82.

[84] Sen C., Kim M., Chen D., *et al.* 'Assessing the impact of thermal profiles on the elimination of light- and Elevated-Temperature-Induced degradation'. *IEEE Journal of Photovoltaics.* 2019;**9**(1):40–8.

[85] Winter M., Walter D., Bredemeier D., Schmidt J. 'Light-induced lifetime degradation effects at elevated temperature in Czochralski-grown silicon beyond boron-oxygen-related degradation'. *Solar Energy Materials and Solar Cells.* 2019;**201**(July):110060.

[86] Sperber D., Furtwängler F., Herguth A., Hahn G. 'On the stability of dielectric passivation layers under illumination and temperature treatments'. Proceedings of the 32nd European Photovoltaic Solar Energy Conference and Exhibition; 2016. pp. 523–6.

[87] Sperber D., Herguth A., Hahn G. 'Instability of dielectric surface Passivation quality at elevated temperature and illumination'. *Energy Procedia.* 2016;**92**(1):211–7.

[88] Sperber D., Graf A., Skorka D., Herguth A., Hahn G. 'Degradation of surface passivation on crystalline silicon and its impact on light-induced degradation experiments'. *IEEE Journal of Photovoltaics.* 2017;**7**(6):1627–34.

[89] Niewelt T., Post R., Schindler F., Kwapil W., Schubert M.C. 'Investigation of LeTID where we can control it – Application of FZ silicon for defect studies'. 15th International Conference on Concentrator Photovoltaic Systems (CPV-15); 2019. p. 140006.

[90] Vargas C., Nie S., Chen D., *et al.* 'Degradatio*n* and recovery of *n* type multi-crystalline silicon under illuminated and dark annealing conditions at moderate temperatures'. *IEEE Journal of Photovoltaics.* 2018;**9**(2):355–63.

[91] Renevier C., Fourmond E., Forster M., Parola S., Le Coz M., Picard E. 'Lifetime degradation on N-type wafers with boron-diffused and SiO_2/SiN-passivated surface'. *Energy Procedia.* 2014;**55**(1):280–6.

[92] Phang S.P., Macdonald D. 'Direct comparison of boron, phosphorus, and aluminum gettering of iron in crystalline silicon'. *Journal of Applied Physics.* 2011;**109**(7):073521.

[93] Vargas C., Kim K., Coletti G., Payne D., Chan C., Wenham S. 'Influence of silicon nitride and its hydrogen content of carrier-induced degradation in multicrystalline silicon'. Proceedings of the 33rd European Photovoltaic Solar Energy Conference and Exhibition; 2017. p. 561.

[94] Sperber D., Schwarz A., Herguth A., Hahn G., Bulk H.G. 'Bulk and surface-related degradation in lifetime samples made of czochralski silicon passivated by plasma-enhanced chemical vapor deposited layer stacks'. *Physica Status Solidi.* 2018;**215**(24):1800741.

[95] Sperber D., Furtwängler F., Herguth A., Hahn G. 'Does LeTID occur in c-Si even without a firing step?' AIP Conference Proceedings; 2019. p. 140011.

[96] Vargas C., Kim K., Coletti G., *et al.* 'Carrier-Induced degradation in multicrystalline silicon: dependence on the silicon nitride Passivation layer and hydrogen released during firing'. *IEEE Journal of Photovoltaics.* 2018;**8**(2):413–20.

[97] Varshney U., Abbott M.D., Liu S., Chen D., Kim M., Sen C. 'Influence of dielectric passivation layer thickness on LeTID in multicrystalline silicon'. IEEE 7th World Conference on Photovoltaic Energy Conversion (WCPEC) (A Joint Conference of 45th IEEE PVSC, 28th PVSEC & 34th EU PVSEC); 2018.

[98] Jensen M.A., Wieghold S., Poindexter J.R., *et al.* 'Solubility and diffusivity: important metrics in the search for the root cause of light- and elevated temperature-induced degradation'. *IEEE Journal of Photovoltaics.* 2018;**8**(2):448–55.

[99] Nampalli N. Characterisation and Passivation of Boron-Oxygen Defects in P-type Czochralski Silicon [PhD Thesis]UNSW Sydney; 2017.

[100] Schmidt J., Bredemeier D., Walter D.C. 'On the defect physics behind light and elevated temperature-induced degradation (LeTID) of multicrystalline silicon solar cells'. *IEEE Journal of Photovoltaics*. 2019;**9**(6):1497–503.

[101] Chen D2020Elucidating the Mechanics Behind Light- and Elevated Temperature-Induced Degradation in Silicon Solar Cells [PhD Thesis]KensingtonUNSW

[102] Bredemeier D., Walter D.C., Schmidt J. 'Possible candidates for impurities in mc-Si Wafers responsible for light-induced lifetime degradation and regeneration'. *Solar RRL*. 2018;**2**(1):1700159.

[103] Eberle R., Kwapil W., Schindler F., Glunz S.W., Schubert M.C. 'Firing temperature profile impact on light induced degradation in multicrystalline silicon'. *Energy Procedia*. 2017;**124**(1):712–7.

[104] Luka T., Turek M., Großer S., Hagendorf C. 'Microstructural identification of Cu in solar cells sensitive to light-induced degradation'. *Physica Status Solidi (RRL) – Rapid Research Letters*. 2017;**11**(2):1600426.

[105] Sachdeva R., Istratov A.A., Weber E.R. 'Recombination activity of copper in silicon'. *Applied Physics Letters*. 2001;**79**(18):2937–9.

[106] Lindroos J., Boulfrad Y., Yli-Koski M., Savin H. 'Preventing light-induced degradation in multicrystalline silicon'. *Journal of Applied Physics*. 2014;**115**(15):154902.

[107] Savin H., Yli-Koski M., Haarahiltunen A. 'Role of copper in light induced minority-carrier lifetime degradation of silicon'. *Applied Physics Letters*. 2009;**95**(15):152111.

[108] Yli-Koski M., Serué M., Modanese C., Vahlman H., Savin H. 'Low-temperature dark anneal as pre-treatment for LeTID in multicrystalline silicon'. *Solar Energy Materials and Solar Cells*. 2019;**192**:134–9.

[109] Vahlman H., Haarahiltunen A., Kwapil W., Schön J., Inglese A., Savin H. 'Modeling of light-induced degradation due to Cu precipitation in P-type silicon. *II. Comparison of simulations and experiments*'. *Journal of Applied Physics*. 2017;**121**(19):195704.

[110] Zuschlag A., Skorka D., Hahn G. 'Degradation and regeneration analysis in mc-Si'. Proceedings of the IEEE Photovoltaic Specialists Conference; 2016. pp. 1051–4.

[111] Skorka D., Zuschlag A., Hahn G. 'Firing and gettering dependence of effective defect density in material exhibiting LeTID'. AIP Conference Proceedings; 2018. p. 130015.

[112] Chakraborty S., Huang Y., Wilson M., Aberle A.G., Li J.B. 'Mitigating light and elevated temperature induced degradation in multicrystalline silicon Wafers and PERC solar cells using phosphorus diffusion gettering'. *Physica Status Solidi*. 2018;**215**(13):1800160.

[113] Sperber D., Herguth A., Hahn G. 'Investigating possible causes of light induced degradation in boron-doped float-zone silicon'. Proceedings of the 33rd European Photovoltaic Solar Energy Conference and Exhibition; 2017. pp. 565–8.

[114] Bredemeier D., Walter D.C., Heller R., Schmidt J. 'Impact of hydrogen-rich silicon nitride material properties on light-induced lifetime degradation in multicrystalline silicon'. *Physica Status Solidi (RRL) – Rapid Research Letters*. 2019;**13**(8):1900201.

[115] Winter C., Zuschlag A., Skorka D., Hahn G. 'Influence of dielectric layers and thermal load on LeTID'. *AIP Conference Proceedings*; 1999. p. 130020.

[116] Varshney U., Abbott M., Ciesla A., *et al*. 'Evaluating the impact of SiN_x thickness on lifetime degradation in silicon'. *IEEE Journal of Photovoltaics*. 2019;**9**(3):601–7.

[117] Walter D.C., Bredemeier D., Falster R., Voronkov V.V., Schmidt J. 'Easy-to-apply methodology to measure the hydrogen concentration in boron-doped crystalline silicon'. *Solar Energy Materials and Solar Cells*. 2019;**200**(14):109970.

[118] Jensen M.A., Zuschlag A., Wieghold S., *et al*. 'Evaluating root cause: the distinct roles of hydrogen and firing in activating light- and elevated temperature-induced degradation'. *Journal of Applied Physics*. 2018;**124**(8):085701.

[119] Bourret-Sicotte G., Hamer P.G., Tweddle D., Bonilla R.S., Wilshaw P.R. 'Hydrogen related defects in float zone silicon investigated using a shielded hydrogen plasma'. IEEE 7th World Conference on Photovoltaic Energy. Conversion (WCPEC 2018). (A Joint Conference of 45th. IEEE PVSC, 28th PVSEC); 2018. pp. 298–302.

[120] Chen D., Kim M., Shi J., *et al*. 'Defect engineering of p-type silicon heterojunction solar cells fabricated using commercial-grade low-lifetime silicon wafers'. *Progress in Photovoltaics: Research and Applications*. 2019:1–15.

[121] Lucovsky G., Tsu D.V. 'Plasma enhanced chemical vapor deposition: differences between direct and remote plasma excitation'. *Journal of Vacuum Science & Technology A: Vacuum, Surfaces, and Films*. 1987;**5**(4):2231–8.

[122] Hamer P., Bourret-Sicotte G., Martins G., Wenham A., Bonilla R.S., Wilshaw P. 'A novel source of atomic hydrogen for passivation of defects in silicon'. *Physica Status Solidi (RRL) – Rapid Research Letters*. 2017;**11**(5):1600448.

[123] Sah Chih-Tang., Sun Jack Yuan-Chen., Tzou Joseph Jeng-Tao. 'Deactivation of the boron acceptor in silicon by hydrogen'. *Applied Physics Letters*. 1983;**43**(2):204–6.

[124] Zundel T., Mesli A., Muller J.C., Siffert P., Festkrperforschung M. 'Boron neutralization and hydrogen diffusion in silicon subjected to low-energy hydrogen implantation'. *Applied Physics A Solids and Surfaces*. 1989;**48**(1):31–40.

[125] Mchedlidze T., Weber J. 'Location and properties of carrier traps in mc-Si solar cells subjected to degradation at elevated temperatures'. *Physica Status Solidi*. 2019;**216**(17):1900142–10.

[126] Varshney U., Abbott M., Liu S., Chen D., Kim M., Sen C. 'Studying degradation in multicrystalline silicon by modulating the thickness of dielectric layer'. 2012:5–6.

[127] Kang M.H., Ryu K., Upadhyaya A., Rohatgi A. 'Optimization of Sin Ar coating for Si solar cells and modules through quantitative assessment of

optical and efficiency loss mechanism'. *Progress in Photovoltaics: Research and Applications*. 2011;**19**(8):983–90.

[128] Sharma R., Aberle A.G., Li J.B. 'Optimization of belt furnace anneal to reduce light and elevated temperature induced degradation of effective carrier lifetime of p-type multicrystalline silicon wafers'. *Solar RRL*. 2018;**2**(9):1800070.

[129] Hamer P., Chan C., Bonilla R.S., *et al.* 'Hydrogen induced contact resistance in PERC solar cells'. *Solar Energy Materials and Solar Cells*. 2018;**184**(April):91–7.

[130] Hallam B., Herguth A., Hamer P., *et al.* 'Eliminating light-induced degradation in commercial P-type Czochralski silicon solar cells'. *Applied Sciences*. 2017;**8**(1):10.

[131] Hallam B.J., Hamer P.G., Wang S., *et al.* 'Advanced hydrogenation of dislocation clusters and Boron-oxygen defects in silicon solar cells'. *Energy Procedia*. 2015;**77**(10):799–809.

[132] Hamer P., Hallam B., Abbott M., Wenham S. 'Accelerated formation of the boron-oxygen complex in P-type Czochralski silicon'. *Physica Status Solidi (RRL) – Rapid Research Letters*. 2015;**9**(5):297–300.

[133] Sen C., Chan C., Hamer P., *et al.* 'Annealing prior to contact firing: a potential new approach to suppress LeTID'. *Solar Energy Materials and Solar Cells*. 2019;**200**(March):109938.

[134] Skorka D., Zuschlag A., Hahn G. Spatially resolved degradation and regeneration kinetics in mc-Si. Proceedings of the 32nd European Photovoltaic Solar Energy Conference and Exhibition; 2016. pp. 643–6.

[135] Bredemeier D., Walter D.C., Pernau T., Romer O., Schmidt J. 'Production compatible remedy against LETID in high-performance multicrystalline silicon solar cells'. 35th European Photovoltaic Solar Energy Conference and Exhibition; 2018. pp. 406–9.

[136] Sharma R., Chong A.P., Li J.B., Aberle A.G., Huang Y. 'Role of post-metallization anneal sequence and forming gas anneal to mitigate light and elevated temperature induced degradation of multicrystalline silicon solar cells'. *Solar Energy Materials and Solar Cells*. 2019;**195**(4):160–7.

[137] Yli-koski M., Pasanen T.P., Heikkinen I., Serué M. 'Low-T anneal as cure for LeTID in Mc-Si PERC cells low-T Anneal as cure for LeTID in Mc-Si PERC cells'. AIP Conference Proceedings; 2019. p. 140013.

[138] Varshney U., Hallam B., Hamer P., *et al.* 'Controlling light- and Elevated-Temperature-Induced degradation with thin film barrier layers'. *IEEE Journal of Photovoltaics*. 2020;**10**(1):19–27.

[139] Wang Z., Shen H., Hu D., *et al.* 'Influence of double current injection annealing on anti-LID effect in mono-like cast silicon PERC solar cells'. *Journal of Materials Science: Materials in Electronics*. 2020;**31**(4):3221–7.

[140] Wilking S., Ebert S., Herguth A., Hahn G. 'Influence of hydrogen effusion from hydrogenated silicon nitride layers on the regeneration of boron-oxygen related defects in crystalline silicon'. *Journal of Applied Physics*. 2013;**114**(19):194512.

[141] Krugel G., Wolke W., Geilker J., Rein S., Preu R. 'Impact of hydrogen concentration on the regeneration of light induced degradation'. *Energy Procedia*. 2011;**8**:47–51.
[142] Bredemeier D., Walter D., Schmidt J. 'Light-induced lifetime degradation in high-performance multicrystalline silicon: detailed kinetics of the defect activation'. *Solar Energy Materials and Solar Cells*. 2017;**173**(August):2–5.
[143] Fung T.H., Kim M., Chen D., Samadi A., Chan C.E., Hallam B.J. 'Influence of bound hydrogen states on carrier-induced degradation in multi-crystalline silicon'. *AIP Conference Proceedings*; 2018. p. 130004.
[144] Rougieux F.E., Forster M., Macdonald D., Cuevas A., Lim B., Schmidt J. 'Recombination activity and impact of the boron–oxygen-related defect in compensated N-type silicon'. *IEEE Journal of Photovoltaics*. 2011;**1**(1):54–8.
[145] Herguth A., Schubert G., Kaes M., Hahn G. 'A new approach to prevent the negative impact of the metastable defect in boron doped CZ silicon solar cells'. 2006 IEEE 4th World Conference on Photovoltaic Energy Conference. IEEE; 2006. pp. 940–3.
[146] McQuaid S.A., Binns M.J., Newman R.C., Lightowlers E.C., Clegg J.B. 'Solubility of hydrogen in silicon at 1300 °C'. *Applied Physics Letters*. 1993;**62**(14):1612–4.
[147] Voronkov V.V., Falster R. 'Formation, dissociation, and diffusion of various hydrogen dimers in silicon'. *Physica Status Solidi*. 2017;**254**(6):1600779.
[148] Wilking S., Beckh C., Ebert S., Herguth A., Hahn G. 'Influence of bound hydrogen states on BO-regeneration kinetics and consequences for high-speed regeneration processes'. *Solar Energy Materials and Solar Cells*. 2014;**131**(10):2–8.
[149] Van de Walle C.G., Neugebauer J. 'Hydrogen in semiconductors'. *Annual Review of Materials Research*. 2006;**36**(1):179–98.
[150] Newman R.C., Tucker J.H., Brown A.R., McQuaid S.A. 'Hydrogen diffusion and the catalysis of enhanced oxygen diffusion in silicon at temperatures below 500 °C'. *Journal of Applied Physics*. 1991;**70**(6):3061–70.
[151] Chavali R.V.K., De Wolf S., Alam M.A. 'Device physics underlying silicon heterojunction and passivating-contact solar cells: a topical review'. *Progress in Photovoltaics: Research and Applications*. 2018;**26**(4):241–60.
[152] Masuko K., Shigematsu M., Hashiguchi T., *et al.* 'Achievement of more than 25% conversion efficiency with crystalline silicon heterojunction solar cell'. *IEEE Journal of Photovoltaics*. 2014;**4**(6):1433–5.
[153] Hallam B., Chen D., Shi J., Einhaus R., Holman Z.C., Wenham S. 'Pre-fabrication gettering and hydrogenation treatments for silicon heterojunction solar cells: A possible path to >700 mv open-circuit voltages using low-lifetime commercial-grade p-type czochralski silicon'. *Solar RRL*. 2018;**2**(2):1700221.
[154] Vicari Stefani B., Weigand W., Wright M., *et al.* 'P-type upgraded metallurgical-grade multicrystalline silicon heterojunction solar cells with open-circuit voltages over 690 mV'. *Physica Status Solidi*. 2019;**216**(17):1900319.

[155] Morishige A.E., Jensen M.A., Needleman D.B., Nakayashiki K., *et al.* 'Lifetime spectroscopy investigation of light-induced degradation in p-type multicrystalline silicon PERC'. 2017 IEEE 44th Photovoltaic Specialist Conference, PVSC 2017; 2018.

[156] Peral A., Dastgheib-Shirazi A., Fano V., Jimeno J.C., Hahn G., Del Cañizo C. 'Impact of extended contact cofiring on multicrystalline silicon solar cell parameters'. *IEEE Journal of Photovoltaics*. 2017;**7**(1):91–6.

[157] Chan C., Hamer P., Bourret-Sicotte G., *et al.* 'Instability of increased contact resistance in silicon solar cells following post-firing thermal processes'. *Solar RRL*. 2017;**1**(11):1700129.

[158] Hamer P., Li H., Chan C., Sen C., Bonilla R.S., Wilshaw P. The behavior and transport of hydrogen in silicon solar cells observed through changes in contact resistance. 2018 IEEE 7th World Conference on Photovoltaic Energy Conversion; 2018. pp. 1682–6.

[159] Herring C., Johnson N.M., Van de Walle C.G. 'Energy levels of isolated interstitial hydrogen in silicon'. *Physical Review B*. 2001;**64**(12):125209.

[160] Job R., Fahrner W.R., Ulyashin A.G. 'Electronic device fabrication by simple hydrogen enhanced thermal donor formation processes in P-type Cz-silicon'. *Materials Science and Engineering: B*. 2000;**73**(1):197–202.

[161] Barakel D., Matinuzzi S. 'Donor behaviour of implanted hydrogen ions in silicon wafers damien barakel and santo martinuzzi TECSEN- UMR 6122 - university of aix-marseilles iii - 13397 marseilles – france'. *Materials Research Society.*. 2004;**813**:3–8.

[162] Tavendale A.J., Pearton S.J., Williams A.A. 'Evidence for the existence of a negatively charged hydrogen species in plasma-treated *n*-type Si'. *Applied Physics Letters*. 1990;**56**(10):949–51.

[163] Shumate S.D., Hutchings D.A., Mohammed H., Beilke G., Newton B.S., Young M.G. 'Self aligned hydrogenated selective emitter for n-type solar cells'. Proceedings of the 38th IEEE Photovoltaic Specialists Conference; 2012. pp. 1110–14.

[164] Lavrov E.V., Weber J. 'Evolution of hydrogen platelets in silicon determined by polarized Raman spectroscopy'. *Physical Review Letters*. 2001;**87**(18):185502.

[165] Zhang S.B., Jackson W.B. 'Formation of extended hydrogen complexes in silicon'. *Physical Review B*. 1991;**43**(14):12142–5.

[166] Heyman J.N., Ager I.I.I., Haller E.E., Johnson N.M., Walker J., Doland C.M. 'Hydrogen-induced platelets in silicon: infrared absorption and Raman scattering'. *Physical Review B*. 1992;**45**(23):13363–6.

[167] Nickel N.H., Anderson G.B., Johnson N.M., Walker J. 'Nucleation of hydrogen-induced platelets in silicon'. *Physical Review B*. 2000;**62**(12):8012–15.

[168] Nastasi M., Höchbauer T., Lee J.-K., *et al.* 'Nucleation and growth of platelets in hydrogen-ion-implanted silicon'. *Applied Physics Letters*. 2005;**86**(15):154102–3.

[169] Aspar B., Bruel M., Moriceau H., *et al.* 'Basic mechanisms involved in the Smart-Cut® process'. *Microelectronic Engineering*. 1997;**36**(1–4):233–40.

[170] Hwang J.M., Schroder D.K., Biter W.J. 'Deep levels introduced into silicon during hydrogen plasma annealing'. *Journal of Applied Physics*. 1985;**57**(12):5275–8.

[171] p Johnson NM., Ponce F.A., Street R.A., Nemanich R.J. 'Defects in single-crystal silicon induced by hydrogenation'. *Physical review. B, Condensed matter*. 1987;**35**(8):4166–9.

[172] Johnson N.M., Herring C., Doland C., Walker J., Anderson G.B., Ponce F.A. 'Hydrogen-Induced platelets in silicon: separation of nucleation and growth'. *Materials Science Forum*. 1992;**83-87**:33–8.

[173] Leitch A.W.R., Alex V., Weber J. 'Raman spectroscopy of hydrogen molecules in crystalline silicon'. *Physical Review Letters*. 1998;**81**(2):421–4.

[174] Kim Y.S., Chang K.J. 'Structural transformation in the formation of H-induced (111) platelets in Si'. *Physical Review Letters*. 2001;**86**(9):1773–6.

[175] Chang K.J., Chadi D.J. 'Hydrogen bonding and diffusion in crystalline silicon'. *Physical Review B*. 1989;**40**(17):11644–53.

[176] Leitch A.W.R., Alex V., Weber J. 'H2 molecules in c-Si after hydrogen plasma treatment'. *Solid State Communications*. 1998;**105**(4):215–19.

[177] Muto S., Takeda S., Hirata M. 'Hydrogen-induced platelets in silicon studied by transmission electron microscopy'. *Philosophical Magazine A*. 1995;**72**(4):1057–74.

[178] Nickel N.H., Anderson G.B., Walker J. 'Hydrogen-induced platelets in disordered silicon'. *Solid State Communications*. 1996;**99**(6):427–31.

[179] Höchbauer T., Misra A., Nastasi M., Mayer J.W. 'Physical mechanisms behind the ion-cut in hydrogen implanted silicon'. *Journal of Applied Physics*. 2002;**92**(5):2335–42.

[180] Höchbauer T., Misra A., Nastasi M., Mayer J.W. 'Investigation of the cut location in hydrogen implantation induced silicon surface layer exfoliation'. *Journal of Applied Physics*. 2001;**89**(11 I):5980–90.

[181] Sopori B., Zhang Y., Reedy R., Jones K., Yan Y.F., Al-Jassim M. 'A comprehensive model of hydrogen transport into a solar cell during silicon nitride processing for fire-through metallization'. Proceedings of the 31st IEEE Photovoltaic Specialists Conference; 2005. pp. 1039–42.

[182] Steingrube S., Altermatt P.P., Steingrube D.S., Schmidt J., Brendel R. 'Interpretation of recombination at c-Si/SiNx interfaces by surface damage'. *Journal of Applied Physics*. 2010;**108**(2010):014506.

Chapter 7
Laser doping for rapid diffusion in silicon

Michelle Vaqueiro-Contreras[1], Matthew Wright[1], and Catherine Chan[1]

This chapter will provide an introduction to laser doping, including the benefits of the significant increases in the diffusion coefficients of impurities with increased temperature, particularly when laser doping is performed in the liquid-state, and in general, the higher solubility of impurities with increasing temperature and rapid cooling. It will discuss various methods for laser doping using gas-phase dopant sources, spin-on dopant (SOD) sources, doped dielectric layers and residual dopants from emitter diffusion/phosphosilicate glass (PSG) layers. It then discusses the impact of different laser parameters on doping profiles for laser-doping purposes.

7.1 The origin of laser doping for silicon solar cell manufacturing

The development and use of laser technology have been exponentially increasing in the past few decades. Rapid advances in the understanding and manufacture of lasers have enabled the now wide range of readily available high-efficiency light sources in a plethora of wavelengths, pulse shapes/widths and compactness, which has allowed their implementation in manufacturing at a massive scale. This becomes clear when noticing that nearly every product manufactured these days benefits, in at least one part of its manufacturing cycle, from the use of a laser-based process. From an industrial point of view, the desirability of these somewhat sophisticated light sources resides in the fact that they have evolved to provide a very accurate, reliable, inexpensive, fast and contactless replacement to conventionally used tools for manufacturing.

A little over half-century ago, in the late 1960s, a team of scientists from IBM demonstrated for the first time that a pulsed laser beam directed to a boron-doped silicon substrate covered with a phosphorous-containing film effectively created a shallow (~1 μm) p-n junction in the subsurface of the silicon crystal. This, without

[1]The School of Photovoltaic and Renewable Energy Engineering, The University of New South Wales, Sydnew, NSW, Australia

raising the global temperature of the substrate or affecting the rest of the material [1]. Not long after these results were published, laser–semiconductor interactions became a topic of worldwide interest, leading to the development of what we now identify as laser-doping techniques. In this chapter, an introduction to the laser-doping technology used for solar cell fabrication, its benefits over traditional doping methods and the impact of laser parameters in cell manufacture will be reviewed.

7.2 Dopant diffusion in crystalline silicon

P-n junction formation is the building block of semiconductor device manufacturing. This has led to the development of various methods for dopant diffusion and activation. In the case of photovoltaic devices, a typical commercial front-junction solar cell contains a p-type substrate with a diffused phosphorus-doped layer usually referred to as the emitter [2]. From an electrical point of view, the emitter plays a key role in the determination of the three key electrical parameters affecting solar cell performance overall through the collection probabilities of photogenerated carriers [3]: first, the open-circuit voltage (V_{OC}), which is directly impacted by the emitter saturation current density (J_{0e}) and driven by recombination; second, the short-circuit current density (J_{SC}), which is directly proportional to the external quantum efficiency and thus highly dependent on absorption and surface passivation; and last, the fill factor (FF), which is affected by lateral resistive losses from the emitter and the contact interfaces.

A typical emitter is a few hundred nanometres thick and contains a high peak level of dopants in the order of $1 \times 10^{20}/cm^3$ [4, 5]. These high doping levels are necessary in thin emitters to achieve acceptable lateral conductivity, screen carriers from the relatively high surface recombination velocity, as well as to facilitate the formation of sufficiently low contact resistance with Ag screen-print pastes. Nevertheless, such high concentration of dopants results in major losses in diffusion lengths which go down to a few hundreds of nanometres due to Auger recombination. Until recently, this suboptimal emitter design had been acceptable due to significantly higher recombination losses appearing at the rear of the full-area aluminium back surface field (Al-BSF) of commercial solar cell architectures dominating the dark saturation current. However, the rapid transition to passivated emitter and rear cell (PERC) technology has made the emitter saturation currents a limiting factor for the achievement of the 24–25% efficiency cells, which have already been demonstrated in the laboratory. The application of laser-doping methods has therefore emerged as an attractive alternative for the realisation of more adequate designs that can tackle the limitations imposed by full area emitters and conventional doping methods, all this without compromising production cost or throughput. This section describes the established diffusion techniques for emitter formation and outlays the fundamental differences between them and laser-induced diffusion.

7.2.1 Conventional doping methods

Conventional methods for doping of semiconductors such as ion implantation and thermal (gas phase) diffusion have been largely investigated throughout the years and are mature technologies. On the one hand, ion implantation, the potential of which was first identified in 1954, has raised to be one of the most important technologies for today's integrated circuit (IC) manufacture, without which, IC devices would simply not exist [6]. Ion implantation has allowed the formation of very precise, reliable and repeatable doping profiles in semiconductors, together with high tunability in terms of doping concentration and depth. However, challenges imposed by the precise control of low-energy ion beams required for shallow and sharp junction formation have represented a significant disadvantage for its application [7]. Similarly, inherent damage caused by highly energetic dopants and additional thermal processing required for damage removal and dopant activation may lead to precipitation, undesired dopant redistribution, wafer contamination and surface degradation [8]. The combination of these factors has thus impeded the adoption of this technology for mainstream solar cell manufacture, which overall also requires low-cost and high-throughput processes.

Thermal diffusions, on the other hand, have been an industrial standard for emitter formation in silicon solar cells since the 1970s, more specifically, the use of phosphorus oxychloride ($POCl_3$) diffusion for electron collector formation on p-type silicon wafers and boron tribromide (BBr_3) on n-type silicon wafers. Both of which have been dominating the market due to their low costs, good stability and high throughput, particularly for the case of $POCl_3$ diffusions [4]. At present, thermal processing at an industrial level is reaching throughput levels in the excess of 7 000 wafers/h, which equals other manufacture processes such as metallisation and is not far behind the chemical processes. However, and even though some evolution in throughput is expected in the following decade, it is expected that thermal processing will be lagging behind the latter two processes [9].

Together with the fact of having to use some rather hazardous chemicals such as BBr_3 and $POCl_3$ for thermal diffusion, there is the need to carefully control the diffusion depth during processing and activate the dopants to avoid the formation of the so-called dead layer. Generally speaking, during gas-phase diffusion, a phosphorus-rich or boron-rich layer commonly termed PSG and borosilicate glass (BSG), respectively, is formed. Depending on the dopant species, 'drive-in' processes are used after the dopant layer formation. Due to the differences in dopant diffusion coefficients, as shown in Figure 7.1, Boron sources require higher temperature and more than 80 minutes for junction formation at a temperature of 900 °C to 1 000 °C [11]. Phosphorus diffusions, on the other hand, are usually performed at lower temperatures in the range of 800 °C to 900 °C for 50 minutes or so [12]. Upon annealing, this rich source of dopants then introduces large quantities of phosphorus or boron into the silicon subsurface, some of which will occupy substitutional lattice sites in the crystal and become electrically active dopants. Others will diffuse to form dopant clusters, precipitates or occupy interstitial sites, which can lead to severe Shockley-Read-Hall (SRH) recombination losses and decreased collection

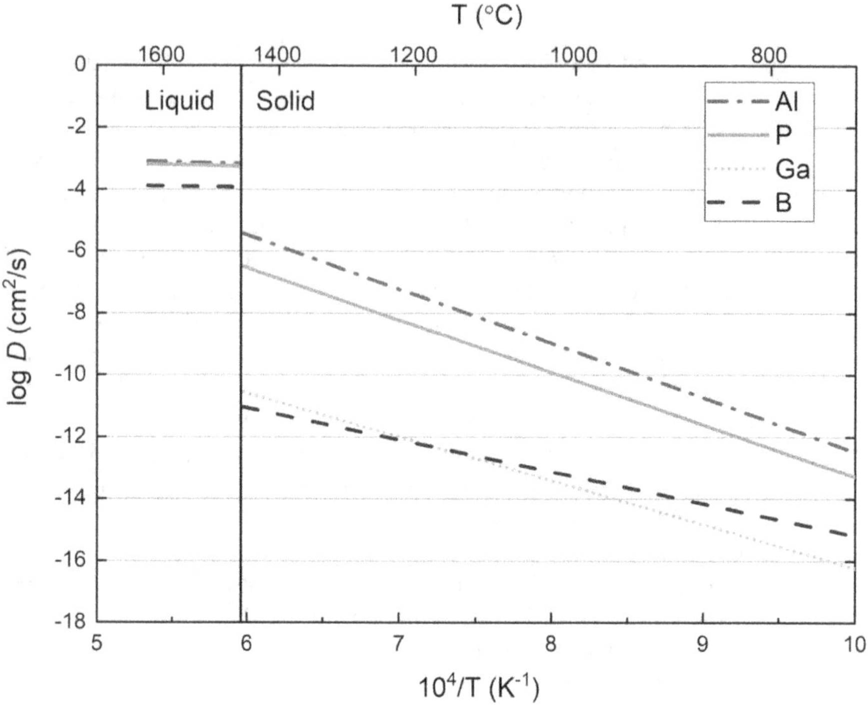

Figure 7.1 *Diffusivity of dopants in silicon of key relevance for silicon solar cells as extracted from figures 12.11 and 12.12 in [10]. Reprinted by permission from Springer Nature: Springer Nature, Thermochemical and Kinetic Databases for the Solar Cell Silicon Materials. Kai Tang* et al. *Copyright (2009).*

efficiencies of the photogenerated carriers [13]. These dopants are inactive and do not contribute to the conductivity of the material. The difference between the dopant atoms occupying substitutional sites and the rest then results in two different depth profiles, an electrically active profile creating the p-n junction and the total dopant concentration profile including both active and inactive dopants. Process optimisation techniques such as reduction of $POCl_3$ flow rates have been shown to decrease the concentration of inactive P; however, a minimal concentration of these impurities is also required for the formation of low-resistivity contacts [14].

Figure 7.2 (a) and (b) shows the differences between phosphorus profiles on c-Si samples diffused using different $POCl_3$:N_2 flow rate ratios, with Figure 7.2 (a) showing the total phosphorus concentrations (active and inactive dopants) from secondary ion mass spectroscopy (SIMS) measurements and Figure 7.2 (b) showing the electrically active phosphorus concentrations from electrochemical capacitance-voltage

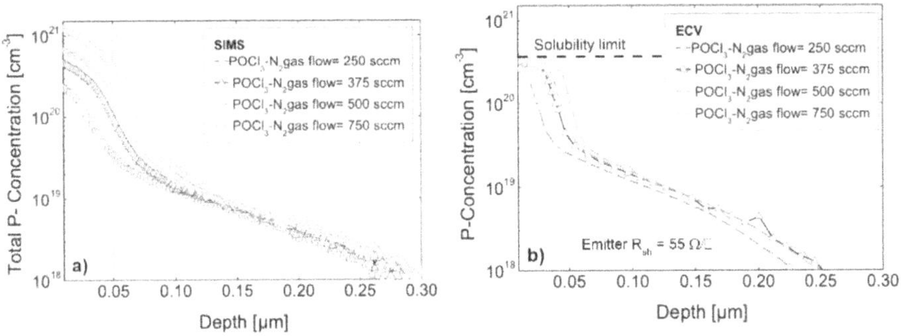

Figure 7.2 *Phosphorus concentration profiles from gas-phase dopant diffused c-Si as a function of $POCl_3$:N_2 ratios. (a) SIMS spectrum showing the total phosphorus concentration. (b) Electrically active phosphorus measured with the ECV method. Reprinted from [14], Copyright (2003), with permission from Elsevier.*

(ECV) measurements.* It can be seen that inactive forms of phosphorus are present at a higher than the solubility limit concentrations, particularly near the surface. Possible forms of such inactive phosphorus atoms have been related to phosphorus clusters and precipitates, which can be enhanced by oxygen diffusion from the same PSG layer [14].

In general, highly doped layers on lightly doped base material have been key in the minimisation of carrier recombination at the contacts by coulombic repulsion of carriers and, in the bulk, by effective defect gettering as discussed in Chapter 4. Yet, it is largely acknowledged that the gas-phase diffusion of phosphorus or boron into a silicon wafer generates stress on the silicon lattice as a result of atomic radius mismatch [15]. If not carefully designed and controlled, these stresses can exceed the critical resolved shear stress, resulting in the introduction of intrinsic point defects i.e. diffusion-induced dislocations and vacancies. Additionally, thermal diffusion with $POCl_3$ gas inevitably results in large thermal budgets and the diffusion of dopants all around the wafer. The former may result in unwanted defect activation whereas the latter results in a diffused region on the front, rear and edges of the wafer, which creates a low-resistance path for photo-generated carriers to freely flow from the front to the back of the cell, also severely impacting the performance of the cell. Hence an extra step in cell manufacture for edge junction isolation is required to remove the conductive path between the contacts and, hence, effectively increase the shunt resistance of the device. Otherwise, an inclusion of diffusion barriers on the back of the cell is required. Some applied approaches for the isolation include sawing, grinding, plasma etching, laser ablation and rear wet etching, with the latter being the most widely used currently, which also removes the diffused layer from the rear of the device [16, 17]. While all these techniques are effective at removing the resulting shunt, they can cause

*For more details on ECV/SIMS, see Section 7.4.1.2.

losses in cell performance and/or experience manufacture shortcomings in terms of cost and yield. Common performance losses include surface damage from plasma treatments, which increases J_0, reduction of active cell area and hence short-circuit current density by laser isolation, and partial emitter and front texture removal as the result of the wrap-around effect from the wet-etching processing. However, newer generation rear wet-etching tools have fewer issues with wrap-around.

7.2.2 Laser doping

The application of laser beam radiation for semiconductor processing has been known for many decades. Significant advances in the understanding of its theoretical and experimental base were heavily reported in the years preceding the 1990s by authors such as Young et al. [18], whose findings have served as building blocks for current laser-processing developments. Most of these studies arose from the need to reduce the inherent displacement damage produced by the energetic dopant ions during ion-implantation and for dopant activation for which additional annealing steps were commonly applied [19]. As a consequence, nomenclature such as laser annealing, laser thermal annealing and laser thermal processing was commonly used in literature in the early days. However, as the technology developed and interest grew in the application of lasers for dopant diffusion and activation, terms such as laser-induced diffusion and laser doping have become the most widely used terminologies within the photovoltaic research community.

The principle of laser doping is the resulting large energy transfer that occurs when a focused light beam is directed to a silicon crystal leading to an almost instantaneous (~ps) raise of lattice temperature to over 1 000 °C (or above 1 414 °C when in the molten state), which is followed by a similarly fast quenching process (~10^{10} K/s). Such processes, when sufficiently high energies are used, result in the simultaneous fast melting and recrystallisation of laser exposed areas in the silicon crystals, with the key advantage of a 'near-zero' thermal budget in the rest of the volume due to the selectivity of the exposure and the speed of the process. Manipulation of the laser fluence, wavelength and pulse length during the doping processes can then be used to finely tune the width and depth of melted regions in any given sample for an effective diffusion or activation of dopants.

In the early days, laser application research was devoted to laser annealing of ion-implanted semiconductors, largely because it was found that it allowed the concentration of active dopants to easily exceed the solubility limits, which is not possible with gas-phase diffusions. This is related to the fact that melting velocities occurring during pulsed laser anneal are in the order of 4 m/s, which are far from equilibrium conditions [20–22], and hence the equilibrium segregation coefficients do not apply. This results in substantially higher effective segregation coefficients and, therefore, higher concentrations of the impurities being incorporated into the solid. In the case of crystalline silicon, the surface concentration of phosphorus for an unlimited source is given by a solubility limit in the range of 3–6 × 10^{20}/cm^3 between 800 °C and 900 °C, which is the average diffusion temperature in industrial manufacture [23]. These active dopant limits, on the contrary, are easily surpassed

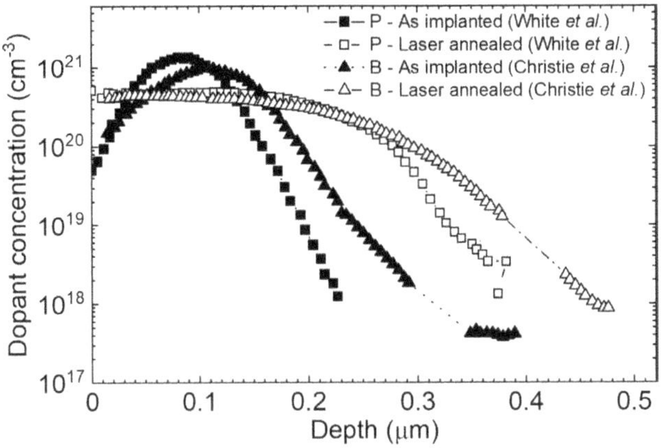

Figure 7.3 *SIMS profiles of phosphorus- and boron-implanted silicon before and after laser annealing. Data from [24] and [25].*

by laser irradiation reaching concentrations in the order of $1 \times 10^{21}/cm^3$. Figure 7.3 shows SIMS profiles of a ^{11}B and ^{31}P-implanted silicon samples before and after laser irradiation. Laser annealing of the initially clustered and disordered dopant atoms, which are responsible of high levels of recombination at the surface, the so-called dead layer, induces the recrystallisation of the material through instantaneous melting. The resulting rearrangement of the atoms, as is shown in the figure, will have a slightly deeper dopant profile but more laterally localised, less recombination active and more conductive.

At first, most applications of laser doping to solar cells focused on the formation of a homogeneous emitter after ion implantation in an attempt to avoid conventional thermal diffusions and the high thermal budgets that such processes impose on the substrates. This, while leveraging its ability to tailor the emitter profile, will be discussed in Section 7.3. Historically, however, the performance of such devices with full area laser-doped emitters has generally been inferior to the equivalent devices subjected to thermal diffusions. Potential reasons for such poor performance have been related to contact resistance issues due to the lack of heavy diffusions, low surface passivation, lack of effective anti-reflection coatings and the introduction of laser-induced defects [26, 27]. Only a few reports using this technology have shown good results since 2006, including the work from Esturo-Breton *et al.* [28] demonstrating efficiencies of 15.4% and Eisele *et al.* reporting efficiencies of 18.9% on p-type monocrystalline wafers using laser-doped emitter cell structures with evaporated contacts [29]. While these results demonstrate the potential of the approach, the technological complexity for the achievement of homogeneous doping, including the required laser powers and the fast processing, makes the adoption of this technology improbable.

As will be discussed in Chapters 8 and 9, there is much more promise and applicability of laser doping in some advanced solar cell structures which require

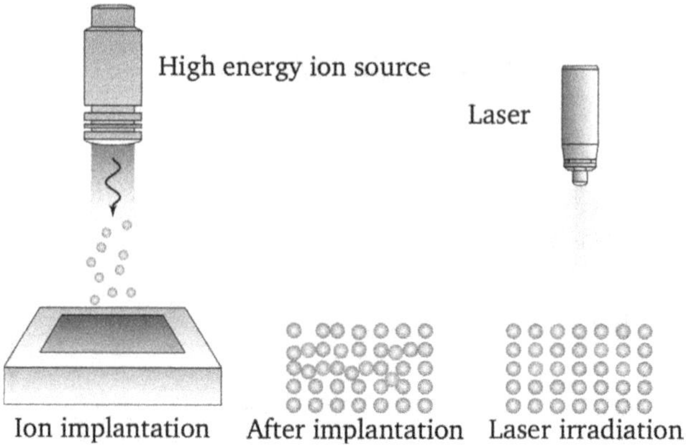

Figure 7.4 *Illustration of a laser annealing post-implantation process and recrystallisation*

localised doping for improved conduction at the contacts and reduced overall surface recombination. These replicate aspects of the record-breaking PERL cell, namely the emitter and localised back-surface field regions.

7.3 Approaches for laser doping

The earliest laser-doping methods for dopant activation involved laser annealing of ion-implanted silicon. In this process, ion implantation was followed by laser beam exposure to remove crystallographic defects by melting the surface and allowing it to epitaxially resolidify. During this process, sufficient energy is provided by the irradiated photons to the lattice (a few eV) which promotes the redistribution of the dopant atoms to substitutional sites. Once in the substitutional position, these atoms are easily ionised at room temperature aided by lattice vibrations (a few hundredths of eV). A schematic illustration of laser thermal annealing post-implantation is shown in Figure 7.4. This technique is still widely used in the MOS-device industry; however, it has not reached mainstream solar cell production as of yet, situation that may change as more advanced concepts enter the market.

In more recent years, and as this technology developed, a wide variety of dopant sources became available in gas, liquid or solid form. Doping from gaseous sources, also known as gas immersion laser doping (GILD), is based on laser-induced melting of silicon after surface adsorption of the dopants from a gaseous species. Most commonly used gases include PH_3 [30] and PCl_3 [31]. As an alternative option, laser irradiation to produce dopant molecules by gas-phase photolysis while holding the substrates at elevated temperatures [32], or using gas discharge implantation and XeCl annealing for the junction formation, has also been reported [33]. An illustration of the adsorbed dopant species to the clean silicon surface after the gas exposure and subsequent laser-induced diffusion is shown in Figure 7.5.

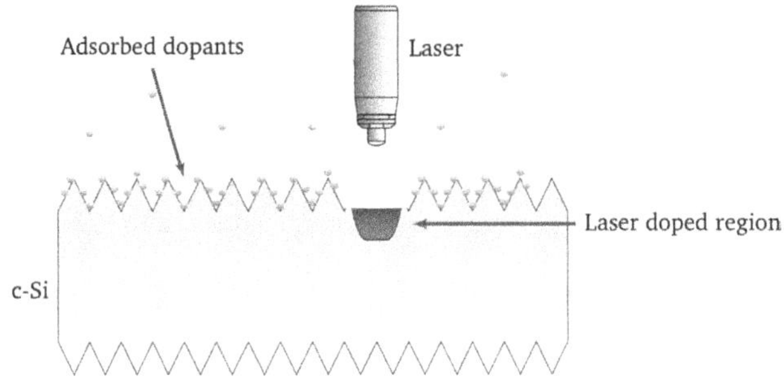

Figure 7.5 *Cross section of silicon device showing the adsorption of the dopant species after dopant gas exposure (GILD process). The dopant is incorporated into the silicon upon laser annealing.*

In the liquid form, dopants can be introduced via scanning the laser over the silicon while it is immersed in a solution as reported in the early 1980s by Stuck *et al.* [34], or via laser chemical processing (LCP) as it was more recently developed by Willeke and Kray [35, 36] and depicted in Figure 7.6. The principle of operation of this method is based on a laser light-guiding water jet developed in the mid-1990s, in which a laser beam is focused in a nozzle carrying pressurised water from a chamber [37]. Total internal reflection at the water–air interface, similarly to light in a glass fibre, allows the laser beam to be directed to the sample. Unlike dry-laser-doping techniques, in this wet-laser-doping technique there is no extra step required

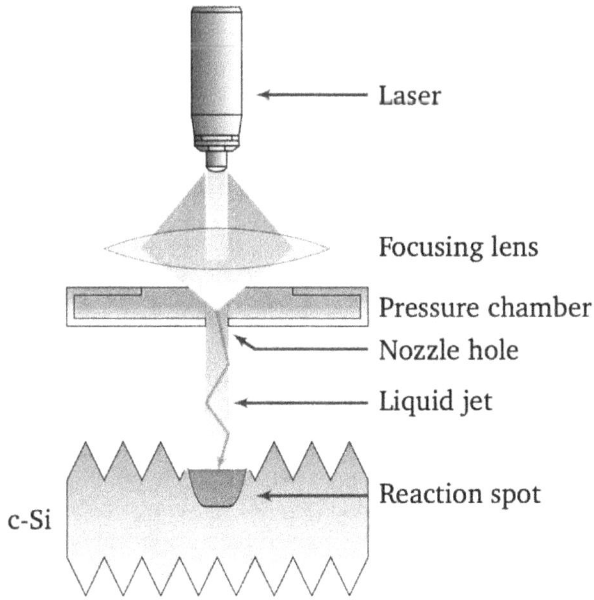

Figure 7.6 *Illustration of LCP for junction formation on a silicon device*

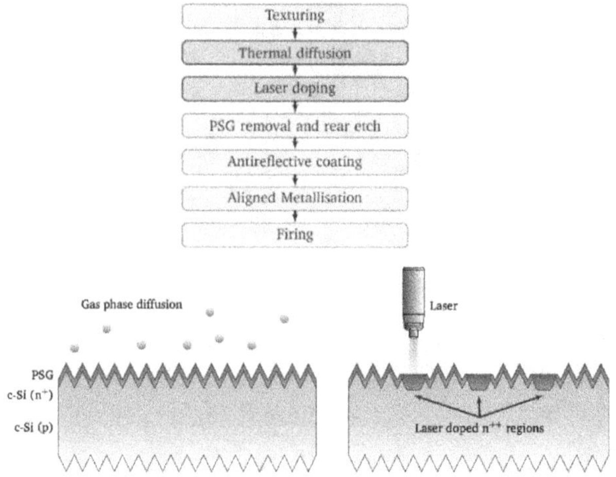

Figure 7.7 *Solar cell manufacturing sequence including laser doping (top). Cross section illustration of a LIMPID process using $POCL_3$ emitter diffusion and laser doping from PSG layer (bottom).*

to apply dopant sources onto the sample. Nevertheless, given the intrinsic complexity of laser beam reflection within the chemical solution jet, the laser beam reaching the sample normally consists of multiple freckles rather than a single focused spot, which may result in discontinuous laser-doped regions. Still, and even though its initial application focused on wafer cutting using water, changes in the liquid composition under the same principle have been used for crystal doping, and efficiencies in excess of 20% have been reported [36, 38].

If the dopant source is solid, the laser-doping process is typically termed laser-induced melting of pre-deposited impurity doping (LIMPID). In LIMPID processes, the dopant source may already be present in the wafer, as in the ion implantation case [39, 40], or in the PSG or BSG glass layer formed during gas-phase thermal diffusion [41–43]. Advanced cell designs such as the selective emitter concept, which will be described in Chapter 8, have been shown to successfully take advantage of these pre-diffused layers for full emitter formation to reduce contact resistance and maximise carrier collection. Figure 7.7 illustrates a cross section of a textured silicon wafer after thermal diffusion and subsequent localised laser doping, together with a manufacturing sequence including selective emitter formation. In the industrial approach, high-power and speed box-like laser beams are used for high yield and to retain the surface texturing.

Alternatively, a dopant source can be deposited onto the silicon surface through vacuum deposition [44, 45], sputtering [29, 46], spraying [47] or via SOD sources [48–50]. The latter two methods have gained the most attention in recent years as these are low-cost- and high-yield-processing methods which have already led to high efficiencies using a variety of substrates and dopants. In both cases the dopant source is solution-based and is deposited at room temperature onto the silicon

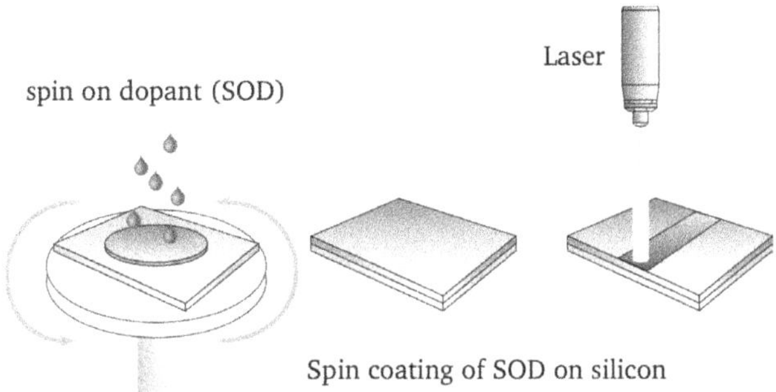

Figure 7.8 Illustration of a LIMPID process using a SOD method for deposition

surface. As the name suggests, spray coating involves atomisation of the solution using an ultrasonic nozzle and a steady substrate, whereas SOD requires of a less advanced dispenser such as a pipette, nonetheless, with the need of a spinner to rotate the substrate and distribute the solution. Once on the surface, the solvents are left to dry for subsequent laser exposure (see Figures 7.8 and 7.9). Although flexible, these methods require the application of an extra step, which reduces the appeal over the more straightforward PSG/BSG LIMPID process.

A major drawback of these approaches is the need for aligned metallisation when cell designs such as the selective emitter concept are to be applied. In this case either an extra step using photolithography for area masking or precise alignment with the screen-printing pattern is required. An alternative to this is to use a self-aligned method such as that developed at UNSW in the early 2000s [51]. In this method the dopant precursor, such as a SOD, is applied onto the dielectric film. The laser is then used to ablate the dielectric and diffuse the dopants in a single step. Posterior metallisation is then achieved selectively on the laser-exposed regions having the dielectric serving as a mask and anti-refection coating. This method can

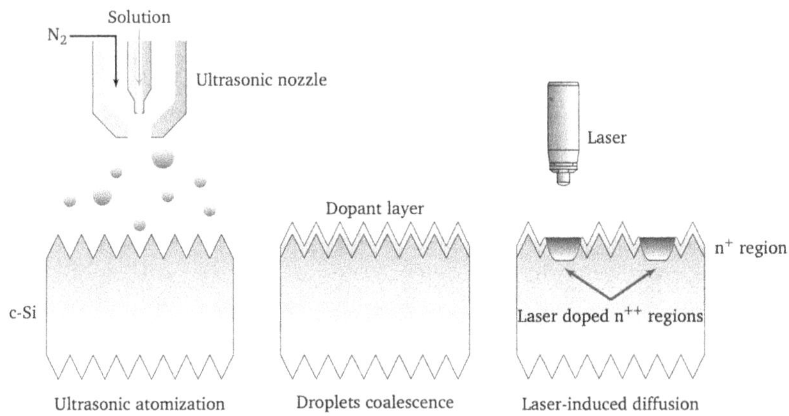

Figure 7.9 Illustration of a LIMPID process using ultrasonic spray deposition

be applied for the fabrication of screen-printed cells as well as more novel metallisation techniques such as metal plating [52] as will be discussed in Chapter 8. In a less common approach, the dopants are present in an overlying dielectric film such as aluminium oxide [53–55] or a phosphorus- or boron-doped silicon nitride [56] as will be discussed in detail in Chapter 9.

7.4 Impact of laser parameters on doping profiles

The process of laser doping involves a highly localised melting of the silicon. The molten silicon region allows for the impurity atoms to be driven from the surface into the silicon wafer. One of the most important parameters that determines the doping profile is the diffusion coefficients of the dopant materials in silicon. Importantly, these diffusion coefficients are significantly higher, by up to six to ten orders of magnitude, in liquid silicon compared to in the solid form at 977 °C [10]. The influence of temperature on the diffusion coefficients on common dopants is displayed in Figure 7.1. The fact that laser doping utilises these extremely high diffusion coefficients means that diffusion processes that would take several hours in the solid state at temperatures in the range of 800 °C to 1 000 °C can be achieved on a timescale of nano- to microseconds using laser doping. This relies on the molten region penetrating into the silicon, the depth of the molten region heavily dictates the achievable junction depth. Several other factors influence the diffusion properties during laser doping: the segregation coefficient, which describes the ratio between the diffusion coefficient in the solid- and liquid-silicon phase, plays a large role in determining the diffusion profile, while the concentration of diffused atoms will depend on the solubility limit. Control of these diffusion properties strongly depends on the type of laser system used.

This chapter details the influence of the primary parameters that influence the achievable junction properties in the laser-doping process, including:

- The type of laser system used (Q-switched vs continuous wave (C))
- Duration of exposure (Number of pulses and/or scan speed)
- Pulse length
- Repetition frequency
- The choice of laser wavelength
- Beam optics.

Before discussing the influence of these parameters, we detail the methods used to characterise laser-doped regions.

7.4.1 Characterisation of laser-doped regions

There are three major characterisation approaches to analyse the doping profile produced during laser doping; these include:

- Sheet resistance measurements

- ECV and/or SIMS
- Scanning electron microscopy (SEM) and/or electron beam-induced current (EBIC).

Each approach provides a different view of the effectiveness of laser doping.

7.4.1.1 Sheet resistance

The most basic metric to describe the doping is to measure the sheet resistance. This is performed by measuring the resistivity on the surface of the doped region, typically using a four-point probe arrangement. The sheet resistance of the diffused layer can be defined as

$$\rho_{sheet} = \frac{\rho}{t} \tag{7.1}$$

where × is the resistivity of the layer and t is the thickness of the diffused layer. The ρ is inversely proportional to the doping density, so measuring p_{sheet} gives a direct indication of the doping concentration. For the purpose of laser doping to form contacts in silicon solar cells, it is preferable to form a heavily diffused layer, as characterised by a low sheet resistance under the contacts. Due to the small widths of the laser-doped lines, the measurement of sheet resistance for laser-doped regions is more complicated than for a homogeneous emitter formed via thermal diffusions. If the spacing between measurement probes is large, the measurement requires the formation of a 2D box of laser-doped area on the surface, often with approximate dimensions of 1 cm × 1 cm. This is achieved by laser-doping parallel lines. The interface of these lines overlap, meaning that some regions will have an artificially high doping concentration. The overlapping lines could also potentially impact the dopant source in surrounding regions, meaning that these measurements may not give a true representation of the sheet resistance for a narrow finger with a single pass of the beam (e.g. 20–100 μm).

The sheet resistance provides information about how heavy or light the diffusion is. However, it does not show the depth profile of how the dopants are distributed from the surface.

7.4.1.2 ECV/SIMS

Accurate knowledge of the doping profiles is required for an understanding of both the J_0 value in simulations as well as for the contact resistivity. To do this, ECV profiling or SIMS can be used. During ECV profiling, a Schottky diode is formed at the interface of the semiconductor surface and an electrolyte, which forms a depletion region. Due to the ions formed in the depletion region, charge is stored, like in a capacitor. Analysis of the electrical characteristics of the diode can lead to the extraction of the number of electrically active dopants at a specific depth; however, the accuracy greatly depends on the accuracy of the contact area. The area used for these measurements is also in the range of 1 cm × 1 cm. The semiconductor is etched electrolytically to provide depth resolution of the desired dopant atoms.

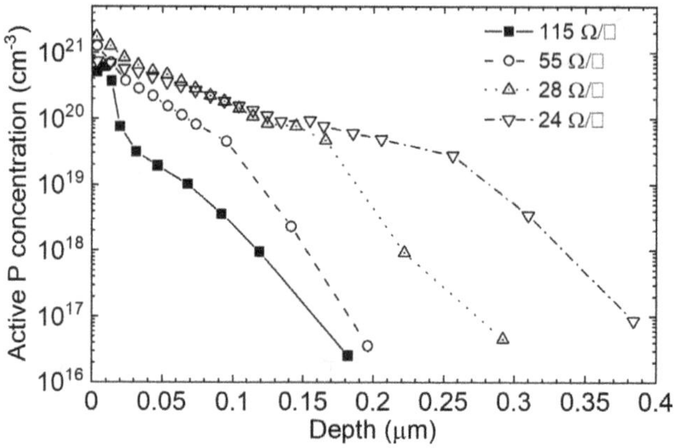

Figure 7.10 *ECV profiles of areas doped using a Q-switched frequency-doubled solid-state laser. The profile for four different regions, with varying sheet resistances, is shown. The sheet resistance was reduced by increasing the pulse energy density of the laser-doped region. Reproduced (digitised) from [58] under Creative Commons CC BY-NC-ND. Original image © 2011. Published by Elsevier Ltd. https://doi.org/10.1016/j.egypro.2011.06.129.*

For the SIMS approach, the surface is bombarded with primary ions via sputtering. These ions interact with the surface, causing the removal of the surface atoms, which are the 'secondary ions' [57]. The composition of these secondary ions is analysed using a mass spectrometer.

As the sputtering process physically etches the surface, the technique allows for an assessment of the concentration of the dopants atoms as a function of depth. Although both techniques provide a similar depth profile, there is one fundamental difference: SIMS shows the concentration of all dopant atoms whereas ECV only shows the concentration of electrically active dopants. If all dopants incorporated during laser doping are electrically active, the ECV and SIMS profile will be identical. Figure 7.10 shows a representative example of ECV profiles for laser-assisted phosphorous diffusions on p-type wafers [58]. The doping profiles were performed using a Q switched, frequency-doubled solid-state laser (λ = 532 nm). The sheet resistance of the emitter was decreased from 115 Ω/□ to 24 Ω/□ by increasing the pulse energy density of the laser from 3 J/cm^2 to 4.8 J/cm^2. Four ECV depth profiles are displayed for varying sheet resistance laser-doped emitters, which illustrates the impact of the laser parameters on the diffusion profile.

As discussed, with sheet resistance measurements, depth profiles obtained from either ECV or SIMS are limited by the physical dimension of the measurement. As the formation of a series of laser-doped lines are required on a macroscale, microscale inhomogeneities are not captured by the measurement. This spatially averaged measurement area is not sensitive to the nuances of doping density variations across the laser-doped line, the nature of which depends on the beam profile. A deeper understanding of the 2D-doping distribution in a laser-doped line is crucial to better understand the potential efficiency improvements achievable by laser doping in silicon solar cells. To investigate the potentially misleading results from ECV, Weber *et al.* used scanning spreading resistance microscopy to understand the doping profiles of laser-doped phosphorous selective emitters [59]. This approach led to better resolution of the laser-doping profiles. Three key pieces of understanding, previously unobtainable using ECV, resulted from this work:

1. The doping close to the surface (<0.2 μm) is largely homogeneous.
2. The monotonic reduction in doping as a function of depth seen in ECV measurements is a simplification, the true profile varies spatially and can contain step changes and plateaus.
3. The doping near the p-n junction can be inhomogeneous.

This improved methodology will provide a more accurate determination of the specific contact resistance and the J_0 resulting from recombination at the interface of the emitter and metal. This improved understanding will allow for more accurate inputs for simulations.

7.4.1.3 SEM/EBIC

To provide an understanding of the spatial distribution of the laser-doping region, SEM and EBIC measurements are used. SEM images provide a view of the cross-sectional area, while the nature of the diffused layer is uncovered using EBIC. For EBIC, the electron beam used during SEM is absorbed at the surface, which generates electron–hole pairs. This generates an electron beam current, which is converted to a signal. Each pixel in the SEM can then be converted to a 'brightness', corresponding to the magnitude of the EBIC signal. Cleaving of the sample during preparation means the recombination rate at the surface is very high, implying that the only contribution to the EBIC signal is due to absorption within one diffusion length of the p-n junction. This allows for a clear spatial depiction of the junction properties and can provide an accurate measure of the junction depth. Figure 7.11 shows an example of an SEM/EBIC of a laser-doped sample.

7.4.2 Type of laser

Early work on laser doping used Q-switched lasers, which produce a pulsed output beam. The pulsed operation means that the laser does not emit light at a constant intensity, rather, the output occurs during emission of light in short pulses, with a certain pulse length. This approach allows for the generation of higher power densities compared to lasers operated in the CW mode. Initial attempts at laser doping in silicon used Q-switched lasers with a pulse length in the range of 10–30 ns. This short duration only allowed for junction depths in the range of 200–300 nm [48, 60]. This work showed that when using a pulsed laser, the junction depth did not depend on the number of laser pulses, rather, it was dependent on the depth of the molten silicon region. Increasing the number of pulses for a given energy density would

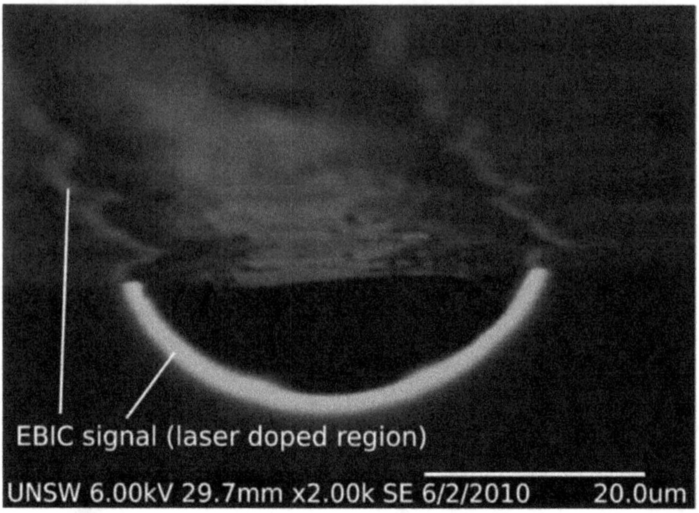

Figure 7.11 *Combined SEM/EBIC image* with *×2 000 magnification showing a laser-doped region. The bright EBIC signal shows the location of the p-n junction. Reprinted from [60], Copyright (2013), with permission from Elsevier.*

simply lead to a redistribution of the dopants within the defined junction depth [48]. The depth of the molten region in silicon depends heavily on the pulse energy density and pulse length. Figure 7.12 displays the relationship between the pulse energy density, pulse duration and melt depth [61]. The melt depth in the silicon is maximised when the surface of the silicon reaches the boiling temperature of 2 900 K. This is more easily achieved when the pulse length is short – it becomes more difficult as the pulse length is increased, as less of the laser energy is absorbed at the surface. As such, increasing the length of the pulse requires a larger pulse energy density to achieve the same melt depth.

By increasing the pulse length to 60 ns, White *et al.* were able to produce a junction depth of about 0.7 μm [39]. Hameiri *et al.* fabricated single side laser-doped solar cells using laser doping to form the contacts [62]. A Q-switched Nd:YVO_4 diode laser (λ = 532 nm) with a repetition rate of 100 kHz and scanning speed of 10 mm/s was used for the laser doping. A phosphorous SOD was used to supply the dopant atoms. Figure 7.13 shows the SIMS measurements of the doping profile of the laser-doped lines. Three different diode currents, which are representative of the pulse energy density, were used. This indicates that the junction depth can be increased by increasing the energy of the laser pulse, and a maximum junction depth of about 1.5 μm was achieved. The authors showed that the junction depth depends on the rate-limiting process. Initially, the limiting process is the laser pulse energy density; however, as the threshold energy density is reached, the dopant source becomes rate limiting. Although increasing the energy density can increase the junction depth, it is important to note that if the energy density is too high, the

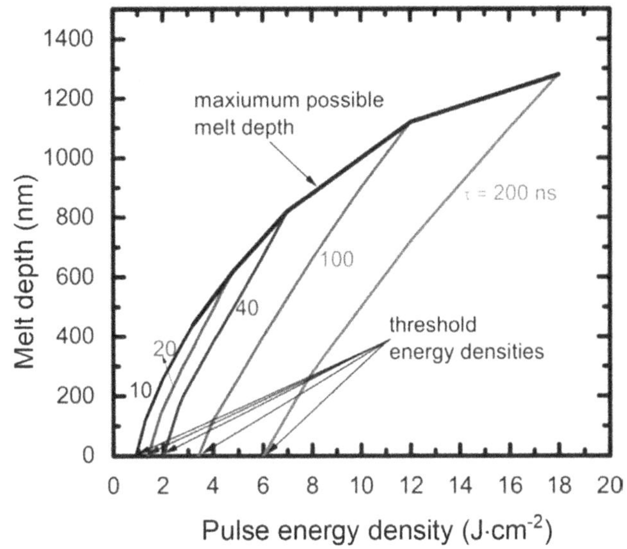

Figure 7.12 *Simulated depth of the molten silicon as a function of the pulse energy density. The laser pulse duration in the model is varied from 10 ns to 200 ns. Reproduced with permission from [61].*

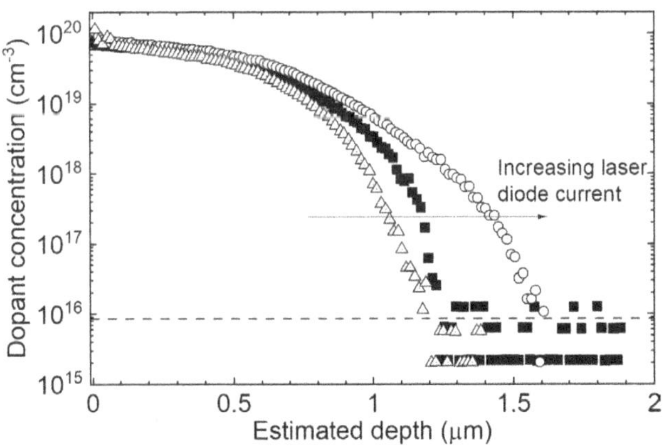

Figure 7.13 *Junction depth profiles measured using SIMS, for laser-doped lines formed using a Q-switched Nd:YVO$_4$ diode laser (l = 532 nm) with a repetition rate of 100 kHz and scanning speed of 10 mm/s. Three different diode currents were used, indicating the relationship between laser energy density and junction depth. The dashed line represents the background boron-doping level. Reprinted from [62] Copyright (2011), with permission from Elsevier.*

process can cause unwanted ablation of the silicon. This represents a fundamental limitation for the use of low-frequency Q-switched lasers.

Even with increased pulsed length and applying multiple pulses, the achievable junction depth for low-frequency Q-switched pulse lasers is limited to <2 μm [63]. Additionally, pulsed lasers can cause more laser-induced damage. When the laser pulse is switched off, the silicon undergoes recrystallisation. When the silicon undergoes many pulses, the thermal cycling induced by the melting and subsequent recrystallisation of the same volume of silicon is one of the primary causes of laser-induced defects. For these reasons, research interest has shifted to the use of CW lasers [64, 65]. CW lasers are fundamentally different to pulsed lasers as the energy output from the laser is uninterrupted. For the purpose of laser doping, the long effective pulse duration of CW lasers, determined by the uninterrupted energy supply, allows for the molten region to penetrate deeper into the silicon, allowing for much deeper junction depths. Subsequently, the uninterrupted supply from CW lasers means that the thermal conductivity in the surrounding silicon can be used to enhance the depth of the molten region, which is not possible with a Q-switched laser with fast switching. Mode-locked quasi-continuous wave (Q-CW) lasers, which use a much higher pulse repetition frequency than Q-switched lasers, can have a similar effect. The avoidance of the thermal cycling also leads to significantly less laser-induced defects, which is advantageous for the efficiency of laser-doped silicon solar cells. Hameiri *et al.* showed that the recrystallisation of silicon from the molten state occurs epitaxially according to the orientation of the substrate [66]. The authors showed that after recrystallisation, there were no observable disorientation of the crystalline structure and thus laser-induced defects occur primarily at the surface rather than in deeper regions. Thus, the use of CW to form deep junctions does not lead to significant defects that penetrate deep into the substrate, which is beneficial for cell operation.

7.4.3 Diffusion profile in laser-doped regions

Conventional solid-state thermal diffusions typically produce a doping profile which closely resembles a Gaussian distribution. The concentration of diffused dopants as a function of depth and time can be modelled using the following expression:

$$C(x,t) = \frac{Q}{\sqrt{\pi Dt}} exp\left(\frac{-x^2}{4Dt}\right) \tag{7.2}$$

where Q is the total dose of dopants, D is the diffusion coefficient at that temperature and t is the time of the process. For this case, the diffusion occurs entirely in the solid state. During laser doping, the diffusion of dopant atoms occurs in the liquid state. As the diffusion coefficient is orders of magnitude higher than in the solid state, the depth of the liquid–solid interface becomes crucial in determining the depth and also the shape of the dopant profile. For low-frequency Q-switched lasers, the depth of the molten region is shallow. In this case, the diffusion of dopants is limited by the shallow depth of the molten region and the diffusion profile can vary significantly from the typical standard Gaussian structure. Such shallow profiles are often

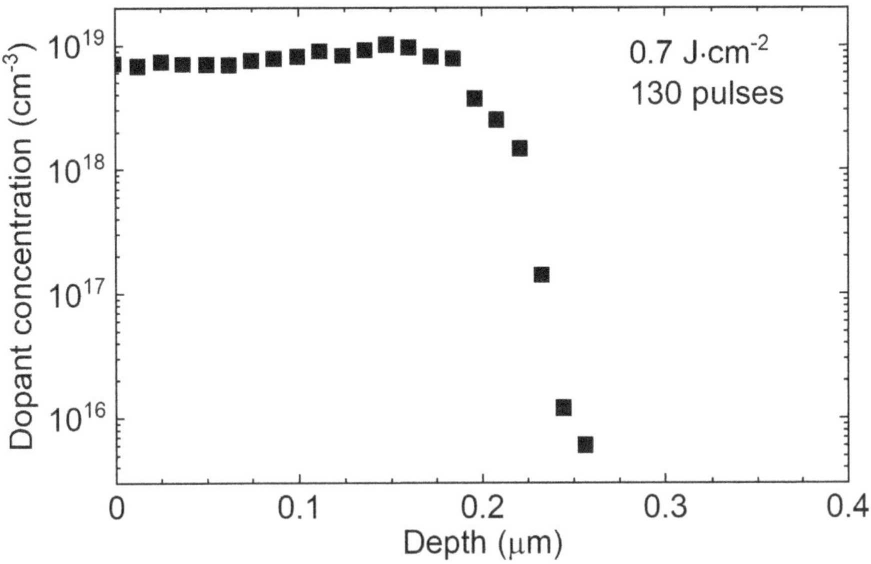

Figure 7.14 *Box-like diffusion profile formed using gas-immersed laser doping. Reprinted from [48]. Copyright (1993), with permission from Elsevier.*

described as 'box-like' [48, 67]. Figure 7.13 shows an example of a box-like diffusion profile with a very shallow depth of about 0.2 μm.

In contrast, CW lasers lead to a much deeper molten region. In this case, depending on the speed, the solid–liquid interface does not limit the dopant depth. In these cases, the dopant profile can be modelled as Gaussian for most cases, even for junction depths up to 5 μm. Figure 7.14 displays the fit for a Gaussian distribution for CW laser processing using a variety of laser speeds (2–10 m/s), indicating that the melt depth is much larger [68]. However, for a sufficiently slow speed, such as 0.5 m/s, the profile is more box-like, indicating that the diffusion is again limited by the depth of the molten region (Figure 7.15). In this example (see Figure 7.16), the diffused region extends 10 μm into the silicon.

An alternative approach to either the low-frequency Q-switched laser or a CW laser is a Q-CW laser source. This can be achieved by operating a pulsed laser with a very high pulse repetition frequency, in the order of tens of MHz (e.g. 80 MHz or 120 MHz). Another common property of Q-CW lasers is the short pulse duration (on the order of tens of picoseconds) and very high output power. A schematic comparison of the power output for a CW and Q-CW laser is shown in Figure 7.17. The continuous, uninterrupted output of the CW laser is shown by the horizontal line. For the Q-CW laser, the power emission occurs in short pulses that are very close together. The peak power output for the Q-CW is much higher than the constant output of the CW laser; however, as the pulses are so closely spaced, the average power output from the two lasers can be comparable. This is displayed schematically in

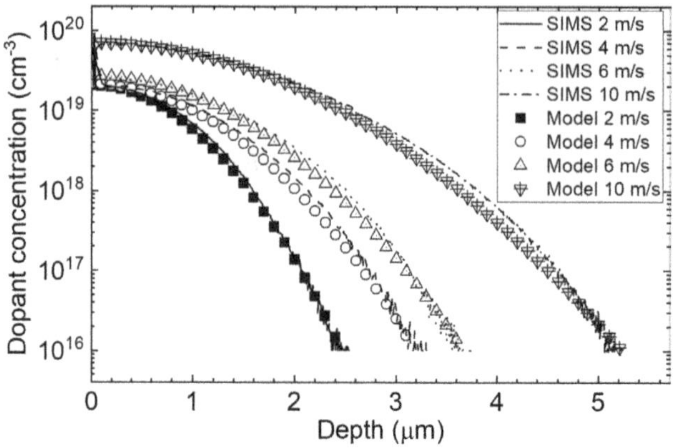

Figure 7.15 Comparison of SIMS doping profile with Gaussian fit for laser-doped regions formed using a 532-nm CW laser. Reproduced from [68].

the image (although the integrated power is not to scale). The total power output of the CW laser over a unit time, shown in cross hatching, is comparable to the average output from the pulsed Q-CW laser over the same duration, represented by the dark grey regions. The higher peak power, combined with the short pulse duration, means that Q-CW lasers can rapidly heat and melt the silicon, possibly with a higher efficiency than for CW lasers. This may allow for reduced processing speeds. Although the pulses are very short, the accumulation effect facilitated by the very fast switching allows for an expansion of the molten region. This allows for much larger junction depth compared to low-frequency Q-switched lasers. Although the junction depth achieved using Q-CW lasers can be comparable to that of CW lasers, the doping profile deviates from a Gaussian distribution. This deviation from a Gaussian-shaped profile is caused by the increased impact of the solid–liquid interface, as discussed earlier. This is likely due to the reduced duration in the molten state for the deepest molten regions, with a partial solidification in between successive pulses [68].

7.4.4 Impact of laser speed

For CW lasers, the junction depth can be simply controlled by the laser speed. Hallam *et al.* explored the impact of the laser speed on the laser-doped profile using a 532-nm CW laser with a maximum power of 13 W [69]. Combined SEM/EBIC images shown in Figure 7.16 provide a clear picture of the impact of the laser speed on the laser lines. When the samples are cleaved during sample preparation, the SRV is very high, meaning that only electron–hole pairs within a diffusion length of the junction contribute to the measured EBIC signal, allowing for a precise spatial determination of the junction properties. As such, the bright regions in the images represent the junction. The series of images displays a very clear trend. For the 1 m/s case, a clearly defined deep junction, with junction depth of about 10 μm, is

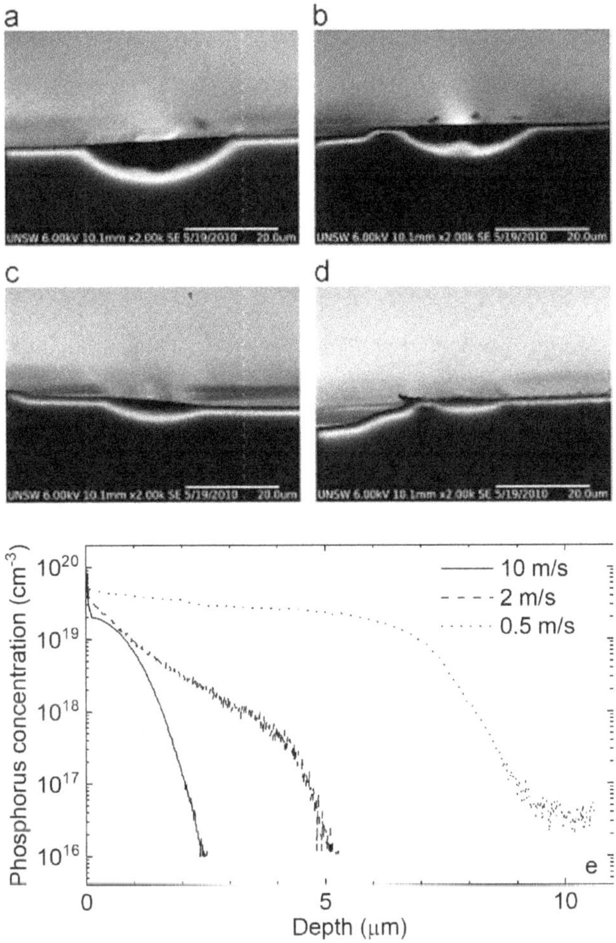

Figure 7.16 *Combined SEM/EBIC images of n-type laser doping performed at a laser speed of (a) 1 m/s, (b) 2 m/s, (c) 5 m/s and (d) 10 m/s. A 532-nm CW laser was used. The bright line in the image corresponds to the p-n junction. (right) SIMS-doping profiles of n-type laser doping performed at various scanning speeds. Reprinted from [69], Copyright (2013), with permission from Elsevier.*

observed. Increasing the laser-doping speed systematically reduces the depth of the junction. This result matches the theory, as the faster scanning speeds will reduce the time of laser illumination on the surface, which reduces the amount of time during which the silicon is in a molten state.

The observation in these EBIC images is confirmed by the doping profiles measured using SIMS, shown in Figure 7.16. For the slow laser speed (0.5 m/s), a junction depth of about 10 μm is observed. Increasing the speed, thus reducing the

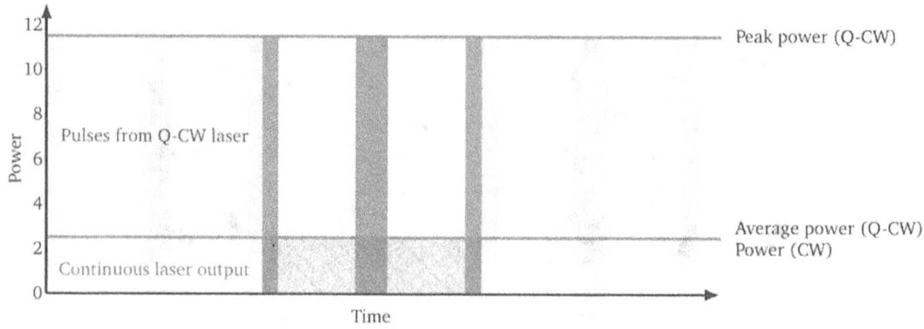

Figure 7.17 *Comparison of laser emission for CW and Q-CW. The CW laser output remains constant over time, as indicated by horizontal line. Q-CW lasers emit high-energy pulses over a short period of time, with a very high switching frequency. As the switching frequency is very fast, the average power emitted from a Q-CW laser can be comparable to that of a CW laser. This is illustrated schematically in the figure; the amount of power emitted by the Q-CW shown in the dark grey regions is equal to the continuous output of the CW laser at lower power, shown by the cross hatch. The output of the laser is drawn as a square pulse for simplicity.*

amount of time the silicon is in a molten state, vastly shifts the diffusion profile. The 10 m/s case produces a much shallower junction profile, with a junction depth of about 2 μm. The very deep junction achieved using slow scanning speeds is particularly advantageous for laser doping for contact formation in silicon solar cells, as it allows for very low sheet resistivity values and assists in shielding minority carriers from the metal–silicon interface.

Hallam *et al.* extended this knowledge to fabricate laser-doped selective emitter industrial PERC (i-PERC) solar cells with an efficiency of 20.5% [70]. A thorough explanation of the selective emitter concept is detailed in Chapter 8. In fabricating this selective emitter solar cell, the authors explored the impact of the SOD layer on the sheet resistance of the emitter. A phosphorous-containing SOD was coated on top of the existing emitter, which had a sheet resistance of approximately 100 Ω/□. The sheet resistance is plotted as a function of laser-processing speed for three different SOD conditions in Figure 7.18. For the case with no SOD (black squares), the sheet resistance was reduced compared to the initial emitter profile. For this case, where no additional dopants are introduced, the laser process is simply redistributing the dopants from the existing emitter region and possibly from the dead layer at the surface. Changing the laser speed displays a clear trend for the sheet resistance measurements. For high speeds, 10 m/s, the laser process only slightly reduced the sheet resistance, to about 70 Ω/□. At this laser speed, the silicon is only molten for a short time. Decreasing the laser speed, which increases the time the silicon is molten, led to a reduction in sheet resistance. A value as low as 40 Ω/□ was achieved

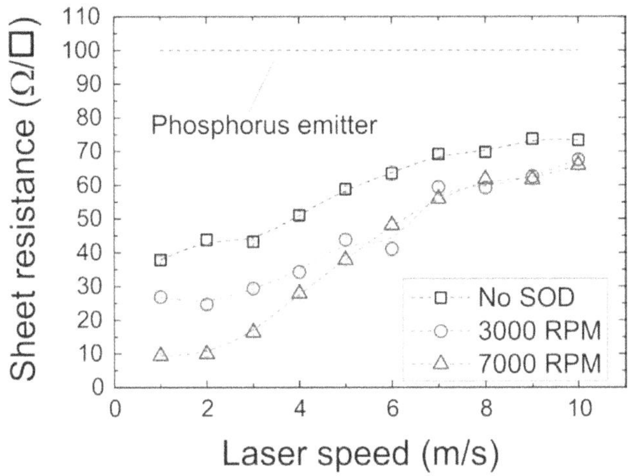

Figure 7.18 *Sheet resistance caused by dopant distribution during laser doping, plotted as a function of laser-processing speed. Reprinted from [71], Copyright (2015), with permission from Elsevier.*

with a laser-doping speed of 1 m/s. Modelling using PC1D [72] suggested that this reduction in sheet resistance was related simply to a redistribution of dopants.

The impact of two different SOD films on the sheet resistance was compared to the control case. The films were spin coated at two different speeds, 3 000 rpm and 7 000 rpm. The spin-coating speed impacts the film thickness, whereby faster spin-coating speeds lead to thinner films and hence, less phosphorous can be incorporated. Conventional wisdom would suggest that the thicker films would lead to lower sheet resistance values, due to the increased number of dopants available for diffusion into the silicon. Surprisingly, the thinner SOD film, spin coated with a speed of 7 000 rpm, leads to the biggest reduction in sheet resistance, with a minimum value of 10 Ω/□. This result highlights the complicated nature of laser doping, which depends on the interactions between the laser beam and the dopant source, dielectric layers and the silicon surface. The authors suggest that the thicker SOD film changes the optics of the laser beam due to increased absorption and reflection, which makes it more difficult to melt and thus redistribute dopant atoms in the underlying silicon.

7.4.5 *Choice of wavelength*

Another important parameter is the wavelength of the laser. The absorption coefficient of silicon at 300 K (~26 °C) varies by approximately five orders of magnitude across the absorption range [73], which means that the absorption depth of the laser beam will depend heavily on the wavelength chosen. The wavelength of a laser is a material property. Diode-pumped solid-state lasers used for industrial applications typically involve wavelengths generated by neodymium-doped media and frequency-conversion optics [74]. The fundamental emission from such lasers occurs at 1 064 nm. This wavelength is close to the band edge, the absorption in

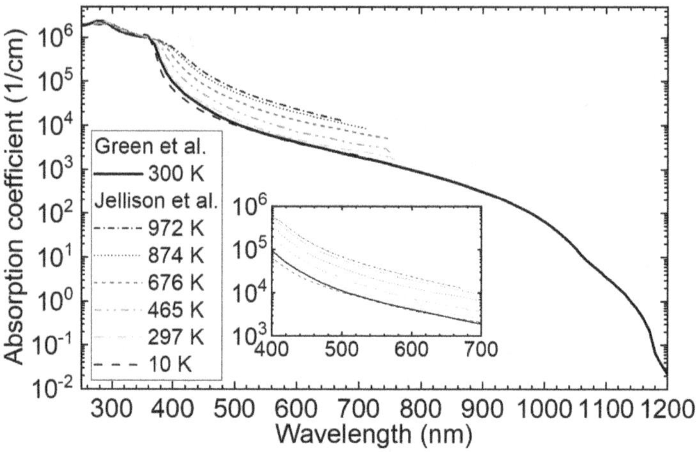

Figure 7.19 Absorption coefficient of silicon at different temperatures. Data from [73, 77].

silicon at this wavelength is very weak, with a penetration depth on the order of 100 μm [75]. This is approaching the thickness of the wafer and is thus too deep for solar cell applications at room temperature. Using a laser with this fundamental frequency has also shown increased laser-induced defects due to cracking. The third harmonic of these lasers emits at a wavelength of 355 nm. Silicon is strongly absorbing at this wavelength, so laser light would be absorbed at the surface. The penetrated depth would be expected to be in the range of 10–100 nm. This depth of the molten region is likely too shallow for the most laser-doping applications for silicon solar cells. The second harmonic of neodymium-doped laser occurs in the visible spectrum at approximately 532 nm, which corresponds to green colour. This corresponds to an absorption depth in the order of microns, which is more appropriate for solar applications. It is well known that the absorption coefficient of silicon is dependent on temperature [76]. Therefore, all 1 064 nm, 532 nm and 355 nm wavelengths can be used for laser doping. By systematically analysing the absorption coefficient as a function of temperature, Jellison *et al.* showed that the absorption coefficient can change by as much as a factor of five in the temperature range from 24 °C to 700 °C [77]. Figure 7.19 displays the absorption coefficient of silicon measured at temperatures ranging from 10 K to 972 K (−263 °C to 699 °C). This illustrates the large impact temperature has on the optical properties of silicon, which is particularly pronounced at the wavelength of interest ($\lambda = 532$ nm). The increased absorption at higher temperature causes more melting, which can increase the chances of undesirable ablation of the silicon. Further complicating the properties of the molten region is the fact that the absorption coefficient is different in the molten and solid state. Jellison *et al.* showed that the absorption coefficient can be more than two orders of magnitude higher in liquid silicon compared to the solid state at room temperature [78]. In fact, at the crucial wavelength of $\lambda = 532$ nm, the absorption coefficient in molten silicon is ~150 times higher than at room temperature. This means that the laser light will be absorbed in a very shallow region if the silicon undergoes a phase change from solid to liquid. The much larger absorption coefficient for molten silicon can cause problems, as the increased absorption/heating can lead to unwanted ablation of the silicon. It is thus important to judiciously choose the laser properties for laser doping, to match the desired junction properties.

7.4.6 Beam optics

Most laser beam profiles are Gaussian in nature. This means that the laser irradiance reduces from the centre of the beam to the edges, meaning that the 2D output profile is non-uniform. This profile is not ideal for laser-doping applications, as the edges or 'wings' of the Gaussian profile mean that the doping is shallower on the edges of the line and also that some areas outside of the line receive undesirable laser illumination. This can lead to undesirable recombination, which could reduce the V_{OC}. This is particularly true for laser doping using high laser powers. An alternative beam profile is referred to as a 'top hat' profile. In this case, the beam output is much more constant across the spot diameter and falls away sharply at the edges. The top hat beam profile can be achieved by either simply truncating the Gaussian profile to create a flat top or the application of beam-shaping optics to convert the Gaussian beam profile to flat top. Beam-shaping optics allow for a more efficient use of the laser energy; however, they are more costly.

Figure 7.20 displays a comparison of the 2D laser irradiation from both a Gaussian and top hat beam profile. The constant output in the desired area and abrupt reduction in irradiance on the edges means that top hat beam profiles allow for a higher efficiency and a more precise control over the area of silicon that is molten. This beam output will have an impact on the 2D-doping profile of the laser-doped regions. A thorough comparison of the difference doping profiles achieved using Gaussian or top hat beam optics is currently missing from literature. In particular, a deeper understanding of the impact of the different profiles at the outside fringes of the laser-doped line is needed. Further optimisation of this property could allow for further advancements in emitter design for industrial silicon solar cells.

7.5 Chapter summary

Laser processing has become popular in a wide range of industrial applications including electronic devices, medical tools and aerospace technology. The first demonstration of using laser doping to form a p-n junction in silicon occurred in 1968; however, the use of laser processing in the photovoltaics industry only became widespread in the early 2000s, as researchers were focused on implementing higher efficiency solar cell designs in production. In the solar cell industry, diffusion processes have typically involved a thermal diffusion at high temperatures. Nevertheless, these processes are lengthy, introduce a large thermal budget to the materials and can induce the formation of defects in large concentrations. Hence, laser doping raises as a very attractive alternative to commonly applied methods. Amongst the key advantages it provides are very precise control of the diffusion profile, minimal

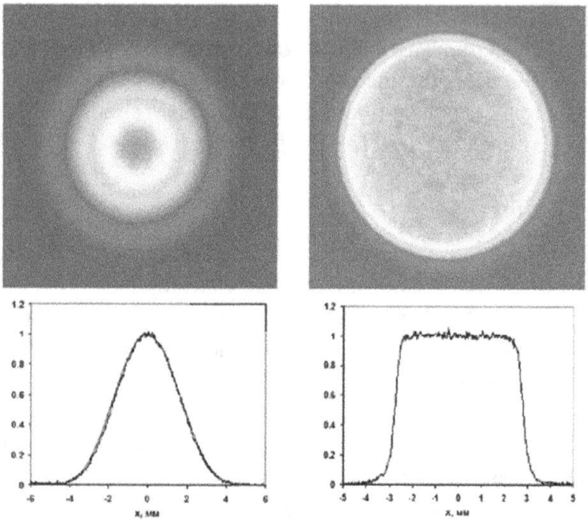

Figure 7.20 Comparison of the beam profile of a Nd:YAG laser. The intensity map and 2D intensity distribution on the left side shows the original Gaussian beam profile, the profile on the right is after the top hat beam-shaping optics, illustrating the more uniform distribution. Reproduced from [79] with permission. © (2016) COPYRIGHT Society of Photo-Optical Instrumentation Engineers (SPIE).

unwanted thermal budgets and defect formation, while maintaining the fast processing required in large scale manufacture. It also highlights the ability to form deep heavily doped regions with junction depths of up to 10 μm when using CW lasers, to overcome limitations of using nanosecond pulse length lasers that result in the complete solidification of the silicon between successive pulses. To fully utilise the possible benefits of lasler doping, however, it is important to understand the advantages and disadvantages of the available approaches and the parameters involved during laser processing. This chapter provided an overview of the background of laser doping for solar cell devices, a technology that is on its way to becoming more prominent as industry continues to move towards high-efficiency cell designs. It also highlights key aspects of laser doping in silicon solar cells that are currently missing from literature, in particular, the importance of gaining a deeper understanding of potential variations in the 2D-doping profile across a thin laser-doped line, which will lead to more accurate predictions for the future potential of this approach.

References

[1] Fairfield J.M., Schwuttke G.H. 'Silicon diodes made by laser irradiation'. *Solid-State Electronics*. 1968;**11**(12):1175–6.

[2] Green M.A. 'The passivated emitter and rear cell (PERC): from conception to mass production'. *Solar Energy Materials and Solar Cells*. 2015;**143**(2):190–7.

[3] Green M.A. Solar Cells: Operating Principles, Technology, and System Applications [online]. 1981. Available from http://www.amazon.com/Solar-Cells-Applications-Prentice-Hall-electronics/dp/0138222703.

[4] Li H., Kim K., Hallam B., Hoex B., Wenham S., Abbott M. '$POCl_3$ diffusion for industrial Si solar cell emitter formation'. *Frontiers in Energy*. 2017;**11**(1):42–51.

[5] Suh D. 'Efficient implementation of multiple drive-in steps in thermal diffusion of phosphorus for PERC solar cells'. *Current Applied Physics*. 2018;**18**(2):178–82.

[6] Rubin L., Poate J. 'Ion implantation in silicon technology'. *The Industrial Physicist*. 2003;**9**(3):12–15.

[7] Liu T.M., Oldham W.G. 'Channeling effect of low energy boron implant in'. *IEEE Electron Device Letters*. 1983;**4**(3):59–62.

[8] Narayan J., Young R.T., White C.W. 'A comparative study of laser and thermal annealing of boron-implanted silicon'. *Journal of Applied Physics*. 1978;**49**(7):3912–7.

[9] VDMA. *International Technology Roadmap for Photovoltaic (ITRPV). 11th edition*; 2020.

[10] Tang K., Øvrelid E.J., Tranell G., Tangstad M. 'Thermochemical and kinetic databases for the solar cell silicon materials' in Nakajima K., Usami N. (eds.). *Crystal Growth of Si for Solar Cells*. Berlin, Heidelberg: Springer; 2009. pp. 219–51. Available from https://doi.org/10.1007/978-3-642-02044-5_13.

[11] Basnet R., Phang S.P., Samundsett C., *et al.* '22.6% efficient solar cells with polysilicon passivating contacts on n-type solar-grade wafers'. *Solar RRL*. 2019;**3**(11):1900297–5.

[12] Li H., Hallam B., Wenham S., Abbott M. 'Oxidation drive-in to improve industrial emitter performance by $POCl_3$ diffusion'. *IEEE Journal of Photovoltaics*. 2017;**7**(1):144–52.

[13] Min B., Wagner H., Dastgheib-Shirazi A., Kimmerle A., Kurz H., Altermatt P.P. 'Heavily doped Si:P emitters of crystalline Si solar cells: recombination due to phosphorus precipitation'. *(RRL) – Rapid Research Letters*. 2014;**8**(8):680–4.

[14] Dastgheib-Shirazi A., Steyer M., Micard G., Wagner H., Altermatt P.P., Hahn G. 'Relationships between diffusion parameters and phosphorus precipitation during the $POCl_3$ diffusion process'. *Energy Procedia*. 2013;**38**:254–62.

[15] Cousins P.J., Cotter J.E. 'The influence of diffusion-induced dislocations on high efficiency silicon solar cells'. *IEEE Transactions on Electron Devices*. 2006;**53**(3):457–64.

[16] Hahn G., Spiegel M., Hauser A., *et al.* 'Comparison of different techniques for edge isolation'. Proceedings of the 17th European Photovoltaic Solar Energy Conference; Munich, Germany; 2001. pp. 1739–42.

[17] Reinders A., Verlinden P.J., van sark W., Freundlich A. *Photovoltaic Solar Energy: From Fundamentals to Applications*. Chichester, UK: John Wiley and Sons, Ltd.; 2017.

[18] Wood R.F., Willardson R.K., White C.W., Young R.T. *Pulsed Laser Processing of Semiconductors*. Academic Press; 1984Orlando, USA.

[19] Young R.T., Wood R.F., Narayan J., White C.W., Christie W.H. 'Pulsed laser techniques for solar cell processing'. *IEEE Transactions on Electron Devices*. 1980;**27**(4):807–15.

[20] White C.W., Wilson S.R., Appleton B.R., Young F.W. 'Supersaturated substitutional alloys formed by ion implantation and pulsed laser annealing of group-III and group-V dopants in silicon'. *Journal of Applied Physics*. 1980;**51**(1):738–49.

[21] Poate J.M. 'Laser annealing of semiconductors'. *Journal of Soviet laser research*. 1982;**6**:395–404.

[22] Nickel N.H. *Laser Crystallization of Silicon*. **3**. San Diego, CA, USA: Elsevier Inc; 2003. pp. 54–67.

[23] Trumbore F.A. 'Solid solubilities of impurity elements in germanium and Silicon*'. *Bell System Technical Journal*. 1960;**39**(1):205–33.

[24] White C.W., Narayan J., Young R.T. 'Laser annealing of ion-implanted semiconductors'. *Science*. 1979;**204**(4392):461–8.

[25] Christie W.H., White C.W. 'Sims depth-profiling studies: the effect of laser annealing on the distribution of ion-implanted boron in silicon'. *Surface Science*. 1980;**100**(1):43–52.

[26] Abbott M.D. *Advanced Laser Processing and Photoluminescence Characterisation of High Efficiency Silicon Solar Cells PhD Thesis*. Sydney, Australia: The University of New South Wales; 2006.

[27] Carlsson C., Esturo-Breton A., Ametowobla M., Kohler J.R., Werner J.H. 'Laser doping of textured monocrystalline silicon wafers'. The 22nd European Photovoltaic Solar Energy Conference2007. pp. 1593–5.

[28] Esturo-Breton A., Ametowobla M., Carlsson C., Köhler J., Werner J. '15.4 % Efficiency silicon solar cells with laser doped emitter'. The 22nd European Photovoltaic Solar Energy Conference; 2006. pp. 1247–9.

[29] Eisele S.J., Röder T.C., Köhler J.R., Werner J.H. '18.9% efficient full area laser doped silicon solar cell'. *Applied Physics Letters*. 2009;**95**(13):133501–4.

[30] Turner G.B., Tarrant D., Pollock G., Pressley R., Press R. 'Solar cells made by laser-induced diffusion directly from phosphine gas'. *Applied Physics Letters*. 1981;**39**(12):967–9.

[31] Pirozzi L., Besi-Vetrella U., Salza E. 'Innovative applications of laser technology in photovoltaics'. Laser Applications in Microelectronic and Optoelectronic Manufacturing II; 1997. pp. 119–28.

[32] Deutsch T.F., Fan J.C.C., Turner G.W., Chapman R.L., Ehrlich D.J., Osgood R.M. 'Efficient Si solar cells by laser photochemical doping'. *Applied Physics Letters*. 1981;**38**(3):144–6.

[33] Young R.T., van der Leeden G.A., Sandstrom R.L., Wood R.F., Westbrook R.D. 'High-efficiency Si solar cells by beam processing'. *Applied Physics Letters*. 1983;**43**(7):666–8.

[34] Stuck R., Fogarassy E., Muller J.C., Hodeau M., Wattiaux A., Siffert P. 'Laser-induced diffusion by irradiation of silicon dipped into an organic solution of the dopant'. *Applied Physics Letters*. 1981;**38**(9):715–17.

[35] Willeke G.P. 'Thin crystalline silicon solar cells'. *Solar Energy Materials and Solar Cells*. 2002;**72**(1–4):191–200.

[36] Kray A. 'Laser-doped silicon solar cells by laser chemical processing (LCP) exceeding 20% efficiency'. 2008:7–9.

[37] Richerzhagen B. 'Complete model to simulate the thermal defocusing of a laser beam focused in water'. *Optical Engineering*. 1996;**35**(7):2058.

[38] Kluska S., Granek F. 'High-efficiency silicon solar cells with boron local back surface fields formed by laser chemical processing'. *IEEE Electron Device Letters*. 2011;**32**(9):1257–9.

[39] White C.W., Christie W.H., Appleton B.R., Wilson S.R., Pronko P.P., Magee C.W. 'Redistribution of dopants in ion-implanted silicon by pulsed-laser annealing'. *Applied Physics Letters*. 1978;**33**(7):662–4.

[40] Young R.T., White C.W., Clark G.J., *et al*. 'Laser annealing of boron-implanted silicon'. *Applied Physics Letters*. 1978;**32**(6):139–41.

[41] Carlsson C., Esturo-Bretón A., Ametowobla M., Kohler J.R., Werner J.H. 'Laser doping for selective silicon solar cell emitter'. Proceedings of the 21st European Photovoltaic Solar Energy Conference; Dresden, Germany; 2006. pp. 938–40.

[42] Poulain G., Blanc D., Focsa A., *et al*. 'Selective laser doping from boron silicate glass'. *Energy Procedia*. 2012;**27**(December):455–9.

[43] Jäger U., Okanovic M. *'Matthias Hörtheis AG and RP. Selective emitter by Laser Doping from Phosphosilicate'*. European Photovoltaic Solar Energy Conference; Hamburg, Germany; 2009.

[44] Narayan J., Young R.T., Wood R.F., Christie W.H. 'P-N junction formation in boron-deposited silicon by laser-induced diffusion'. *Applied Physics Letters*. 1978;**33**(4):338–40.

[45] Fogarassy E., Stuck R., Grob J.J., Siffert P. 'Silicon solar cells realized by laser induced diffusion of vacuum-deposited dopants'. *Journal of Applied Physics*. 1981;**52**(2):1076–82.

[46] Dahlinger M., Eisele S.J., Lill P.C., Köhler J.R., Werner J.H. 'Full area laser doped boron emitter silicon solar cells'. IEEE Photovoltaic Specialists Conference; Seattle, USA; 2011. pp. 1029–31.

[47] Ametowabla M., Esturo-Breton A., Kohler J.R., Werner J.H. 'Laser processing of crystalline silicon solar cells'. Conference Record of the Thirty-first IEEE Photovoltaic Specialists Conference, 2005; Lake Buena Vista, USA; 2005. pp. 1277–80.

[48] Wong Y.W., Yang X.Q., Chan P.W., Tong K.Y. 'Excimer-laser doping of spin-on dopant in silicon'. *Applied Surface Science*. 1993;**64**(3):259–63.

[49] Hameiri Z. 'Laser-doped selective emitter and local back surface field solar cells with rear Passivation'. *Centre for Photovoltaic Engineering*. 2010.

[50] VC L., Wong Y.W., Cho H.C., Chen Y.Q., SM H., Chan P.W. 'Excimer laser assisted spin-on doping of boron into silicon'. *Semiconductor Science and Technology*. 1996;**11**(9):1285–90.

[51] Wenham S., Green M. *A Self Aligning Method for Forming a Selective Emitter and Metallization in a Solar Cell*. U.S. Patent US6429037B1; 2002.

[52] Lennon A., Yao Y., Wenham S. 'Evolution of metal plating for silicon solar cell metallisation'. *Progress in Photovoltaics: Research and Applications*. 2013;**21**(7):1454–68.

[53] PHD L., Lin D., Wang X., Lennon A., Wenham S. 'Laser doping through anodic aluminium oxide silicon solar cell'. *Solar Energy Materials & Solar Cells*. 2016;**145**:349–57.

[54] Harder N.P., Larionova Y., Brendel R. 'Al+-doping of Si by laser ablation of Al_2O_3/SiN passivation'. *Physica Status Solidi (A) Applications and Materials Science*. 2013;**210**(9):1871–3.

[55] Martin I., Ortega P., Colina M., Orpella A., López G., Alcubilla R. 'Laser processing of Al_2O_3/a-SiCx: H stacks: a feasible solution for the rear surface of high-efficiency P-type c-Si solar cells'. *Progress in Photovoltaics: Research and Applications*. 2013;**21**(5):1171–5.

[56] Paviet-Salomon B., Gall S., Monna R., *et al*. 'Laser doping using phosphorus-doped silicon nitrides'. *Energy Procedia*. 2011;**8**(7):700–5.

[57] Benninghoven A. 'Analysis of submonolayers on silver by negative secondary ion emission'. *Physica Status Solidi*. 1969;**34**(2):K169–71.

[58] Germershausen S., Bartholomäus L., Seidel U., *et al*. 'Investigation of emitter homogeneity on laser doped emitters'. *Energy Procedia*. 2011;**8**:232–7.

[59] Weber J., Brand A. 'A scanning spreading resistance microscopy study on a laser-doped selective phosphorous emitter'. *IEEE Journal of Photovoltaics*. 2020;**10**(5):1313–18.

[60] Sameshima T., Usui S., Sekiya M. 'Laser-induced melting of predeposited impurity doping technique used to fabricate shallow junctions'. *Journal of Applied Physics*. 1987;**62**(2):711–13.

[61] Köhler J., Grabitz P., Eisele S.J., Röder T.W.J.H. 'Laser doped selective emitters yield 0.5% efficiency gain'. *Solar Energy*. 2009;**September**:1847–50.

[62] Hameiri Z., Mai L., Puzzer T., Wenham S.R. 'Influence of laser power on the properties of laser doped solar cells'. *Solar Energy Materials and Solar Cells*. 2011;**95**(4):1085–94.

[63] Röder T., Grabitz P., Eisele S., Wagner C., Kohler J.R., Werner J.H. '0.4% absolute efficiency gain of industrial solar cells by laser doped selective emitter'. Proceedings of the 34th IEEE Photovoltaic Specialists Conference; Philadelphia, USA; 2009. pp. 871–3.

[64] Ogane A., Hirata K., Horiuchi K., Kitiyanan A., Uraoka Y., Fuyuki T. 'Feasible control of laser doping profiles in silicon solar cell processing using multiple excitation wavelengths'. Conference Record of the IEEE Photovoltaic Specialists Conference; San Diego, USA; 2008. pp. 5–8.

[65] Sugianto A., Bovatsek J., Wenham S., Tjahjono B., Xu G., Yao Y. '18.5% laser-doped solar cell on CZ p-type silicon'. The 35th IEEE Photovoltaic Specialists Conference; 2010. pp. 689–94.

[66] Hameiri Z., Puzzer T., Mai L., Sproul A.B., Wenham S.R. 'Laser induced defects in laser doped solar cells'. *Progress in Photovoltaics: Research and Applications*. 2011;**19**(4):391–405.

[67] Ishida E., Sigmon T.W., Weiner K.H., Frost M.R. 'Ultra-shallow box-like profiles fabricated by pulsed ultraviolet-laser doping process'. *Journal of Vacuum Science & Technology B: Microelectronics and Nanometer Structures*. 1994;**12**(1):399–404.

[68] Hallam B. *High Efficiency Laser-Doped Silicon Solar Cells with Advanced Hydrogenation [PhD Thesis]*. Sydney, Australia: University of New South Wales; 2014.

[69] Hallam B., Chan C., Sugianto A., Wenham S. 'Deep junction laser doping for contacting buried layers in silicon solar cells'. *Solar Energy Materials and Solar Cells*. 2013;**113**(6):124–34.

[70] Hallam B., Urueña A., Russell R., *et al.* 'Efficiency enhancement of i-PERC solar cells by implementation of a laser doped selective emitter'. *Solar Energy Materials and Solar Cells*. 2015;**134**(4):89–98.

[71] Hallam B., Urueña A., Russell R., *et al.* 'Efficiency enhancement of i-PERC solar cells by implementation of a laser doped selective emitter'. *Solar Energy Materials and Solar Cells*. 2015;**134**(4):89–98.

[72] Basore P.A. 'Numerical modeling of textured silicon solar cells using PC-1D'. *IEEE Transactions on Electron Devices*. 1990;**37**(2):337–43.

[73] Green M.A., Keevers M.J. 'Optical properties of intrinsic silicon at 300 K'. *Progress in Photovoltaics: Research and Applications*. 1995;**3**(3):189–92.

[74] Bovatsek J.M., Sugianto A., Hallam B., Tjahjono B., Wenham S., Patel R.S. 'High-speed fabrication of laser doping selective emitter solar cells using 532nm continuous wave (CW) and modelocked quasi-cw laser sources'. Proceedings of the 26th European Photovoltaic Solar Energy Conference Exhibit; Hamburg, Germany; 2011. pp. 895–900.

[75] Sze S.M., Kwok K.N. *Physics of Semiconductor Devices*. John Wiley & Sons; 2006.

[76] Nguyen H.T., Rougieux F.E., Mitchell B., Macdonald D. 'Temperature dependence of the band-band absorption coefficient in crystalline silicon from photoluminescence'. *Journal of Applied Physics*. 2014;**115**(4):43710.

[77] Jellison G.E., Modine F.A. 'Optical absorption of silicon between 1.6 and 4.7 eV at elevated temperatures'. *Applied Physics Letters*. 1982;**41**(2):180–2.

[78] Jellison G.E., Lowndes D.H. 'Measurements of the optical properties of liquid silicon and germanium using nanosecond time-resolved ellipsometry'. *Applied Physics Letters*. 1987;**51**(5):352–4.

[79] Laskin A., Kaiser P., Laskin V., Ostrun A. 'Laser beam shaping for biomedical microscopy techniques'. *Biophotonics Photonic Solut Better Heal Care V*. 2016;**9887**:98872E.

Chapter 8

Laser-doped selective emitter formation and the passivation of laser-induced defects

Arman Mahboubi Soufiani[1], Catherine Chan[1], Jingjia Ji[1], and Brett Hallam[1]

This chapter will introduce the concept of selective emitters for silicon solar cells as a pathway for increased solar cell efficiencies. It will briefly discuss the record 25% p-type UNSW passivated emitter, rear locally diffused (PERL) cell, as well as several industrial approaches for selective emitter formation including the buried contact solar cell (BCSC) and inkjet masking-based methods, along with the strengths and weakness of each approach. Then the various laser-doping methods for selective emitter formation will be discussed, which can overcome many of the limitations of non-laser-doping-based methods. One such benefit that will be discussed is the ability to form the selective emitter from the residual phosphosilicate glass (PSG) after emitter diffusion, which avoids the need for masking the silicon surface prior to selective emitter formation and avoids the need to apply a separate doping source, as is currently used in the industry. The ability to form a self-aligned metallisation with a laser-doped selective emitter (LDSE) structure will also be presented, as was used for Suntech's Pluto technology that was scaled to 500 MW of production. The challenges of the laser-doping process will also be discussed in terms of defect generation, with hydrogen passivation again playing a vital role in the passivation of the formed defects. In particular, it will discuss how hydrogen passivation processes for laser-induced defects can be incorporated into the solar cell manufacturing process, leading to efficiency enhancements of up to $1.5\%_{abs}$ [1]. This includes the natural passivation of defects for the industrial LDSE process for passivated emitter and rear cells (PERC) by performing laser doping before dielectric deposition enabling efficiencies approaching 24% [2]. It also discusses the ability to incorporate passivation with sintering for nickel-plated contacts, or a dedicated hydrogen passivation process, which can potentially be used simultaneously to treat light-induced degradation. It also highlights the importance of assessing laser-doping parameters used in conjunction with thermal processes for solar cell fabrication for defect passivation.

[1]The School of Photovoltaic and Renewable Energy Engineering, The University of New South Wales, Sydney, NSW, Australia

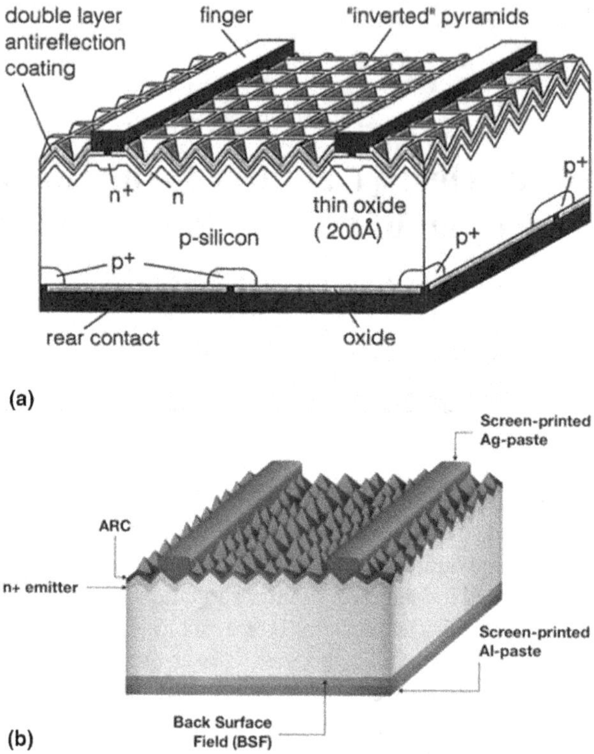

Figure 8.1 *(top) PERL solar cell architecture using front-side selective emitter and rear locally diffused contacts. Reprinted from [6]. Copyright (1996), with permission from Elsevier. (bottom) Screen-printed Al-BSF solar cell with a front heavily doped homogeneous emitter.*

8.1 The selective emitter concept

A large number of solar cell architectures have been explored since the first silicon solar cell was fabricated in 1941, using epitaxially grown junctions, with a reported power conversion efficiency of 1% [3]. Among these, the UNSW developed PERL configuration with its first reported efficiency of 24.2% in 1990 [4] stood as the record-holding single-junction structure for about 25 years, the highest reported efficiency for this structure after correction for the illumination spectrum is 25% by UNSW [5]. One of the key high-efficiency attributes of the PERL cell is the selective emitter structure implemented on its front side, contributing to its high open-circuit voltage (V_{OC}) and short-circuit current density (J_{SC}) (Figure 8.1 (top)). In contrast to homogeneous emitters in p-n junction silicon solar cells with a uniform, heavily doped emitter across the full front surface (Figure 8.1 (bottom)), in selective emitter structures, the heavily doped regions are localised to areas in the immediate vicinity of the metal contacts, with a more lightly diffused emitter formed over the rest of the wafer surface.

The other most notable feature of the PERL cell architecture was its low-area-coverage of heavily doped rear point contacts, which allow the majority of the rear surface to be passivated, thereby allowing much higher V_{OC} values to be achieved compared to the conventional full-area Al-BSF design (Figure 8.1 (bottom)) where the entire rear forms the metal/Si interface. The rear surface design of the PERL cell will be discussed further in Chapter 9.

Despite the benefits of a selective emitter being known for a long time, homogenously diffused screen-printed solar cells dominated the silicon PV market for many years due to the relative simplicity of production compared that for selective emitter formation with additional processing steps required, potentially also requiring precise alignment. Additionally, the benefits associated with implementing a selective emitter were not fully realised while full-area Al-BSF cells dominated the industry, as the performance of solar cells was largely limited by the rear of the device rather than the front surface. Further, as silver paste and printing technology improved, rapidly enabling narrower printed lines that could contact more lightly doped homogeneously diffused emitters, the benefits of using selective emitter technology were often not cost-effective. However, with the introduction of a PERC rear design to the industry, which now dominates the industry, the front surface design has become increasingly important. In 2019, selective emitter technologies were used on >30% of industrial p-type cells of which almost all are formed via laser doping [7]. LDSEs are expected to become mainstream in 2024, surpassing homogenous emitter technology, and their share predicted to rise to almost 70% of industrial p-type cells by 2030 [7].

8.1.1 Trade-offs leading to the use of a selective emitter

When designing solar cells, many trade-offs exist where opposing effects need to be balanced. One of the most notable examples is the doping level of the front-side emitter diffusion, where the sheet resistance and surface doping concentration must be carefully chosen to maximise the performance of the solar cell. If the doping is too heavy, the solar cell will have high emitter dark saturation current density (J_{0e}) due to Auger recombination, therefore reducing V_{OC}, and reduced quantum efficiency (QE) at short wavelengths resulting in a reduced J_{SC}. On the other hand, if it is too lightly doped, the solar cell will have high contact resistivity and poor lateral conductivity leading to high series resistance losses, as well as high recombination at the metal/Si interface, also increasing J_{0e} and decreasing V_{OC}. Table 8.1 summarises the desirable and undesirable effects of a heavily doped versus a lightly doped emitter. A detailed loss analysis of a PERC-type solar cell with homogeneous emitter by Saint-Cast *et al.* revealed that the front metal/Si interface and bulk component of the J_{0e} were responsible for more than 50% of its V_{OC} loss, which can be largely circumvented with a selective structure [8]. Selective emitter solar cells allow for the decoupling and separate optimisation of the emitter doping at the metallised and non-metallised regions to overcome these trade-offs. Each of these effects is described in Subsections 8.1.2 to 8.1.4.

Table 8.1 Effect of front homogeneous emitter doping on solar cell performance: desirable =√ undesirable =×

Heavily doped surface, low sheet resistance emitter	Lightly doped surface, high sheet resistance emitter
× High emitter recombination due to Auger recombination	√ Low emitter recombination due to Auger recombination
× Higher SRV (when passivated)	√ Lower SRV (when passivated)
× Low QE at short wavelengths	√ High QE at short wavelengths
√ Low contact resistivity	× High contact resistivity
√ Low metal/Si interface recombination	× High metal/Si interface recombination
√ Low lateral resistance	× High lateral resistance

QE, quantum efficiency;SRV, surface recombination velocity.

8.1.2 Emitter recombination, surface recombination velocity and quantum efficiency

The doping level and profile of the emitter can greatly impact solar cell performance. High levels of emitter doping lead to increased Auger recombination leading to a high J_{0e}, generally a higher surface recombination velocity (SRV) when passivated for instance by SiO_2, as well as band-gap narrowing, which reduces the absorption of shorter wavelengths of light. Thus, a heavily doped emitter with a high surface doping concentration is often referred to as having a 'dead layer', and this can be observed through a reduction in the cell's QE. The QE of a solar cell refers to the ratio of carriers collected by a solar cell to the number of photons incident on the cell under short-circuit condition. That is, it represents how effectively photons of each wavelength are absorbed by the solar cell, are transported to the contacts and therefore contribute to the short-circuit current (J_{SC}) of the device. However, the internal quantum efficiency (IQE) does not take into account photons that are reflected or transmitted through the solar cell (this is referred to as the external quantum efficiency (EQE) or parasitically absorbed in the device). If a solar cell has high emitter recombination or a high front SRV, minority carriers generated close to the front surface or in the emitter region are more likely to recombine before being collected by the p-n junction and being able to contribute to the J_{SC}. This causes a drop in IQE (and hence also EQE) of photons with shorter wavelengths, which are absorbed in the first few tens of nanometres of the device, therefore affecting the blue response of the device. Thus, in non-metallised, light-exposed regions of a solar cell, it is ideal to have a lightly doped emitter with low surface concentration $<2 \times 10^{19}/cm^3$, well below the surface concentration of conventional screen-printed solar cells with a homogenous emitter ($>1 \times 10^{20}/cm^3$). Figure 8.2 shows an example of experimentally measured IQE gain at short wavelengths of less than 600 nm for a selective emitter cell compared to a standard homogeneous emitter cell [9]. The blue response of the solar cell is becoming increasingly important with the development of more transparent encapsulant materials.

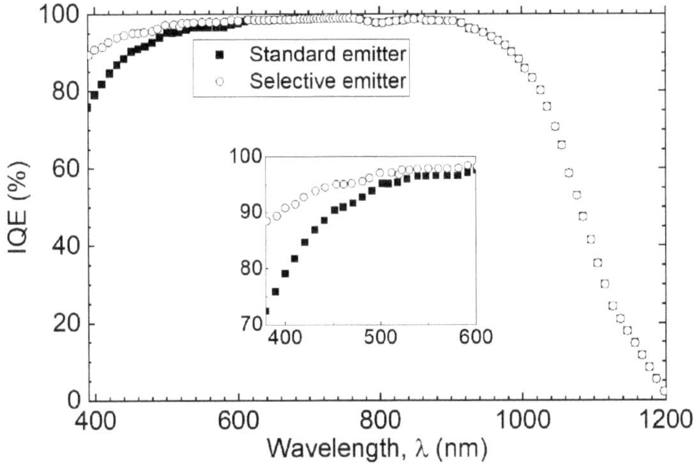

Figure 8.2 *Improvement in the IQE at short wavelengths of a silicon solar cell with a selective emitter compared to a cell incorporating a homogeneous emitter. Reproduced from [9] with permission from John Wiley and Sons. Copyright © 2010 John Wiley & Sons, Ltd.*

The impact of doping profile on Auger recombination and the J_{SC} of a solar cell is further demonstrated in the simulation results of Figure 8.3. A range of phosphorous dopant profiles with different surface concentrations and depths (see Figure 8.3 (a)) was created using a complementary error function (ERFC) diffusion profile, which describes the n-type emitter profile, assuming a p-type base with bulk resistivity of

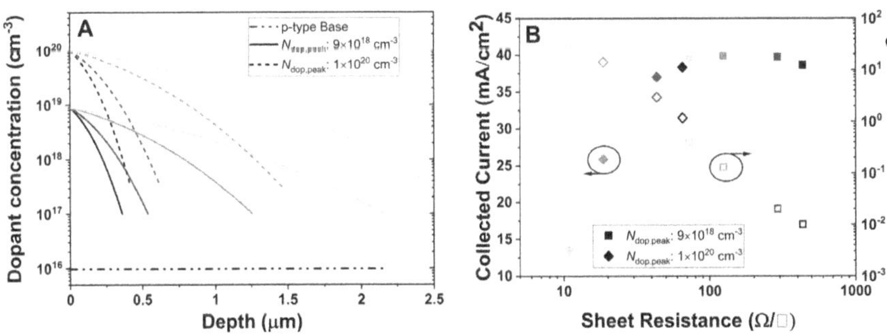

Figure 8.3 *(a) Modelled n-type emitter dopant profiles on a 1.5 Ω·cm p-type wafer. The doping profiles were created using a ERFC with different depths. (b) Simulated total extracted photo-generated current (filled symbols) for monochromatic 400 nm incident light with an illumination intensity corresponding to 40 mA/cm² short-circuit current generation under 100% EQE. Also shown are the corresponding Auger recombination current density (hollow symbols). Simulations were performed using EDNA2 [10].*

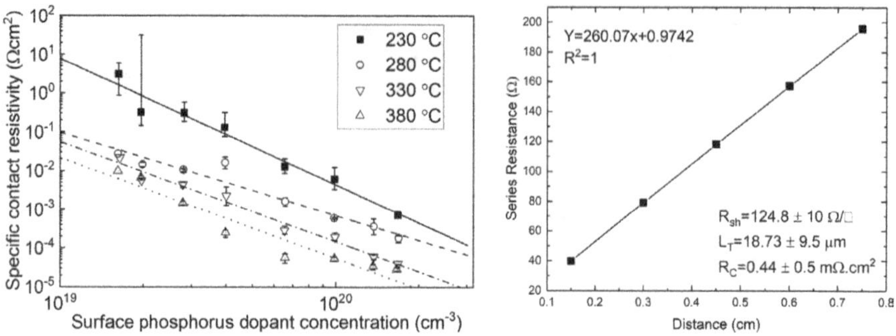

Figure 8.4 *(left) Impact of surface dopant concentration on contact resistivity of phosphorous-doped profiles and evaporated aluminium. The diffused profiles were created using a high-temperature quartz furnace tube with $POCl_3$ as the dopant source. The temperatures indicate the temperature at which the sintering was performed for 30 minutes in forming gas. © 2015 IEEE. Reprinted, with permission, from [11]. (right) Example transmission line measurement (TLM) data from a screen-printed Ag finger on an n-type LDSE region for PERC.*

1.5 Ω·cm. Simulations were performed using a freely available EDNA2 calculator on PV Lighthouse [10]. The Auger recombination current associated with each profile, as well as the collected current density for an incident 400 nm monochromatic light source, is shown as a function of emitter sheet resistance (Figure 8.3 (b)). To highlight the impact of Auger recombination current on the total collected current, the minority carrier (i.e. holes in the case of an n-type emitter), the SRV was set to a low value of 1 cm/s. The impact of Auger related J_{0e} on the collected current density is apparent, with increasing Auger recombination reducing collection efficiency at short wavelengths and therefore reducing current collection. For homogeneously heavily doped, low sheet resistance emitters, this highlights that large losses to the J_{SC} of the device can occur.

8.1.3 Contact resistivity and metal/Si interface recombination

A key requirement for a high-efficiency solar cell is having a low resistive loss at the metal/Si interface. To keep contact resistance low, high surface doping is required. Figure 8.4 (left) shows an example of the impact of the surface dopant concentration on contact resistivity for the case of phosphorus-diffused silicon being contacted with evaporated aluminium [11]. This used an etch-back technique to vary the surface dopant concentration. For screen-printed contacts, which penetrate several tens of nanometres into the silicon, a deep junction is also desirable to ensure that the dopant concentration remains high at the contact interface, as well as to allow for a larger processing window for firing and avoid shunts caused by metal penetrating to depths in the vicinity of the junction [9, 12]. However, more recent pastes are self-limiting and tend not to shunt devices if over fired. A roadmap study by

Min *et al.* used specific contact resistivities of 2 mΩ·cm² and 0.1 mΩ·cm² assuming a laser-doped n⁺⁺ phosphorous emitter profile with surface dopant density of approximately $2 \times 10^{20}/cm^3$ and sheet resistance of 74 Ω/□ when contacted with screen-printed Ag-paste and plated Ni/Cu fingers, respectively [13]. More recent measurements of an industrial screen-printed LDSE PERC solar cell using transmission line measurements have indicated a specific contact resistivity of 0.44 ± 0.5 mΩ·cm² using a fire-through Ag paste (see Figure 8.4 (right)). Hence, more recent pastes are capable of reaching specific contact resistivities below 1 mΩ·cm².

Additionally, the SRV at a metal/Si interface is near infinite (usually assumed to be ~10^7 cm/s). To reduce the impact of this high SRV on cell performance, it is desirable for the emitter to be both deep and heavily doped to keep minority carriers away from the highly recombination-active contact by way of field-effect passivation [14]. Deep, heavily doped profiles increase the resilience to high SRV values such as at the metal/Si interface. However, for the reasons described in Section 8.2.1, a deep, heavily doped emitter with high surface doping concentration across the entire solar cell is not ideal. Lightly doped emitters are capable of much lower J_{0e} values, provided the SRV is sufficiently low, such as for a very well passivated surface with an SRV of <1 000 cm/s. Thus, in a selective emitter solar cell, deep, heavily doped regions are restricted to only being used in the immediate vicinity of the metal/Si interface contact areas. This enables a low contact resistivity and reduces recombination at the metal/Si interface, while also reducing the localised contribution of Auger-related J_{0e} in the heavily doped region to the total J_{0e} of the device. Figure 8.5 demonstrates this effect, with the impact of the different emitter

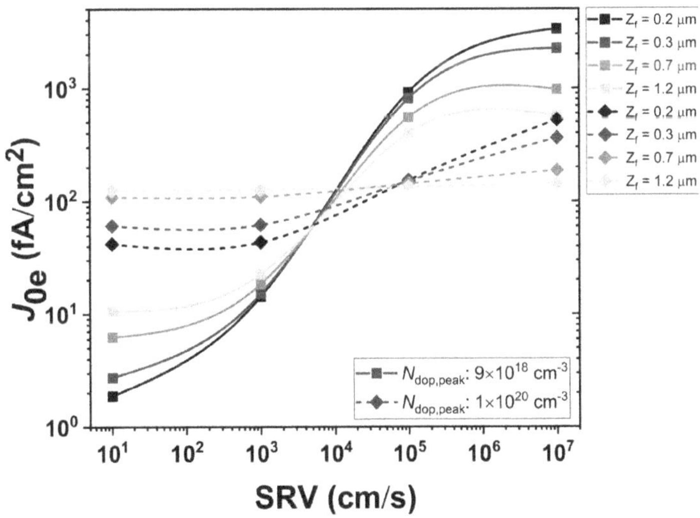

Figure 8.5 The calculated total emitter recombination current density, J_{0e}, *as a function of minority carrier SRV for different phosphorous dopant density profiles provided in panel (a) of Figure 8.3. Simulations performed with EDNA2 [10].*

doping profiles from Section 8.2.1 on the total J_{0e} for a wide range of minority carrier SRVs. At high SRVs, as is expected at a metal/Si interface, deeper emitters with high surface concentration are favourable, shielding the minority carriers from the high non-radiative recombination rates occurring at the interface. As the SRV is reduced beyond a threshold (~10^4 cm/s), Auger recombination becomes the dominant pathway for minority carrier loss for such profiles and the trend inverts, where shallower emitters with lower surface concentrations become favourable (i.e. lower J_{0e}).

The J_{0e} of the metal/Si interface regions of industrial LDSE PERC solar cells with aligned screen-printed contacts is in the vicinity of 300–500 fA/cm^2 (e.g. see [13]). However, with an area fraction of typically 3% for the metal/Si interface, this results in a J_{0e} addition by the metal/Si interface of approximately 9–15 fA/cm^2.

Another benefit of the selective emitter structure is the higher surface passivation quality via the deposited dielectrics on the lightly diffused emitter layer in between the selective emitter lines. Lighter diffusion profiles have shown greater passivation quality through SiO_2 deposition [12]. This benefit can be also realised by considering the trend in J_{0e} for the low SRV range related to SiO_2 passivation, presented in Figure 8.5.

8.1.4 Lateral emitter resistance

In a high-efficiency solar cell design, it is also important to ensure that the resistance losses for carriers travelling laterally through the emitter are sufficiently low such that carriers generated some distance away from a metal finger can travel to the contact with ease. As the emitter sheet resistance is increased for a more lightly doped emitter, these lateral resistive losses increase with a linear dependence on the sheet resistance, leading to a reduction in cell FF. To compensate for this loss, the metal finger spacing can be reduced due to the quadratic dependence of the relative power loss on the finger spacing. However, this has the adverse impact of increasing the shading losses, reducing the J_{SC}, as well as increasing the metal/Si interface area and therefore the total metal/Si interface recombination. Thus, there is a trade-off between finger spacing and emitter sheet resistance that must be carefully optimised. The interplay between the two power losses due to shading and lateral resistance is clearly illustrated in the contour map of Figure 8.6. For selective emitter solar cells with a lightly diffused emitter in non-metallised regions, thin finger widths are therefore ideal, allowing fingers to be spaced more closely together, without high shading losses. Generally, as the finger widths of screen-printed solar cells become narrower, the fingers are spaced more closely, resulting in similar shading losses in the fingers of 2–3%, while reducing emitter series resistance losses and also benefiting from reduced series resistance losses in the fingers. The relative power loss due to emitter series resistance for a PERC solar cell is typically in the vicinity of 1%.

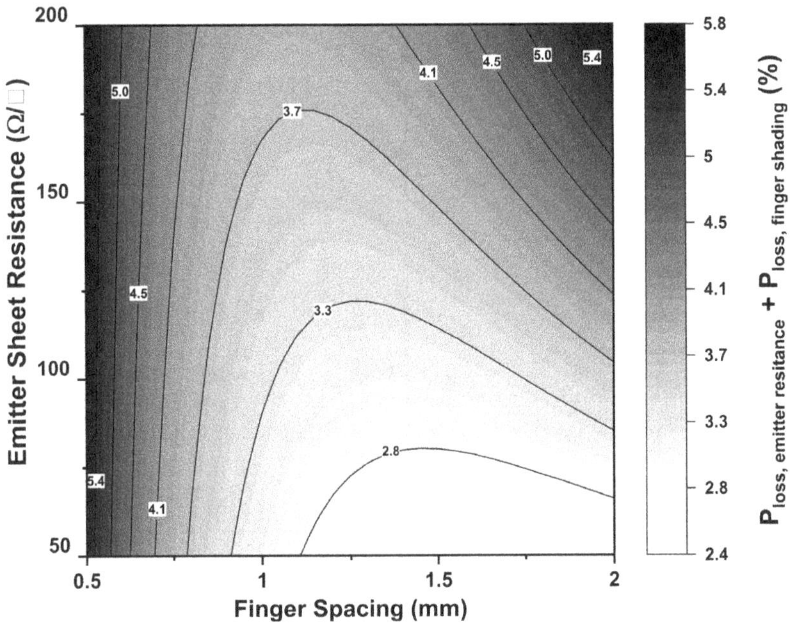

Figure 8.6 *Contour map illustrating the total power loss due to combined effects of emitter resistive losses and finger shading as a function of emitter sheet resistance and spacing between the fingers. The modelling was performed using Sentaurus TCAD assuming a maximum power point voltage (V_{MP}) and current (J_{MP}) of 583 mV and 38.56 mA/cm^2, respectively.*

8.2 Techniques for selective emitter formation

Many techniques to form selective emitters for high-efficiency cell structures have been developed over the years. The earliest methods involved using multiple thermal diffusion steps through oxide-masking layers as used in the UNSW PERL cell [6] and BCSC [15]. The use of multiple high-temperature diffusion steps to form the selective emitter was not only cost- and energy-intensive but also incompatible with lower-quality wafer types such as multi-crystalline silicon that is known to degrade at high temperatures [16, 17] and can risk the growth of oxygen precipitates in Cz-grown material [18]. In addition, one of these thermal diffusion steps needed to be performed through an oxide-masking layer formed via high-temperature oxidation and then patterned for subsequent selective diffusion, adding more complexity to the fabrication process. A number of methods have been used to pattern the oxide layer, an overview of the most common is presented in this section.

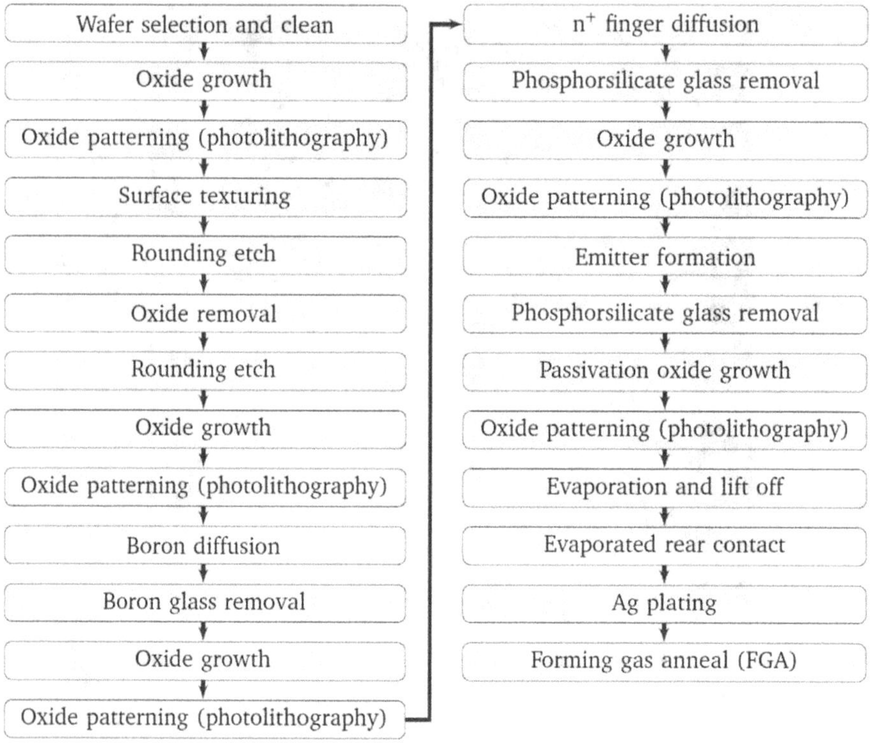

Figure 8.7 Processing sequence for the UNSW PERL cell. Adapted from [19].

8.2.1 Dielectric patterning using photolithography

The world-record UNSW PERL cell used photolithography to pattern its masking oxides not only for selective emitter formation but also for the formation of the rear locally diffused contacts, the inverted pyramid texturing scheme and defining metal contact openings. The complex processing sequence for this cell structure is shown in Figure 8.7.

Although all five instances of oxide patterning by photolithography are depicted as single step in Figure 8.7, this process actually involves multiple complex steps [20] using expensive equipment:

1. application of photoresist to the surface dielectric
2. pre-baking of the photoresist
3. exposure to UV light through a patterned mask
4. development Of the photoresist
5. post-baking of the photoresist
6. removal of the area of the photoresist exposed to UV light
7. chemical etching of the exposed dielectric layer
8. removal of the photoresist.

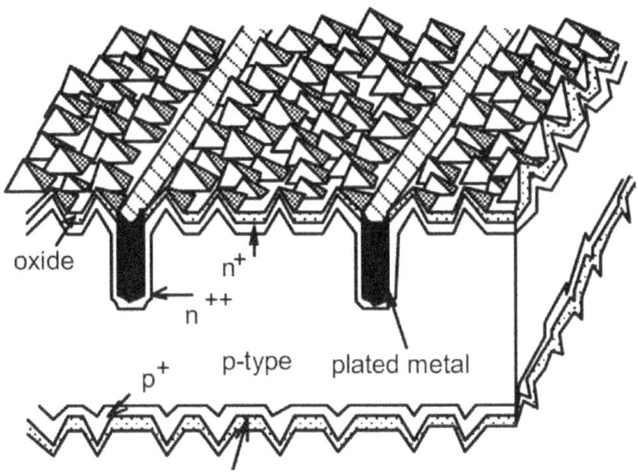

Figure 8.8 *Schematic of the UNSW BCSC architecture. Image reproduced from [15] with permission. © 1993 by John Wiley & Sons, Ltd.*

The patterns are also not particularly easy to modify. A new photolithography mask must be manufactured for each new pattern. Thus, while photolithography is advantageous for its extremely precise features and hence very narrow openings (e.g. ~5 μm) facilitating low metal/Si interface area and narrow plated contacts with minimal shading losses, it is prohibitively slow and expensive for use in commercial solar cell production where throughput and cost are paramount. In addition, photolithography is a particularly wasteful process, with large amounts of photoresist, fluoride-based silicon etchants and solutions to remove the photoresist layers being required and subsequently needing disposal. Photolithography is therefore reserved for high-efficiency laboratory devices. Combined with the need for the multiple high-temperature steps the PERL cell was not a commercially viable device architecture at the time.

8.2.2 Dielectric patterning through laser ablation

Another method for creating a patterned oxide-masking layer is through the use of laser ablation. Laser processing has the advantages of being high throughput, creating high-resolution features, and being easier to change patterns whilst also being a much leaner process in terms of waste when compared to photolithography. A key example of a solar cell structure using laser ablation to pattern its dielectric layer and form a selective emitter is the UNSW BCSC [15] shown schematically in Figure 8.8.

The selective emitter for the BCSC was formed via a two-step diffusion process in a similar manner to the PERL cell. However, instead of photolithography, laser ablation was used to pattern the oxide layer, which simultaneously formed deep (~50 μm) and narrow (~20 μm) grooves in the silicon surface. The grooves were then chemically etched to remove laser damage and heavily diffused via a second thermal diffusion prior to PSG removal and subsequent plating with Ni and Cu. As its name suggests, the resulting front contacts were essentially buried in these narrow, deep grooves, resulting in low shading loss and excellent adhesion. The BCSC technology was commercialised by BP Solar as Saturn solar cells, achieving cell efficiencies of 17.5–18% in production on commercial grade Cz wafers, an approximate $30\%_{rel}$ performance gain than that achieved via their standard screen-printed cell process at the time [21]. This was the first time that selective emitter solar cells were successfully mass-produced. Although the process was significantly less complex than that of the laboratory PERL cell, a second high-temperature diffusion step was still required to form the heavily doped emitter contact regions, again prohibitive to the use of low-quality multi-crystalline silicon wafers.

A more recent industrial approach for creating a selective emitter structure using laser ablation was developed by Centrotherm, whereby only one thermal diffusion was required, making it more attractive for commercial use [22]. In this approach, a thin dielectric layer (e.g. oxide) was first grown on the wafer at temperatures well below the subsequent diffusion temperature, followed by laser scribing of the layer to form 300 μm wide openings to the underlying bare silicon wafer. It is noted that the wet chemically etched porous silicon layer has also been used as the inhibitive dielectric layer [22]. Following a wet chemical etch to remove laser damage, the wafer then underwent a single thermal diffusion to form the selective emitter. This method relied on phosphorus dopants diffusing through the thin oxide-masking layer, albeit at a slower rate than at the exposed unmasked regions. Sheet resistances of 45 Ω/ϒ and 110 Ω/ϒ were achieved in the exposed and masked regions, respectively, thereby creating a selective emitter with only one diffusion step. This approach was then combined with an aligned screen-print metallisation process.

While the use of lasers to pattern dielectric layers has enabled the simplification of selective emitter formation, a disadvantage of this process is the creation of laser-induced defects in the silicon which can propagate further into the wafer during high-temperature processing [23]. These can, however, typically be removed after the ablation step using a wet chemical silicon etch such as NaOH or KOH [23]. In addition, the high-temperature oxidation increases the thermal budget, posing risks for low-quality material as well as ensuring control of the thickness of the SiO_2 layer to enable sufficient diffusion in the passivated regions.

8.2.3 Dielectric patterning through inkjet or aerosol jet printing

Inkjet and aerosol jet printing (also known as aerosol jet etching) can be used in place of photolithography or laser ablation to pattern dielectric layers. Initial work using inkjet printing involved the application of a polymer resist coating onto the dielectric layer [24]. A plasticiser solution was then inkjet printed in a specific pattern onto the polymer resist which altered the permeability of the resist coating to a subsequent hydrofluoric acid (HF) dip. The HF would permeate through the printed regions only and etch the underlying dielectric. After removal of the polymer resist in piranha etching solution, a subsequent masked diffusion could then be performed to form a selective emitter. This approach was denoted ‘indirect’ inkjet patterning.

In the 'direct' approach [25], a water-soluble polymer such as polyacrylic acid was first applied to the dielectric layer and then an inactive etching component such as NH_4 was inkjet-printed onto this layer where it would react with the polymer to form an active etchant such as HF. The localised HF would then etch the dielectric layer directly underneath the regions that were printed. This approach had the advantage of less waste than the indirect approach, as it removed the need for complete wafer immersion in HF. The feature sizes achievable with the inkjet approach were limited to approximately 30–50 μm, due to the limitations of the size of the inkjet-printed droplets. The replacement of the inkjet printer with an aerosol jet printer in a process, termed aerosol-jet etching, reduced this down to approximately 25 μm with potential for 10 μm [26], making it a more effective approach to forming the fine features required on high-efficiency devices. Both the inkjet and aerosol jet approaches have the advantage over laser ablation of dielectrics in that they do not induce damage to the underlying silicon wafer. However, achieving high throughputs via these approaches can be a key challenge.

8.2.4 Use of doped silicon inks

A novel approach for forming a selective emitter was developed by Innovalight using doped silicon inks in their selective emitter 'Cougar' cells [27]. In this method, heavily doped silicon nanoparticles were screen-printed onto the surface of a silicon wafer in regions where metal contacts were to be printed. The wafers then underwent a single step phosphorus diffusion to form heavily doped 30–50 Ω/□ regions at areas covered by the silicon ink and a lightly doped 80–100 Ω/□ emitter over the rest of the wafer surface. After SiN_x deposition, an aligned screen-printing step (aligned to the pattern of the doped silicon ink) was used to form the front metal grid. This process was used by JA Solar, demonstrating peak cell efficiencies of more than 19% in production [27]. An attractive feature of this technology is the compatibility with the widely used screen-printing technology and not requiring oxide-masking. However, it requires the addition of a printing step.

8.2.5 Ion implantation

Forming selective emitters via ion implantation has been another technique of commercial interest. A process developed by Varian involved the use of an in situ proximity mask to form a selective emitter with a single ion implantation step [28]. Ion implantation is a single-side-doping technique; therefore, this approach eliminates the need for the edge isolation step required by other selective emitter formation approaches involving tube furnace diffusions. However, the implanted dopants require activation, which can be performed in an oxide tube furnace where a beneficial high-quality passivating SiO_2 layer is formed on the surface, simultaneously annealing the implanted local regions at high temperatures, 800–900 °C for a duration of 10–60 min [29], typically performed in a tube furnace. Additionally, no PSG layer is formed during this dopant activation step, thereby eliminating altogether the need for the PSG removal step (typically combined with wet chemical edge isolation for industrial solar cells). Subsequently, screen-printed contacts can be aligned to the heavily doped regions. The dry nature of the ion implantation method is an advantage over most of the techniques briefly reviewed here, for which this wet chemical step can be eliminated, removing the associated waste. This high-precision technique allows simultaneous control of the depth, width, and density of the dopant, which in turn makes this controllable approach highly repeatable. The tool developed by Varian Semiconductor Equipment Associates was capable of processing >1 000 wafers per hour [28]. However, the industry has strongly preferred $POCl_3$-based diffusion processes due to the reduced equipment requirements, reduced cost and simplicity of the process.

8.2.6 Etch-back technique

Another industrial method for forming a selective emitter structure was developed by the University of Konstanz and commercialised by Schmid GmbH [30, 31]. In the inSECT (inline Selective Emitter ConcepT) approach, a heavily doped homogeneous emitter was first formed via inline diffusion. Subsequently, areas intended for metallisation were masked with an etch-resistant barrier applied via screen-printing or inkjet printing and the wafers underwent a wet chemical silicon etch to etch back the dead layer of the heavily doped emitter on the unmasked areas not intended for metallisation. Approximately 10–20 nm of silicon was removed in a controllable manner via the formation of porous silicon and its subsequent removal. This process was particularly attractive for the industry as only one additional tool was required to apply the etch resistant mask as the wet chemical processes could be performed in additional baths of the wet bench already used for rear etching edge isolation [12]. This method was used in production by many companies including Sunrise Global Solar in Taiwan, where 19.2% efficiency was achieved using this approach [32]. The heavily diffused regions were wider than the width of the screen-printed contacts to prevent contact misalignment. Therefore, as with other non-self-aligning selective emitter approaches, there was a larger percentage of heavily doped silicon than actually required, limiting the device performance.

8.3 Laser doping for selective emitter formation

As discussed in section 8.2, many techniques have been proposed for selective emitter formation, some requiring multiple high-temperature processes (including typically two thermal diffusions to form the two levels of doping selective emitter), expensive photolithography, wet chemical etching steps or additional materials.

Another method to form a selective emitter is through the use of laser doping, a technique that has gained significant commercial interest over the past decade owing to its cost-effectiveness, fast throughput and ability to eliminate high-temperature diffusion processes, with an option to use the by-product of emitter-forming diffusion step, PSG or borosilicate glass (BSG), as the dopant source, eliminating the need for an additional source for selective emitter formation. Furthermore, as laser doping only heats the silicon locally where required, it is compatible with multicrystalline and low-quality silicon substrates that are susceptible to degradation during the long and high-temperature processes [16, 17] required by other selective emitter formation techniques.

Laser doping allows tunable, locally formed dopant profiles to be created (refer to Chapter 7) without the complexity involved in other abovementioned techniques in Section 8.2. The creation of locally heavily doped regions through laser doping is commonly practised in two main ways:

1. Prior to the deposition of a passivating dielectric
2. After the deposition of a passivating dielectric where the laser doping simultaneously ablates the dielectric and dopes the underlying silicon.

Both approaches can be implemented on both front [12, 33] and rear surfaces of the solar cell [34–36]. In the first approach, local contact openings must be made through the dielectric layer after the laser-doping step. This can be performed via an aligned photolithography via a more industrially appropriate means of aligned laser ablation [35] or aligned screen-printing as will be described in Section 8.3.1.

8.3.1 Industrial LDSE PERC solar cells with aligned screen-printed contacts

The current method for forming a LDSE for industrial solar cells uses aligned screen-printed fire-through metallisation process cells [37]. In this approach, directly after phosphorus diffusion, the un-passivated, diffused silicon wafer is laser-doped using the PSG layer as the dopant source (note that in an n-type base silicon cell with a p-type emitter laser doping is performed through the BSG formed after BBr_3 diffusion process [38]). The laser doping is performed in narrow selectively heavily doped regions with a width of approximately 100 μm to allow suitable alignment tolerance for the screen-printed fingers. After PSG removal and edge junction isolation, the surface passivation/anti-reflection coating dielectric(s) are deposited onto the wafer [7, 39]. Then, an aligned screen-printing process is used, followed by a co-firing step, during which the metal contact paste penetrates through the dielectrics to contact the underlying laser-doped silicon region [37]. Simultaneously, the firing process enables the passivation of laser-induced defects as will be discussed in Section 8.6. Currently, printing widths of approximately 40 μm are used, which allows an alignment tolerance of approximately 30 μm [7] of the fire-through screen-print metal electrodes on the narrow selectively heavily doped regions. The advantage of this approach is that it removes the need for a separate process for local contact opening of the dielectric layers in the laser-doped region or the addition of a dopant source, thereby only adding one process step. The solar cell structure for an industrial LDSE PERC solar cell with aligned screen-printed contacts is shown in Figure 8.9 (a). The corresponding process flow for industrial LDSE PERC solar cells is shown in Figure 8.9 (b). Figure 8.9 (c–f) shows scanning electron microcopy (SEM) and electron beam-induced current (EBIC) images of an industrial LDSE PERC solar cell with an aligned screen-printed contact [40]. Using the industrial LDSE process, the junction depth is relatively shallow, as the laser doping is performed in the solid state to ensure that the pyramid

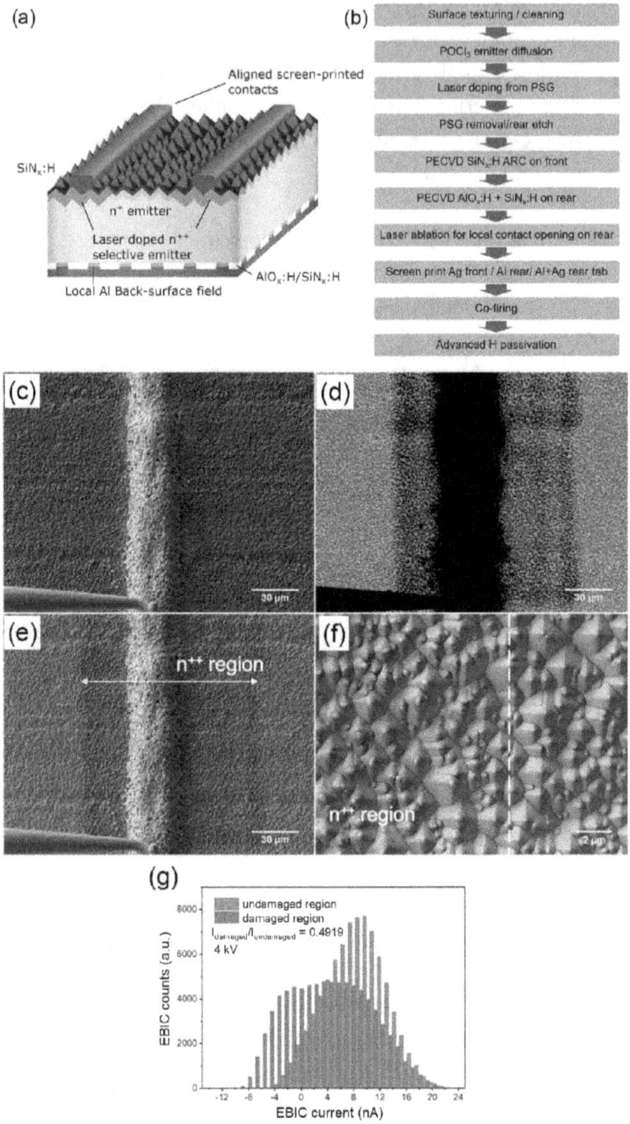

Figure 8.9 *(a) Schematic of an industrial LDSE solar cell with aligned screen-printed contacts and (b) the corresponding process sequence. (c) SEM image, (d) EBIC image, (e) combined SEM/EBIC image at low magnification and (f) combined SEM/EBIC image at high magnification in the vicinity of a Ag screen-printed laser-doped line. (g) Histogram of the current collected from the undamaged region and the regions damaged due to laser doping. (c–g) Reprinted from [40], Copyright (2020), with permission from the International Solar Energy Society.*

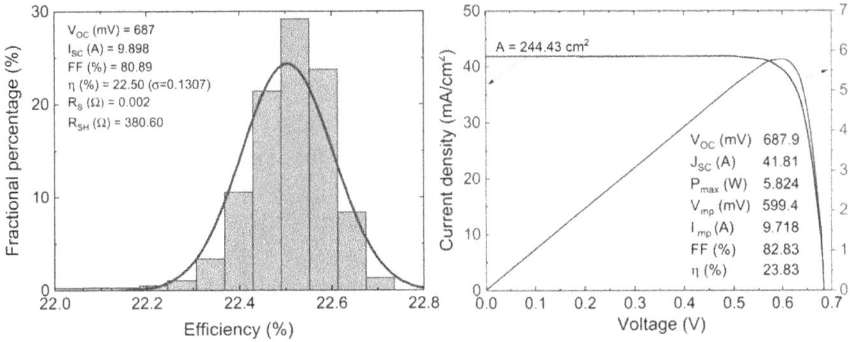

Figure 8.10 *(left) Efficiency distribution for 40 228 cells from an industrial LDSE PERC production line with aligned screen-printed contacts from LONGi, treated with advanced hydrogenation. (right) Certified* J–V *and power–voltage curves for high-efficiency LDSE PERC champion cell. Figures reproduced from [2] under Creative Commons CC BY. © 2020 The Authors. Progress in Photovoltaics: Research and Applications published by John Wiley & Sons Ltd. https://doi.org/10.1002/pip.3243.*

texture is retained on the surface for optical properties in the non-metallised region (see Figure 8.9 (f)). This limits the diffusion of phosphorus into the silicon compared to that in the liquid state, hence resulting in a reduced junction depth, although still deeper than the conventional emitter diffusion in the surrounding region. This is indicated by a slight weakening of the EBIC signal in the laser-doped region. Note that the lower current collection and thus EBIC signal at the tip of the pyramids as compared to pyramid valleys in both the undamaged and damaged parts of Figure 8.9 (f) could be due to (i) deeper junction, (ii) greater laser damage or presence of inactive dead layer or (iii) topography artefact as reported in [40].

This process has led to the achievement of peak efficiencies of 23.83% and an average of 22.5% efficiency from a batch of more than 40 000 cells (cell area 244.43 cm^2) by LONGi [2]. These cells also incorporate advanced hydrogenation with illumination to eliminate light-induced degradation using a laser-based tool. The efficiency histogram of the cells is shown in Figure 8.10 (left) along with the certified current density–voltage (*J–V*) measurement by the Wuxi National Quality Inspection Center in Figure 8.10 (right). This solar cell technology represents a low-cost implementation of the PERL solar cell. Impressively, efficiency for the large area device is within 1.2% absolute of the record 25% efficiency small area device, which was fabricated using expensive float-zone (FZ) silicon and photolithographic processes [5].

8.3.2 LDSE solar cells with self-aligned plated contacts

The alternative method, developed at UNSW [41], applies lasing directly through the stack of dielectrics deposited on the homogeneously diffused silicon wafer (with preferred emitter sheet resistance in the range of 80–120 Ω/□) [42]. Simultaneous local contact opening throughout the dielectric stack, melting and dopant incorporation into the local area of the underlying silicon is implemented in this method. The dopant source in this scenario can be externally deposited through either a spin-on-dopant (SOD) source, ultrasonic spaying [43] or doped silicon nanoparticle ink [35]. Alternatively, dielectric layers such as phosphorous-doped SiN_x deposited on the emitter underneath the SiN_x anti-reflection coating [44] can be used, or a passivating AlO_x layer [45] in the case of a selective emitter for an n-type solar cell (or local back surface field formation in PERL cells as will be discussed in Chapter 9). Alternatively, it may merely be through redistribution of the dopants in the emitter itself [42].

When laser doping is performed through the dielectric layer, it can be accompanied by self-aligned light-induced- or electro-plating [46] of Ni/Cu/Ag contacts. Here, the SiN_x dielectric layer is used as the self-aligned mask whereby plating only occurs in the exposed laser-doped region. In this instance, the Ni is used as a seed layer and to prevent diffusion of the Cu into the silicon after the formation of a nickel silicide (NiSi) layer, and the Ag capping layer is used to prevent oxidation of the copper. The use of plating enables very fine metallisation lines (<20 μm wide) only on the heavily doped selective emitter regions, while also removing the need for contact alignment. Suntech Power commercialised this technology in their Pluto cells, achieving 500 MW in production in 2011 [47] and was used to fabricate the world's first industrial p-type Cz solar cell with an efficiency of over 20%, which also incorporated advanced hydrogenation, in this instance to passivate laser-induced defects as will be discussed in Section 8.6 [47]. Figure 8.11 (a) shows the solar cell structure of an LDSE solar cell with self-aligned plated contacts and a full area Al-BSF. The corresponding process flow diagram is shown in Figure 8.11 (b). Figure 8.11 (c) shows an example cross-sectional SEM focused-ion beam image of a plated LDSE contact, highlighting the high crystallinity of the plated metal. An advantage of plated copper is the low line resistivity of the finger, with a value of approximately 1.6 μΩ·cm for pure copper, compared to that of a screen-printed silver finger of the order of 3 μΩ·cm [13, 46].

Similar to the UNSW LDSE method, the laser chemical processing (LCP) method developed at Fraunhofer ISE, as described in Chapter 7, is usually implemented on the passivating dielectric layer enabling self-aligned metallisation. This method requires one less processing step than the former method since the liquid dopant source deposition is not a separate step from that of the laser-doping process. LCP has been used for both n-type phosphorous doping [49, 50] and p-type boron doping [51], and efficiencies beyond 20% have been demonstrated for both LCP-based front [49] and rear selective emitter [52] structures.

An advantage of the self-aligned metallisation approaches for LDSE solar cells is that a much lower laser-doped coverage area (~1–2%) can be used compared

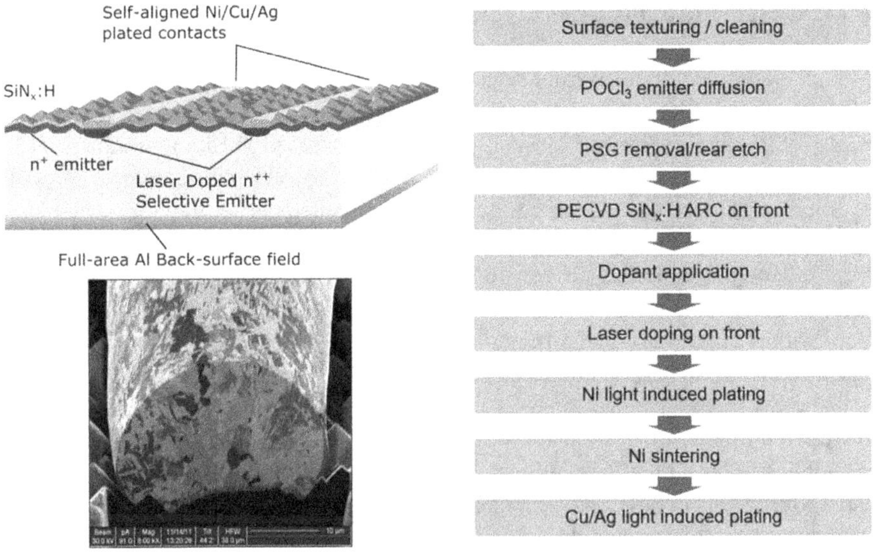

Figure 8.11 *(top left) Schematic of an LDSE Al-BSF solar cell with self-aligned Ni/Cu/Ag-plated contacts. Adapted from [46]. (bottom left) Contact focused ion beam image on a laser-doped finger with high aspect ratio self-aligned Ni/Cu/Ag light-induced plating. The image has been stretched in the vertical direction to correct for the angle. Reproduced from [48]. (right) Corresponding process flow diagram for an LDSE Al-BSF solar cell with self-aligned plated contacts.*

to the current industrial process with laser doping from PSG layers with aligned screen-printed contacts (~10%). This can reduce the contribution of the laser-doped region to the total J_{0e} of the device. Another advantage is the high conductivity of the plated contacts compared to that of screen-printed contacts, and reduced contact resistance, on the order of 0.1 mΩ·cm^2 [13].

However, plating is much more challenging than screen-printing, particularly in a production environment. The relatively complicated and delicate processing steps for plating involve a large number of processing parameters and conditions that require fine optimisation. Two of the main challenges with plating for metallisation are contact adhesion, as well as ensuring the formation of an effective copper diffusion barrier in the contact area to avoid potential shunting of the devices [53]. Plating is also a significant deviation away from mainstream technology used for industrial silicon solar cells, and the tools, particularly for light-induced plating, require very large footprints. Another key challenge is the potential issue of managing liquid metallic wastes [54]. At the same time, the constant improvements of screen-printing technology including narrower fingers, improved alignment capabilities and reduced Ag consumption have reduced the financial benefits of shifting to Cu-based plated contacts. As such, LDSE solar cells continue to use screen-printed contacts.

8.4 Laser-induced defect generation

Although attractive to form front selective emitter structures and local back surface fields in high-efficiency silicon solar cells as successfully demonstrated in various PERC/PERL and passivated emitter, rear totally diffused (PERT) structures [42, 52, 55–57], the implementation of laser doping has its own shortcomings. As will be briefly discussed below, the choice of laser parameters (i.e. fluence, speed and repetition rate in the case of pulsed laser) for optimal selective emitter formation requires careful fine-tuning. The major downside of laser doping is that it can create a large number of potentially recombination-active defects at the solid/liquid interface, within the recrystallised region and at silicon/dielectric interfaces, even for optimised laser processing conditions and dielectric layers [23, 58].

The detrimental impact of laser-induced defects was identified as the performance limiting factor even in the pioneering work of Fairfield and Schwuttke on the application of lasers for p-n junction formation in 1968 [59]. The laser-induced damage reduces the bulk and surface minority carrier lifetimes, therefore reducing solar cell efficiency compared to the ideal case of a damage-free, selectively, heavily doped region. Earlier studies proposed that the frequent melting and recrystallisation processes caused by the pulsed lasing can introduce impurities such as oxygen, carbon and nitrogen into the laser-doped region, forming new recombination pathways within the forbidden gap, which in turn degrades the minority carrier lifetime [60]. This further indicates the significance of using a high-purity dopant source especially when it comes to SOD sources. Instead, PSG/BSG are relatively pure and clean dopant-rich layers created during the high-temperature tube furnace diffusion process, although still containing oxygen. As already mentioned, PSG is currently the preferred dopant source for n-type laser doping in industrial PERC solar cells.

In theory, a high degree of laser pulse overlap (i.e. a longer dwell time on a region hit by the laser) is preferred to obtain deep laser-doping profiles with low sheet resistances [61]. This can be achieved by reducing the laser scan head speed and/or increasing the frequency of laser pulses. Increasing the laser dwell time, such that areas receiving overlapping pulses do not have time to re-solidify between pulses, can reduce the likelihood of inducing structural defects into the laser-doped regions and allows for sufficient time for the dopants to distribute themselves adequately through the molten region [42, 62]. However, this may require very high pulse frequencies or significant temporal overlap between pulses to avoid recrystallisation between successive pulses. On the other hand, reducing the pulse frequency and laser pulse energy has been shown to lead to poorer characteristic features of the laser-doped contact region, such as a higher contact resistivity and generally a narrower width of the doped region, which can create exposed un-passivated edges acting as a high recombination and shunting sites [63] (see Section 8.2.6 for further discussion). Therefore, continuous-wave (CW) and high-frequency quasi-CW (Q-CW) lasers with frequencies in the range of 80–100 MHz typically induce less damage to the carrier lifetime in the laser-doped regions [61, 64].

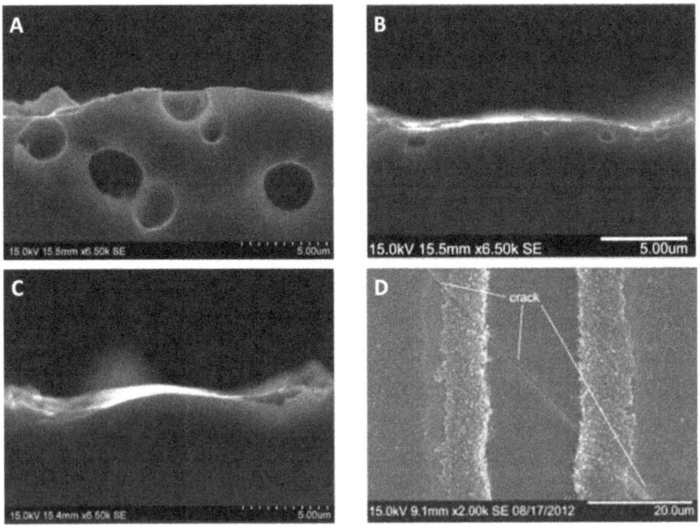

Figure 8.12 *Cross-sectional SEM images of the laser-doped region for samples processed with a 532-nm CW laser and 13 W optical power to dope through H_3PO_4 SOD source using (a) 0.5 m/s, (b), 2 m/s and (c) 5 m/s laser scan speed. The images illustrate the voids created in the laser-doped region with clearly larger ones formed at low laser speeds. (d) Top SEM image of a crack formed along crystal planes after laser doping at 0.01 m/s. Images adapted from [42, 48].*

The formation of voids within the bulk of the melted region has been demonstrated for slow laser scan speeds when using a CW laser with their concentration and size increasing at lower speeds [48], as indicated by the SEM images in Figure 8.12. The density of such voids is further dependent on the dielectric layers and the deposited dopant source. In addition, large concentrations of crystallographic defects such as dislocations [65, 66] can be formed at relatively high laser fluences due to melting and recrystallisation of a large amount of silicon, inducing extensive thermal stress [42]. However, dislocations seem not to be formed in the deep regions of the laser-doped area, close to the junction, possibly because of its similar crystallographic orientation as the silicon substrate [67] and the fact that this is the first part of the molten region to re-solidify. Although free from dislocation defects, a decrease in the electrical performance can be still experienced in these regions. Electrically active point defects with their density depending on the recrystallisation velocity were identified as the main cause of the performance degradation in these deep zones [68]. It is also possible to create micro-cracks [66] when performing laser doping, resulting in junction discontinuity and possible shunting; the chances increase at higher laser powers [68]. An example of a formed crack, and extended beyond laser-doped region, is provided in the SEM image of Figure 8.12 (d). Nevertheless, in some cases, laser-induced defects can actually be beneficial as they have been shown to be effective as gettering sites for bulk impurities [69]. This concept will be further discussed in Chapter 9.

8.4.1 Characterisation of laser-induced defects

Identification and characterisation of laser-induced defects go back to the 1970s [58]. With the ongoing advancements in the available characterisation techniques, the development of the methodologies, and the interest in industrial implementation of laser doping for silicon solar cell fabrication, there has been renewed interest in improving the understanding of laser-induced defects, as suppression of these defects is critical to achieving high cell efficiencies. In Sections 8.4.1.1 to 8.4.1.4, methods for characterising laser-induced defects are discussed along with the current understanding of the root cause of the laser-induced defects gained from using these various methods.

8.4.1.1 Electron beam microscopy-based characterisation

Dislocation formation at different depths of the laser-doped region has been investigated using transmission electron microscopy (TEM). In an earlier study, Hameiri *et al.* demonstrated that dislocations are likely to form near the surface and not in the deeper epitaxially recrystallised layers [68]. As will be discussed later in this section, this depth-dependent trait is consistent with the findings of other optical and electrical characterisation methods. From depth-dependent deep-level transient spectroscopy (DLTS) characterisation of the laser-melted region, an exponential-like drop in laser-induced defect concentration was suggested (see Section 8.5) [70]. In addition, a high-energy excitation wavelength has allowed clear detection of luminescence signal from the laser-induced damage within the fully molten region, whereas excitation using low-energy photons, with a deeper optical penetration depth in silicon, did not reveal an observable emission signal at the corresponding laser-induced defect peak (see Section 8.5) [65].

TEM has provided further important findings regarding laser line width threshold below which defect-free recrystallisation of the molten region can occur [71]. It has been shown that the maximum of this line width, however, depends on the crystallographic orientation of the surface of the silicon wafer. Wider laser lines were allowed for (100)-oriented silicon wafers, the standard orientation of wafers used for silicon solar cells, as opposed to (111)-oriented wafers before defects start to form. However, silicon recrystallised epitaxially in all studied cases, without the formation of amorphous silicon and/or twin crystals. An example of TEM image from the study is provided in Figure 8.13.

8.4.1.2 Combined QSSPC and photoluminescence-based characterisation

Measurements of the injection-level dependent minority carrier effective lifetime (τ_{eff}) of a silicon sample before and after laser doping can provide valuable information regarding laser-induced defects. One of the common characterisation techniques

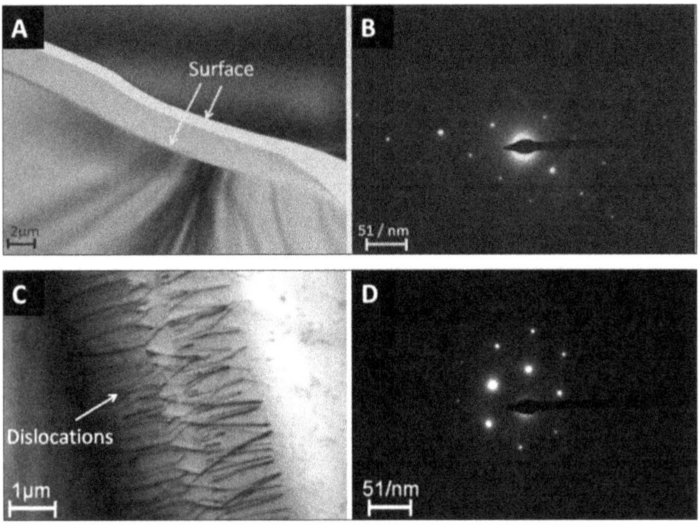

Figure 8.13 *(Top) TEM results of silicon with a (100)-oriented surface, irradiated with a laser beam focused to a line with a pulse energy density of 2.5 J/cm², 28 ns pulse length, 8 μm line width at full width at half maximum (FWHM). (A) Two surfaces are joined, allowing an analysis of two cross sections perpendicular to these surfaces at a time. The darker areas are bending contours evolving due to the furling of the thinned silicon. (B) Diffraction pattern. The positions of the maxima correspond to the maxima of a (100)-oriented crystal. The silicon is epitaxially regrown with no defects/dislocations. (Bottom) Corresponding TEM results of silicon with a (111)-oriented surface, irradiated with a pulse energy density of 3.7 J/cm², 62 ns pulse length, 15.5 μm line width at FWHM. (C) TEM image shows a high defect concentration in some areas. (D) Diffraction image indicating a (111)-oriented crystal. The silicon is epitaxially regrown. © 2011 IEEE. Reprinted, with permission, from [71].*

to monitor this is via quasi-steady-state photoconductance (QSSPC), which was described briefly in Chapter 3.

In Figure 8.14 (a), the experimentally measured inverse effective minority carrier lifetime of a symmetrical p-type lifetime test structure with diffused phosphorus emitter of 103 ± 3 Ω/□, passivated with hydrogenated SiN_x is shown in its as-deposited state. The sample then underwent rapid firing with a peak temperature of 680 °C (triangle symbols in panel (b)). Using a 532 nm CW laser, laser doping was performed through phosphoric acid as the SOD to pattern a 50 mm × 50 mm area with 500 μm laser line spacing (circle symbols; unpublished data). After the firing process, the effective lifetime enhanced significantly with a clear reduction

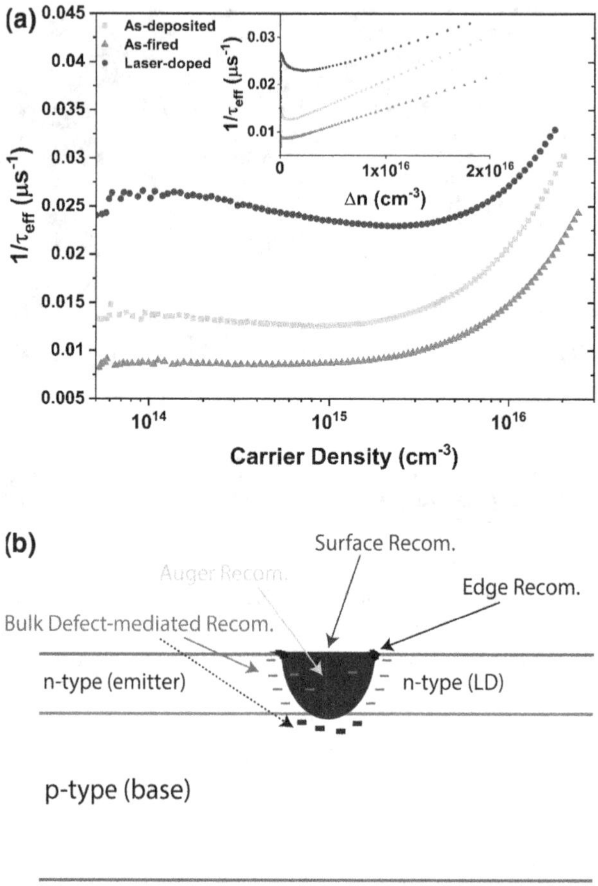

Figure 8.14 *(a) Injection-dependent inverse lifetime curves of a symmetrical p-type lifetime test structure with n-type emitter passivated with hydrogenated SiN_x in the as-deposited start (square symbols), after firing (triangle symbols) and after laser doping (circle symbols). The inset uses a linear scale to demonstrate the impact of each step on high-injection level slope associated with J_{0e}. (b) Schematic of possible recombination pathways in the laser-doped region and its vicinity.*

in the emitter recombination current density (J_{0e}). The latter is understandable considering the reduction in the slope of the inverse effective lifetime curve versus minority carrier density, in the linear scale at high injection levels (see the inset of Figure 8.14 (a)). It is noted that the above argument and the below discussion can be understood considering the equation for recombination rate (R) describing surface-diffused samples (i.e. with emitter) which encompasses the recombination parameter J_{oe}, where n is the electron concentration, W is the width of the sample,

τ_{bulk} is the bulk lifetime, N_{dop} is the doping density, q is the elementary charge and n_i is the intrinsic carrier concentration [72]:

$$R_{tot,\ nr} = \frac{n\,W}{\tau_{bulk}} + 2J_{0e}\frac{n\,(n + N_{dop})}{qn_i^2} \tag{8.1}$$

Degradation in the minority carrier effective lifetime of a laser-doped sample, and thus its implied open-circuit voltage (iV_{OC}), can be caused by several different mechanisms that can be difficult to separate. This is schematically illustrated in Figure 8.14 (b). The effective lifetime deterioration measured experimentally, such as that presented in Figure 8.14 (a), can be due to degradation in the bulk lifetime, caused by defects extending beyond those within the molten region (i.e. heat-affected zone [73]). This can be observed in the low-injection part of the inverse effective lifetime curve in Figure 8.14 (a) (circle symbol curve). This is further exemplified by the thick dashed lines in the p-type base of the test structure shown in Figure 8.14 (b). Most of the remaining losses imposed on the charge-carrier lifetime are commonly lumped into the increased total J_{0e}. For example, this can be partly realised by the change in the slope of the effective lifetime curves at high injection levels in Figure 8.14 (a). It should be noted that this slope may also be influenced by the change in the Auger recombination lifetime in the heavily doped laser-doped region, likely here due to a short spacing between the laser lines (i.e. 500 μm). In addition to the slope change, increase J_{0e} due to laser-induced damage, causes a constant increase in the $1/\tau_{eff}$ at low injection levels (note (8.1)). As indicated in Figure 8.14 (b), such contributions to the lifetime deterioration include defect-mediated non-radiative recombination in the molten region, surface recombination of the laser-doped region, edge recombination (at the edge of the molten region surrounding the laser-doped region), and the increased Auger recombination in the heavily doped laser-doped region. The separation of these effects accumulated in the J_{0e} is a nontrivial task, which is discussed further in the following paragraph. A proper quantitative analysis through modelling of the injection-dependent effective lifetime data including all such defects is currently (at the time of preparing this chapter) missing in the literature.

Recently, Ernst *et al.* decomposed the lumped averaged J_0 of the laser-doped region (i.e. total non-radiative recombination saturation current density induced by laser-doping process) into its components for samples fabricated using a two-step combined laser doping/laser contact opening (LCO) contact formation route. Contributions from the edges and the fully molten central region were disentangled and discussed within the recombination current density context [74]. This was achieved by fitting the averaged J_0 data as a function of the edge area fraction. The former values were obtained for a range of laser fluences and spot sizes using photoluminescence (PL) intensity images, calibrated by QSSPC. To vary the edge fraction, the laser spot size was varied while keeping the total area fraction of the laser-doped region constant, which required simultaneous changing of the laser pitch with the laser spot size. The modelling showed that the edge areas around the central fully laser-melted region are the major contributors to the increased charge-carrier recombination in the laser-doped region, particularly for high laser fluences. Although the contribution of the fully molten regions comes second, it is not insignificant [74]. Estimated J_0 values reported for the edge areas were at least three times larger than the recombination current density of the fully molten region.

8.4.1.3 Spectral photoluminescence-based characterisation

The combination of a spectral PL measurement with confocal microscopic signal detection allows probing the recombination properties of spatially non-uniform defects at microscopic levels. This technique is known as micro-PL. In this technique, a focused monochromatic light illuminates the sample to optically excite the electrons from valence band to the conduction band, enabling PL emission from the semiconductor. The PL signal emitted from the sample, which needs to be measured, passes through a confocal pinhole to eliminate the out-of-focus background signal and therefore localise the detection volume to the vicinity of the excitation volume centre. This detection technique is particularly useful for probing non-uniform localised defects for narrow laser-doped regions, as compared to large-area detection methods such as steady-state PL imaging described in Chapter 4, quasi-steady-state PL and QSSPC. These latter techniques are more affected by spatial inhomogeneity in the charge-carrier bulk lifetime. The confocal spectral micro-PL system has been used for many studies related to defects in silicon [75] such as characterising D-lines in multi-crystalline silicon wafers [76], and for microscopic transient carrier lifetime analysis [77].

Through detailed analysis of the characteristic sub-bandgap luminescence spectral signal using micro-PL, a large number of dislocations in the edge area of the laser-doped region have been identified. This was observed to a lesser extent in the heavily doped fully molten region, possibly formed during the solidification phase [65] (see Figure 8.15). The detection of PL signal from laser-induced dislocations requires low-temperature measurements to avoid being obscured by the thermal broadening of the silicon and the heavily doped region band-to-band emission signals, and to prevent charge carriers from de-trapping back to their corresponding band-edge. The emergence of the associated PL peak by lowering the silicon temperature is illustrated in Figure 8.15 (b). It also needs to be measured with a high-energy excitation light-source with shallow penetration depth, which further reduces at higher temperatures for silicon, to enable detection of the defective sites formed close to the surface (see Figure 8.15 (c)).

The PL peak energy position associated with the laser-induced defect matches well with the energy offset of sub-bandgap D4 peak of the D-line series in multi-crystalline silicon with respect to the un-doped band-to-band peak [65, 76]. This emission has a wavelength between 1 250 nm and 1 300 nm, with a broad signature peak emission at about 100 meV lower than the heavily doped band-to-band emission peak. The temperature-dependent PL intensity behaviour of the laser-induced defect was observed to mostly follow that of D3 and D4, which are commonly attributed to the intrinsic properties of dislocations in silicon [65, 78, 79]. These, in addition to the similar activation energy for laser-induced defects and D4, have led

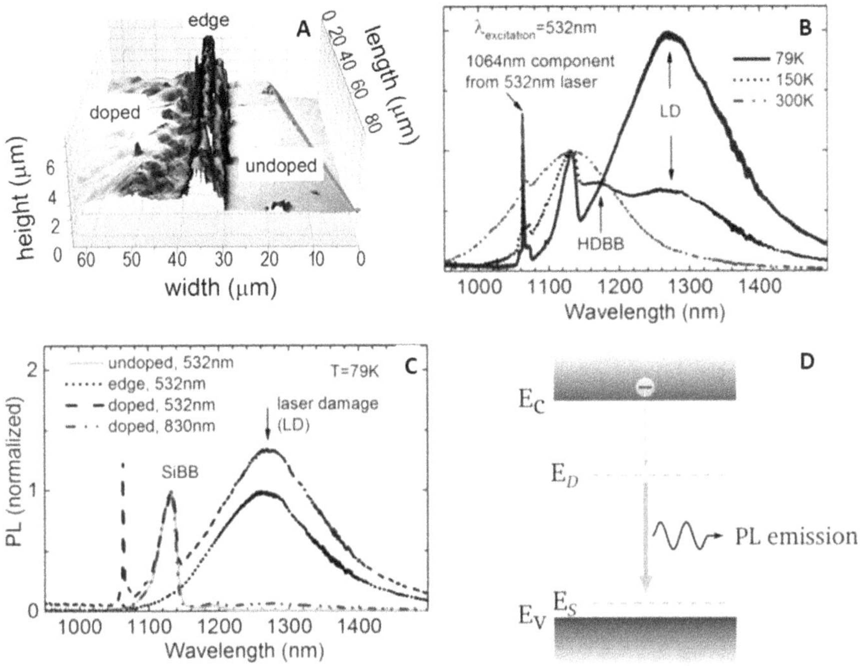

Figure 8.15 *(a) Optical surface image of a severe laser damaged area, using high-fluence 248 nm KrF pulsed laser (25 ns pulse duration), with the three components of the laser-doped region labelled. (b) PL spectra of the doped region, excited with a 532 nm CW laser, measured at different temperatures. (c) PL spectra detected from the three regions identified in panel (a). The impact of excitation beam energy on the dislocation-induced PL signal is also illustrated (dashed line vs dash-dotted line). The emitted PL signal from the locally excited edge regions is very strong compared to the signal from the doped region only for 532 nm excitation wavelength and, therefore, its spectra are normalised. (a–c) Reprinted from [65], with the permission of AIP Publishing. (d) Schematic of the shallow and deep states with respect to the band-edges, which contribute to the PL emission from laser-created dislocations.*

to the postulation of D3 and D4, being the emissive components of the broad laser-induced defect PL peak. Due to a clear temperature-dependent characteristic of the dislocation-induced PL intensity caused by laser doping, resembling that of its postulated elements, D3 and D4, the laser-induced defects are described with a shallow state close to one of the band-edges. Furthermore, the relatively large peak position energy difference with respect to that of bandgap-narrowing-induced heavily doped band-to-band emission, the laser-induced defects are also described as a deep state distant from the other band-edge. These are schematically shown in Figure 8.15 (e).

It is still unclear whether or not the generated defects and their corresponding luminescence behaviour is universal and relevant to a variety of silicon substrate types, dopant sources and laser systems [58].

These understandings of the defect states associated with the postulated dislocations and/or any other defect precursor created by laser doping can be further explored for validation with characterisation techniques that allow direct identification of the active defect state(s), their concentration and capture cross sections. Examples of such techniques are temperature- and injection-dependent lifetime spectroscopy and DLTS.

8.4.1.4 DLTS-based characterisation

DLTS has been used to distinguish and understand the role of dopant source and laser irradiation itself on the nature of the defects created within the laser-melted (and subsequently epitaxially re-crystallised) region. Laser irradiation was only conducted on bare silicon wafers in the absence of an external dopant source or stack of dielectric layers [70, 80]. It is noteworthy that upon laser doping through a dielectric stack, or when dopant source is used, new recombination active sites within the forbidden gap may be introduced, modifying the energy levels and/or carrier trapping behaviour (i.e. capture cross section) associated with the deep defect levels.

A variety of pulsed laser systems including solid-state Nd-YAG and ruby lasers and gas-state ArF excimer laser with different irradiation wavelengths were used to laser-melt both non-diffused n- and p-type silicon. These earlier studies provided insights into the potential defect states created upon laser melting and re-crystallisation of silicon; multiple deep electron/hole trap states were formed in the n-type/p-type silicon wafers, with energy levels ranging from a couple of hundred meV from the valence/conduction band-edges all the way to the mid-gap energy level. The laser-induced defect features detected by DLTS have been primarily associated with vacancy clusters and vacancy-impurities complexes [70, 80]. The energy levels and capture cross sections of the identified laser-induced defects were proposed to be independent of the laser type, as long as the processing conditions and parameters are such that the depth of the molten region and the melting-cooling rate (i.e. velocity of solidification) remain similar [70]. Nevertheless, detailed DLTS as well as micro-PL-based characterisation of laser-induced defects formed using CW laser seem to be missing in the literature.

The existence of deep defect levels characterised by DLTS is consistent with the proposition made for the energy level of laser-induced defects through the analysis of temperature- and illumination intensity-dependency of micro-PL signal emitted from laser-doped region of n-type FZ silicon. However, there seems to be a discrepancy between the proposed origins of the laser-induced defect; structural imperfections, or dislocations, have been postulated as the origin of laser-induced defect luminescence [65]. It is also likely that not all the deep defect energy levels identified by DLTS are emissive, which could potentially result in radiative recombination of the captured charge carriers and be detected through PL measurements.

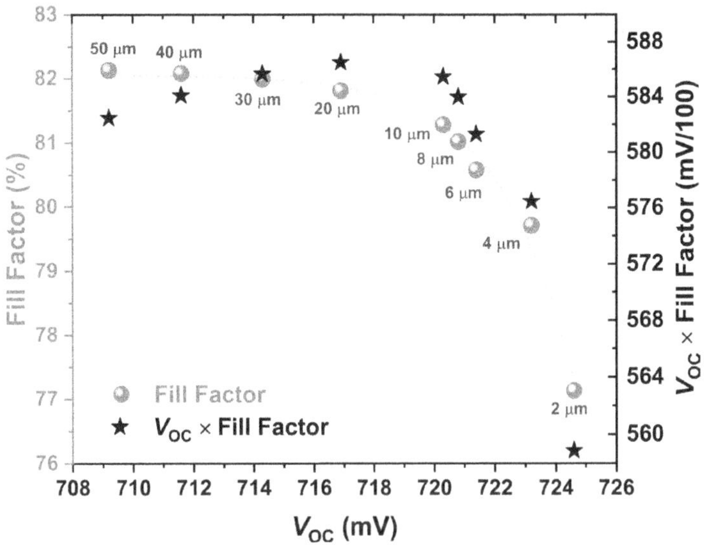

Figure 8.16 *Impact of laser-doped line width on the solar cell FF and* V_{OC}. *Laser-doped line width is considered here to be equal to the metal finger line width.*

8.5 Techniques to minimise the impact of laser-induced defects

8.5.1 Reduction in laser-doping coverage and metal/Si interface area

To minimise the impact of high recombination rates within the laser-doped region on the V_{OC} of the device, which can also influence fill factor (FF) [81], one can reduce the area fraction of the selectively laser-processed region and/or metal/Si interface area [55]. Although promising, care must be taken due to the trade-off between recombination losses primarily affecting V_{OC} and resistive losses primarily affecting FF. Reducing the area fraction of laser doping can lead to slight increases in the total resistance of the device. For example, if the distance carriers have to travel in the lightly diffused region increases, this can increase the lateral resistance between fingers. In addition, if there is a significant reduction in the contact area, it can lead to substantial increases in contact resistance due to the inverse proportionality of resistive loss with the contact area, which in turn causes a decrease in the solar cell FF. Note that this argument applies to any selective emitter formation technique while the nature of the recombination process and the extent of defects formed can be different. Figure 8.16 illustrates modelling results exemplifying this behaviour assuming a hypothetical laser-doped region with a perfectly aligned metal/Si interface contact region of the same width as the laser-doped region. Here, on the one hand, a reduction of the laser-doped line width increases V_{OC}, a drop in width below 10 μm causes a substantial drop in FF. As such it creates a trade-off. Here, an

Table 8.2 *Baseline parameters used as an input to the software Sentaurus TCAD to simulate the impact of laser-doped line width/finger width on LDSE PERC solar cells as presented in Figure 8.16*

Parameters	Values
Substrate thickness	180 μm
Substrate resistivity	p-type 2 Ω·cm
SRH lifetime (e, h)	$\tau_e = 2.4$ ms, $\tau_h = 28.9$ ms
n region (passivated)	ρ = 380 Ω/□, $J_0 = 7.3$ fA/cm^2
n^+ region (metallised)	ρ = 74 Ω/□, SRV = 10^7 cm/s, $J_0 = 585$ fA/cm^2
p region (passivated)	$S_p = 38.5$ cm/s, $S_n = 5.47 \times 10^3$ cm/s $Q = -3\times10^{12}$/cm^2, $J_0 = 6.2$ fA/cm^2
p^+ region (contact)	SRV = 10^7 cm/s, $J_0 = 210$ fA/cm^2
Finger pitch	1 mm
LDSE width	Varied
Finger width	Varied

optimum range in terms of maximising the product of V_{OC} and FF exists between 10 μm and 30 μm. The device parameters considered for the purpose of this modelling, performed in Sentaurus TCAD, are given in Table 8.2. Changes in the metal/Si interface area through changing the width of metal contacts can also affect the J_{SC} of the device. In addition, changes to the width of the laser-doped region and metal contact width could potentially impact yield, if the laser doping width is too small for the required alignment tolerance of a subsequent process like screen-printing. Changes to the width of the laser-doped area beyond the metallised region can also potentially impact the J_{SC} through the formation of a 'dead layer'.

8.5.2 Laser doping prior to dielectric deposition

A powerful approach to reduce the generation of laser-induced defects is to perform laser doping prior to dielectric deposition [82–84]. This is the current method for industrial LDSE PERC solar cells [2]. In this approach, after dielectric deposition, an aligned screen-printing process with fire-through silver paste penetrates through the dielectric to form the metal/Si contact. In this method, the metal finger width must be much narrower than the laser-doped line width. This ensures that the metal only makes contact with the heavily diffused laser-doped region (see Figure 8.17 (a)), to avoid deterioration to the fill factor and V_{OC} of the device that would occur if the metal finger was misaligned and contacted the lightly diffused (non-laser-doped) emitter region (see Figure 8.17 (b)).

This method requires high-precision screen-print alignment, while benefiting from the prominent reduction in SRV of the non-contacted area on the laser-doped region and passivating the edge regions. In general, the laser doping is performed in a stripe approximately 100–150 μm wide, with an aligned screen-printed finger with a width of 40–50 μm, providing an alignment tolerance of approximately

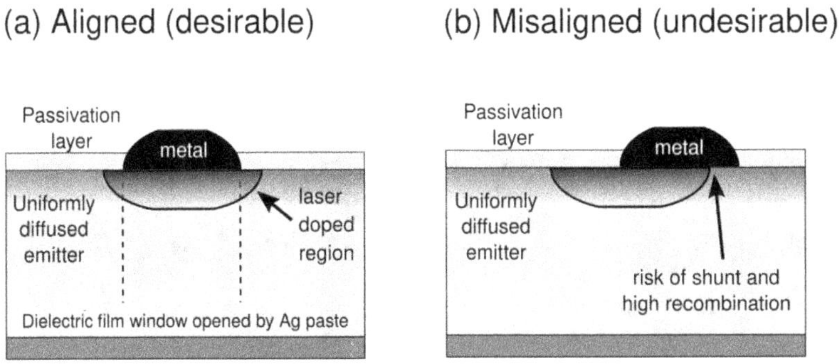

Figure 8.17 Schematic of (a) wide laser-doped line with aligned screen-printed fire-through finger in the middle (desirable) depicting the current implementation for industrial LDSE PERC, and (b) misaligned screen-print finger touching the undoped silicon (undesirable)

± (30–55) μm. It should be noted that this approach of edge passivation can become ineffective if the corresponding local dopant profile is such that recombination becomes SRV independent and only Auger limited (refer to Figure 8.5).

When this process flow is adopted, apart from a lower extent of laser-induced defect formation due to the performing laser doping prior to dielectric deposition, the subsequent dielectric deposition processes and firing process can enable natural hydrogen passivation of the process-induced defects during cell fabrication as discussed in Chapter 4 [83]. Hydrogen passivation of laser-induced defects will be discussed in detail in Section 8.6.

Another approach is a two-step laser-doping/LCO contact formation route. After laser doping and the subsequent dielectric deposition, an LCO process can be used to open a region of the dielectric layer above the laser-doped area. Here, the LCO region must again be smaller than the laser-doped area and precisely aligned to ensure that the opened region is fully contained within the laser-doped area. In this method, after LCO a self-aligned plating metallisation could be used, or an aligned screen-printing process using a non-fire-through paste. This method benefits from the extended dielectrics covering over the recombination-active defects at the edge of the laser-doped regions, tackling the associated local high recombination rates around the edge of laser-doped regions, particularly if forming part of the metal/Si interface region.

An advantage of performing laser doping prior to dielectric deposition is that dielectric layers deposited on top of the laser-doped region enable chemical and field-effect passivation of the laser-doped area in the non-metallised region. Full coverage of the laser-doped region with a thin dielectric layer containing fixed charges with the appropriate polarity (e.g. SiN_X on n^+ laser-doped regions and AlO_X on p^+ laser-doped region) can significantly reduce the recombination parameter J_0 of the laser-doped region.

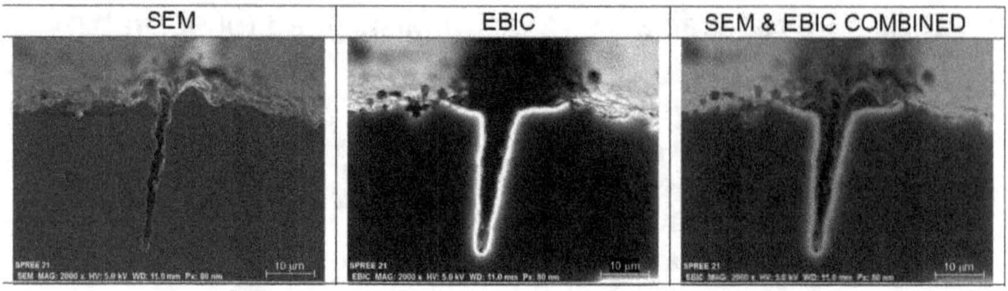

Figure 8.18 *Example with cross-sectional SEM image, corresponding EBIC image and combined SEM/EBIC image of a typical laser-doped groove silicon wafer. Reproduced from [85] Copyright (2017) with permission of Elsevier.*

8.5.2.1 Simultaneous laser doping and grooving

An alternative approach developed by Wang *et al.* to allow laser doping to be performed prior to dielectric deposition to reduce defect formation, yet still allow a self-aligned metallisation scheme, is to finely tune laser parameters to allow the formation of heavily doped grooves [85]. This also overcomes the adhesion issue of the self-aligned metal-plated contacts in the conventional LDSE structure when laser doping is performed through dielectric layers.

In this technique, laser doping can be implemented using the PSG from the emitter diffusion process as a dopant source or using an applied SOD source. For the latter case, laser doping can be performed prior to emitter diffusion; however, this results in additional processing steps and care must be taken for the propagation of laser-induced defects during the high-temperature emitter diffusion process.

With finely tuned laser conditions, very narrow (3–5 μm wide) and deep (10–15 μm) grooves can be formed whereby the groove walls are heavily doped. The grooves form due to the reduced density of silicon in the molten state than the solid state, thereby creating the groove during cooling. Figure 8.18 shows an example with cross-sectional SEM and EBIC image of a laser-formed groove highlighting the p-n junction formed during the laser-doping process [85]. The dielectric layer is then deposited on the laser-doped wafer, which acts also as a passivating layer, minimising the impact of defects, particularly edge defects, created during laser processing. Due to the conformal nature of plasma-enhanced chemical vapour deposition (PECVD) processes, to a large extent, the narrow groove prevents the deposition of the dielectric layer into the groove, leaving most of the groove walls exposed to allow nucleation of the plated metal. This approach also reduces laser-induced defect formation caused by thermal stress, in contrast to when laser doping is conducted through the dielectric stack. In addition, a substantially lower laser-doped coverage area can be used compared to that for the current industrial process. This technology has achieved efficiencies over 19% for a solar cell with a full area Al-BSF [85], indicating further efficiency enhancements is possible by optimisation of the rear of the device such as by adopting a PERC cell structure.

8.5.3 Localised passivated contact structures on laser-doped regions

Attempts have also been made to incorporate passivated contact structures on laser-doped regions to reduce the lumped emitter J_0 after laser doping by depositing ultrathin passivating contact layers on the laser-doped regions. For example, SiO_2, hydrogenated amorphous silicon (a-Si:H), Al_2O_3 and bilayer stacks of SiO_2/a--Si:H and Al_2O_3/a-Si:H have been applied on very lightly doped n^+ (190 Ω/Υ) or p^+ (196 Ω/Υ) regions [86]. Without a passivated contact, the very lightly doped regions provided little protection against recombination at a metal/Si interface, hence very high initial J_0 values of approximately 2 000 fA/cm^2, substantially higher than that for industrial LDSE PERC solar cells (~300–500 fA/cm^2). As such, with the implementation of the passivated contacts, the J_0 reduced substantially to 60–70 fA/cm^2 and approximately 1 000 fA/cm^2 after dielectric stack deposition for n-type and p-type laser-doped regions, respectively.

These structures were implemented on n-type PERL solar cells fabricated on FZ silicon. Here, laser doping was used to form localised BSF regions with the LCP process used to perform laser doping through the rear SiNx layer. Subsequently, passivated contacts were formed, with a full area a-Si:H layer and evaporated Al contacts. This resulted in a promising power conversion efficiency enhancement of 1.5%$_{abs}$ with respect to a non-passivated reference cell with 4.5% rear contact area (efficiency improvement from 19.4% to 20.9%) [52]. However, undesirable, the addition of the passivated contact structure increased the specific contact resistance by a factor of up to three, although specific contact resistivities below 10^{-4} Ω·cm^2 were achievable for optimal annealing conditions [52, 86].

To date, efficiencies achieved are well below that for industrial LDSE PERC solar cells, with sheet resistance values in the laser-doped regions in the range of 60–80 Ω/□. The implementation of passivated contacts with laser doping would be of reduced benefit for LDSE PERC due to the relatively effective shielding of the heavily doped regions and low coverage area, which have already demonstrated efficiencies of almost 24% [2], and essentially capable of 25% efficiency based on the record PERL cell [87]. The addition of passivated contact structures would also greatly increase processing complexity. Furthermore, the use of a-Si:H-based passivated contacts would affect the ability to use high-temperature screen-printed pastes due to the thermal instability of such passivated contact structures above 350 °C [88]. In addition, if implemented on the front surface, increased parasitic absorption would be expected, and much smaller drops in the J_0 would be expected using the current doping profiles. The J_0 values are also substantially higher than that achieved for industrial TOPCon solar cells [89], in which a lightly diffused region naturally forms underneath the tunnel oxide layer during fabrication.

8.5.4 Performing laser doping through dielectric layers

When laser doping is performed through the dielectric stack and into the emitter such as LDSE technologies with self-aligned plating metallisation scheme, the induced damage is expected to be worse [68] compared to the above contact formation

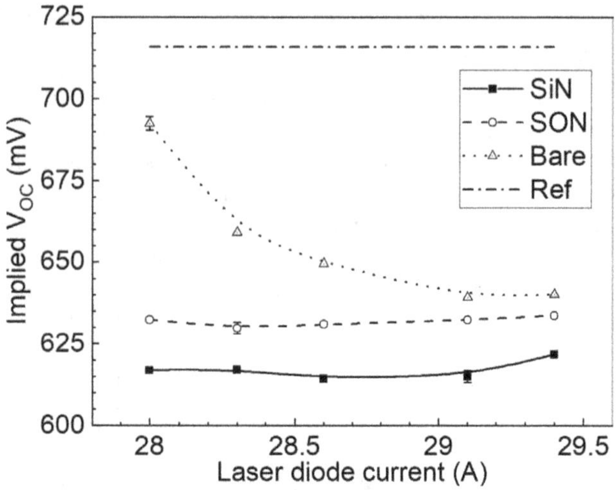

Figure 8.19 *The impact of laser conditions in the presence (square data points) and absence (circle data points) of a thin silicon oxide layer underneath the silicon nitride (SiO_2/SiN, denoted 'SON') on implied V_{OC}. Implied V_{OC} of laser doping performed on bare wafers is shown with triangle data points, with the least drop. Reproduced from [68]. Copyright © 2010 John Wiley & Sons, Ltd.*

approaches. This can be understood considering that the smallest implied V_{OC} drop was obtained when laser doping was performed on a bare-diffused wafer as demonstrated in Figure 8.19 (triangle data points). To alleviate the deteriorating impact of laser doping when performed through the dielectric layer(s), one strategy has been to optimise the type and thickness of the dielectric layers deposited on the fully diffused emitter of the silicon cell precursors. The dielectric stack has a noticeable impact on the extent of laser-induced defect formation. It has been demonstrated, for a range of laser conditions, that degradation in the minority carrier lifetime and thus the implied V_{OC} of the laser-doped samples is reduced when a thin layer of SiO_2 is deposited underneath the SiN_x, together acting as the anti-reflecting coating [33, 68]. This is reflected in the implied V_{OC} data shown in Figure 8.19 (square data points). The effectiveness of this tactic has been proposed to be due to the formation of lower defect density in the case of using SiO_2/SiN_x stack. Reduced thermal-induced stress due to an improved consistency between the thermal expansion coefficients of the silicon substrate and the dielectric stack has been proposed as a reason for this development [63, 90]. The use of a thin SiO_2 layer also has additional benefits of improved surface passivation.

The thickness of the dielectric stack can also have an impact on the extent of defects formed during laser-doping process. A study by Xu *et al.* has verified that for optimal laser processing conditions and SiN_x thickness, with a thin SiO_2 layer underneath the SiN_x layer, proper coverage of the edges with the passivating

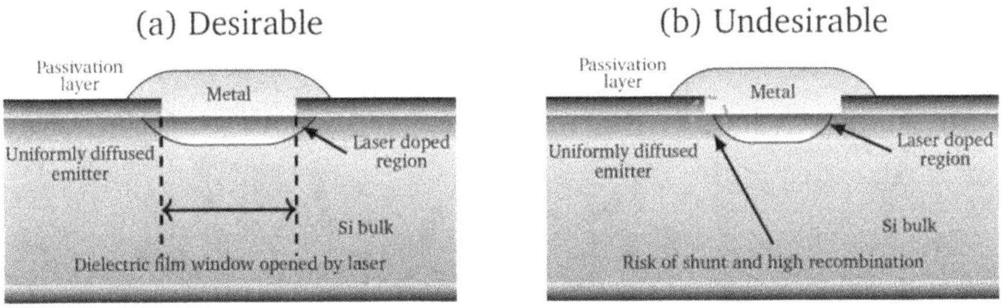

Figure 8.20 *Schematic illustrations of possible scenarios after laser doping for selective emitter formation. Panel (a) depicts a scenario in which implementation of optimised dielectric layers' types and thicknesses allow LDSE technology with preferred coverage of the laser-induced edges coverage (i.e. laser-doped region extends laterally under the dielectric opening) to reduce its detrimental effect. Panel (b) the exposed, un-doped and un-passivated regions indicated in the 'undesirable' case are prone to high recombination rates and possible shunting upon metallisation. Adapted from [63].*

dielectric layers is achievable, which potentially allows partial suppression of laser-induced edge recombination [63]. Figure 8.20 schematically compares the favourable silicon-dielectric stack interfacial configuration (panel **a**) to an undesirable one (panel **b**) formed after the laser-doping process. The undesirable configuration can induce high carrier recombination rates due to the lightly doped exposed surface and likely to further cause shunting upon metallisation [63]. This is essentially the same for a two-step laser-doping/LCO process as described before.

8.5.5 Deep-junction laser doping

As discussed in Chapter 7, the laser parameters can greatly impact defect generation. For example, using a CW or Q-CW laser to avoid the repeated melting and solidification of the silicon, thereby enabling very deep molten regions to be formed. One advantage of this approach is that the doping can extend several microns beyond the opening in the dielectric layer, thereby ensuring that the opened region is fully enclosed within the laser-doped region, similar to that with a multi-stage laser-doping/LCO method. As such, using a CW laser, a self-aligned metallisation scheme can be used without the risk of shunting. Figure 8.21 (a) shows a combined SEM/EBIC image of an n-type laser-doped region processed at 0.2 m/s highlighting the p-n junction, with a junction depth of more than 10 μm in the central region, and extends more than 5 μm beyond the opening in the dielectric layer [91]. A downside of this approach is the complete melting of the silicon, resulting in the loss of the pyramid structure (see Figure 8.21 (b) as adapted from [92]). As such, the use of such laser parameters is not suitable for the current implementation of LDSE PERC solar cells whereby the laser-doped region is substantially wider than the screen-printed finger. From an optical point of view, the approach with a CW laser and

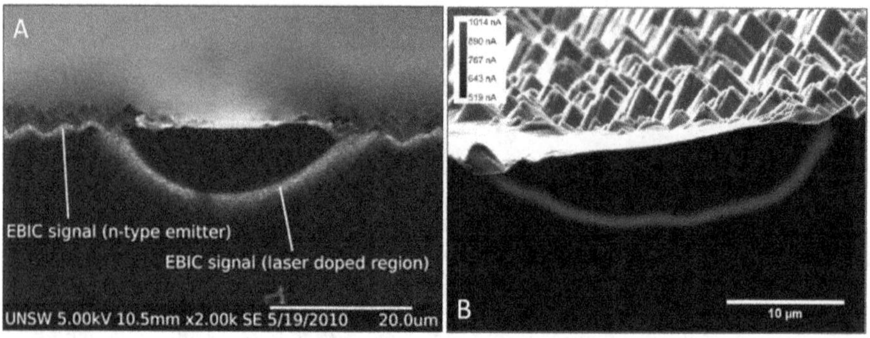

Figure 8.21 *(a) Combined SEM/EBIC images at ×2 000 magnification of an n-type laser-doped line processed at 0.2 m/s on a textured surface using a CW laser. Adapted from [91]. (b) Combined SEM/EBIC image of a laser-doped region processed with a CW laser highlighting the loss of pyramids in the laser-doped region, collected at 5 kV accelerating voltage. Adapted from [92].*

flattened surface profile is very well suited to plated contacts, where the flattened region lies underneath the metal contact. However, higher processing speeds can partially maintain the surface texture, which may assist with improved adhesion.

This approach was used in a number of publications to achieve efficiencies over 20% for p-type devices such as Suntech's Pluto technology [42, 46, 93] and as high as 22.5% for n-type devices [1, 94]. However, these high efficiencies were critically dependent on the passivation of laser-induced defects, as will be discussed in Section 8.6.

8.5.6 Laser- and thermal-annealing for localised repair of laser damage

Another approach for passivating the defects created by laser doping is the implementation of a laser- or thermal-annealing process. For example, a second pass of lasing is run over the laser-doped region at much lower laser power to anneal the damage induced in the first high-powered laser-doping pass [66, 95]. Alternatively, laser-induced defects can be annihilated through furnace annealing at 800 °C for 30 minutes.

Both approaches were observed to largely recover the minority carrier lifetime of the laser-doped regions as reflected in the enhancement in the band-to-band PL intensity after the treatments (Figure 8.22 (a)) and simultaneous reduction in the PL signal originated from the so-called laser-induced dislocations (Figure 8.22 (b)) [95].

These two treatments were further applied on IBC cells at different stages of cell fabrication with intriguing findings. Laser annealing applied immediately after the laser-doping step mainly improves V_{OC} and increases the shunt resistance, whereas its implementation after contact formation boosts the FF due to a reduction in series

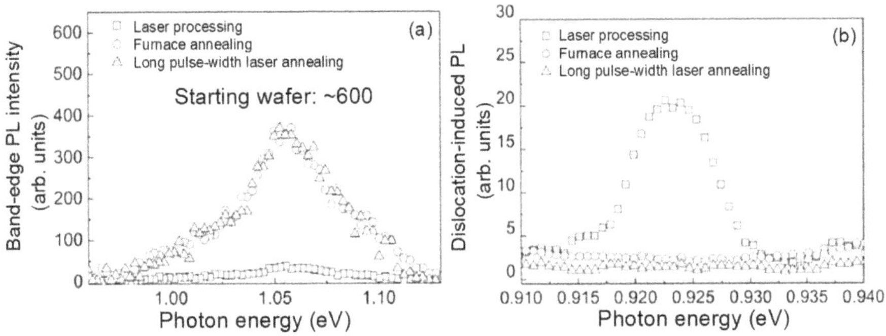

Figure 8.22 *The impact of furnace firing at 800 °C and long-pulsed laser annealing on (a) the increase in the band-to-band PL signal after laser processing a bare p-type FZ silicon wafer and (b) decrease in the dislocation-induced PL signal. Reprinted from [95], with the permission of AIP Publishing. © 2018 Author(s).*

resistance. The implementation of a two-stage laser-annealing treatment during the cell fabrication process sequences outperformed that of the two-step furnace-annealing counterpart for decreasing non-radiative recombination losses and contact resistance, providing the better of the two worlds with a lower thermal budget.

8.6 Hydrogen passivation of laser-induced defects

Hydrogen can also be used to passivate laser-induced defects. For example, an early study identified that a low-energy ion implantation process could be used to reduce the recombination activity associated with laser-induced defects [96]. Even in an ideal scenario where the combination of laser-processing parameters and the passivating dielectric layers is cautiously selected and implemented to minimise the generation of laser-induced defects, hydrogen can also be used to passivate the remaining laser-induced defects to reduce recombination activity. Such reductions may occur for non-radiative recombination pathways within the bulk as well as the J_0. Note that in Figure 8.14 (a), this latter J_0 incorporates:

- the effect of laser-induced defect-mediated non-radiative recombination in the bulk of the fully molten region,
- surface recombination in the laser-doped region with an SRV in the order of 10^7 cm/s, if exposed to air or forming a metal/Si interface and
- recombination associated with the laser-created edge area.

8.6.1 Performing laser doping prior to firing

One of the most common, effective, and industrially relevant approaches to mitigate the impact of laser-induced defects is to apply a post thermal treatment on the laser-processed sample. This occurs naturally for the current industrial process for a LDSE PERC solar cells where the laser doping is performed using the PSG layer as a dopant source prior to dielectric deposition and has been used to demonstrate cell efficiencies approaching 24% [2]. Given that the PECVD SiN_x:H dielectric layers used for PERC solar cells are typically deposited at temperatures of around 400 °C in the presence of hydrogen-containing gases, the PECVD process itself can be considered as a hydrogen-plasma treatment, as well as providing a surface passivation layer. This mild thermal process will likely have some effect on passivating laser-induced defects introduced during laser doping.

In addition, the fast-firing process used for the formation of screen-printed contacts can lead to a passivation of laser-induced defects. The fast-firing process with a peak temperature in the range of 700–900 °C allows a release of hydrogen from the dielectric layers to enable a simultaneous passivation of laser-induced defects. As such, when this process flow is adopted, apart from a reduced generation of laser-induced defects by performing laser doing prior to dielectric deposition, a certain level of passivation of the process-induced defects (as discussed in Chapter 4) can automatically occur during the cell fabrication process [83].

Similarly, the passivation of laser-induced defects may also occur if laser doping is performed after dielectric deposition, but prior to thermal processes used for forming the metal contacts, particularly when using a high-temperature firing process. This method was used, for example, in a study by Hallam *et al.* comparing laser ablation and laser doping for forming the front n-type contact of LDSE PERC solar cells together with a self-aligned plated metallisation scheme, where the laser processing was performed before the firing step for the formation of the local Al-BSF region [97]. Although the passivation of laser-induced defects was not explicitly explored in that work, a peak efficiency of 20.5% was achieved for the LDSE solar cell with a FF of 79.7% and V_{OC} of 662 mV. In addition, the pseudo-FF of 82.9% suggested minimal impact of laser-induced defects in the finished device. This efficiency was approximately 0.4%$_{abs}$ higher than that of the cell with a homogenous emitter with laser ablation for contact definition (20.1% efficiency with a corresponding V_{OC} of 656 mV, FF of 78.9% and pseudo-FF of 82.8%) [97]. This clearly highlights the benefit of using an LDSE structure to improve series resistance as indicated by a significant increase in FF, as well as improved V_{OC} for a more effective shielding of minority carriers and the n-type metal/Si interface, despite both cells featuring narrow plated contacts. However, it is possible that some of the benefits with firing for the annealing of laser-induced defects are purely thermal. Further studies in this area should aim to clearly separate thermal from hydrogen-related effects. Nevertheless, the impact of the firing process on passivation of the laser-induced defects can be assessed by performing a pre-firing step, before laser-doping step, to help isolate the effects of the passivation of laser-induced defects from that of other bulk/surface defects. This is depicted through the injection-dependent inverse minority carrier lifetime ($1/\tau_{eff}$) curves of Figure 8.23.

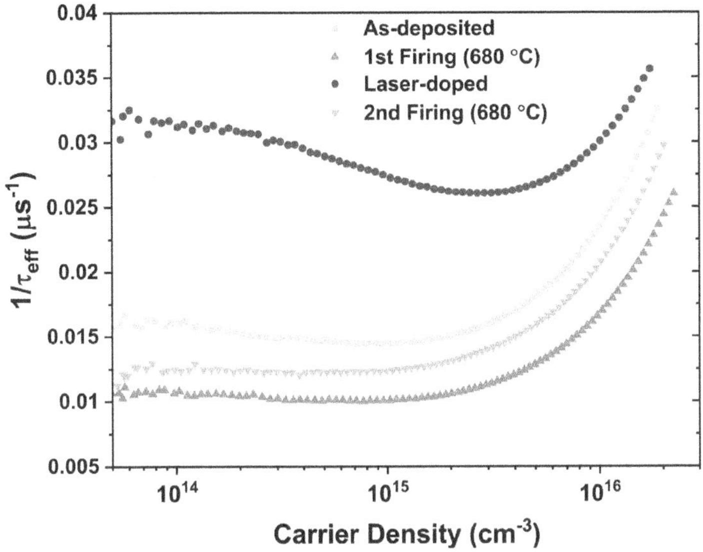

Figure 8.23 *Injection-level-dependent effective minority carrier lifetime curves for p-type Cz wafers, 169 ± 1 μm thick, with a 102 ± 3 Ω/ϒ n-type emitter and laser doping at 4 m/s with a 532 nm wavelength CW laser with an optical power of 14 W. A fast-firing process in a belt furnace was performed at 680 °C before laser doping. Laser-induced defect passivation was performed through a second firing step under identical conditions as the first firing step.*

The initial firing process improved surface passivation with a reduction in J_{0e} from 252 fA/cm^2 to 178 fA/cm^2 and resulted in an improvement in bulk lifetime from 930 μs to 2.8 ms. After the laser-doping process, a significant reduction in the lifetime is observed for all injection levels, due to a substantial increase in the J_{0e} to 279 fA/cm^2 and reduction in bulk lifetime to 80 μs. The subsequent firing process slightly decreased the J_{0e} to 255 fA/cm^2 and increased the bulk lifetime to 1.3 ms through the passivation of laser-induced defects. It is noted that it is likely the first firing step has not passivated all the defects in the sample, prior to the laser-doping process. These results indicate the potential of furnace firing on the passivation of laser-induced defects and the subsequent impact on both the J_{0e} and bulk lifetime (see Table 8.3).

8.6.2 Additional low-temperature belt furnace-annealing processes

On the other hand, if laser doping is performed after the typical hydrogenation process in the solar cell production sequence, such as after the co-firing process, an additional hydrogenation process may be required for passivation of the laser-induced defect. For example, it has been demonstrated that low-temperature belt furnace annealing (BFA) for less than 5 minutes after the implementation of the

Table 8.3 *Lifetime parameters for the lifetime curves measured on the samples that went through similar process sequences as those in Figure 8.23. The iV_{OC} is extracted at 1-sun and the τ_{eff} value is extracted at a minority carrier density $10^{15}/cm^3$. J_{0e} is the total emitter recombination current density of both surfaces*

Sample type	Processing stage	iV_{OC} (mV)	τ_{eff} (μs)	J_{0e} (fA/cm^2)	τ_{bulk} (μs)
n/p/n	As-deposited	659.7	70	252	930
	1st 680°C firing	669.0	101	178	2820
	Laser-doped	641.4	37	279	80
	2nd 680°C firing	660.2	71	255	1300

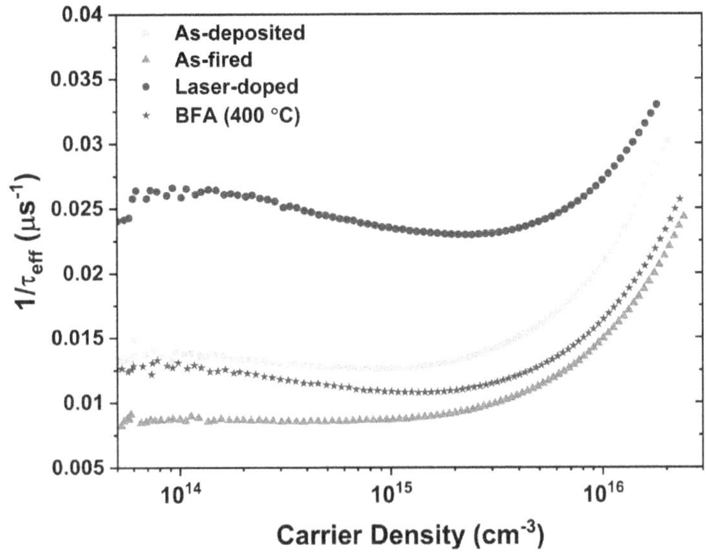

Figure 8.24 *Injection-level-dependent effective minority carrier lifetime curves for p-type Cz wafers, 167±2 μm thick, with a 102±3 Ω/ϒ n-type emitter and laser doping at 4 m/s with a 532-nm wavelength CW laser at an optical power of 14 W. Firing for a duration of several seconds in a firing furnace at 680 °C was performed before laser doping to isolate the effects of the passivation of laser-induced defects from that of other bulk/surface defects. Laser-induced defect passivation was performed at approximately 400 °C for approximately 4 minutes in the same belt furnace. No SiO_2 passivation was used for the lifetime test structure.*

laser-doping process can noticeably reduce the recombination associated with laser-induced defects. Figure 8.24 shows the minority carrier lifetime recovery of a symmetrical n/p/n lifetime test structure passivated by PECVD SiN_x:H after various processes, including directly after dielectric deposition (as-deposited), after a high-temperature firing process with a set peak temperature of 680 °C in a CAMiNI firing furnace (as-fired), after laser doping, and after a subsequent hydrogenation process for laser-induced defect passivation at 400 °C.

The short, relatively low temperature approximately 400 °C inline BFA process improves the effective lifetime with respect to the laser-doping state, with a substantial reduction in the J_{0e} from 305 fA/cm^2 to 194 fA/cm^2 (see Table 8.4). It is noted that this is slightly lower than the value after firing before laser doping at 205 fA/cm^2, which may indicate a slight change in surface passivation between firing and annealing. In addition, τ_{bulk} increased from 60 μs to 530 μs, a factor of three lower than that before laser doping. It is noted that as opposed to the 680 °C firing laser-doping passivation approach (see results in Table 8.3), where large recovery of τ_{bulk} is realised, the BFA passivation approach results in slightly improved surface

Table 8.4 *Lifetime parameters for the lifetime curves measured on the samples that went through similar process sequences as those in Figure 8.24. The* iV_{OC} *is extracted at 1-sun and the* τ_{eff} *value is extracted at a minority carrier density* $10^{15}/cm^3$. J_{0e} *is the total emitter recombination current density of both surfaces.*

Sample type	Processing stage	iV_{OC} (mV)	τ_{eff} (µs)	J_{0e} (fA/cm^2)	τ_{bulk} (µs)
n/p/n	As-deposited	655.1	61	289	740
	As-fired	665.3	87	205	1470
	Laser-doped	635.8	32	305	60
	400°C BFA	664.9	82	194	530

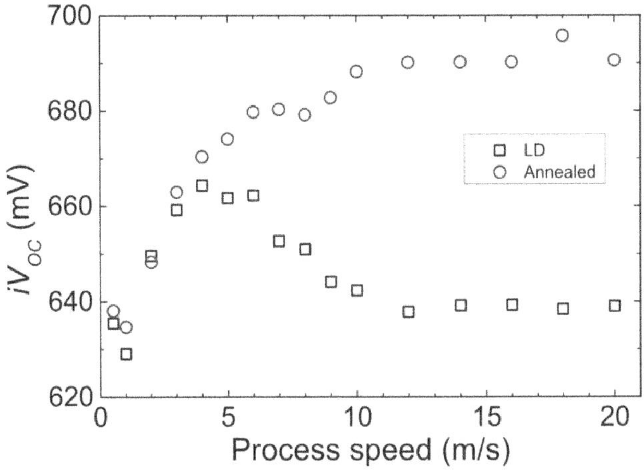

Figure 8.25 *Implied* V_{OC} *as a function of laser scan speed for p-type laser doping from* AlO_x *layers directly after laser doping, and after a subsequent hydrogenation process for the passivation of laser-induced defects (annealed). ©2014 IEEE. Reprinted, with permission, from [94].*

passivation compared to firing. This furnace-annealing approach has been proved to be as effective as the long (~30 minutes) and costly forming gas annealing (FGA) approach [1], which is commonly executed in the research labs. It is noted, however, that FGA typically has a dual purpose; to decrease contact resistivity [74] as well as reducing J_0 of the laser-doped region [98], although improvements in contact resistance and J_0 would also be expected for a BFA process.

However, in addition to the generation of laser-induced defects greatly depending on the choice of laser parameters and subsequent doping profile, the impact of annealing can also greatly depend on the laser-doping conditions used. For example, Hallam *et al.* performed p-type laser doping using AlO_x passivation layers (which will be discussed in detail in Chapter 9) for a range of processing speeds (Figure 8.25) [94]. A key observation was that directly after laser doping, the highest iV_{OC} of approximately 660 mV was obtained from a laser scan speed of 4 m/s. In particular, higher speeds reduced the iV_{OC} before it saturated at approximately 640 mV for processing speeds above 10 m/s. In contrast, after hydrogen passivation of the laser-induced defects, a very different trend was observed. For low speeds of 3 m/s and below, almost no change in the iV_{OC} was observed. In contrast, speeds higher than 3 m/s displayed significant the iV_{OC} enhancements increases with increasing laser scan speed. For speeds higher than 10 m/s, an improvement in iV_{OC} of approximately 40–50 mV was observed. As a result, despite these speeds causing the lowest iV_{OC} directly after laser doping, they were responsible for the highest values after hydrogenation. This clearly highlights the importance of assessing laser doping directly not only after the laser-doping process but also after the expected thermal processing for solar cell fabrication and attempts to passivate the defects.

Optimising laser-doping processes

The extent of laser-induced defect passivation and thus τ_{eff} improvement can depend heavily on the laser-doping processing conditions used. For laser-doping process optimisation, consideration should be given to the response of laser-doped regions to the hydrogenation processes that will naturally occur during cell fabrication, as well as the potential for dedicated hydrogenation processes for laser-induced defect passivation.

The use of a low-temperature BFA process to passivate laser-induced defects can have a significant impact on the electrical performance of laser-doped solar cells where laser doping is performed after the conventional high-temperature firing process. For example, an efficiency gain of $0.35\%_{abs}$ has been demonstrated on conventional LDSE solar cells with a full area rear Al-BSF structure and front self-aligned light-induced plated contacts [94]. For this solar cell structure, a high-temperature firing step was required to form the Al-BSF after screen-printing of the full area rear Al layer, which simultaneously releases hydrogen from the SiN_x:H passivation layer into the bulk. Subsequently, laser doping is performed, followed by a low-temperature BFA step to passivate the laser-induced defects. Figure 8.26 illustrates both the impact of the laser-doping process in reducing the iV_{OC} of the sample through the generation of laser-induced defect, as well the improvement in the iV_{OC} by the subsequent BFA process [94]. Current density–voltage (*J–V*) measurements are shown in Table 8.5.

The potential efficiency gain is even larger for higher efficiency device structures such as PERC and n-PERT solar cells, due to the larger impact defects can have on the overall performance. As shown in Table 8.5, efficiency gains of up to

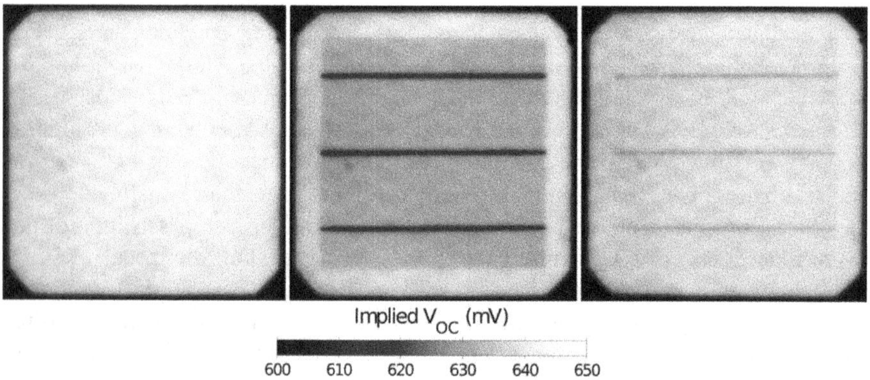

Figure 8.26 *Implied V_{OC} images of a laser-doped Al-BSF cell (a) after firing for Al-BSF formation, (b) after laser doping and (c) after laser-induced defect passivation by low-temperature BFA. ©2014 IEEE. Reprinted, with permission, from [94].*

Table 8.5 *Impact of hydrogenation processes on the passivation of laser-induced defects for LDSE solar cells. P-type Al-BSF and n-type PERT data from [94]. P-type PERC data from [57].*

Cell type	Stage	V_{OC} (mV)	J_{SC} (mA/cm^2)	FF (%)	Efficiency (%)
P-type Al-BSF	Before anneal	627.4 ± 0.1	37.8 ± 0.1	77.4 ± 0.6 0.611	18.35 ± 0.10
	After anneal	636.7 ± 0.7	37.7 ± 0.4	77.9 ± 0.7	18.71 ± 0.09
N-type PERT	Before anneal	648.4 ± 1.0	39.2 ± 0.1	78.0 ± 0.1	19.81 ± 0.03
	After anneal	664.1 ± 1.8	39.5 ± 0.0	78.3 ± 0.2	20.54 ± 0.11
P-type PERC	Before anneal	679.1	39.5	68.9	18.5
	After anneal	691.0 ± 1.2	40.3 ± 0.3	71.6 ± 0.6	19.9 ± 0.4

0.7%$_{abs}$ for n-PERT solar cells have been demonstrated [94]. In addition to passivating laser-induced defects, this BFA process was also observed to improve rear surface passivation, leading to an improved spectral response for long wavelengths (1 000–1 200 nm). This again highlights the potential for a single hydrogenation process to cause changes in the recombination activity of multiple recombination mechanisms. Optimised thermal processing by pre-firing of the dielectrics and subsequent BFA at 500 °C of laser-induced defects has facilitated 23% efficiency achievement for IMEC with the n-PERT structure [55].

For solar cells with a low-temperature metallisation scheme such as plating, the contacts generally require a sintering process, such as with nickel seed layers to form a copper-diffusion barrier of NiSi, and also improve contact resistance. Depending on the metallisation approach, this may occur in the range of 350–400 °C. Such processes have been used to demonstrate at least partial passivation of laser-induced defects in n-PERT solar cells, although a more effective level of passivation was achieved in that study using a dedicated annealing process at a higher temperature of 450 °C rather than 350–400 °C [94]. Another study by Ciesla *et al.* for p-type Pluto solar cells with a LDSE observed more effective passivation of the laser-induced defect during the nickel sintering process at 350–400 °C, likely benefited by the prior firing process to ensure effective release of hydrogen from the dielectric layers and enable the availability of hydrogen for passivation during the sintering process. This resulted in efficiency enhancements of up to 1.5%$_{abs}$ and V_{OC} over 690 mV for the LDSE PERC-like solar cells (see Table 8.5). These solar cells also featured anchor points to assist with metal adhesion [57].

Particularly, if the annealing process is performed as a dedicated processing step rather than as part of an integrated part of the cell fabrication process, such as for the firing of the metal contacts or sintering of plated contacts, the additional cost of performing such a step must be considered. Depending on the performance improvements gained, including a single BFA step for laser-induced defect passivation may not necessarily be justifiable from cost point of view for high throughput cell fabrication and the implications on yield by adding a process to the fabrication sequence. However, for a 22.5% efficient PERC solar cell with a wafer-to-cell conversion cost of approximately $0.33/piece [99], an additional $0.0089/piece can be justified if the efficiency gain is 0.2%$_{abs}$ to result in the same fabrication cost in terms of $/W. In addition, such a change would result in a reduced fabrication cost at the module level.

8.6.3 *Simultaneous laser-induced defect passivation and light-induced degradation mitigation*

Another approach for laser-induced defect passivation that involves hydrogen, is illuminated annealing [94, 100], as was used in Chapters 4–6 to improve bulk defect passivation due to the manipulation of hydrogen charge states. The use of illuminated annealing for passivating laser-induced defects creates the possibility for a simultaneous elimination of light-induced degradation (LID) in p-type Cz material

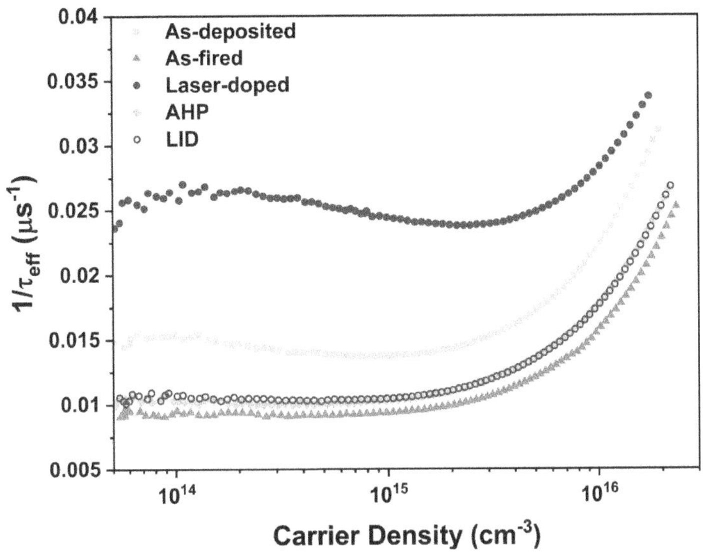

Figure 8.27 *Injection-level-dependent effective minority carrier lifetime curves for p-type Cz wafers, 170 ± 2 μm thick, with a 102 ± 3 Ω/ϒ n-type emitter and laser doping at 4 m/s with a 532 nm wavelength CW laser at an optical power of 14 W. A firing for a duration of several seconds in a firing furnace at 680 °C (filled triangles) was performed before laser doping (filled circles). Laser-induced defect passivation was performed through a rapid 10-second laser-based AHP (filled diamonds). To test for LID (open circles), an accelerated degradation test was performed using the same commercial laser system (DR laser) as for AHP with the exception that the temperature of the sample stage was held at 150 °C during the 10 seconds laser illumination.*

through the passivation of B-O defects, as well as for the passivation of the laser-induced defects.

Figure 8.27 and corresponding Table 8.6 show a clear example of the simultaneous passivation of laser-induced defects and the elimination of B-O LID in p-type Cz material using a laser-based advanced hydrogenation process (AHP). This shows essentially the effective passivation of laser-induced defects and the complete elimination of B-O LID. Here, a drop in effective lifetime of only 10 μs from 94 μs to 84 μs was observed from after firing to after laser doping, advanced hydrogenation and the subsequent LID test. The AHP process also recovered all of the J_{0e} introduced during laser doping; however, light soaking led to a slight increase.

The effectiveness of laser-induced defect passivation and LID mitigation can be impacted by the laser process used, along with the chosen passivation process. In Figure 8.28 (a), the effectiveness of different annealing processes for the passivation of laser-induced defects are compared. Specifically, the impact of a low-temperature

Table 8.6 *Lifetime parameters for the lifetime curves measured on the samples that went through similar process sequences as those in Figure 8.27. The* iV_{OC} *is extracted at 1-sun and the* τ_{eff} *value is extracted at a minority carrier density* $10^{15}/cm^3$. J_{0e} *is the total emitter recombination current density of both surfaces.*

Sample type	Processing stage	iV_{OC} (mV)	τ_{eff} (µs)	J_{0e} (fA/cm^2)	τ_{bulk} (µs)
n/p/n	As-deposited	656.4	64	276	780
	As-fired	667.9	94	193	2020
	Laser-doped	640.6	36	283	80
	300 °C AHP	665.5	~84	194	560
	After LID	664.1	~84	217	1900

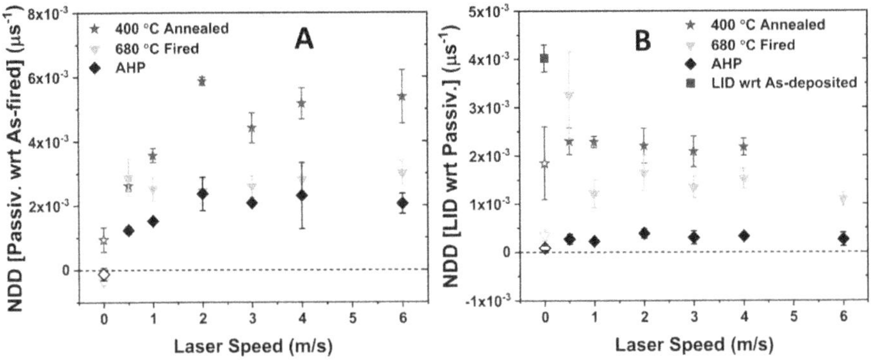

Figure 8.28 *(A) NDD values reported at minority carrier density of $10^{15}/cm^3$ of the laser-doped samples calculated for the passivated state with respect to the initial fired state for various CW laser speeds. The values reported at 0 m/s are from the non-laser-doped control samples. (B) NDD values for the control- and laser-doped samples calculated for the LID state with respect to the passivated state. The LID test was performed using the same conditions as in Figure 8.27.*

BFA at 400 °C, a fast-firing process with a peak temperature of 680 °C, as well as a rapid high-intensity 10-second- illuminated annealing process at 300 °C on passivating the laser-induced defects for n^+ laser-doped regions on mono-crystalline p-type Cz wafers are presented in the context of normalised defect density (NDD). Here, AHP was the most effective process for passivating laser-induced defects for all speeds investigated. It is also noted that the NDD after passivation, showed a dependence on processing speed, with higher values at higher speeds.

The impact of these passivation processes for their effectiveness in simultaneously reducing susceptibility to LID is also demonstrated (see Figure 8.28 (b)). This clearly highlights the effective elimination of B-O LID using AHP with a reduction of NDD by more than 90% compared to that in the as-deposited state, while firing and a low-temperature BFA only reduced the NDD by approximately 50%, likely due to the thermal path for B-O defect mitigation during firing as discussed in Chapter 5.

Future work should more closely assess the impact of different annealing processes such as belt furnace firing and illuminated annealing for the passivation of laser-induced defects using a range of different laser types and laser processing conditions for the industrial LDSE approach.

8.7 Conclusion

Selective emitter solar cell structures offer significant benefits over their homogenous emitter counterparts, as was used for the record 25% efficient PERL cell. Laser doping is an elegant way to achieve such a structure and has become the method of choice for selective emitter formation for industrial PERC solar cells. This approach avoids many complexities or additional processing steps of other selective emitter approaches. For LDSE, laser doping can be performed prior to dielectric deposition, as is done in the industry. In this approach, the PSG layer from the emitter diffusion is used as a dopant source. An advantage of this approach is that it reduces laser-induced defect generation and enables a passivation of laser-induced defects naturally during solar cell fabrication with the metallisation firing process. This has been used by LONGi to fabricate PERC solar cells with efficiencies of more than 23.8%. The downside of this approach is that it requires relatively large laser-doped coverage areas, of around 10% with lase-doped stripes approximately 100 μm wide to provide sufficient alignment tolerance for the subsequently aligned fire-through Ag screen-print metallisation with fingers approximately 40 μm wide, resulting in a metal/Si interface area of approximately 3%.

On the other hand, laser doping can be performed through a dielectric layer to provide a self-aligned metallisation scheme in conjunction with light-induced plating. An advantage of this approach is that it can enable narrow contact regions 10–20 μm wide, to result in metal/Si interface areas of 1–2%. In addition, copper is substantially cheaper than silver, more abundant, and has a much lower resistance than screen-printed silver. On the downside, this approach results in a more significant generation of laser-induced defects. As such, a CW or Q-CW laser is preferred to reduce thermal cycling, and also enable heavily doped regions up to 10 μm deep to protect against recombination. However, this method should be used in conjunction with a hydrogen passivation process to passivate laser-induced defects, which can improve efficiency by as much as $1.5\%_{abs}$. This method was used by Suntech for its Pluto technology to fabricate the world's first industrial 20% efficient p-type Cz solar cell and efficiencies of 22.5% have been achieved for n-PERT as a front-surface field. The passivation of defects can be achieved naturally during nickel sintering, or through a dedicated belt furnace anneal, or even using an AHP incorporating minority carrier injection to simultaneously eliminate B-O light-induced degradation. A critical finding is that laser-doping process conditions should not be optimised to maximise lifetime directly after laser doping. Rather they should be optimised in conjunction with hydrogen passivation processes.

Another approach for laser doping is performing laser doping prior to dielectric deposition using a precise control of laser-doping parameters to form very narrow grooves approximately 3–5 μm wide. Due to the conformal coating of PECVD layers, a self-aligned plated metallisation can be used, while still benefiting from a reduced generation of laser-induced defects by performing laser doping on bare silicon.

However, to date, there is no such LDSE for industrial TOPCon solar cells. A potential method is presented in Chapter 9 using AlO_x as a dopant source, thereby also allowing an elegant solution that does not require an additional dopant-application step.

References

[1] Urueña A., Aleman M., Cornagliotti E., *et al.* 'Progress on large area N-type silicon solar cells with front laser doping and a rear emitter'. *Progress in Photovoltaics: Research and Applications*. 2016;**24**(8):1149–56.

[2] Chen R., Tong H., Zhu H., *et al.* '23.83% efficient mono-PERC incorporating advanced hydrogenation'. *Progress in Photovoltaics: Research and Applications*. 2020;**28**(12):1239–47.

[3] Ohl R.S. Light-Sensitive electric device including silicon US2443542A. US patent US2443542A; 1948.

[4] Zhao J., Wang A., Green M.A. 24% efficient PERL structure silicon solar cells. IEEE Conference on Photovoltaic Specialists; 1990. pp. 333–5.

[5] Green M.A. 'The path to 25% silicon solar cell efficiency: History of silicon cell evolution'. *Progress in Photovoltaics: Research and Applications*. 2009;**17**(3):183–9.

[6] Zhao J., Wang A., Altermatt P.P., Wenham S.R., Green M.A. '24% efficient perl silicon solar cell: Recent improvements in high efficiency silicon cell research'. *Solar Energy Materials & Solar Cells*. 1996;**41**(42):87–99.

[7] Eleventh E. *International Technology Roadmap for Photovoltaic [online]*. 2020. Available from https://itrpv.vdma.org/viewer/-/v2article/render/48393879.

[8] Saint-Cast P., Werner S., Greulich J., Jäger U., Lohmüller E., Höffler H. "Analysis of the losses of industrial-type PERC solar cells". *Physica Status Solidi*. 2017;**214**(3):1600708.

[9] Röder TC E.S.J., Grabitz P., Wagner C., Kulushich G K.J.R. "Add-on laser tailored selective emitter solar cells". *Progress in Photovoltaics: Research and Applications*. 2010;**18**(7):505–10.

[10] 'PV Lighthouse. EDNA 2'. 2014.

[11] Fong K.C., Kho T.C., Fell A., *et al.* 'Contact resistivity of evaporated Al contacts for silicon solar cells'. *IEEE Journal of Photovoltaics*. 2015;**5**(5):1304–9.

[12] Hahn G. 'Status of selective emitter technology'. 25th European Photovoltaic Solar Energy Conference and Exhibition; Valencia, Spain; 2010.

[13] Min B., Müller M., Wagner H., *et al.* 'A roadmap toward 24% efficient PERC solar cells in industrial mass production'. *IEEE Journal of Photovoltaics*. 2017;**7**(6):1541–50.

[14] Aberle A.G. 'Surface passivation of crystalline silicon solar cells: a review'. *Progress in Photovoltaics: Research and Applications*. 2000;**8**(5):473–87.

[15] Wenham S. 'Buried-contact silicon solar cells'. *Progress in Photovoltaics: Research and Applications*. 1993;**1**(1):3–10.

[16] Hanke I., Apel M., Schroter W. 'Influence of annealing conditions on recombination properties and diffusion length in cast multicrystalline silicon for solar cells'. Proceedings of the 14th European Photovoltaic Solar Energy Conference; 1997. pp. 735–8.

[17] McDonald D., Cuevas A. 'The trade-off between phosphorus gettering and thermal degradation in multicrystalline silicon'. Proceedings of the 16th European Photovoltaic Solar Energy Conference; 2000.

[18] Voronkov V.V., Falster R. 'Grown-in microdefects, residual vacancies and oxygen precipitation bands in Czochralski silicon'. *Journal of Crystal Growth*. 1999;**204**(4):462–74.

[19] Wang A. *High Efficiency Perc and Perl Silicon Solar Cells*. Sydney: University of New South Wales; 1992.

[20] Sze S.M., KK N. *Physics of Semiconductor Devices*. Hoboken, New Jersey, USA: John Wiley & Sons; 2006.

[21] Green M.A. Chapter I-2-B - High-Efficiency Silicon Solar Cell Concepts. In: Kalogirou SABT-MH of P (Third E, editor. Academic Press; 2018. p. 95–128. Available from http://www.sciencedirect.com/science/article/pii/B9780128099216000057.

[22] Esturo-Breton A., Binaie Masouleh F., Breselge M., Frieß T., Geiger M., Holbig E. 'Crystalline si solar cells with selective emitter for industrial mass production'. 24th European Photovoltaic Solar Energy Conference; Hamburg, Germany; 2009. pp. 1068–71.

[23] Abbott M., Cousins P., Chen F., Cotter J. 'Laser-induced defects in crystalline silicon solar cells'. Conference Record of the Thirty-first IEEE Photovoltaic Specialists Conference, 2005; 2005. pp. 1241–4.

[24] Utama R.Y. Inkjet printing for commercial highefficiency silicon solar cells[PhD thesis]. Sydney, Univ New South Wales; 2009.

[25] Lennon A.J., Ho-Baillie A.W.Y., Wenham S.R. 'Direct patterned etching of silicon dioxide and silicon nitride dielectric layers by inkjet printing'. *Solar Energy Materials & Solar Cells*. 2009;**93**(10):1865–74.

[26] Lennon A. Dielectric Patterning for Commercial High Efficiency Silicon Solar Cells[PhD thesis .]Sydney, The University of New South Wales; 2010.

[27] Antoniadis H., Jiang F., Shan W., Liu Y. 'All screen printed mass produced silicon ink selective emitter solar cells'. 2010 35th IEEE Photovoltaic Specialists Conference; 2010. pp. 1193–6.

[28] Low R., Gupta A., Bateman N., Ramappa D., Sullivan P., Skinner W. 'High efficiency selective emitter enabled through patterned ion implantation'. 2010 35th IEEE Photovoltaic Specialists Conference; 2010. pp. 1440–5.

[29] Dubé C.E., Tsefrekas B., Buzby D., Tavares R., Zhang W., Gupta A. 'High efficiency selective emitter cells using patterned ion implantation'. *Energy procedia*. 2011;**8**:706–11.

[30] Dastgheib-Shirazi A., Haverkamp H., Raabe B., Book F., Hahn G. 'Selective emitter for industrial solar cell production: a wet chemical approach using a single side diffusion process'. 23rd European Photovoltaic Solar Energy Conference, EU PVSEC; 2008. pp. 1197–9.

[31] Lauermann T., Dastgheib-Shirazi A., Book F., Raabe B., Hahn G., Haverkamp H. 'InSECT: an inline selective emitter concept with high efficiencies at competitive process costs improved with inkjet masking technology'. 24th

European Photovoltaic Solar Energy Conference; Hamburg, Germany; 2009. pp. 1767–70.

[32] Osborne M. '*Schmid's selective emitter technology supports Sunrise Global Solar's record 19.2% cell efficiencies*' [online]. 2011. Available from https://www.pv-tech.org/news/schmids_selective_emitter_technology_supports_sunrise_global_solars_record#disqus_thread.

[33] Tjahjono B.S., Hameiri Z., Mai L., Sugianto A., Wang S., Wenham S.R. 'JHG high efficiency solar cell structures through the use of laser doping'. 22nd European Photovoltaic Solar Energy Conference; Milan, Italy; 2007.

[34] Duttagupta S., Nandakumar N., Padhamnath P., Buatis J.K., Stangl R., Aberle A.G. 'MonoPolyTM cells: Large-area crystalline silicon solar cells with fire-through screen printed contact to doped polysilicon surfaces'. *Solar Energy Materials & Solar Cells*. 2018;**187**:76–81.

[35] Ernst M., Walter D., Fell A., Lim B., Weber K. 'Efficiency potential of P-Type Al_2O_3/SiN passivated perc solar cells with locally laser-doped rear contacts'. *IEEE Journal of Photovoltaics*. 2016;**6**(3):624–31.

[36] Ernst M., Fell A., Franklin E., Weber K.J. 'Characterization of recombination properties and contact resistivity of laser-processed localized contacts from doped silicon nanoparticle ink and spin-on dopants'. *IEEE Journal of Photovoltaics*. **2017**;7(2):471–8.

[37] Miao F., Zhang S., Lian W., Zhao B., Wei Q. Improvement of PERC solar cell efficiency based on laser-doped selective emitter. AIP Conference Proceedings; 2019. pp. **2147**(1)–40011.

[38] Fernandez-Robledo S., Jäger U., Lohmüller E., Nekarda J. Laser doping from borosilicate glass for metallization of boron emitters. Proceedings of the 28th European Photovoltaic Solar Energy Conference and Exhibition; Paris, France; 2013.

[39] Rohatgi A., Meier D.L., McPherson B., Y-W O., Upadhyaya A.D., Lai J.-H. 'High-Throughput Ion-Implantation for low-cost high-efficiency silicon solar cells'. *Energy Procedia*. 2012;**15**:10–19.

[40] Yu M., Zhou R., Hamer P., Chen D., Zhang X., Altermatt P.P. 'Imaging and quantifying carrier collection in silicon solar cells: a submicron study using electron beam induced current'. *Solar Energy*. 2020;**211**:1214–22.

[41] Wenham S.R., Green M.A. 'Self aligning method for forming a selective emitter and metallization in a solar cell'. *Google Patents*. 2002.

[42] Hallam B., Urueña A., Russell R., Aleman M., Abbott M., Dang C. 'Efficiency enhancement of i-PERC solar cells by implementation of a laser doped selective emitter'. *Solar Energy Materials & Solar Cells*. 2015;**134**:89–98.

[43] Li T., Wang W., Zhou C., Song Y., Duan Y., Li Y. 'Laser-doped solar cells exceeding 18% efficiency on large-area commercial-grade multicrystalline silicon substrates'. *Progress in Photovoltaics: Research and Applications*. 2013;**21**(6):1337–42.

[44] Norouzi M.H., Saint-Cast P., Lohmüller E., Weber J., Gutscher S., Bartsch J. 'Development and characterization of multifunctional PECVD SiNX:P

layers for laser-doped selective emitters'. *Aip Conference Proceedings*. 2018;**1999**(1):110005.

[45] Xiao B., Mai L., Lin D., Hallam B., Chong C.M., Lennon A. 'Study of p-type laser doping using ALD AlOx as a dopant source'. IEEE 40th Photovoltaic Specialist Conference (PVSC); Denver, USA; 2014. pp. 3032–5.

[46] Hallam B., Wenham S., Sugianto A., *et al.* 'Record large-area p-type CZ production cell efficiency of 19.3% based on ldse technology'. *IEEE Journal of Photovoltaics*. 2011;**1**:43–8.

[47] Wang Z., Han P., Lu H., *et al.* 'Advanced PERC and PERL production cells with 20.3% record efficiency for standard commercial p-type silicon wafers'. *Progress in Photovoltaics: Research and Applications*. 2012;**20**(3):260–8.

[48] Hallam B. 'High efficiency laser-doped silicon solar cells with advanced hydrogenation'. *School of Photovoltaic and Renewable Energy Engineering*. University of New South Wales; 2014.

[49] Kray D., Aleman M., Fell A., Hopman S., Mayer K., Mesec M. '*Laser-doped silicon solar cells by laser chemical processing (LCP) exceeding 20% efficiency*'. 2008 33rd IEEE Photovoltaic Specialists Conference; 2008. pp. 1–3.

[50] Fell A. *Modelling and Simulation of Laser Chemical Processing (Lcp) for the Manufacturing of Silicon Solar Cells*. München: Dr. Hut; 2010.

[51] Kluska S., Granek F. 'High-Efficiency silicon solar cells with boron local back surface fields formed by laser chemical processing'. *IEEE Electron Device Letters*. 2011;**32**(9):1257–9.

[52] Yang X., Bullock J., Bi Q., Weber K. 'High efficiency N-type silicon solar cells featuring passivated contact to laser doped regions'. *Applied physics letters*. 2015;**106**(11):113901.

[53] Hördt W., Kluska S., Geisler C., Hopman S., Bartsch J., Mondon A. 'Failure mechanisms for solar cells with laser doped selective emitter and plated Ni-Cu metal contacts'. 29th European PV Solar Energy Conference and Exhibition; Amsterdam, the Netherlans; 2014.

[54] Lennon A., Colwell J., Rodbell K.P. 'Challenges facing copper-plated metallisation for silicon photovoltaics: insights from integrated circuit technology development'. *Progress in Photovoltaics: Research and Applications*. 2019;**27**(1):67–97.

[55] Aleman M., Uruena A., Cornagliotti E., Choulat P., John J., Russell R. 'How to achieve 23% efficient large-area Cu plated n-PERT cells?' 2017 IEEE 44th Photovoltaic Specialist Conference (PVSC); Washington DC, USA; 2017. pp. 2227–31.

[56] Wu W., Zhang Z., Zheng F., Lin W., Liang Z., Shen H. 'Efficiency enhancement of bifacial PERC solar cells with laser-doped selective emitter and double-screen-printed al grid'. *Progress in Photovoltaics: Research and Applications*. 2018;**26**(9):752–60.

[57] Ciesla A., Chen R., Wang S., Ji J., Shi Z., Mai L. 'High-Voltage P-type PERC solar cells with anchored plating and hydrogenation'. *Progress in Photovoltaics: Research and Applications*. 2018;**26**(6):397–401.

[58] Mooney P.M., Young R.T., Karins J., Lee Y.H., Corbett J.W. 'Defects in laser damaged silicon observed by DLTS'. *Physica Status Solidi*. 1978;**48**(1):K31–4.
[59] Fairfield J.M., Schwuttke G.H. 'Silicon diodes made by laser irradiation'. *Solid-state electronics*. 1968;**11**(12):1175–6.
[60] Ametowobla M., Bilger G., Köhler JR W.J.H. 'Laser induced lifetime degradation in P-type crystalline silicon'. *Journal of applied physics*. 2012;**111**(11):114515.
[61] Tjahjono B.S. 'Laser doped selective emitter solar cells'. *Photovoltaics & Renewable Energy Engineering*. Sydney: University of New South Wales; 2010.
[62] Sugianto A. Development of commercial high-efficiency solar cells incorporating selective emitters and rear Passivated contacts[PhD Thesis]Sydney Australia, The University of New South Wales; 2011.
[63] Xu L., Weber K., Fell A., *et al*. 'The impact of sio_2/sin_x stack thickness on laser doping of silicon solar cell'. *IEEE Journal of Photovoltaics*. 2014;**4**(2):594–600.
[64] Bovatsek J.M., Sugianto A., Hallam B., Tjahjono B., Wenham S., Patel R.S. 'High-speed fabrication of laser doping selective emitter solar cells using 532 nm continuous wave (CW) and modelocked quasi-cw laser sources'. Proceedings of the 26th European Photovoltaic Solar Energy Conference; Hamburg, Germany; 2011. pp. 895–900.
[65] Nguyen H.T., Han Y., Ernst M., Fell A., Franklin E., Macdonald D. 'Dislocations in laser-doped silicon detected by micro-photoluminescence spectroscopy'. *Applied Physics Letters*. 2015;**107**(2):022101–22101.
[66] Sun Z., Gupta M.C. 'Laser annealing of silicon surface defects for photovoltaic applications'. *Surface science*. 2016;**652**:344–9.
[67] Young R.T., Narayan J.R., Westbrook D., Wood R.F. 'Transmission electron microscopy and electrical properties measurements of laser doped silicon and GaAs'. AIP Conference Proceedings; 1979. pp. 579–84.
[68] Hameiri Z., Puzzer T., Mai L., Sproul A.B., Wenham S.R. 'Laser induced defects in laser doped solar cells'. *Progress in Photovoltaics: Research and Applications*. 2011;**19**(4):391–405.
[69] Chen D., Edwards M., Wenham S., Abbott M., Hallam B. 'Laser enhanced gettering of silicon substrates'. *Front Energy*. 2017;**11**(1):23–31.
[70] Barhdadi A., Hartiti B., Muller J.-C. 'Active defects generated in silicon by laser doping process'. *African Review of Physics*. 2011;**6**:229–38.
[71] Ohmer K., Weng Y., Köhler J.R., Strunk H.P., Werner J.H. 'Defect formation in silicon during laser doping'. *IEEE Journal of Photovoltaics*. 2011;**1**(2):183–6.
[72] Cuevas A., Macdonald D. 'Measuring and interpreting the lifetime of silicon wafers'. *Solar energy*. 2004;**76**(1):255–62.
[73] Hopman S., Fell A., Mayer K., Fleischmann C., Drew K., Kray D. Study on laser parameters for silicon solar cells with LCP selective emitters. In: Proceedings if the 24th European photovoltaic solar energy conference,

edited by: WC Sinke, HA Ossenbrink, and P Helm (WIP, Munich, Germany, 2009); 2009.

[74] Ernst M., Huyeng J.D., Walter D., Fong K.C., Blakers A. 'Unravelling the origins of contact recombination for localized laser-doped contacts'. 2018 IEEE 7th World Conference on Photovoltaic Energy Conversion (WCPEC) (A Joint Conference of 45th IEEE PVSC, 28th PVSEC & 34th EU PVSEC); 2018. pp. 2195–9.

[75] Mchedlidze T., Seifert W., Kittler M., Blumenau A.T., Birkmann B., Mono T. 'Capability of photoluminescence for characterization of multi-crystalline silicon'. *Journal of applied physics*. 2012;**111**(7):73504.

[76] Chin R.L., Pollard M., Zhu Y., Hameiri Z. 'Detailed analysis of radiative transitions from defects in N-type monocrystalline silicon using temperature- and light intensity-dependent spectral Photoluminescence'. *Solar Energy Materials & Solar Cells*. 2020;**208**:110376.

[77] Heinz F.D., Kasemann M., Warta W., Schubert M.C. 'Microscopic charge carrier lifetime in silicon from a transient approach'. *Applied physics letters*. 2015;**107**(12):122101.

[78] Sauer R., Weber J., Stolz J., Weber E.R., Küsters K.H., Alexander H. 'Dislocation-related photoluminescence in silicon'. *Applied Physics A*. 1985;**36**(1):1–13.

[79] Kveder V.V., Steinman E.A., Shevchenko S.A., Grimmeiss H.G. 'Dislocation-related electroluminescence at room temperature in plastically deformed silicon'. *Physical review. B*. 1995;**51**(16):10520–6.

[80] Hartiti B., Slaoui A., Muller J.C., Siffert P. 'Electrically active defects in silicon after excimer laser processing'. *Journal of applied physics*. 1989;**66**(8):3934–7.

[81] Green M.A. 'Accurate expressions for solar cell fill factors including series and shunt resistances'. *Applied physics letters*. 2016;**108**(8):81111.

[82] Kray D., Fell A., Hopman S., Mayer K., Willeke G.P., Glunz S.W. 'Laser Chemical Processing (LCP)–A versatile tool for microstructuring applications'. *Applied Physics A*. 2008;**93**(1):99.

[83] Hallam B.J., Ciesla A.M., Chan C.C., *et al*. 'Overcoming the challenges of hydrogenation in silicon solar cells'. *Australian Journal of Chemistry*. 2018;**71**(10):743–52.

[84] Jäger U., Okanovic M., Hoerteis, A M., Grohe R.P. 'Selective Emitter by Laser Doping from Phosphosilicate Glass'. 24th European Photovoltaic Solar Energy Conference; Munich, Germany; 2009. pp. 1740–3.

[85] Wang S., Mai L., Wenham A., Hameiri Z., Payne D., Chan C. 'Selective emitter solar cell through simultaneous laser doping and grooving of silicon followed by self-aligned metal plating'. *Solar Energy Materials & Solar Cells*. 2017;**169**:151–8.

[86] Yang X., Bullock J., Xu L., Bi Q., Surve S., Ernst M. 'Passivated contacts to laser doped P+ and N+ regions'. *Solar Energy Materials & Solar Cells*. 2015;**140**:38–44.

[87] Green M.A. 'The path to 25% silicon solar cell efficiency: History of silicon cell evolution'. *Progress in Photovoltaics: Research and Applications.* 2009;**17**(3):183–9.

[88] Bullock J., Yan D., Cuevas A., Demaurex B., Hessler-Wyser A., Wolf S.D. 'Passivated contacts to n+ and p+ silicon based on amorphous silicon and thin dielectrics'. 2014 IEEE 40th Photovoltaic Specialist Conference (PVSC); 2014. pp. 3442–7.

[89] Chen D., Chen Y., Wang Z., Gong J., Liu C., Zou Y. '24.58% total area efficiency of screen-printed, large area industrial silicon solar cells with the tunnel oxide passivated contacts (i-TOPCon) design'. *Solar Energy Materials and Solar Cells : An International Journal Devoted to Photovoltaic, Photothermal, and Photochemical Solar Energy Conversion.* 2019;**2020**(206):1–8.

[90] Hu SM. 'Stress-related problems in silicon technology'. *Journal of applied physics.* 1991;**70**(6):R53–80.

[91] Hallam B., Chan C., Sugianto A., Wenham S. 'Deep junction laser doping for contacting buried layers in silicon solar cell'. *Solar Energy Materials & Solar Cells.* 2013;**113**:124–34.

[92] Zhou R., Yu M., Tweddle D., Hamer P., Chen D., Hallam B. 'Understanding and optimizing EBIC pn-junction characterization from modeling insights'. *Journal of applied physics.* 2020;**127**(2):24502.

[93] Wang Z., Han P., Lu H., *et al.* 'Advanced PERC and PERL production cells with 20.3% record efficiency for standard commercial p-type silicon wafers'. *Progress in Photovoltaics: Research and Applications.* 2012;**20**(3):260–8.

[94] Hallam B., Sugianto A., Mai L., *et al.* 'Hydrogen passivation of laser-induced defects for laser-doped silicon solar cells'. *IEEE Journal of Photovoltaics.* 2014;**4**(6):1413–20.

[95] Sun Z., Gupta M.C. 'A study of laser-induced surface defects in silicon and impact on electrical properties'. *Journal of applied physics.* 2018;**124**(22):223103.

[96] Slaoui A., Barhdadi A., Muller J.C., Siffert P. 'Passivation of laser induced defects in silicon by low energy hydrogen ion implantation'. *Applied Physics A.* 1986;**39**(3):159–62.

[97] Hallam B., Urueña A., Russell R., *et al.* 'Efficiency enhancement of i-PERC solar cells by implementation of a laser doped selective emitter'. *Solar Energy Materials and Solar Cells.* 2015;**134**(4):89–98.

[98] Aleman M., Uruena A., Choulat P., Hallam B., Dang C., Russell R. 'Reducing front recombination losses to improve the efficiency of rear junction Cu-plated n-Si cells'. 29th European Photovoltaic Solar Energy Conference and Exhibition; Paris; 2014. pp. 528–31.

[99] Chang N.L., Ho-Baillie A., Wenham S., Woodhouse M., Evans R., Tjahjono B. 'A techno-economic analysis method for guiding research and investment directions for c-Si photovoltaics and its application to Al-BSF, PERC, LDSE and advanced hydrogenation'. *Sustainable energy & fuels.* 2018;**2**(5):1007–19.

[100] Herguth A., Schubert G., Käs M., Hahn G., Kaes M., Hahn G. 'A new approach to prevent the negative impact of the metastable defect in boron doped Cz silicon solar cells'. 4th IEEE World Conference on Photovoltaic Energy Conversion. IEEE; 2006. pp. 940–3.

Chapter 9

Applications of laser doping

Catherine Chan[1] and Brett Hallam[1]

This chapter discusses further applications of laser doping for silicon solar cells. First, it will cover the application of laser doping for high-efficiency solar cell fabrication. This includes the formation of localised heavily doped p^{++} regions using either applied boron spin-on-dopants (SODs) or the rear surface aluminium oxide (AlO_X) dielectric layer in industrial passivated emitter and rear cell (PERC) solar cells as a dopant source to form a low-cost implementation of the rear surface of the passivated emitter rear locally (PERL) diffused cell. Other high-efficiency solar cells, such as the laser-doped semiconductor finger cell and interdigitated back-contact (IBC) solar cells, will also be discussed. The ability for laser doping to form heavily diffused contact regions that can penetrate through emitter surface layers to contact the underlying bulk silicon will be reviewed, with applications in novel solar cell structures and edge junction isolation to avoid resistance-limited recombination mechanisms at low carrier concentrations. Finally, the application of laser doping for enhanced gettering of bulk impurities is discussed.

9.1 Local back surface field formation for PERC-type cell structures

The simultaneous localised ablation of surface dielectrics and doping of the underlying silicon makes laser doping ideal for any polarity of local contact opening in high-efficiency solar cell structures. In Chapter 8, the use of laser doping to form a front-side n-type selective emitter was discussed. Laser doping can also be implemented on the rear side of solar cell devices to form the local back surface field (LBSF) contact regions of the PERC family of cell structures. Note that although the term PERC was originally used by UNSW to denote devices without heavily doped silicon under the rear contacts, it is now being used by the industry to describe a broader range of devices with localised rear contacts and passivated rear surfaces

[1]The School of Photovoltaic and Renewable Energy Engineering, The University of New South Wales, Sydney, NSW, Australia

that would have originally been classified as PERL and PERT (passivated emitter rear totally diffused) [1].

Initial implementations of laser doping for LBSF formation appeared in bifacial cells utilising laser doping combined with self-aligned copper plating for both front and rear contacts [2, 3], greatly simplifying the usual high-temperature furnace diffusion, patterning and masking approach used to fabricate similar structures such as UNSW's double-sided buried contact cell [4]. Other approaches have utilised an aligned screen-printing step to the laser-doped back surface field (BSF) regions to form the rear metal contacts for bifacial cells [5]. In the most common implementation of the PERC structure, a full-area aluminium contact was deposited or screen-printed onto the rear surface to contact the laser-doped regions. Laser doping for BSF formation has been performed using SOD sources through a passivation layer [3, 6–8], and in some cases implemented using a doped rear passivation layer [9–13], the laser chemical processing (LCP) method [14] or even a screen-printable-doped silicon nanoparticle paste [15].

A summary of some key efficiencies achieved for PERL cells with laser-doped (LD) rear contacts is shown in Table 9.1.

9.1.1 *Laser doping as an alternative to the Al/Si alloying process for PERC*

The commercial implementation of the original world-record-holding UNSW PERL cell design [20], now generally referred to as PERC, consists of a passivated emitter and screen-printed front Ag metallisation, and a passivated rear surface with local heavily doped p^{++} back contact regions. These heavily doped p^{++} regions, required for low contact resistance and field-effect passivation, are typically formed via selective laser ablation of the rear surface dielectric layer and subsequent screen-printing and firing of a homogeneous layer of aluminium paste over the patterned dielectric to form local aluminium/silicon (Al/Si) alloyed BSF regions when the cell is fired at a high temperature [21, 22]. During the firing step in an industrial fast firing furnace (typically 700–900 °C), an Al/Si eutectic layer is formed when the temperature exceeds 577 °C. During the cooldown, the silicon recrystallises epitaxially and a p^{++} Al-doped layer is formed. The doping concentration of this p^{++} layer is dictated by the temperature at which the recrystallisation takes place, determined from the Al/Si phase diagram. A limitation of this technique is the relatively low-doping concentrations achievable, with a peak dopant concentration of around $3 \times 10^{18}/cm^3$ due to the relatively low solid solubility of aluminium in silicon [23–25] compared to elements such as boron (B) (refer to Figure 9.1a). To increase the doping concentrations achievable via this process, boron, which has a much higher solid solubility in silicon compared to aluminium, is often added to the aluminium paste (in concentrations ~1% of the aluminium concentration), resulting in Al-B co-doping. This can increase the doping concentrations achieved by one order of magnitude to approximately $3 \times 10^{19}/cm^3$ [23], hence improving field-effect passivation and reducing the J_0. An example of doping profiles achieved by this method, with and without boron additives in the paste, is shown in Figure 9.1b, along with estimations of the

Table 9.1 Efficiencies of various PERL cells utilising laser-doped rear contacts

Reference	Year	Area (cm^2)	Wafer type	Front contact	Rear contact	J_{SC} (mA/cm^2)	V_{OC} (mV)	FF (%)	Eff (%)
Kluska *et al.* [14]	2011	4	p-type float-zone (FZ)	Photolithography; Ti/Pd/Ag evaporation	Boron LD (LCP); Al evaporation	38.6	675	80	20.9
Wang *et al.* [16]	2012	155	p-type Cz	n-type laser doping; Ni/Cu plating	Boron LD; Al sputtering	40.9	665	74.4	20.3
Ortega *et al.* [17]	2012	4	p-type FZ	Photolithography; Ti/Pd/Ag e-beam	Al LD from Al_2O_3/SiC_X stack; Al e-beam	38.6	678	81.9	21.4
Li *et al.* [7]	2013	13	p-type Cz	n-type laser doping; Ni/Cu plating	Boron LD; Al evaporation	40.2	667	75.6	20.3
Lin *et al.* [8]	2014	6	p-type Cz	n-type laser doping; Ni/Cu plating	Boron LD; Al screen-print	40.2	655	76.1	20
Benick *et al.*[18]	2014	4	n-type Cz	Photolithography; Ti/Pd/Ag evaporation	SiC_X:P laser doping; Al evaporation	41.3	699	80.5	23.2
Benick *et al.*	2014	4	n-type Cz	Photolithography; Ni/Cu/Sn plating	SiN_X:P laser doping; Al evaporation	39.6	677	81.2	21.7
Jaeger *et al.* [19]	2014	238	p-type MCz	Screen-printed Ag	B-doped Si nanoparticle paste laser doping; evaporated Al	38.4	654	79.4	20
Tomizawa *et al.* [15]	2015	25	p-type Cz	Screen-printed Ag	B-doped Si nanoparticle paste laser doping; Al screen-print	37.9	656	79.1	19.7
Cornagliotti *et al.* [12]	2015	238	p-tpe MCz	Laser ablation; Ni/Cu/Ag plating	Al LD from Al_2O_3 layer; Al sputtering	39.4	657	79.9	20.7
Ernst *et al.* [13]	2016	238	p-type Cz	Screen-printed Ag	Al LD from Al_2O_3 layer; Al evaporation	38.9	641	79.7	19.9

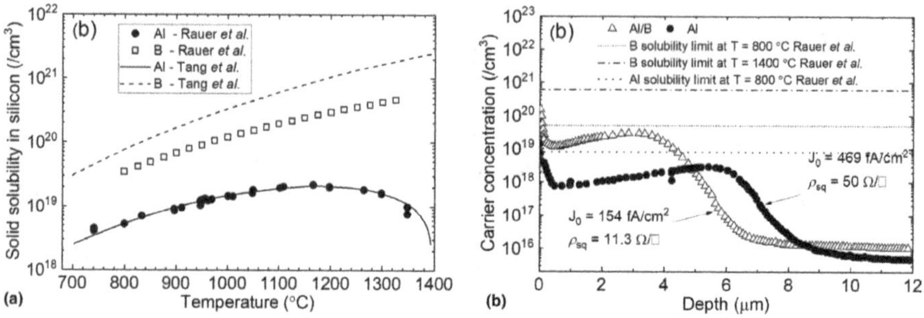

Figure 9.1 *(a) Solid solubilities of Al and B in crystalline Si as a function of temperature under thermal equilibrium conditions from Rauer* et al. *[24] and Tang* et al. *[25]. (b) Electrochemical capacitance-voltage (ECV) measurements of the doping profiles formed by the Al-Si alloying process (at 850 °C) with standard Al paste and Al paste with B additive from Lölgen* et al. *[23]. Sheet resistance and J_0 values simulated using EDNA2 [25] using assumed values of S_{n0}, S_{p0} = 1×10^7 cm/s, τ_{n0}, τ_{p0} = 100 μs.*

localised sheet resistance and corresponding J_0 values as simulated using EDNA2 on PV lighthouse [26].

When the Al/Si alloying method is used for LBSF formation on PERC structures, an air gap can sometimes form instead of a solid eutectic layer, known as a Kirkendall void [27, 28]. These voids form due to the much larger solubility of silicon in aluminium compared to aluminium in silicon, resulting in significant amounts of silicon spreading far from the local contact region into the aluminium layer during the high-temperature process, and the inability of the silicon to return quickly enough during cooling [27]. Voids are undesirable as they can reduce electrical contact and increase recombination in the area if the LBSF does not form adequately. The firing temperature for LBSF formation in this way is selected based on a trade-off between minimising voids, which tend to form at higher firing temperatures combined with rapid cooling [27], and firing at a high enough temperature to ensure the LBSF is sufficiently thick, and with a higher-doping concentration due to the higher solid solubility with increasing temperature. Figure 9.2 shows scanning electron microscope (SEM) images of four different scenarios that can occur at the Al/Si contact: (a) an ideal Al/Si contact with sufficiently thick (~6 μm) LBSF; (b) a Kirkendall void where no eutectic layer or LBSF is visible; in this case, there is poor electrical contact leading to high series resistance as well as high recombination leading to low V_{OC} due to the exposure of the non-passivated silicon bulk to the air gap; (c) an Al/Si contact where the LSBF formation is negligible, again leading to poor electrical contact and high recombination; and (d) a Kirkendall void where an LBSF has formed, thus shielding minority carriers from recombining at the surface, however, still resulting in poor electrical contact. To minimise the formation of voids, specialised Al-LBSF pastes with silicon additives have been developed for PERC cells.

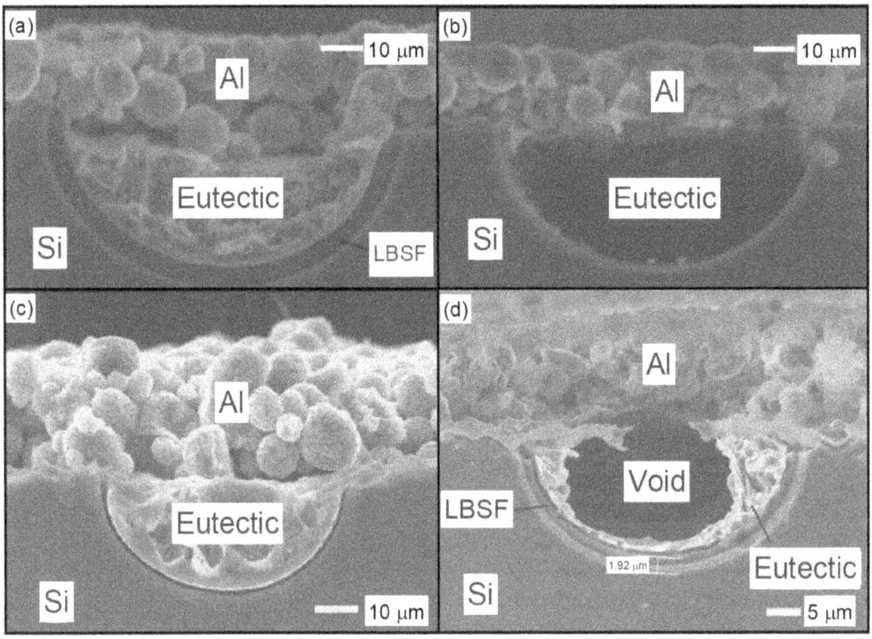

Figure 9.2 *SEM cross-sectional images of (a) an Al/Si contact with LBSF; (b) a Kirkendall void with no LBSF; (c) an Al/Si contact with negligible LBSF and (d) an Al/Si contact with LBSF protecting against a void. Reproduced from reference [29].*

This provides a background concentration of silicon within the aluminium paste, which reduces the diffusion of silicon into the aluminium layer from the bulk of the wafer [30], ensuring sufficient silicon is available at the solid–liquid interface for the formation of the p^+ region upon cooling. However, these pastes are typically less conductive than standard Al-BSF pastes due to the presence of silicon in the paste and also adversely decrease internal reflectance [31], thereby reducing the J_{SC}.

As an alternative to the Al/Si alloying process, laser doping, typically with boron, can be used to form the deep p^{++} regions [8, 32]. Note that this method should not be confused with laser-fired contacts, whereby a laser is used to fire a metallic layer through a dielectric to alloy it with the underlying silicon [33].

The use of laser doping to form the LBSF regions has several distinct advantages over this process; these are further explained in the paragraphs below:

1. higher-doping concentrations in the LBSF region
2. removes the trade-off between minimising Kirkendall voids and maintaining sufficient LBSF thickness
3. allows the use of regular Al-BSF paste without Si additives (higher conductivity than Al-LBSF paste)

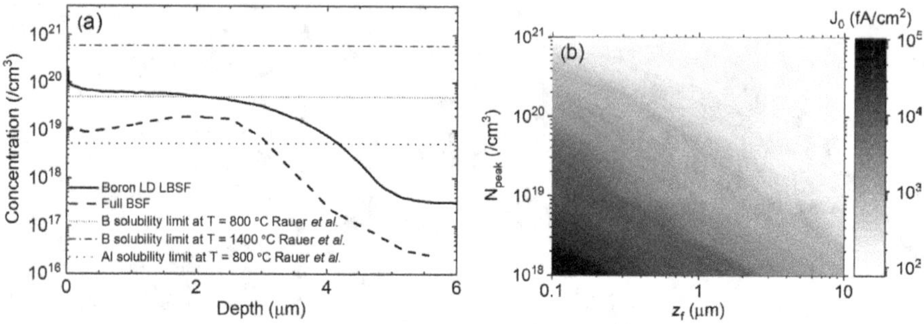

Figure 9.3 *(a) Boron ECV profiles for laser-doped regions formed using a boron SOD source and a 355-nm Q-switched laser with a scan speed of 0.5 m/s and power of 5 W. Adapted from Li* et al. *[7]. (b) Simulated value of J_0 as a function of JD (z_f) and peak concentration (N_{peak}) with a complementary error function profile using EDNA2 [25].*

4. allows alternative metallisation methods other than screen-print paste to be used (e.g. sputtering/evaporation/plating).

First, laser doping can heat the Si in localised regions to temperatures much higher than achieved through the liquid phase epitaxial growth of Al-doped Si during Al/Si alloying, meaning higher solid solubilities for the concentrations of dopants such as B in the Si when B is used as the dopant source (refer to Figure 9.1a). Temperatures achieved via laser doping are above the melting point of silicon (1 414 °C), compared to approximately 700–900 °C for the Al/Si alloying process. For B, this can increase the solid solubility limit over an order of magnitude from approximately $3 \times 10^{19}/cm^3$ at 800 °C to approximately $5 \times 10^{20}/cm^3$ for temperatures approaching 1 400 °C. It should also be noted that this is sufficiently higher than the solid solubility of B at a temperature of 1 000 °C ($\sim 1 \times 10^{20}/cm^3$), which is used in conventional thermal oxidation for the drive-in after tube-furnace boron diffusions. This enables more heavily doped LBSF regions to be formed, facilitating better shielding of minority carriers from the back-contact regions as well as a lower contact resistance. Example ECV profiles of a B-doped p^{++} region formed using a Q-switched 355 nm laser is shown in Figure 9.3a, demonstrating the higher dopant concentrations achievable (boron LD LBSF) compared to the Al/Si alloying process (full Al-BSF) [7]. It is noted that in this case, the Al-BSF is most likely an Al/B paste due to the concentration well above the aluminium solubility limit but similar to that observed in Figure 9.1, although it was not explicitly noted in the manuscript. Higher peak-doping concentrations, as well as deeper junctions, are favourable for BSFs, resulting in lower J_0 in these regions due to an improved shielding of minority carriers. Simulated J_0 and sheet resistance values as a function of peak-doping concentration and junction depth (JD) assuming an error function profile are shown in Figure 9.3b. Simulations were performed using EDNA2 [25].

Second, by pre-forming the LBSFs via laser doping, lower firing temperatures can be used to minimise Kirkendall voids independently of LBSF thickness.

Third, as Kirkendall void formation is minimised, regular aluminium paste used for full-area BSF cells may be used, therefore improving line conductivity of the metal and reducing the undesirable decrease in internal reflectivity for PERC LBSF pastes [31].

Finally, as the Al/Si alloying process is not required to form the LBSF, alternative metallisation methods can be used for the rear contacts whereby metal is deposited directly onto the laser-doped LBSF regions. This could be performed via sputtering, evaporation, plating or potentially low-temperature screen-print pastes. One example of where this has been implemented is Suntech Power's 'next-generation Pluto-PERL' cells, where a record 20.3% efficiency was achieved in 2011 on commercial-grade p-type Cz wafers using boron laser doping for LBSF formation coupled with Al sputtering [16]. This was the first demonstration of an industrial p-type Cz solar cell with an efficiency over 20%. In these cells, peak active doping concentrations for the LBSF regions formed by laser doping approached $2 \times 10^{20}/cm^3$ (as measured by ECV) in comparison with the equivalent Al/Si alloyed LBSFs (formed with Al-B paste), which were an order of magnitude lower.

Lin *et al.* [8] demonstrated several of the advantages described above through the fabrication of PERC cells with boron laser-doped LBSFs, formed through the application of a boron SOD source to the rear passivation stack prior to laser processing, and contacted with screen-printed aluminium paste. SEM cross-sectional images of the contact region showing the boron-doped BSF (a) before and (b) after Al/Si alloying are shown in Figure 9.4. It was shown that PERC structures with LD formed LBSF and standard aluminium paste fired at reduced temperature (700 °C) could achieve the same efficiencies as standard PERC cells with LBSF formed through Al/Si alloying using specific LBSF paste fired at 800 °C. It was also found that the deep laser-doped LBSF had the potential to shield regions where Kirkendall voids did form, reducing recombination at these voids. Thus, a wider processing window and increased fabrication robustness using the laser-doping technique was demonstrated.

Laser doping for LBSF formation was originally implemented using a line contact geometry. However, point contacts as used in the world record PERL cell [20] have the advantage of a reduced area of localised heavy doping, reduced metal–Si interface area, both of which reduce the contribution of the contact to the total J_0 of the device, and also resistive losses through a reduced average distance that carriers have to travel throughout the bulk of the silicon to reach the heavily doped region. Laser-doped point contacts have been demonstrated by several groups [19, 34]. The use of a CW laser to form deep junctions is advantageous for point contacts, to ensure adequate LBSF thickness to shield minority carriers from the laser-induced defects. It also increases the collection area of majority carriers from the bulk through a 3D high–low junction structure, as opposed to a more 2D structure from conventional shallow diffused regions, potentially reducing the impact of current crowding, which can be of concern for small area contacts [35]. Hallam [34] demonstrated junctions >8 μm deep (Figure 9.5) by scanning a CW laser perpendicularly to a physical mask

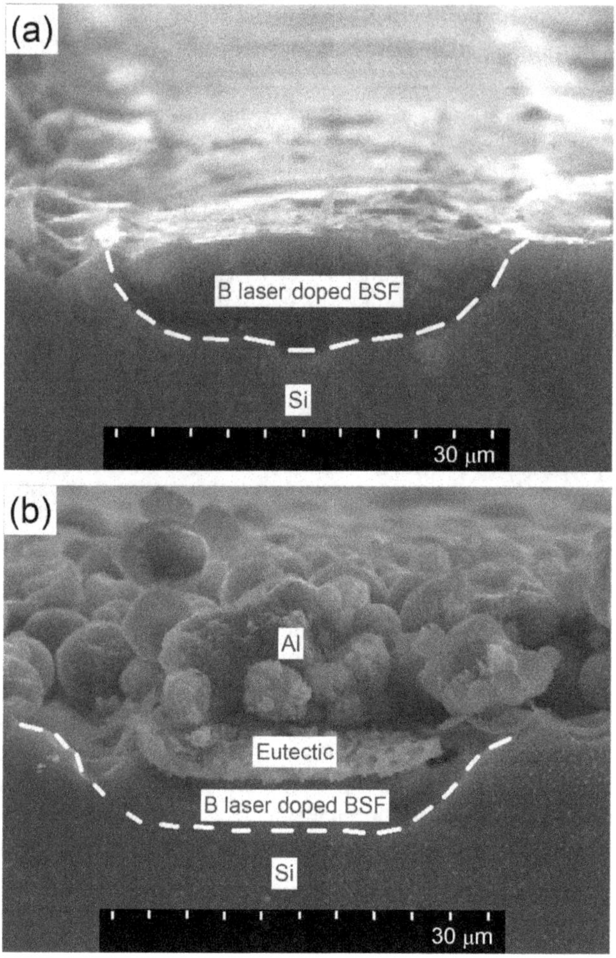

Figure 9.4 *SEM cross-sectional images of (a) a boron laser-doped LBSF region and (b) an ideal Al/Si contact to a boron laser-doped LBSF where the eutectic region is protected by a thick LBSF. Reprinted from [8], Copyright (2014), with permission from Elsevier.*

with small line-shaped openings to form the point-like structure, but this could also be done with an optical shutter. A key issue for small point contacts if implemented with a scanning procedure is the potential non-uniform laser doping along the length of point-like contact, resulting in the generation of defects at the starts and ends of the points through a reduced exposure time to laser, and possibly insufficient doping. As shown in Figure 9.5, the Marangoni effect can also result in a mound appearing at the last portion of the point contact to be processed as the laser is rapidly removed from the silicon wafer [37]. It is unclear what potential impact this region could have on the electrical performance of a solar cell. For other implementations where

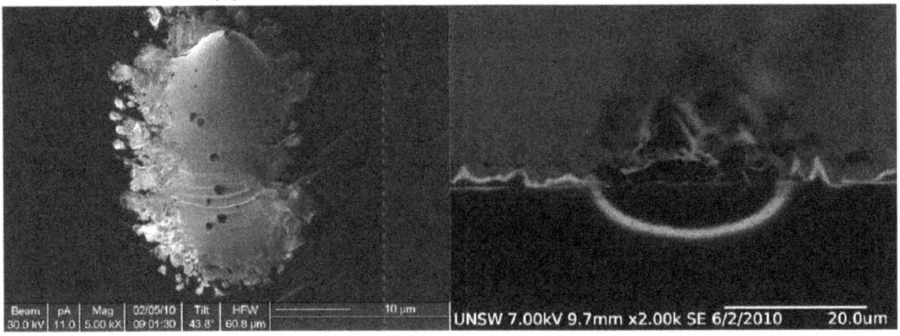

Figure 9.5 *(a) Focused ion beam (FIB) image of the surface of a laser-doped point contact on a textured silicon surface using a 532-nm CW laser; (b) combined SEM/electron beam-induced current (EBIC) image of the cross section of the laser-doped point contact. Reprinted from [36], Copyright (2013), with permission from Elsevier.*

the laser is stationary over the given region or using nanosecond pulse-length lasers where the scanner is essentially stationary on the wafer's surface with respect to the pulse length, this effect is likely minimised.

9.1.2 Aluminium oxide as a dopant source for p^{++} silicon formation

Although much simpler than forming the same structure using photolithography or other patterning methods and multiple thermal diffusion steps, one drawback of the self-aligned laser-doping process (laser doping through dielectric layers) is the typical requirement for the additional step of applying the dopant-containing film (e.g. phosphoric acid (H_3PO_4) or a boron SOD) prior to laser doping and the subsequent removal of the film after the laser-doping process. Alternatively, if a dielectric stack containing an appropriate dopant is used as discussed in Chapter 8, these extra steps can be eliminated.* The rear side of a typical industrial PERC cell utilises a rear dielectric scheme comprising a stack of AlO_X capped with SiN_X, where the AlO_X can double as both an excellent surface passivation layer (due to its high density of fixed negative charge causing an accumulation effect on p-type silicon surfaces [38, 39] as discussed in Chapter 3) and also fortuitously as a rich source of aluminium, a p-type dopant, for forming heavily doped p^{++} regions in the underlying silicon as demonstrated by several research groups [34, 40–43]. Thus, p^{++} doped regions can be formed by laser doping on the rear side of industrial PERC cells, which can then subsequently be contacted through evaporation/sputtering of aluminium or through screen-printing of aluminium paste. For this reason, laser doping using AlO_X layers as the dopant source has become an area of interest in the past

*A chemical preparation step such as a HF dip may still be required to remove any residual oxide layers and ensure a clean interface between the heavily doped silicon and overlying metal.

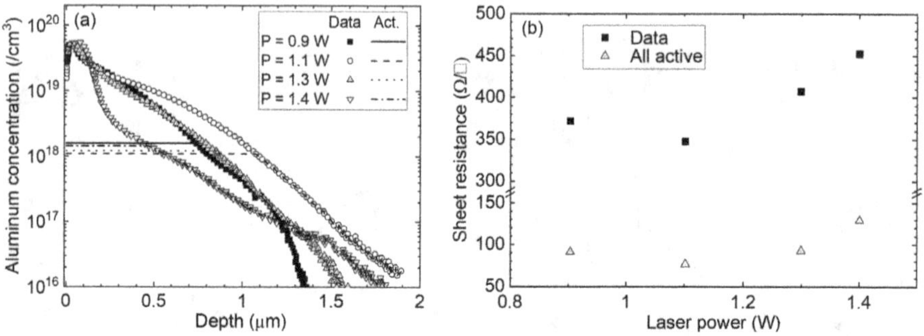

Figure 9.6 *(a) SIMS profiles of aluminium dopants for laser-doped regions formed with a 1 064 nm Nd:YAG laser for various laser powers. Data reproduced from [43] along with estimated electronically active dopant concentrations based on sheet resistance measurements and EDNA2 simulations [26]. (b) The measured four-point-probe sheet resistance of the profiles in (a) reproduced from [43] (closed squares) as well as simulated sheet resistance if all dopants were electrically active as performed using EDNA2 (open triangles) [26].*

few years. As discussed in Section 9.1.1, silicon reaches much higher temperatures during laser doping than those normally achieved through the liquid phase epitaxial growth of Al-doped Si during the Al/Si alloying process. However, in contrast to boron dopants, the solid solubility of aluminium in silicon peaks at a temperature of approximately 1 100–1 200 °C ($2 \times 10^{19}/cm^3$) and subsequently reduces with increasing temperature to below $1 \times 10^{19}/cm^3$ at a temperature of 1 350 °C. As such, the laser doping process using aluminium dopants may not provide the same benefit as when using other dopants such as boron. In particular, the peak active doping concentration using AlO_X-based laser doping is lower than that of Al-B pastes using conventional screen-printing process. However, in addition to simplifying the process by not requiring a boron SOD source, the avoidance of using a boron-based dopant source may help to avoid the formation of boron-oxygen defects in the laser-doped region, where high levels of oxygen are typically found [44].

The formation of a p^{++} Al-LBSF region was demonstrated in 2011 by Martín *et al.* [43], where a 1 064 nm Nd:YAG laser with 100 μs pulse length was used to form local contact openings in a dielectric stack of atomic layer deposited (ALD) Al_2O_3 and amorphous silicon carbide. Using this technique, peak aluminium concentrations of approximately $5 \times 10^{19}/cm^3$ and JDs of 1–2 μm were obtained as measured by SIMS, as shown in Figure 9.6a. The peak concentration of aluminium incorporated into the silicon is well above the solid solubility limit shown in Figure 9.1a. However, the corresponding sheet resistances for these profiles were reported to be in the range of 300–400 Ω/□ (shown in Figure 9.6b), which are significantly higher than expected if all aluminium atoms in the silicon are electrically active as also shown in Figure 9.6b from simulations performed using EDNA2 [25].

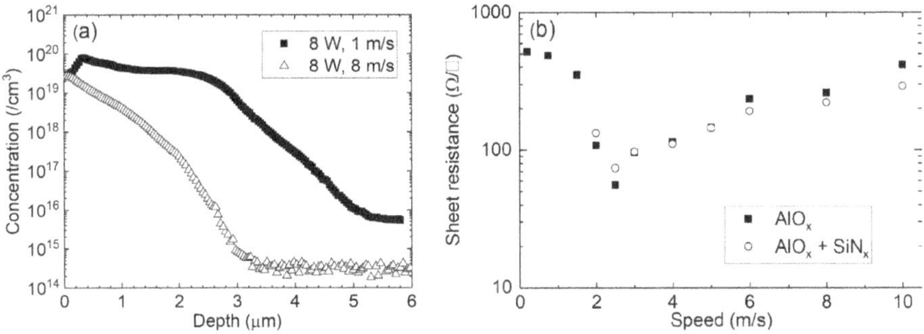

Figure 9.7 *(a) SIMS profiles of Al concentration for laser-doped regions formed through a dielectric stack of 10 nm ALD Al_2O_3 capped with 120 nm PECVD SiN_X using a 355-nm Q-CW laser with speeds of 8 m/s and 1 m/s. Reprinted from [12] Copyright (2015) with permission from Elsevier. (b) Sheet resistance of laser-doped regions as a function of laser speed for the same laser as (a) when performed through the 10 nm ALD Al_2O_3 layer with and without a capping layer of 120 nm PECVD SiN_X:H.*

Additional simulations in ENDA2 matching the measured sheet resistance for each profile suggest a solid solubility limit of approximately 1–1.5 × $10^{18}/cm^3$ for the laser-doping process.

Despite a significant amount of aluminium in the silicon being electrically inactive, efficiencies over 21% were achieved on small-area cells using FZ substrates, where this process was used to form the rear LBSF regions onto which a 2-μm layer of aluminium was deposited as the rear contact [17]. Fill factors over 82% indicate effective shielding of minority carriers at the localised rear contacts and hence the suitability for a rear contact scheme.

With a solid solubility limit of approximately 1–1.5 × $10^{18}/cm^3$, given a substantial amount of aluminium, is incorporated into the silicon in an electrically inactive state during the laser-doping process, deeper laser-doped regions could potentially allow for more effective utilisation of the dopants. This would also reduce the fraction of inactive dopants with an unknown influence on minority carrier lifetime through the generation of defects. Substantially deeper LBSF regions for AlO_X-based laser doping have been achieved through the use of a 355 nm mode-locked Q-CW laser or using a 532 nm CW laser [12, 34]. Using the 355 nm laser, Imec achieved 20.7% efficiency on a full-size Cz PERL cell in 2015 with Cu plated front contacts, using laser doping through the rear dielectric stack consisting of 10 nm ALD Al_2O_3 capped with 120 nm of PECVD SiN_X [12]. The laser-doped regions were contacted with a layer of sputtered aluminium that was then annealed at 450 °C. Figure 9.7a shows the SIMS profiles for the laser-doped regions for laser speeds of 8 m/s (speed used for laser doping of the rear contacts for the 20.7% cell) and 1 m/s. A JD and hence LBSF thickness of approximately 5 μm was achieved for the 8 m/s laser speed. The deep JD achieved with only a 10-nm-thick layer of ALD

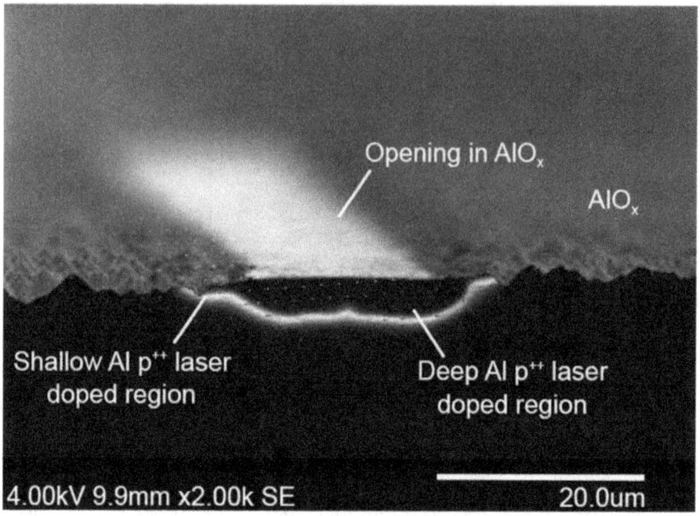

Figure 9.8 *Combined SEM/EBIC images of Al p^{++} laser doping on an n-type wafer using PECVD AlO_X as a dopant source. The p-n junction appears as a bright line, revealing a deep laser-doped region directly beneath the opening in the AlO_X layer, as well as a shallow-doped region extending laterally beneath the AlO_X layer. Image reproduced from [34].*

Al_2O_3 as the dopant source was enabled by the high repetition frequency (80 MHz) of the Q-CW laser, which prevents the solidification of the laser-doped region between successive pulses, despite each pulse only lasting 12.5 ps. Given the solid solubility limit for aluminium, little benefit of a reduced sheet resistance value is expected using thicker AlO_X layers, unless deeper molten regions can be formed. Figure 9.7b demonstrates that sheet resistances as low as approximately 55 Ω/□ can be achieved with this approach.

Using a 532 nm CW laser, JDs of up to 8 μm have been demonstrated [34], along with sheet resistances as low as 50 Ω/□. A key feature that has been noted for AlO_X-based laser doping using CW lasers using EBIC and SEM analysis is that the p^{++} regions formed in this way extend not only deep into the substrate but also laterally beyond the opening in the AlO_X layer (Figure 9.8). This is in contrast to results obtained when performing an identical laser-doping process with a SiN_X passivation layer coated in a boron SOD source. The extended doping beneath the AlO_X layer was hypothesised to be due to the natural incorporation of dopants in the AlO_X enabling thermal diffusion of the dopants from the AlO_X layer into the silicon in the solid state. In the case of the boron SOD on SiN_X layers, dopants are blocked from diffusing into the silicon by the SiN_X. The extended lateral doping is likely to be beneficial in shielding minority carriers from any laser-induced defects that are typically formed at the edges of the laser openings, as discussed in Chapter 8. In addition, the use of a CW laser with high scan speeds results in a melting threshold

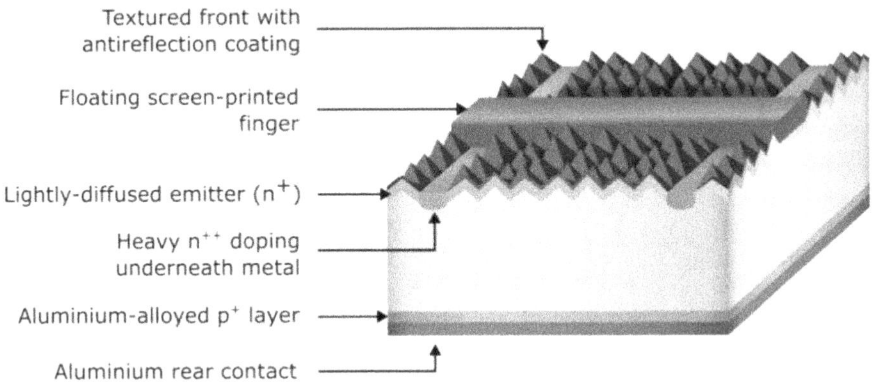

Figure 9.9 *Schematic of the SFSC structure with heavily doped channels running perpendicularly to the front metal fingers. The metal fingers only contact the Si at the intersections with the heavily doped channels and are insulated from the silicon by the front dielectric layer elsewhere.*

region narrower than the full width half maximum beam diameter of the laser. This has enabled ultra-narrow openings and doped regions, as little as 4-μm wide [34], which is on the order of that achievable with photolithography for record PERL cells, without the need for complicated optics. This could greatly reduce metal–Si interface areas for future cell designs.

9.2 Semiconductor finger solar cells

As discussed in Chapter 8, selective emitter solar cells allow for a decoupling of the optimisation of the emitter for contact resistance and short-wavelength response. To be compatible with screen-printing technology, the heavily doped n^{++} regions need to be wide enough to allow enough tolerance for the aligned screen-printing of the metal grid to these regions. In addition, the lightly doped emitter region still needs to be heavily doped enough to be able to transport current laterally to the fingers, or the fingers need to be placed close enough together to allow for a lighter diffusion. This, of course, impacts the amount of optical reflection and hence the short-circuit current of the device.

Introduced in 2006 as a way to implement selective emitter structures in industrial production, the semiconductor finger solar cell (SFSC) (Figure 9.9) is a form of selective emitter device that addresses these trade-offs by making use of narrow heavily doped channels on the surface of the solar cell running perpendicular to the metal fingers [45]. The semiconductor fingers provide lateral conductive paths between the metal fingers, greatly enhancing the effective conductivity of the emitter, enabling the front metal fingers to be placed further apart. The structure boasts all the advantages of a standard screen-printed selective emitter solar cell, such as a near-unity IQE by virtue of the lightly diffused and well passivated front surface, low contact resistivity due to the heavy doping at the metal–Si interface, with the added advantages of:

1. less optical shading due to a wide finger spacing and therefore higher J_{SC}
2. self-aligned metallisation with metal fingers printed perpendicular to semiconductor fingers, which reduces processing complexity
3. reduced metal–Si interface area confined only to the heavily doped regions at the intersection of a metal finger and semiconductor fingers (non-fire-through Ag paste), therefore a reduced device dark saturation current density and higher V_{OC}.

Although optical shading was reduced, a small amount of 'electrical shading' inevitably occurred on such devices due to the heavily doped semiconductor fingers. However, this was generally outweighed by the benefits of the structure if the geometries were appropriately optimised. It was reported by Mai *et al.* [45] in 2006 that the SFSC had approximately a $10\%_{rel}$ performance advantage over the conventional screen-printed solar cells (SPSCs) produced using the same equipment at Suntech Power. The SFSC was put into production by Suntech Power in 2007 with efficiencies reported of 18% compared to the 14–15% average efficiency of conventional screen-printed cells from other manufacturers at the time [46]. However, no details were publicly disclosed of how the semiconductor fingers were formed. While the benefits described above gave clear efficiency advantages over conventional screen-printed cells using homogeneously diffused emitters, drawbacks of the front semiconductor finger (SF) design include the higher resistive losses associated with the highly localised front metal–Si interface (only at the intersection of Ag fingers and semiconductor fingers when using non-fire-through Ag pastes). This makes the structure less suitable for high-efficiency designs such as PERC, where resistive losses can have a larger impact on overall device performance.

The semiconductor fingers were originally formed in a similar way to the heavily doped region in the UNSW buried-contact solar cell device [47], as described in Chapter 8. This comprised complex fabrication processes involving, first, forming the light emitter diffusion, a thermal oxidation step to form a masking barrier for the subsequent heavy diffusion, laser ablation of the oxide layer and silicon to form deep grooves, chemical etching of the deep grooves to remove laser damage and heavy diffusion through oxide masking layer to dope the grooves ('groove diffusion'). Both the thermal oxidation and groove diffusion steps were carried out at temperatures in excess of 900 °C thereby largely inhibiting the use of the technology on lower quality substrates such as mc-Si, which is known to degrade at these high temperatures [48, 49]. In the case of the SFSC, grooves were approximately 20 μm wide and were kept to conservative depths of around 10–25 μm to ensure continuity of the metal screen-printed fingers at the cross-over points when printed perpendicularly to the grooves [50]. Using this approach, an efficiency of 18.4% was achieved for a large area p-type Cz device,

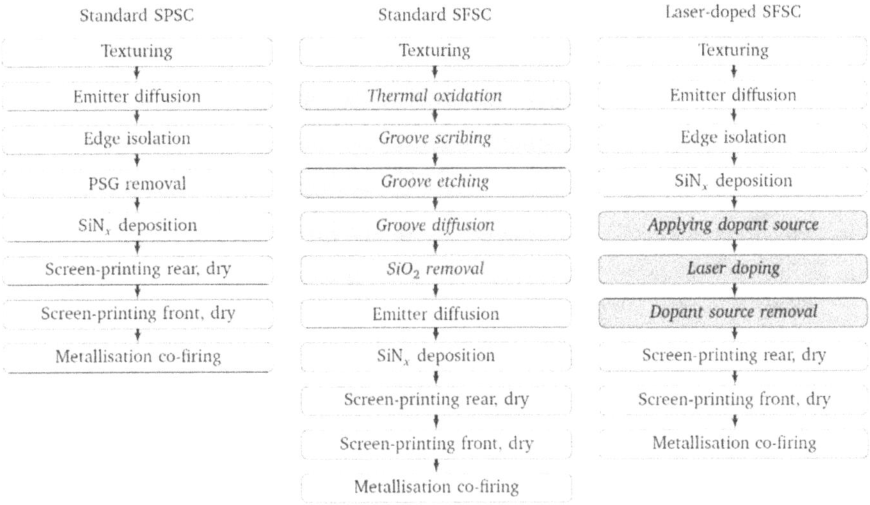

Figure 9.10 *Processing sequence of (left) a standard conventional screen-printed homogeneously diffused solar cell, (middle) a standard SFSC and (right) a laser-doped SFSC. The additional processing steps required for producing the semiconductor fingers compared to conventional SP technology are italicised. Figure adapted from [50].*

with a metal–Si interface area of approximately 0.2%. This was compared to an efficiency of 16.5% on a standard screen-print cell structure using the same equipment [50].

9.2.1 Laser-doped semiconductor finger solar cells

As with the LDSE technology replacing the buried contact cell technology, laser doping was proposed as a simplified method to form the semiconductor fingers [50]. This eliminated the need for the high-temperature oxidation masking step, the thermal groove diffusion and the wet chemical etching for laser damage removal in the grooves – thereby making the technology more commercially viable and also compatible with lower quality multi-crystalline silicon wafers. The processing sequence of a laser-doped SFSC is shown in Figure 9.10 compared to a standard SFSC and standard homogeneously diffused SPSC. Only three extra steps were required over the standard homogeneously diffused SP cell: applying the dopant source, the laser-doping process itself and removal of the dopant source. The advantage of using laser doping as opposed to groove diffusions on the standard SFSC is clear, eliminating the need for the additional two high-temperature steps.

There are several advantages of using laser doping for the semiconductor finger technology:

1. High-temperature oxidation and diffusion processes are avoided (allows the use of lower quality substrates sensitive to thermal degradation or oxygen precipitation).
2. Due to shallow laser-doped lines, problems associated with deep grooves for metal finger continuity are eliminated.
3. Wet chemical steps associated with groove etching and cleaning are eliminated.
4. Wet chemical edge isolation can be performed prior to semiconductor finger formation thereby allowing the doping profile of the semiconductor fingers to be maintained.
5. The ARC layer is not limited in choice to a material that is high-temperature compatible (PECVD SiN_x, for example, can be used).
6. High throughput due to reduced processing and fast laser scanning speeds.
7. Reduced overall front parallel sheet resistance for the same density of semiconductor fingers (due to the heavier doping achieved via laser doping).

Disadvantages include:

1. loss of the gettering step associated with the heavy phosphorus groove diffusion
2. increased surface reflectivity of the laser-doped lines compared to the laser grooves
3. high resistive losses due to poor contact resistance between the screen-printed metal fingers and the laser-doped semiconductor fingers combined with the very low metal–Si interface areas [50].

Early work on the laser-doped SF technology used a 532-nm Q-switched laser with a repetition frequency of 200 kHz [50]. However, at the time, sheet resistances were limited to 30 Ω/ϒ. With a narrow width of approximately 10 μm for the laser-doped region, this did not allow an effective reduction of the emitter sheet resistance, and poor efficiencies of 11.6% were obtained. Subsequent work using a higher laser power and reduced repetition frequency of 100 kHz yielded sheets with much lower resistances of approximately 5 Ω/ϒ [51], with laser-doped regions approximately 1 μm deep and 10 μm wide. However, the major limitation for the laser-doped SF cell structure stemmed from the device fill factor being severely limited by high contact resistance between the screen-printed metal and laser-doped semiconductor fingers, coupled with the very low metal–Si interface area [51]. To achieve high efficiencies with this small metal–Si interface area, a very low contact resistance of approximately 0.1 $m\Omega\cdot cm^2$ or less was required, which was not achievable with screen-printing and laser-doping alone [52]. In particular, low-reactivity Ag pastes with low glass-frit content were required to ensure the metal–Si interface only formed at the intersection of the SF and the screen-printed metal. It was found that the lower the sheet resistance of the laser doping (below 20 Ω/□), the worse the contact resistance became [51]. It was speculated that the commercial Ag pastes used were not compatible with such low sheet resistances and the associated high-surface-doping concentrations of the laser-doped regions. In particular, the heavily doped LD regions were shown to inhibit the formation of Ag crystallites [51], and importantly, this issue was not avoided by applying a HF dip prior to screen-printing. It is noted in [50] that the mechanism by which the LD regions inhibited the formation of the Ag crystallites is not understood and requires further investigation. However, it was speculated that the extremely heavily doped LD regions with concentrations well above the solid-solubility limit may create a resistive layer that hinders the melting of glass-frit.

Subsequent work focused on the use of a 532-nm CW laser for forming LD SF structures, with substantially deeper JDs of up to 10 μm and sheet resistance value as low as 2 Ω/ϒ [52]. However, a striking feature observed in SIMS profiles was the presence of a sharp spike in the phosphorus dopant concentration at the silicon surface to approximately $7 \times 10^{20}/\text{cm}^3$. In that work, an efficiency of 13.0% was achieved for laser-doped SFSC, again limited by contact resistance, with an obtained specific contact resistivity of 2.85 $\text{m}\Omega\cdot\text{cm}^2$, approximately one order of magnitude too high given the low contact area. In comparison, an efficiency of 15.7% was obtained for an identically processed cell using a standard fire-through paste (despite a higher specific contact resistivity of 11.85 $\text{m}\Omega\cdot\text{cm}^2$ on the lightly diffused 100 Ω/ϒ emitter). However, only one process condition (scan speed of 0.7 m/s) was used to fabricate cells or transmission line measurement (TLM) structures. Furthermore, no investigations were performed to test the hypothesis of high doping impacting the ability to nucleate Ag crystallites by etching off the surface layers to contact more moderately doped silicon beneath the spike, with dopant concentrations on the order of approximately $1 \times 10^{20}/\text{cm}^3$. As such, additional work is required to understand the specific contact resistivity for a range of laser-doped profiles using low-reactivity Ag pastes, and whether etching processes can be used to remove the surface region of the laser-doped region to improve contact resistance.

In addition, significant improvement in pastes has occurred in the last decade. It is unclear whether such challenges would be solved by the current generation of pastes, which can achieve specific contact resistivities of approximately 0.5 $\text{m}\Omega\cdot\text{cm}^2$ to laser-doped regions (fire-through Ag pastes for industrial LDSE solar cells), or whether future paste chemistry modifications could be used to reduce contact resistance.[†]

9.2.2 *Advanced laser-doped semiconductor finger cells*

To overcome the limitation of excessively high contact resistances, among others, the advanced SFSC was developed whereby the laser-doped semiconductor fingers were plated with a thin metal layer (e.g. nickel) after the screen-printed fingers were formed [51]. This not only improved the contact resistance to the intersecting screen-printed fingers but also greatly improved the conductivity of the semiconductor fingers so that a much wider screen-printed metal finger spacing was possible,

[†]As measured on an industrial screen-printed LDSE cell at UNSW.

resulting in less optical shading and therefore a higher J_{SC}, as well as reducing Ag consumption. In this way, the benefits of screen-printing and laser-doped-plated contacts were combined, where screen-printing was beneficial for printing thick fingers as well as busbars that were easy to solder to, while laser doping and plating were ideal for forming narrow conductive lines. Using this approach, efficiencies of 18.8% were achieved [53]. Despite the benefits, a barrier for implementation was the requirement for both plating and screen-printing metallisation equipment on the production line.

9.2.3 Future potential of the semiconductor finger technology

The semiconductor finger family of solar cells was particularly attractive when screen-printed fingers were limited to approximately 120 μm in width as was the case in 2010 [50], any narrower being a technical challenge. At that time, emitter series resistance losses played a large role due to the larger finger spacing required to reduce optical shading losses. However, the benefit of SCF cells has reduced over time due to the breakthroughs in screen-printing technology allowing narrower lines to be printed, which are now in the range of approximately 35 μm and are set to reduce even further to 20 μm by 2030 [54], comparable with those achievable via plating. Nevertheless, the advantage of the self-aligning screen-printed metallisation remains an attractive feature of this technology.

In addition, the ability to form narrow, deep, heavily doped channels with very low sheet resistance across an emitter (as opposed to wide, shallower regions) has the potential to greatly reduce the effective sheet resistance while minimising J_{0e} [55], as well as potentially reducing optical shading. This could be increasingly important in the future as the industry moves towards structures with higher sheet resistance emitters.

9.3 Interdigitated back-contact solar cells

IBC solar cells, also known as all back contact or rear contact cells, have both polarities of contacts on the rear side as shown in Figure 9.11. They possess several advantages over conventional solar cells with contacts on both the front and rear surfaces, as detailed below:

1. The lack of contacts on the frontside means zero optical shading loss and, therefore, higher short-circuit currents (J_{SC}) are achievable.
2. Superior front surface passivation can be attained due to the lack of contact openings or heavily doped contact regions required. Instead, a lightly diffused front surface field can be applied over the whole front surface and optimised to give a very low front surface recombination velocity and avoid the formation of a dead layer.
3. Reduced cell series resistance is achievable, as the width and number of rear contacts can be optimised for current collection and current transport without impacting on front surface reflectance. The fact that the optimisation of optical

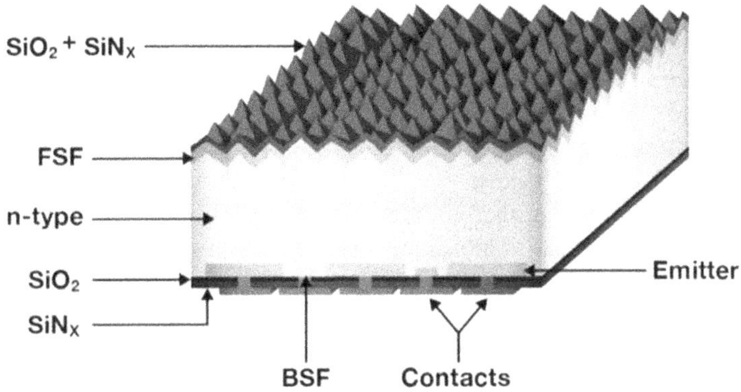

Figure 9.11 *Schematic of a typical IBC solar cell, where both polarities of contacts are on the rear of the device. Figure adapted from [56].*

properties and series resistance can be decoupled is a fundamental reason why IBC cells were initially introduced for concentrator applications [57].

4. The interconnection process is simplified by having both polarities of contact on the rear side, as the connecting tabs run from the rear of one cell to the rear of the adjacent cell rather than front to rear. This means they can be made thicker and therefore carry larger currents, leading to lower series resistance losses at the module level. It also enables cells to be placed closer together to achieve a higher packing density in the module and thus a higher module efficiency.
5. The lack of metal contacts on the front leads to a more uniform appearance which is highly desirable in some cases, particularly architectural building-integrated photovoltaic applications.

In 2004, SunPower commercialised the A300 rear point-contact solar cell concept developed by Stanford University [58] using simplified processing techniques [59], replacing photolithography with screen-printing techniques to form the patterned diffusion masks and achieving 21.5% efficiency in production. Through numerous iterations and technology developments, SunPower now mass-produces cells with efficiencies over 24.5% [60] and can achieve module efficiencies of up to 22.8% [61]. SunPower's technology is often considered a premium product for applications where a high energy density is required (e.g. areas with limited space) and where aesthetics is a primary factor.

Despite the higher efficiencies achievable through the IBC structure, the commercial success of the structure has been limited. In 2019, IBC cell structures accounted for just 2% of the worldwide market share for solar cells, and it is expected that this will increase to no more than 5% by 2030 [54]. This is predominantly due to the extra fabrication complexity of such a structure, requiring multiple aligned masking and high-temperature diffusion (typically three different diffusions required) and oxidation steps which are both costly and incompatible with lower-quality wafer types. The need for alignment, in general for at least two of the processes, also increases the likelihood of yield problems. In addition, IBC cells require silicon substrates with long diffusion lengths to carry the photogenerated carriers (of which the majority are generated at the front of the device) to the rear junction for collection and ensure a high J_{SC}. Ideally, the silicon needs to have a minority carrier diffusion length three times the thickness of the wafer [62]. As a result of the complex processing and the need for expensive, high lifetime substrates, the cost per watt of IBC cells is typically higher than that for conventional front and back contact cell structures, and until the early 2000s, the structure was reserved for niche applications such as solar planes, solar cars and space. Another key challenge faced by IBC solar cells is the transition to bifacial structures for technologies such as PERC and TOPCon. Bifaciality can increase the performance of such technologies by up to 30% relative and is widely seen as the option with the highest potential to increase module power with the lowest additional cost [63], thus making the transition to more complex structures such as the IBC cell less favourable.

9.3.1 Manufacturing challenges of IBC cells

The fabrication of IBC cells presents a complex problem: forms diffused regions of two different polarities on the rear surface and contact each of them with metal while also avoiding shunts. The structure essentially requires contact to be made through the rear emitter diffusion in some areas, while contacting the bulk silicon of opposite polarity to the emitter in others. Prior to the implementation of laser doping for IBC solar cells, there were two main methods to achieve this:

1. Partial removal of a full area emitter to form an opening to the bulk [64–67] or
2. Selective diffusion of the emitter via a diffusion mask, avoiding the covering of the bulk with a diffused layer from the outset as in the passivated emitter rear floating junction (PERF) cell [64, 68].

In the first method, the overlying emitter layer is physically removed locally, through either laser ablation or wet chemical etching through a patterned mask (formed by methods such as photolithography or inkjet printing). One example of this approach is that developed by Stanford University with a mesa-trench whereby after diffusion, the sample is masked and the diffusion is selectively removed to form a trench-like structure, which can potentially enable a self-aligned metallisation scheme, although challenging in mass production to realise [66, 69]. Another example of this is the interdigitated buried back-contact (IBBC) solar cell, where buried groove contacts are formed in a similar manner to the buried-contact solar cell described in Chapter 8, although the fabrication process is far more convoluted due to the IBC structure (see Figure 9.12). First, a light phosphorus diffusion is performed on the n-type substrate to form an n^+ front surface field (FSF), followed by oxidation to mask the FSF from the subsequent boron p^+ emitter diffusion. Photoresist is then applied to the front surface, and wet chemical etching of the rear oxide and unwanted rear phosphorus diffused layer is performed. A boron diffusion is performed to form the p^+ emitter on the rear side, followed by another oxidation

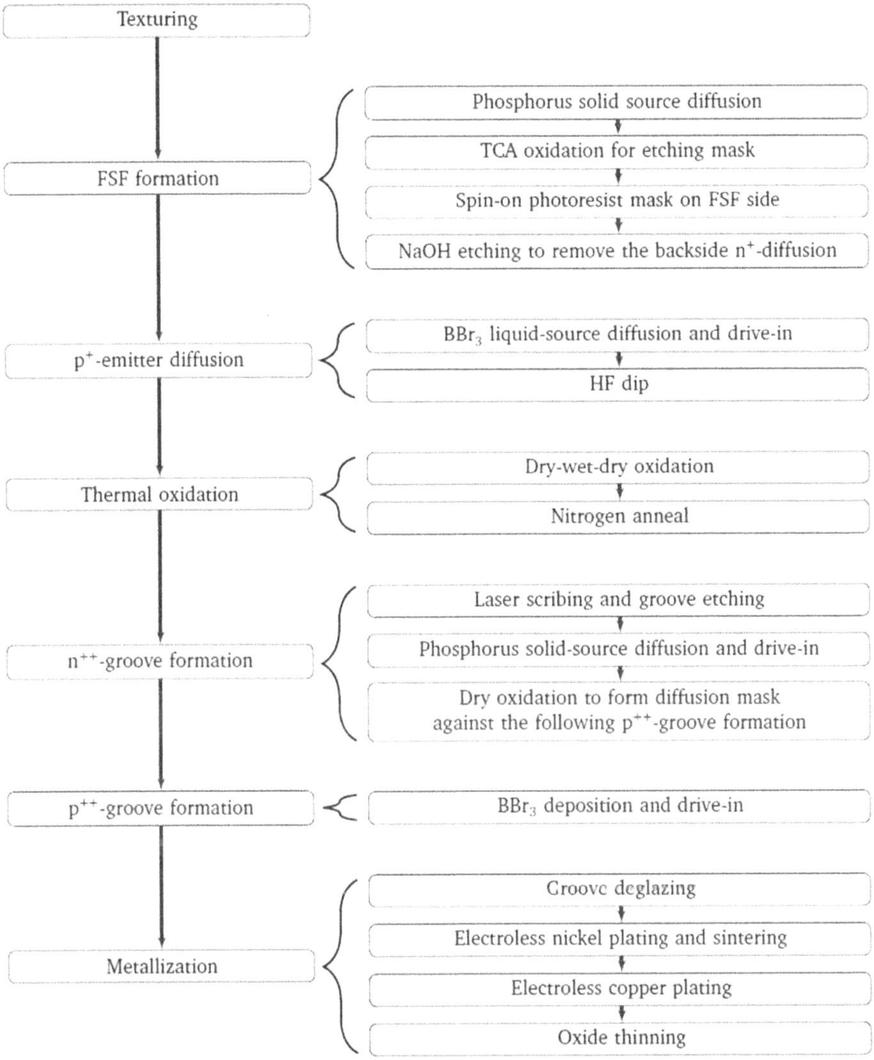

Figure 9.12 Processing sequence of an IBBC solar cell. Figure adapted from [70].

to act as a mask for the subsequent n^{++} base contact phosphorus diffusion and act as a drive-in process for the boron diffusion. This oxide layer is then patterned via laser scribing, which also forms deep grooves in the underlying silicon. After groove etching and cleaning, the heavy phosphorus diffusion is performed for the n^{++} base contacts, followed by yet another oxidation to form a diffusion mask against subsequent p^{++} emitter contact diffusion. The laser grooving and cleaning step is then repeated to form the grooves for the emitter contacts, followed by the final p^{++} emitter contact diffusion. Finally, the grooves are deglazed prior to electroless plating to form the metal contacts, and the oxide layer is thinned [70]. This fabrication sequence is complex and not suitable for mass manufacturing, requiring four thermal diffusions and three thermal oxidation steps. Additionally, the interface between the heavily doped n^{++} groove diffusion and p^{+} emitter diffusion is susceptible to forming a tunnel junction shunt and, therefore, must be carefully optimised and controlled [70], ideally with a region of non-diffused silicon in between the p^{+} and n^{++} regions.

In the second method, the emitter diffusion is performed selectively via the use of a diffusion mask. This method was used to fabricate the PERF junction solar cell (a cell with contacts on both front and rear side) [71] and has since been used to fabricate IBC cells. The PERF cell required three masking steps and two thermal diffusions. First, a thermal oxide was grown onto the rear surface of the wafer, followed by patterning with photolithography to expose regions where the diffusion was to be performed. After thermal diffusion, the oxide was removed and replaced with a new thermally grown oxide layer, again patterned via photolithography to expose areas to be subject to the opposite polarity diffusion, while leaving a small gap between the n^{+} and p^{+} regions. After the second diffusion, a final masking step was performed to mask the entire area apart from contact windows where metal was to contact the silicon [71].

9.3.2 Laser-doped IBC cells

The use of laser doping as a means to form the rear highly doped regions of an IBC cell and simplify its manufacturing sequence has been a significant area of research in recent years. Avoiding multiple high-temperature processes is desirable and can potentially make the cell fabrication process less detrimental to wafers sensitive to oxygen precipitation. Dahlinger *et al.* [72] were able to achieve an efficiency of 22% in 2013 [56] and 23.2% in 2016 using laser doping to form both contact polarities on the rear surface. A schematic of the general process flow is shown in Figure 9.13, and the cell structure was shown previously in Figure 9.11. First, a boron precursor was sputtered onto the rear surface of an n-type wafer, followed by a laser-doping process to form the p^{++} emitter regions. After precursor removal, the wafer underwent $POCl_3$ diffusion to form a lowly doped FSF, which inherently forms on both sides of the wafer. This was followed by another laser-doping step on the rear side which utilised the phosphorus dopants in the PSG layer from the thermal diffusion, to form the n^{++} BSF regions. Thus, a separate dopant source application for the n-type laser-doped regions was not required. After removal of the PSG, a silicon etch was used to remove the first few nanometres of silicon or phosphorus-doped dead layer from the surface. This was particularly important on the rear side because since there was no mask for the light phosphorus diffusion step, the whole rear surface including the p^{++} regions were doped with phosphorus atoms. The silicon etch therefore removed the possible shunt path between the emitter and base. Following this, an oxidation process was performed for drive-in of the diffused regions and surface passivation, followed by deposition of PECVD SiN_X on both sides. Despite eliminating the need for masked diffusions, photolithography was still used to define the contact openings to the heavily doped regions through a thermal oxide and SiN_X rear passivation stack

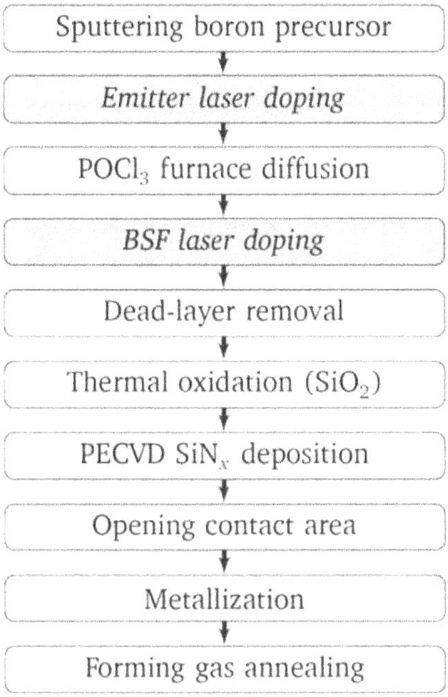

Figure 9.13 Process flow for the University of Stuttgart's laser-doped IBC cell. Adapted from [56].

prior to metallisation for the initial 22% cell. In 2016, this was replaced by laser openings to arrive at the 23.2% result. It was reported that the main loss mechanisms in this cell were associated with bulk resistance and bulk recombination, ahead of the losses associated with the laser-doped layers [72].

Another notable example of laser-doped IBC cells is the work from the Australian National University, where researchers developed an IBC cell that used a 249-nm pulsed excimer laser to dope both polarities of contacts prior to dielectric deposition, followed by local laser contact opening. This achieved 22.8% efficiency in 2017 [73]. However, this proof-of-concept cell still utilised photolithography for contact separation. In addition, work done by the Polytechnic University of Catalonia in recent years includes demonstration of a variety of IBC cell structures with one or both polarities of contacts formed with a pulsed Nd:YAG 1 064 nm laser:

1. An IBC cell on a p-type substrate with light thermal phosphorus diffusion, utilising laser doping from dopants from the dielectric layers for both contacts (ALD Al_2O_3 for p-type and phosphorus-doped PECVD amorphous silicon carbide (a-SiC_x(n)) for n-type) achieving 18.1% in 2015 [74]

2. An n-type IBC cell with the n-type contact formed via laser doping with a-SiC_x(n) and the p-type contact formed by conventional boron diffusion achieving 20.8% efficiency in 2016 [75]
3. An n-type IBC cell fabricated without the need for any high-temperature steps at all in a so-called cold process, again using dopants from these same layers to form both polarities of contacts named the DopLa-IBC (Doped by Laser-IBC) cell achieving 20.4% efficiency in 2017 [76].

9.4 Laser doping through doped surface layers of opposite polarity

Laser doping can also be performed to dope a surface that has already been diffused with the opposite polarity of dopants. In this case, a sufficient concentration of dopants of the opposite polarity to the surface layer (e.g. a thermally diffused emitter) must be present to overcompensate the dopants in that surface layer. It has been shown that with the appropriate laser-doping conditions, a deep p-type laser-doped region could be formed through an n-type-doped emitter layer [77, 78]. In these works, a 532-nm CW laser was used. The process is demonstrated schematically in Figure 9.14. First, a SOD source containing a high concentration of p-type dopants (e.g. boron) is applied to the wafer surface (Figure 9.14a). Laser doping is then performed, ablating the dielectric layer and melting the underlying silicon to incorporate p-type dopants from the SOD. These p-type dopants mix with the n-type dopants from the emitter, of which there are substantially less (Figure 9.14b). As the laser illumination continues, the melt front of the silicon travels further into the wafer, creating a deeper molten region (Figure 9.14c). When the laser illumination is removed, the molten silicon cools and resolidifies, leaving behind a compensated p-type laser-doped region with a certain JD and lateral doping width that acts as a contact to the underlying p-type silicon wafer (Figure 9.14d). Figure 9.15 shows combined SEM/EBIC images of this concept, where it is clear that the laser-doped region is p-type (no p-n junction is visible between the laser-doped region and the p-type silicon base).

The use of laser doping in this way to contact buried layers has the potential to further simplify the fabrication of otherwise complex solar cell structures such as the IBC cells discussed in Section 1.3. One concern with such a technique is the potential for a tunnel junction to form between the heavily doped p-type and n-type region [79]. Although no specific experiments have been performed to investigate the properties of the junction, the high implied V_{OC} obtained via photoluminescence measurements achieved on test structures using this method with boron as the dopant source to overcome a 100 Ω/□ phosphorus-doped emitter indicate minimal shunting behaviour introduced by the laser doping process [80]. While the estimated dopant concentrations at the surface of the laser-doped region and adjacent thermally diffused emitter region are expected to be high enough to lead to a tunnel shunt, it is theorised that due to the laser doping taking place in the molten state and local temperatures, hence, exceeding

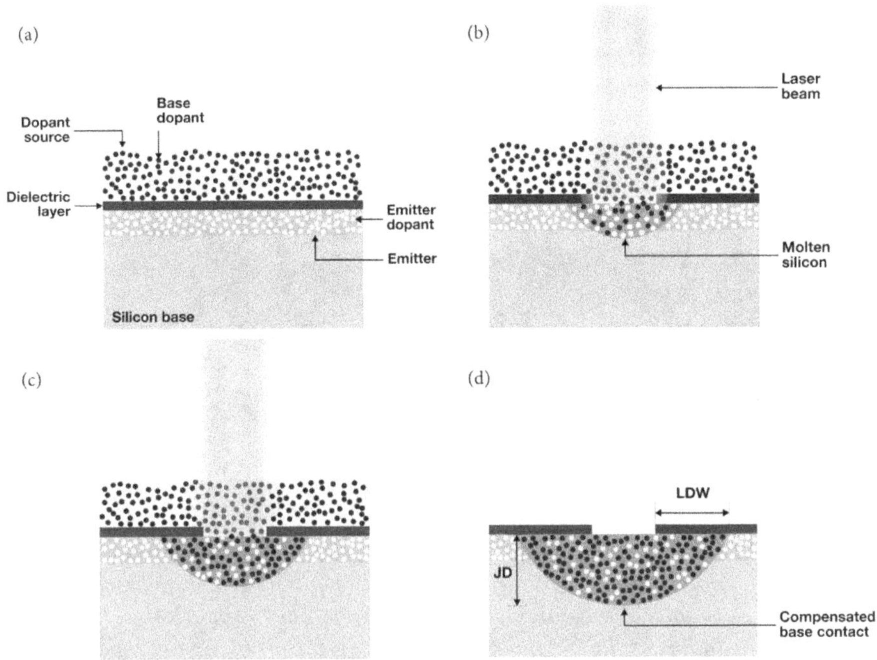

Figure 9.14 *Schematic of the process for forming contacts through an emitter of opposite polarity using laser doping by (a) coating the silicon wafer with a SOD source, (b) illuminating the surface with a laser, ablating the dielectric layer and melting the silicon surface, (c) continuing the laser illumination to expand the molten region and incorporate dopants and (d) the finished structure. Figure adapted from [34, 78].*

1400 °C, dopants from the adjacent emitter are able to diffuse into the molten region, and dopants from the molten region are able to diffuse in the solid state into the adjacent emitter. This would act to reduce the abruptness of the junction, potentially minimising the formation of a tunnel shunt [80]. Another concern with such a structure is the risk of the applied dopants not fully overcompensating the dopants in the existing emitter layer. This was indeed found to be the case for laser processing speeds above 0.5 m/s using the 532 nm CW laser. In this case, a p-n-p transistor structure was formed, as shown in Figure 9.16, where the two p-n junctions are clearly visible as bright regions in the EBIC signal. This is theorised to happen due to the large difference in diffusion coefficients of boron and phosphorus in molten silicon, with phosphorus having a diffusion coefficient almost five times higher than that of boron [25].

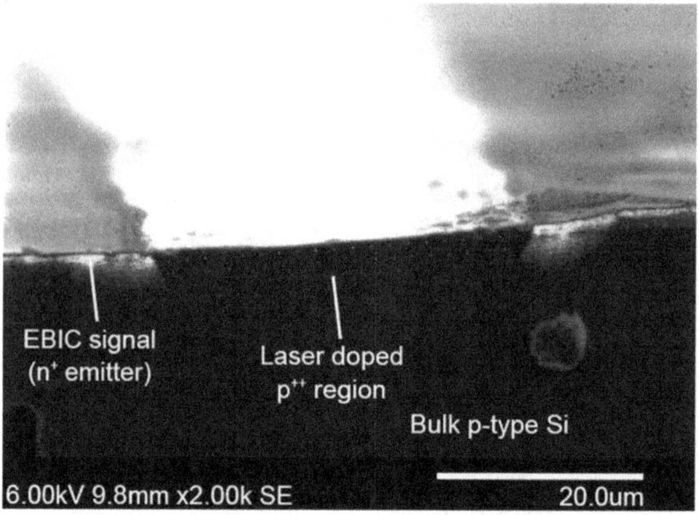

Figure 9.15 *Combined SEM/EBIC image of a cross section of a p-type laser-doped line formed through a 120 Ω/ϒ n-type emitter, making contact to the underlying p-type wafer. The bright EBIC signal indicates the p-n junction near the surface, with no p-n junction visible underneath the laser-doped region indicating full compensation of the n-type dopants in the laser-doped region. Image reproduced from [34].*

9.4.1 *Simplified interdigitated back-contact cells with laser-doped compensated contact*

A demonstration of a simplified laser-doped IBC cell was given in [80], whereby both polarities of contacts were formed via laser doping through a rear phosphorus emitter on a p-type substrate (see Figure 9.17). For this structure, only a single thermal diffusion step was required which simultaneously formed the rear n^+ emitter region and n^+ floating junction on the frontside for surface passivation. Forming the p-type laser-doped contacts directly through the rear n-type emitter enabled the device to have a high emitter coverage fraction – i.e. full emitter coverage except for the localised p-type contacts. A high emitter coverage fraction is advantageous to minimise 'electrical shading'. Although optical shading is eliminated in IBC solar cells due to the lack of front contacts, electrical shading is introduced as the emitter can no longer cover the whole solar cell area [81]; instead, the base contacts and non-diffused gaps between the base contacts and emitter need to exist. At these regions, carriers cannot be collected as there is no p-n junction. Therefore, minority carriers travelling through the base must diffuse further laterally to reach the emitter, and therefore have a higher chance of recombining before contributing to the device current. As a result, the quantum efficiency of areas above these regions is reduced

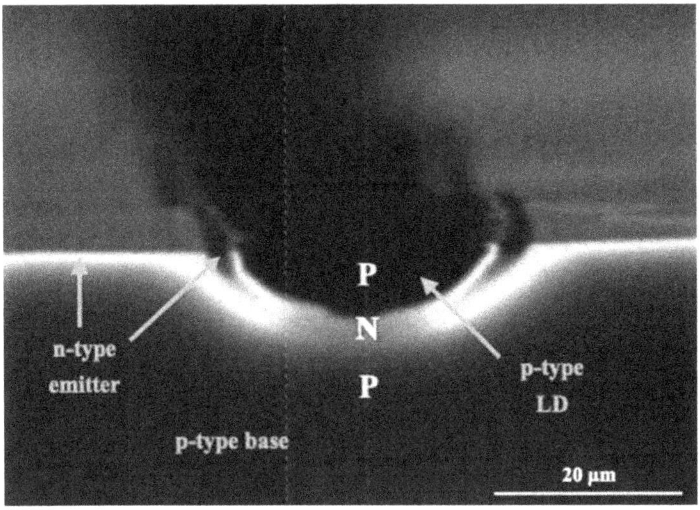

Figure 9.16 *EBIC image of a cross section of a p-type laser-doped line formed through a 100 Ω/Υ n-type emitter. The n-type emitter was not fully compensated by the p-type dopants, leaving behind a semi-circular p-n-p transistor structure indicated by the label 'PNP'. Image adapted from [80].*

where there is a lack of emitter. Hence the emitter coverage can greatly affect the device J_{SC} [82, 83].

Proof of concept devices was fabricated on p-type Cz wafers with a top efficiency of only 14.6%, limited by shunting after metal contact deposition due to pinholes formed at the laser-doped line edge regions in the silicon oxynitride passivating layer. However, modelling suggested that efficiencies of up to 22% could be achieved using this method on commercial p-type substrates.

9.4.2 Isolation of shunts or high recombination areas such as edges

The ability to laser dope through an emitter layer of opposite polarity not only allows contact to the underlying silicon base to be made but also opens up the

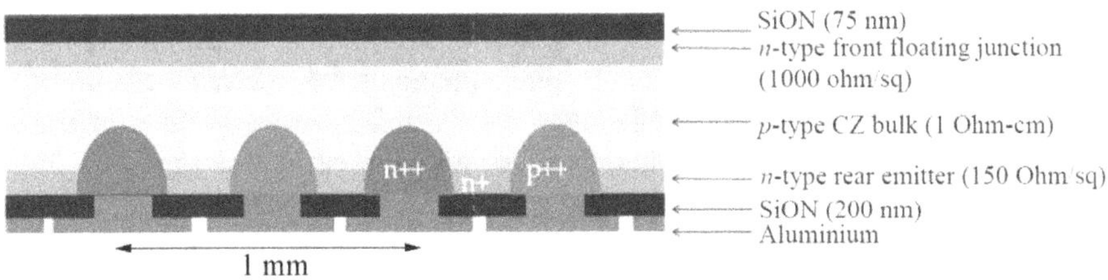

Figure 9.17 *Schematic of simplified laser-doped IBC cell. Reprinted from [77], Copyright (2012), with permission from Elsevier.*

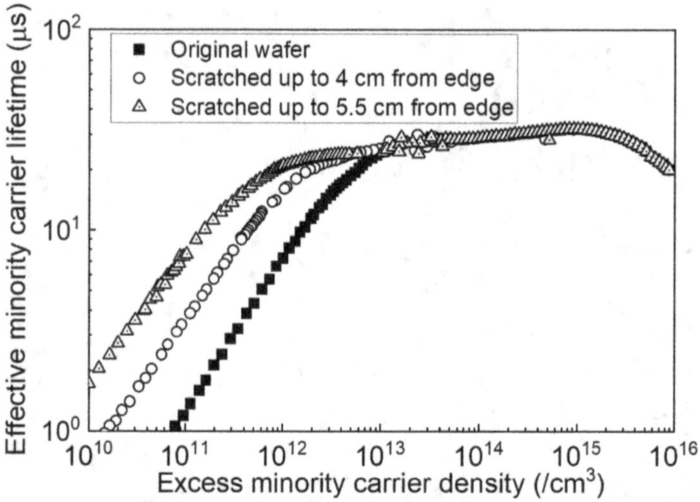

Figure 9.18 *Effective lifetime as a function of excess carrier density as measured by QSS-PL of a symmetrical phosphorus diffused p-type lifetime test structure after an increasing amount of damage is introduced outside the measurement test area. © 2014 IEEE. Reprinted, with permission, from [87].*

opportunity to electrically isolate the regions on either side of the laser-doped line from each other. This concept has been implemented using laser scribing, to locally isolate shunted regions of a solar cell, such as edges of industrial solar cells [84, 85] or other point-like shunts formed across a solar cell [86]. While these shunt regions are highly localised, the lateral conductivity of the emitter and the metal contacts allow carriers from much further away from the shunt or other highly recombination active regions to travel to the region, hence affecting large areas of the cell. This is particularly exacerbated at low-illumination intensities, as demonstrated in Figure 9.18 [87]. This figure shows injection-dependent lifetime data as measured by quasi-steady-state photoluminescence (QSS-PL) of the centre region of a 156 mm × 156 mm p-type lifetime test structure with symmetrical phosphorus diffusion passivated with PECVD SiN_X. Edge damage was increasingly applied in the form of scratches to the wafer surface using a diamond pen outside of the central measurement area. As can be seen from the data, the induced edge damage greatly affects the minority carrier lifetime at injection levels below $1 \times 10^{13}/cm^3$, demonstrating the ability for active recombination areas outside of the active measurement area to drag down the lifetime at low injection. While this has obvious implications for the low-level illumination performance of solar cells with poorly isolated or highly recombination active edges, and this is also not ideal for fundamental studies on lifetime structures where an understanding of the lifetime behaviour at low injection is required.

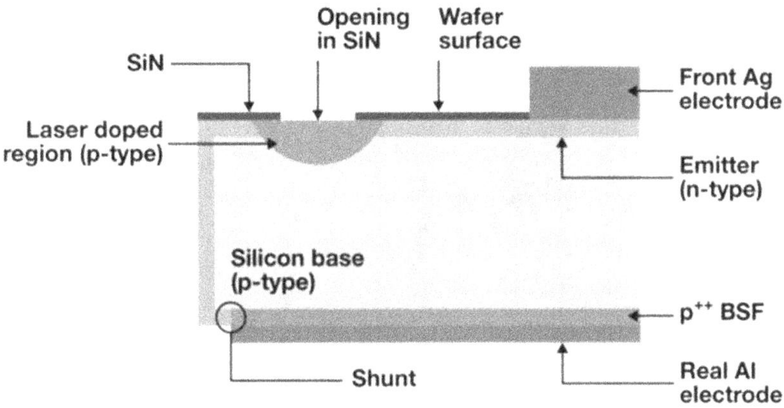

Figure 9.19 Schematic of edge isolation of a solar cell using laser doping. Reprinted from [91], Copyright (2015), with permission from Elsevier.

Edge isolation is an essential step in most solar cell fabrication sequences, removing the conductive path formed around the edges during emitter diffusion. If not removed, this conductive path shunts the cell, resulting in a low fill factor and hence efficiency. Edge isolation can be achieved in a variety of ways: in-line single-side wet chemical etching where the rear n-type surface is etched off [88], laser grooving of an isolation trench close to the perimeter of the solar cell [85], and plasma barrel etching. In laboratory scenarios, the edges of solar cells are commonly isolated through laser scribing from the rear side of the device followed by mechanical cleaving of the edges along the scribe line.

While these methods are very effective at removing the linear shunt path at the edges of the device, they typically introduce varying degrees of damage to the edges of the cell or lifetime structure leading to an increased J_{02}. This is particularly observable under low illumination, when there are fewer generated carriers that can consequently travel with low resistance to the edge regions to recombine [89, 90] (also refer to Figure 9.18). Additionally, all of these methods involve the physical removal of silicon to interrupt the conductive path for carriers to flow between the front and the rear of the device. Laser doping has been demonstrated as an alternative method for achieving isolation with minimal damage, which instead relies on the interruption of the conductive path by an overcompensated region of opposite polarity to the emitter layer [91]. In that work, small-area SPSCs were edge isolated using either laser doping (see Figure 9.19) or the standard laboratory method of laser scribing from the rear with mechanical cleaving. It was demonstrated that not only did the laser doping provide effective edge isolation removing the shunt path around the device (see Figure 9.20) but also the J_{02} associated with edge recombination was reduced by a factor of more than three when using laser doping compared to cleaving, and comparable to the J_{02} values reported when edges are ideally passivated using a thermal oxide.

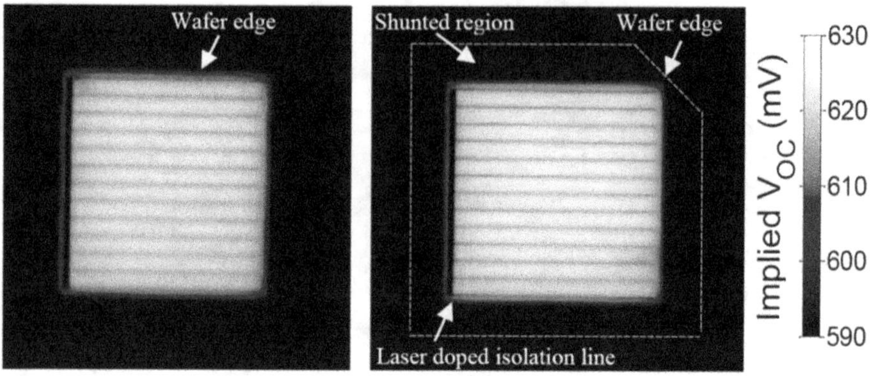

Figure 9.20 *Calibrated PL images showing a spatial map of iV_{OC} of (left) a solar cell with laser cleaved edges and (right) a solar cell within a larger substrate that has been isolated using laser doping. The edge of the host wafer (indicated by the dotted line) is not visible due to the edge shunt. Reprinted from [91], Copyright (2015), with permission from Elsevier.*

9.5 Laser doping for enhanced silicon gettering

One of the lesser explored uses of laser doping in solar cell fabrication is laser-enhanced impurity gettering. Laser-enhanced gettering (LEG) is a form of damage-induced gettering performed within the integrated circuit (IC) industry as a method for bulk silicon purification [92–96]. In this process, a laser beam is scanned across the rear side of the silicon substrate to introduce localised melted regions of elevated strain. This is commonly performed without a dopant source present. Scanning across the rear side allows for IC devices to be later fabricated on the frontside, uninfluenced by the laser damage. Subsequent to laser irradiation, the silicon wafers are exposed to a high-temperature cycle, typically above 700 °C, akin to emitter diffusion and oxidation processes experienced during solar cell manufacturing. According to Pearce *et al.* [93], the exposure to higher temperatures allows the release of stress from the laser-damaged regions. This stress propagates throughout the silicon wafer as a defect array consisting of stacking faults and dislocations, both of which act as effective gettering sites [95]. For most LEG processes, the optimal laser damage depth is typically between 1 μm and 15 μm, which have been observed to subsequently increase the minority carrier lifetime of the gettered wafer.

In solar cell fabrication, however, as the entire silicon substrate is utilised for the device, it is essential that defects are minimised at both surfaces and throughout the silicon bulk in the final solar cell structure. Ametowobla *et al.* observed that upon the introduction of laser damage, the primary reduction in carrier lifetime was due to an increase in the interface defect density. If this was the case, then by removing several microns of silicon from the surface of the wafer, the laser-induced damage could subsequently be removed and the τ_{eff} restored. This was later demonstrated

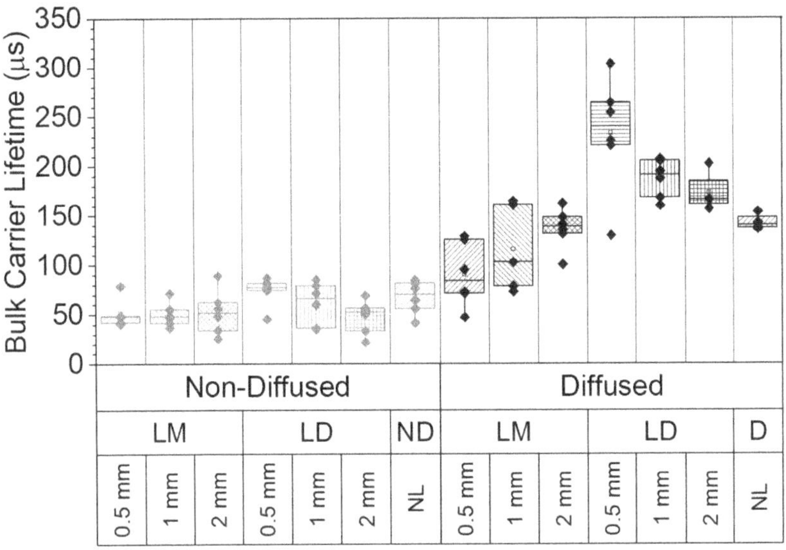

Figure 9.21 *Plot of τ_{bulk} vs. various laser doping or melting conditions with or without a thermal diffusion ($POCl_3$). Image reproduced from [99].*

in the work by Thorstensen *et al.* [97], showing a reversal in τ_{eff} with different etch depth. The impact of rear side LEG was then evaluated by Hayafuji *et al.* [92], showing an enhancement in the bulk carrier lifetime by over two orders of magnitude as the result of defect removal.

In 2017, Chen *et al.* [98] applied LEG to commercially available wafers used for the fabrication of solar cell devices. In that work, a 532-nm CW laser was used to introduce laser damage to the rear side of silicon wafers. Laser-irradiation was applied either directly onto the silicon wafer to create melted regions or through a film of spin-on 85% H_3PO_4 to generate heavily doped n^+ regions. A higher temperature step was then achieved using a conventional phosphorus emitter diffusion process. Once diffused, a re-texturing etch was used to remove the laser damage in addition to the emitters present on both sides of the wafers. Analysis of the τ_{bulk} on subsequently passivated samples showed that both the diffusion process and the presence of a dopant source during laser processing had a significant influence on the quality of the final device, in particular, in the case where samples were doped and underwent a diffusion process, having a greater density of laser-damage present on the surface, resulting in a fivefold improvement in carrier lifetime when compared to non-lasered and undiffused reference cells and a threefold enhancement when compared to an un-lasered but diffused control (see Figure 9.21).

The efficacy of different post-laser thermal annealing and emitter diffusion processes were also evaluated, using either a conventional screen-printed emitter diffusion, a diffusion processes used for lightly doped emitter typically used for selective emitter solar cells, or a dopant-less annealing procedure in N_2. Testing of both mc-Si and Cz-Si wafers suggested that in almost all scenarios, samples with the laser

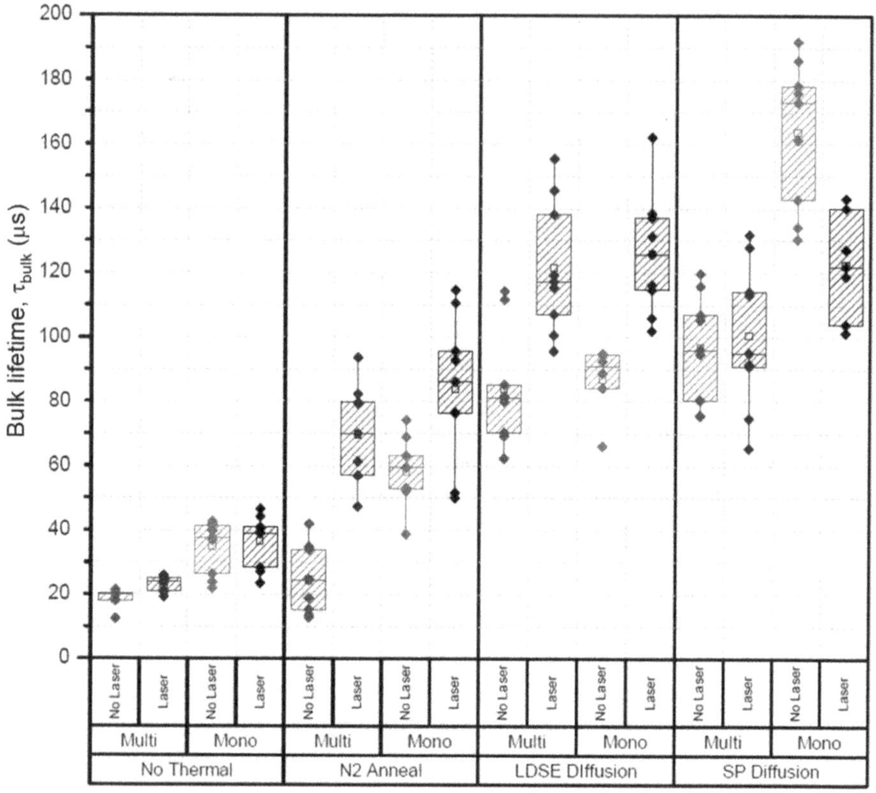

Figure 9.22 *τ_{bulk} comparison for Cz-Si and mc-Si wafers with and without rear side laser doping, undergoing various thermal annealing or $POCl_3$ diffusion processes. Reprinted by permission from Springer Nature Customer Service Centre GmbH: Springer Nature [98], © (2016). https://www.springer.com/journal/11708.*

treatment outperformed those without, with greater enhancements in τ_{bulk} on those with heavier phosphorus diffusions (see Figure 9.22). This would likely be a combination of gettering from the laser damage in addition to conventional phosphorus diffusion gettering (PDG).

Interstitial iron (Fe_i) concentrations were measured as a metric to determine the effectiveness of the gettering. As seen in Figure 9.23, negligible changes in [Fe_i] were recorded if samples were not exposed to high-temperature treatment, irrespective of whether laser damage was applied. Annealing of LEG treated wafers in an N_2 ambient immediately reduces [Fe_i] by a factor of two in mc-Si wafers and a factor of three in Cz-Si wafers. In comparison, on control samples without laser treatment, N_2 annealing increases the overall [Fe_i], likely due to the dissolution of pre-formed iron precipitates.

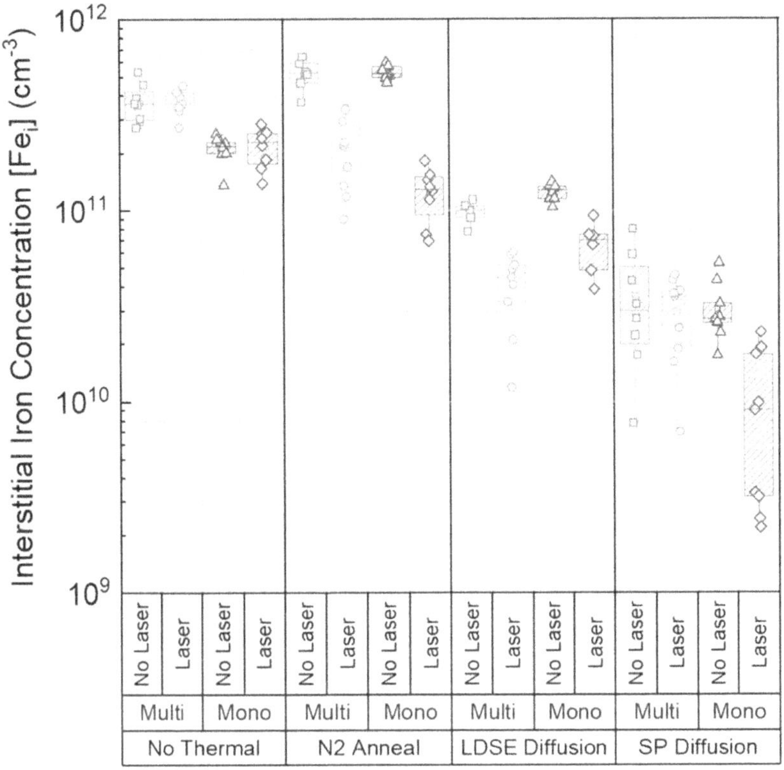

Figure 9.23 *[Fe_i] measured on Cz-Si and mc-Si wafers with and without rear side laser damage either undergoing phosphorus diffusion processing, an anneal in N2 ambient, or not receiving thermal treatment. Reprinted by permission from Springer Nature Customer Service Centre GmbH: Springer Nature [98], © (2016). https://www.springer.com/journal/11708.*

Application of an LDSE diffusion process produces a similar trend in [Fe_i], with almost an order of magnitude reduction in iron on laser-treated mc-Si ([Fe_i] reduction from $3.6 \times 10^{11}/cm^3$ to $4.5 \times 10^{10}/cm^3$). The effectiveness of gettering in a mc-Si wafer using LDSE diffusion conditions appears to be more effective than what is observed on Cz-Si, although this could be attributed to the additional getter sites present in the form of grain boundaries and dislocation clusters. The advantages of the LEG process, however, are less observable when a heavily doped emitter diffusion process is applied. This may be due to the greater potential of gettering as the result of the PDG process, consistent with the work by Phang *et al.*, demonstrating the correlation between peak phosphorus concentration and gettering efficiency. These results indicate that LEG processing may be more applicable for lightly diffused emitters that have recently become more mainstream in the photovoltaics industry.

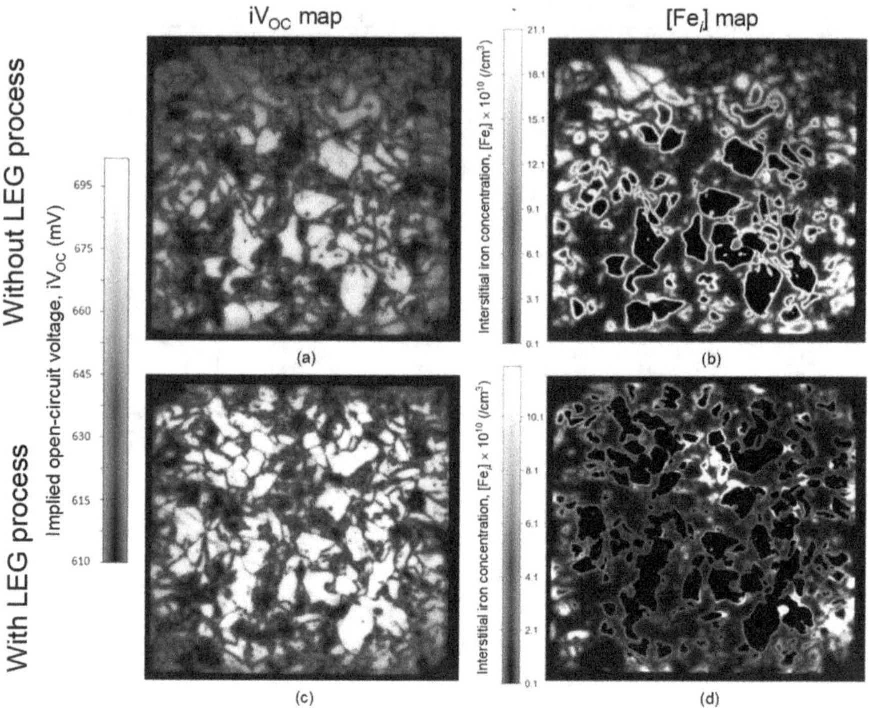

Figure 9.24 *iV_{OC}-calibrated photoluminescence image (left) and [Fe_i] map (right) of mc-Si wafers (a–b) without and (c–d) with a LEG process. Reprinted by permission from Springer Nature Customer Service Centre GmbH: Springer Nature [98], © (2016). https://www.springer.com/journal/11708.*

Both a qualitative and quantitative assessment of the effect of gettering through LEG can be gathered through calibrated PL images as depicted in Figure 9.24. In mc-Si wafers without LEG, large quantities of Fe_i can be seen decorating grain boundaries and some intra-grain regions once activated with a short exposure to illumination (Figure 9.24c). These concentrations are significantly reduced with the laser process (Figure 9.24d). However, to date, no such work has investigated the use of LEG on solar cell performance and how such a process will be integrated into a solar cell fabrication sequence.

9.6 Summary

Laser doping has a number of applications for silicon solar cells, with the potential to significantly simplify otherwise complex fabrication sequences that are too costly in a production environment, or to provide additional benefits to increase solar cell performance. Following on from selective emitter formation described in Chapter 8, this chapter explored the use of laser doping for BSF formation in industrial PERC solar cells and its associated benefits. The concept of using the rear AlO_X passivation layer as a source of p-type dopants for the local BSF laser-doping process was discussed. In particular, it was noted that the solid solubility of aluminium in silicon decreases with increasing temperatures above approximately 1 200 °C, unlike boron, which continues to increase with increasing temperature and, in addition, is generally over an order of magnitude higher than that of aluminium. Thus, the active dopant concentration achieved for aluminium laser doping (taking place at temperatures above the melting point of silicon of 1 414 °C) is limited in comparison to laser doping with boron. Although decent cell efficiencies have been achieved on PERC/PERL cells using aluminium laser doping for rear LBSF formation, future work could explore the potential for increasing the active aluminium dopant concentration in these devices by taking advantage of the higher solid solubility at temperatures in the range of 1 100–1 200 °C.

The use of laser doping to improve the conductivity of a lightly diffused surface was discussed, as in the SFSC devices, which can allow for a self-aligned metallisation scheme with screen-printed contacts. However, cell efficiencies are limited by very low metal–Si interface areas requiring low specific contact resistance values and progress with narrowing of screen-printed finger widths and, therefore, reduced finger spacing has reduced the benefit of the technology. A key area for further technical understanding currently missing from the literature includes a comprehensive study on the contact resistances achievable for both n-type and p-type laser-doped regions with screen-printed contacts for both fire-through and non-fire-through pastes, particularly when laser doping is performed in the liquid state.

Laser doping can also be used to greatly simplify the fabrication process of IBC solar cells, enabling localised FSF and BSF formation and potentially avoid masking and etching processes, or multiple lengthy high-temperature processes for diffusions. Efficiencies of 22.8% have been achieved. Further, the use of laser doping to penetrate a diffused layer of opposite polarity was discussed, with potential application to IBC cells to simplify processing even further. However, the development of bifacial designs for PERC, PERT and tunnel oxide passivated contact (TOPCon) cells are reducing the benefits of transitioning to an IBC structure.

The ability for the laser doping to penetrate through doped surface layers by over-doping to interrupt the conductive path of a diffused layer was discussed, leading to the application of this technique for edge isolation of solar cells or lifetime test structures. Lastly, the application of laser doping to enhance the gettering of bulk impurities was discussed, with little work done in this area for the application of solar cells, but as a potential to remove metallic impurities such as iron. This could become increasingly important with the transition to lighter emitter diffusions that are less suitable for impurity gettering. A key requirement for such processes would be ensuring the removal of the defected regions after the gettering process to ensure the performance of the final device is not affected.

References

[1] Green M.A. 'The Passivated emitter and rear cell (PERC): from conception to mass production'. *Solar Energy Materials and Solar Cells*. 2015;**143**(2):190–7.

[2] Abbott M.D., Cotter J.E., Fisher K. 'N-type bifacial solar cells with laser doped contacts'. 2006 IEEE 4th World Conference on Photovoltaic Energy Conference; Waikoloa, USA; 2006. pp. 988–91.

[3] Hameiri Z., Tjahjono B., Mai L., Sugianto A., Javid S., Wang S. 'Double sided laser doped solar cells'. 18th International Photovoltaic Science and Engineering Conference; Kolkata, India; 2009.

[4] Ebong A.U., Lee S.H., Honsberg C., Wenham S.R. 'High efficiency double sided buried contact silicon solar cells'. *Japanese Journal of Applied Physics*. 1996;**35**(Part 1, No. 4A):2077–80.

[5] Duran C., Eisele S.J., Buck T., Kopecek R., Köhler JR W.J.H. 'Bifacial solar cells with selective B-BSF by laser doping'. *24th European Photovoltaic Solar Energy Conference and Exhibition; Hamburg*. 2009:1775–8.

[6] Wang X., Allen V., Vais V., *et al.* 'Laser-doped metal-plated bifacial silicon solar cells'. *Solar Energy Materials and Solar Cells*. 2014;**131**:37–45.

[7] Li H., Wenham S.R., Shi Z. 'High efficiency PERL cells on CZ P-type crystalline silicon using a thermally stable a-Si:H/SiNx rear surface passivation stack'. *Solar Energy Materials and Solar Cells*. 2013;**117**:41–7.

[8] Lin D., Abbott M., Lu P.H., *et al.* 'Incorporation of deep laser doping to form the rear localized back surface field in high efficiency solar cells'. *Solar Energy Materials and Solar Cells*. 2014;**130**(Part A):83–90.

[9] Suwito D., Jäger U., Benick J., Janz S., Hermle M., Preu R. 'Industrially feasible rear side concept for N-type silicon solar cells approaching 700 mV of VOC'. *Proceedings of the 25th European Photovoltaic Solar Energy and Exhibition. Valencia, Spain*. 2010:1443–5.

[10] Gall S., Manuel S., Lerat J.-F. 'Boron laser doping through high quality Al_2O_3 passivation layer for localized B-BSF PERL solar cells'. *Energy Procedia*. 2013;**38**:270–7.

[11] PHD L., Lin D., Wang X., Lennon A., Wenham S. 'Laser doping through anodic aluminium oxide silicon solar cell'. *Solar Energy Materials and Solar Cells : an International Journal Devoted to Photovoltaic, Photothermal, and Photochemical Solar Energy Conversion*. 2016;**145**:349–57.

[12] Cornagliotti E., Uruena A., Hallam B., *et al.* 'Large area P-type PERL cells featuring local P+ BSF formed by laser processing of ALD Al_2O_3 layers'. *Solar Energy Materials and Solar Cells*. 2015;**138**(6):72–9.

[13] Ernst M., Walter D., Fell A., Lim B., Weber K. 'Efficiency potential of P-type $Al_2\,O_3$/SiN passivated PERC solar cells with locally laser-doped rear contacts'. *IEEE Journal of Photovoltaics*. 2016;**6**(3):624–31.

[14] Kluska S., Granek F. 'High-efficiency silicon solar cells with boron local back surface fields formed by laser chemical processing'. *IEEE Electron Device Letters*. 2011;**32**(9):1257–9.

[15] Tomizawa Y., Imamura T., Soeda M., Ikeda Y., Shiro T. 'Laser doping of boron-doped Si paste for high-efficiency silicon solar cells'. *Japanese Journal of Applied Physics*. 2015;**54**(8S1):08KD06.

[16] Wang Z., Han P., Lu H., *et al.* 'Advanced PERC and PERL production cells with 20.3% record efficiency for standard commercial p-type silicon wafers'. *Progress in Photovoltaics: Research and Applications*. 2012;**20**(3):260–8.

[17] Ortega P., Martín I., Lopez G., *et al.* 'P-Type c-Si solar cells based on rear side laser processing of Al_2O_3/SiCx stacks'. *Solar Energy Materials and Solar Cells*. 2012;**106**(7):80–3.

[18] Benick J., Steinhauser B., Müller R., *et al.* 'High efficiency n-type PERT and PERL solar cells'. 2014 IEEE 40th Photovoltaic Specialist Conference (PVSC); Denver, CO, USA, 08-06-2014; 2014.

[19] Jaeger U., Wolf A., Wufka C., Tomizawa Y., Imamura T., Soeda M. 'Local boron doping for P-type PERL silicon solar cells fabricated by laser processing of doped silicon nanoparticle paste'. 29th European Photovoltaic Solar Energy Conference; Amsterdam, The Netherlands; 2014. pp. 1111–5.

[20] Zhao J., Wang A., Green M.A. '24·5% Efficiency silicon PERT cells on MCZ substrates and 24·7% efficiency PERL cells on FZ substrates'. *Progress in Photovoltaics: Research andApplications*. 1999;**Nov;7**(6):471–4.

[21] Agostinelli G., Szlufcick J., Choulat P., Beaucarne G. 'Local contact structures for industrial PERC-type solar cells'. Proceedings of the 20th European Photovoltaic and Solar Energy Conference; Barcelona, Spain; 2005. pp. 942–5.

[22] Agostinelli G., Choulat P., Dekkers H.F.W., De Wolf S., Beaucarne G. 'Screen printed large area crystalline silicon solar cells on thin substrates'. *Proceedings of the 20th European Photovoltaic and Solar Energy Conference. Barcelona, Spain*. 2005:647–50.

[23] Lölgen P., Sinke W.C., Leguijt C., Weeber A.W., Alkemade P.F.A., Verhoef L.A. 'Boron doping of silicon using coalloying with aluminium'. *Applied Physics Letters*. 1994;**65**(22):2792–4.

[24] Rauer M., Schmiga C., Glatthaar M., Glunz S.W. 'Alloying from Screen-Printed aluminum pastes containing boron additives'. *IEEE Journal of Photovoltaics*. 2013;**3**(1):206–11.

[25] Tang K., Øvrelid E.J., Tranell G., Tangstad M. 'Thermochemical and kinetic databases for the solar cell silicon materials' in Nakajima K., Usami N. (eds.). *Crystal Growth of Si for Solar Cells*Springer Berlin Heidelberg; 2009. pp. 219–51.

[26] PV Lighthouse. EDNA 2 [online]. 2014. Available from www.pvlighthouse.com.au [Accessed 10 September 2020].

[27] Urrejola E., Peter K., Plagwitz H., Schubert G. 'Silicon diffusion in aluminum for rear passivated solar cells'. *Applied Physics Letters*. 2011;**98**(15):153508.

[28] Chen D., Deng W., Sheng J., Zhu H., Zhong M., Wang W. Preventing the formation of voids in the rear local contact areas for industrial-type perc solar cells. 28th European Photovoltaic Solar Energy Conference and Exhibition; Paris, France; 2013. pp. 770–4.

[29] Lin D. *Advanced Metallization for Screen-Printed Solar Cells*. Kensington: UNSW Sydney; 2014.

[30] Rauer M., Woehl R., Ruhle K., *et al.* 'Aluminum alloying in local contact areas on dielectrically passivated rear surfaces of silicon solar cells'. *IEEE Electron Device Letters*. 2011;**32**(7):916–18.

[31] Rauer M., Schmiga C., Woehl R., *et al.* 'Investigation of aluminum-alloyed local contacts for rear surface-passivated silicon solar cells'. *IEEE Journal of Photovoltaics*. 2011;**1**:22–8.

[32] Tomizawa Y., Ikeda Y., Itoh H., *et al.* 'Analysis of contact recombination at rear local back surface field via boron laser doping and screen-printed aluminum metallization on P-type PERC solar cells'. *Energy Procedia*. 2017;**124**:384–91.

[33] Schneiderlöchner E., Preu R., Lüdemann R., Glunz S.W. 'Laser-fired rear contacts for crystalline silicon solar cells'. *Progress in Photovoltaics: Research and Applications*. 2002;**10**(1):29–34.

[34] Hallam B. *High Efficiency Laser-Doped Silicon Solar Cells with Advanced Hydrogenation. PHD thesis*. Kensington: University of New South Wales; 2014.

[35] Zhao J., Wang A., Green M.A. 'Series resistance caused by the localized rear contact in high efficiency silicon solar cells'. *Solar Energy Materials and Solar Cells*. 1994;**32**(1):89–94.

[36] Hallam B.J., Hamer P.G., Wang S., *et al.* 'Advanced hydrogenation of dislocation clusters and boron-oxygen defects in silicon solar cells'. *Energy Procedia*. 2015;**77**(10):799–809.

[37] Bennett T.D., Krajnovich D.J., Grigoropoulos C.P., Baumgart P., Tam A.C. 'Marangoni mechanism in pulsed laser texturing of magnetic disk substrates'. *Journal of Heat Transfer*. 1997;**119**(3):589–96.

[38] Agostinelli G., Delabie A., Vitanov P., *et al.* 'Very low surface recombination velocities on P-type silicon wafers passivated with a dielectric with fixed negative charge'. *Solar Energy Materials and Solar Cells*. 2006;**90**(18):3438–43.

[39] Hoex B., Heil S.B.S., Langereis E., van de Sanden M.C.M., Kessels W.M.M. 'Ultralow surface recombination of c-Si substrates passivated by plasma-assisted atomic layer deposited Al_2O_3'. *Applied Physics Letters*. 2006;**89**(4):042112.

[40] Lu D., Xiao B., Lu Z., Wang K., Lennon A., Wenham S. 'Laser doping through anodic aluminium oxide layers formed on silicon solar cells'. Proceedings of the 21st IEEE Photovoltaic Science and Engineering Conference; Fukuoka, Japan; 2011.

[41] Fell A., Franklin E., Walter D., Suh D., Weber K. 'Laser doping from Al_2O_3 layers'. 27th European Photovoltaic Solar Energy Conference; Frankfurt, Germany; 2012. pp. 706–8.

[42] Harder N.-P., Larionova Y., Brendel R. 'Al+-doping of Si by laser ablation of Al_2O_3/SiN passivation'. *Physica status solidi*. 2013;**210**(9):1871–3.

[43] Martin I., Ortega P., Colina M., Orpella A., López G., Alcubilla R. 'Laser processing of Al_2O_3/a-SiCx: H stacks: a feasible solution for the rear surface of

high-efficiency P-type c-Si solar cells'. *Progress in Photovoltaics: Research and Applications*. 2013;**21**(5):1171–5.

[44] Hameiri Z., Puzzer T., Mai L., Sproul A.B., Wenham S.R. 'Laser induced defects in laser doped solar cells'. *Progress in Photovoltaics: Research and Applications*. 2011;**19**(4):391–405.

[45] Mai L., Wenham S.R., Tjahjono B., Ji J., Shi Z. 'New emitter design and metal contact for screen-printed solar cell front surfaces'. *2006 IEEE 4th World Conference on Photovoltaic Energy Conference*. 2006:890–3.

[46] Australian Technology Improves Solar Panels. *The Generator [online]*. 2007. Available from https://thegenerator.news/australian-technology-improves-solar-panels/ [Accessed 20 May 2021].

[47] Chong C.M., Wenham S.R., Green M.A. 'High-Efficiency, laser grooved, buried contact silicon solar cells'. *Applied Physics Letters*. 1988;**52**(5):407–9.

[48] Hanke I., Apel M., Schroter W. 'Influence of annealing conditions on recombination properties and diffusion length in cast multicrystalline silicon for solar cells'. Proceedings of the 14th European Photovoltaic Solar Energy Conference; Barcelona, Spain; 1997. pp. 735–8.

[49] Macdonald D., Cuevas A. 'The trade-off between phosphorus gettering and thermal degradation in multicrystalline silicon'. 16th European Photovoltaic Solar Energy Conference; Glasgow, UK; 2000. pp. 1707–10.

[50] Mai L. *Overcoming the Performance and Limitations of Commercial Screen-Printed Solar Cells*. Kensington: UNSW; 2010.

[51] Mai L., Mitchell E.J., Wang K.S., Lin D., Wenham S. 'The development of the advanced semiconductor finger solar cell'. *Progress in Photovoltaics: Research and Applications*. 2014;**22**(12):1195–203.

[52] Wang K.S. *From the Laser-doped semiconductor fingers to the advanced semiconductor fingers silicon solar cell [PhD Thesis]*. Sydney:University of New South Wales; 2012.

[53] Wang K.S., Lin D., XR A., Mai L., Mitchell E., Wenham S.R. 18.8 % efficient laser-doped semiconductor fingers screen-printed silicon solar cell with light-induced plating. 2012 38th IEEE Photovoltaic Specialists Conference; Austin, USA; 2012. pp. 1149–53.

[54] VDMA. *International technology roadmap for photovoltaic (ITRPV)*. 11th edition. Frankfurt, Germany: VDMA; 2020.

[55] Zhu L.Q., Gong J., Huang J., *et al.* 'Improving the efficiency of crystalline silicon solar cells by an intersected selective laser doping'. *Solar Energy Materials and Solar Cells*. 2011;**95**(12):3347–51.

[56] Dahlinger M., Bazer-Bachi B., Röder T.C., Köhler J.R., Zapf-Gottwick R., Werner J.H. 'Laser-doped back-contact solar cells'. *IEEE Journal of Photovoltaics*. 2013;**5**(3):812–18.

[57] Verlinden P.J., Swanson R.M., Crane R.A. '7000 high-eficiency cells for a dream'. *Progress in Photovoltaics: Research and Applications*. 1994;**2**(2):143–52.

[58] Swanson R.M. 'Point-contact solar cells: modeling and experiment'. *Solar Cells*. 1986;**17**(1):85–118.

[59] Mulligan W.P., Rose D.H., Cudzinovic M.J., De Ceuster D.M., McIntosh K.R., Smith D.D. Manufacture of solar cells with 21% efficiency. 19th European Photovoltaic Solar Energy Conference; Paris, France; 2004. pp. 387–90.

[60] Cousins P. Lab to GW fab for 25% back-contacted n-type cells. *PV Cell Tech Conference*; Kuala Lumpur, Malaysia; 2016.

[61] SunPower®. *X-Series solar panels [online]*. 2020. Available from https://us.sunpower.com/solar-panels-technology/x-series-solar-panels [Accessed 21 May 2021].

[62] Smith D.D. 'Review of back contact silicon solar cells for low-cost application'. *Proceedings of the 16th European Photovoltaic Solar Energy Conference. Glasgow, Scotland.* 2000:1104–7.

[63] Kopecek R., Libal J. 'Towards large-scale deployment of bifacial photovoltaics'. *Nature Energy*. 2018;**3**(6):443–6.

[64] Harder N.-P., Mertens V., Brendel R. 'Buried emitter solar cell structures: decoupling of metallisation geometry and carrier collection geometry of back contacted solar cells'. *Physica Status Solidi (RRL) – Rapid Research Letters*. 2008;**2**(4):148–50.

[65] Cousins P.J., Smith D.D., Luan H.-C., Manning J., Dennis T.D., Waldhauer A. 'Generation 3: improved performance at lower cost'. Proceedings of the 35th IEEE Photovoltaic Specialists Conference; Honolulu, USA; 2010. pp. 275–8.

[66] Sinton R.A., Swanson R.M. 'Simplified backside-contact solar cells'. *IEEE Transactions on Electron Devices*. 1990;**37**(2):348–52.

[67] Green M.A., Basore P.A., Chang N., *et al.* 'Crystalline silicon on glass (CSG) thin-film solar cell modules'. *Solar Energy*. 2004;**77**(6):857–63.

[68] Altermatt P.P., Heiser G., Dai X., *et al.* 'Rear surface passivation of high-efficiency silicon solar cells by a floating junction'. *Journal of Applied Physics*. 1996;**80**(6):3574–86.

[69] Engelhart P., Harder N.-P., Grischke R., Merkle A., Meyer R., Brendel R. 'Laser structuring for back junction silicon solar cells'. *Progress in Photovoltaics: Research and Applications*. 2007;**15**(3):237–43.

[70] Guo J.-H. *High efficiency N-type laser-grooved buried contact silicon solar cells. [PhD Thesis]*. Kensington: UNSW; 2004.

[71] Wenham S.R., Robinson S.J., Dai X., Zhao J., Wang A., Tang Y.H. 'Rear surface effects in high efficiency silicon solar cells'. *Proceedings of the 18th IEEE Photovoltaic Specialists Conference*. 1994:1278–82.

[72] Dahlinger M., Carstens K., Hoffmann E., Zapf-Gottwick R., Werner J.H. '23.2% laser processed back contact solar cell: fabrication, characterization and modeling'. *Progress in Photovoltaics: Research and Applications*. 2017;**25**(2):192–200.

[73] Ernst M., Franklin E., Fell A., *et al.* 'Fabrication of a 22.8% efficient back contact solar cell with localized laser-doping'. *Physica Status Solidi*. 2017;**214**(11):1700318.

[74] López G., Ortega P.R., Martín I., *et al.* 'Base contacts and selective emitters processed by laser doping technique for P-type IBC c-Si solar cells'. *Energy Procedia*. 2015;**77**(1):752–8.

[75] Masmitja G., Ortega P., Martín I., López G., Voz C., Alcubilla R. 'IBC c-Si(n) solar cells based on laser doping processing for selective emitter and base contact formation'. *Energy Procedia*. 2016;**92**:956–61.

[76] Ortega P., López G., Muñoz D., *et al.* 'Fully low temperature interdigitated back-contacted c-Si(n) solar cells based on laser-doping from dielectric stacks'. *Solar Energy Materials and Solar Cells*. 2017;**169**(6):107–12.

[77] Chan C.E., Hallam B.J., Wenham S.R. 'Simplified interdigitated back contact solar cells'. *Energy Procedia*. 2012;**27**(2):543–8.

[78] Hallam B., Chan C., Sugianto A., Wenham S. 'Deep junction laser doping for contacting buried layers in silicon solar cells'. *Solar Energy Materials and Solar Cells*. 2013;**113**(6):124–34.

[79] Esaki L. 'Discovery of the tunnel diode'. *IEEE Transactions on Electron Devices*. 1976;**23**(7):644–7.

[80] Chan C.E. *Novel Laser Doping Technology for Silicon Solar Cells*. Kensington: The University of New South Wales; 2014.

[81] Hermle M., Granek F., Schultz-Wittmann O., Glunz S.W. 'Shading effects in back-junction back-contacted silicon solar cells'. 2008 33rd IEEE Photovoltaic Specialists Conference; San Diego, USA; 2008. pp. 1–4.

[82] Kim D.S., Meemongkolkiat V., Ebong A., Rounsaville B., Upadhyaya V., Das A. '2D-modeling and development of interdigitated back contact solar cells on low-cost substrates'. 2006 IEEE 4th World Conference on Photovoltaic Energy Conference; Waikoloa, USA; 2006. pp. 1417–20.

[83] Granek F., Hermle M., Reichel C., Schultz-Wittmann O., Glunz S. 'High-efficiency back-contact back-junction silicon solar cell research at Fraunhofer ISE'. Proceedings of the 23rd European Photovoltaic Solar Energy Conference; Valencia, Spain; 2008. pp. 991–5.

[84] Hauser A., Hahn G., Spiegel M., Feist H., Breitenstein O., Rakotoniaia J. 'Comparison of different techniques for edge isolation'. 17th European Photovoltaic and Solar Energy Conference; Munich, Germany; 2001. pp. 1739–46.

[85] Schneiderlöchner E., Grohe A., Glunz S., Preu R., Willeke W. 'Scanning Nd:YAG laser system for industrially applicable processing in silicon solar cell manufacturing'. 3rd World Conference on Photovoltaic Energy Conversion; Osaka, Japan; 2003. pp. 1364–7.

[86] Abbott M.D., Trupke T., Hartmann H.P., Gupta R., Breitenstein O. 'Laser isolation of shunted regions in industrial solar cells'. *Progress in Photovoltaics: Research and Applications*. 2007;**15**(7):613–20.

[87] Chan C.E., Abbott M.D., Juhl M.K., Hallam B.J., Xiao B., Wenham S.R. 'Assessing the performance of surface passivation using low-intensity photoluminescence characterization techniques'. *IEEE Journal of Photovoltaics*. 2014;**4**(1):100–6.

[88] Delahaye F., Löhmann M., Bauer M., Vilsmeier G., Melnyk I., Hauser A. 'Edge isolation: innovative inline wet processing-ready for industrial production'. Proceedings of the 19th European Photovoltaic Solar Energy Conference; Paris, France; 2004. pp. 416–18.

[89] Hermle M., Dicker J., Warta W., Glunz S.W., Willeke G. 'Analysis of edge recombination for high-efficiency solar cells at low illumination densities'. 3rd World Conference on Photovoltaic Energy Conversion; Osaka, Japan; 2003. pp. 1009–12.

[90] Glunz S.W., Dicker J., Esterle M., Hermle M., Isenberg J., Kamerewerd F.J. 'High-efficiency silicon solar cells for low-illumination applications'. Conference Record of the Twenty-Ninth IEEE Photovoltaic Specialists Conference; New Orleans, USA; 2002. pp. 450–3.

[91] Chan C., Abbott M., Hallam B., *et al.* 'Edge isolation of solar cells using laser doping'. *Solar Energy Materials and Solar Cells*. 2015;**132**:535–43.

[92] Hayafuji Y., Yanada T., Aoki Y. 'Laser damage gettering and its application to lifetime improvement in silicon'. *Journal of The Electrochemical Society*. 1981;**128**(9):1975–80.

[93] Pearce C., Zaleckas V. 'Gettering semiconductor wafers with a high energy laser beam'. *United States patent US 4,131,487*. 1978.

[94] Schwuttke G.H. 'High efficiency gettering in silicon through localized superheated melt formation'. *United States patent US 4,257,827*. 1981.

[95] Takemura K., Toyokawa F., Mikami M. *Method of Gettering Semiconductor Wafers with a Laser Beam. EU; 0251280*; 1987.

[96] Yang K.H., Schwuttke G.H. 'Minority carrier lifetime improvement in silicon through laser damage gettering'. *Physica Status Solidi*. 1980;**58**(1):127–34.

[97] Thorstensen J., Foss S.E. 'Investigation of depth of laser damage to silicon as function of wavelength and pulse duration'. *Energy Procedia*. 2013;**38**:794–800.

[98] Chen D., Edwards M., Wenham S., Abbott M., Hallam B. 'Laser enhanced gettering of silicon substrates'. *Frontiers in Energy*. 2017;**11**(1):23–31.

[99] Chen D. *Improved Performance of Low-Cost Silicon Substrates Through Advanced Laser Processing*. Kensington: UNSW; 2016.

Chapter 10
Conclusion and future outlook

Matthew Wright[1], Brett Hallam[1], and Catherine Chan[1]

Addressing the issue of climate change is critical for the future survival of humanity. The significant increase in global mean temperature since the beginning of the industrial revolution is caused primarily by an increase in the amount of CO_2 present in the atmosphere. This increased CO_2 can be attributed to the use of fossil fuels for transportation and the production of electricity. Thus, to reduce CO_2 emissions and limit the impact of climate change, it is important to shift to renewable forms of energy. Among the possible forms of renewable energy, solar energy possesses many advantages. The solar resource is abundant, and the operation of a solar cell in the field does not lead to emissions. As well as the environmental aspect, over the past ten years, the cost of electricity produced using solar power has significantly reduced. This is because the price of solar panels reduced by a factor of ~15 between 2008 and 2018 [1]. In multiple regions, solar energy is becoming more cost-effective than conventional power-generation methods. There are many factors that have driven down the price of solar, among them are the technological improvements made in the process of making commercial silicon solar cells. The rapid reduction in price has skyrocketed the projections of solar energy installed in the future. Multiple reports suggest that harnessing the Sun's energy will not be limited to the generation of electricity. Another industry that may be powered by solar energy is transportation, as we see the electrification of not only private vehicles but also trucks for freight and buses/trains for public transport [2]. It could also be used for the generation of liquid hydrogen [3]. As the scope of solar energy's role in future energy generation diversifies, the demand for industrial silicon solar cells will increase [4], and some experts predict that the amount of solar photovoltaics (PV) installed by 2050 will be in the range of 30–50 TW_p [5]. To achieve this, the solar cell industry will need to adapt to meet this new demand, which increases the importance of industrially relevant processes to improve the performance of current commercial solar cells.

[1]The School of Photovoltaic and Renewable Energy Engineering, The University of New South Wales, Sydney, NSW, Australia

10.1 Importance of hydrogenation and laser doping for silicon solar cells

Over the past 20 years, the structure of industrial silicon solar cells has remained relatively unchanged. Until 2019, the market had been completely dominated by silicon solar cells manufactured using the aluminium-back surface field (Al-BSF) approach. The manufacturing process flow for Al-BSF solar cells is relatively simple, which leads to low manufacturing costs. Decades of optimisation, slight processing tweaks and the ongoing development of materials such as screen-printed pastes have led to progressive improvements in the performance of Al-BSF cells of approximately $0.5\%_{abs}$ per year to achieve efficiencies of almost 20% [6]. However, the efficiency is fundamentally limited by the rear surface due to the direct contact between the silicon and metal over the whole area of the device. The significant surface recombination velocity (SRV) at the metal–Si interface limits the achievable open-circuit voltage (V_{OC}) of this design and hence efficiency.

As such, the efficiencies of industrial Al-BSF solar cells are significantly lower than that of record cells fabricated in the laboratory using more sophisticated cell structures. A key example technology is the passivated emitter, rear locally diffused (PERL) cell developed at UNSW, which held the world record for the highest efficiency single-junction silicon solar cell from 1991 until 2014, with an efficiency of 25%. There are several factors that allow for this cell structure to achieve higher efficiencies than industrial Al-BSF solar cells.

1. Passivation on the rear surface using a dielectric layer with localised openings to reduce the meta–Si interface contact area
2. The local diffused layers at the contact regions, which provide field-effect passivation of the metal–Si interface
3. The selective emitter design on the front surface, which reduces the emitter component of J_0 and improves short wavelength response
4. The use of hydrogenation to passivate defects at the silicon–dielectric interface.

This design represents the state of the art in terms of diffused single-junction solar cells with direct metal–Si interfaces and contains many processing steps that are not amenable with large scale commercial production of silicon solar cells. It would be ideal to incorporate the benefits of the PERL cell design into production; however, some compensations must be made in the industrial setting. In the same family as the PERL cell is the passivated emitter and rear cell (PERC), also developed at UNSW. Although in its first instance, PERC featured a direct localised contact between the rear aluminium layer and a more heavily doped bulk p-type wafer, these days, 'PERC' can be used to represent the family of the traditional PERC solar cell, as well as the passivated emitter, rear totally diffused (PERT) cell and PERL cell. The PERC family, with a passivated rear design, produces much higher efficiency than Al-BSF. However, in the industrial context, the PERC solar cell mimics aspects of the PERL cell, including localised heavily doped Al-BSF regions at the p-type contact, which can be fabricated in a low-cost manner. The significant advantage of PERC solar cells compared to Al-BSF is the reduced metal–Si contact area on the rear surface, which allows for a higher V_{OC}. The process flow for PERC cells requires only minor modifications compared to the conventional Al-BSF structure, which means the manufacturing cost and added capital expenditure (CAPEX) are kept relatively low, and knowledge gained from the fabrication of Al-BSF solar cells can be directly translated for the manufacture of PERC solar cells. The higher efficiency potential and relatively low manufacturing costs have seen PERC solar cells rise to dominance in recent years. It is predicted that the high-efficiency potential and relatively low cost means that PERC solar cells will continue to be the most prominent technology for years to come. Rapid improvements have seen PERC cells in production demonstrated with efficiency exceeding 23%, again with improvements in the efficiency on the range of $0.5\%_{abs}$ per year. However, to date, the majority of industrial PERC solar cells employ a homogeneous emitter.

To push this towards the 25% limit achieved on the record laboratory PERL cell, there are two key aspects that must be addressed. The design of the emitter is crucial in further improving the voltage of commercially produced PERC cells. Efficiency enhancements have largely been driven by the use of a selective emitter design, where the doping density on the front surface is optimised to minimise recombination losses restricting the use of heavily doped regions to the vicinity of the metal contacts, allowing for a more transparent, although more resistive emitter in passivated regions. While several industrially relevant approaches have been developed to form a selective emitter, laser doping has risen to be the most popular technique used in production in recent years [7].

In a commercial setting, PERC cells are predominantly being fabricated using p-type Czochralski (Cz) silicon wafers. This is because p-type Cz wafers are cheaper than the n-type Cz wafers used in several other very-high-efficiency architectures, such as silicon heterojunction (SHJ) solar cells and interdigitated back contact (IBC) cells. However, p-type wafers typically have a higher concentration of bulk defects, which leads to lower minority carrier lifetime. Of particular concern are defects caused by boron-oxygen complexes due to the simultaneous presence of oxygen and boron in the p-type Cz wafers that can induce degradation when exposed to sunlight. This can cause a loss of cell performance of $4–6\%_{rel}$ for PERC solar cells, thereby greatly impacting the levelised cost of electricity generated from photovoltaic modules. Importantly, this degradation can be overcome by using hydrogen to passivate the defects, thereby neutralising the recombination activity.

The future progression of silicon solar cell technologies will likely depend on the successful implementation of laser doping to facilitate the selective emitter design and hydrogenation to passivate both bulk and surface defects. Both of these technologies have been critical in the on-going development of PERC, with the world's first commercial p-type solar cell with an efficiency of over 20% by Suntech in 2011, along with devices approaching 24% efficiency by LONGi achieved in 2019.

10.1.1 Hydrogen passivation

The beneficial role of hydrogen in silicon solar cells and its use for bulk and surface defect passivation has been studied for decades. However, in general, hydrogen diffusion and passivation mechanisms are poorly understood. This is largely due to the difficulty in the direct detection and measurement of hydrogen silicon. As such, often an understanding of hydrogen is indirectly gained and inferred from hydrogen-related reactions. The literature of hydrogen passivation in silicon solar cells is plagued with contradictions, often caused by conclusions drawn from diffusivity data, which can vary by more than five orders of magnitude for a given temperature. A key property of hydrogen is its amphoteric nature with the ability to assume different charge states. Hydrogen in silicon can exist in three charge states, negative (H^-), neutral (H^0) and positive (H^+), depending on the number of electrons held within the atom. As the charge state of hydrogen is dependent on the concentration of electrons available in the lattice, in thermal equilibrium, it is heavily dependent on the materials' dominant conductivity, which results in a shift of the Fermi level, thereby changing the fractional occupation of hydrogen in the different charge states.

The mobility and ability for hydrogen to passivate defects depends heavily on the charge state. Thus, shifting the Fermi level in silicon by carrier injection or elevated temperature, or both, can change the ability of hydrogen to passivate surface and bulk defects. This improved understanding has led to the development of advanced hydrogenation processes incorporating minority carrier injection to manipulate the charge state of hydrogen in silicon solar cells to improve defect passivation [8].

Virtually all silicon solar cells use hydrogen-containing dielectric layers to improve surface passivation. Surface passivation can occur when hydrogen atoms bond with the surface states formed via dangling bonds on the silicon surface. This reduction in the amount of recombination active surface sites is referred to as 'chemical passivation'. Chemical passivation from the dielectric films can be enhanced by subsequent thermal processes, such as forming gas anneal performed using a hydrogen gas, or aluminium annealing (alneal), which incorporates a sacrificial aluminium layer. However, the adoption of hydrogenated plasma-enhanced chemical vapour deposited (PECVD) dielectric layers such as hydrogenated silicon nitride, aluminium oxide and amorphous silicon (SiN_x:H, AlO_x:H and a-Si:H) have led to the ability to use hydrogen from the dielectric layers for chemical surface passivation, rather than requiring an additional hydrogen source. These films can often benefit from a subsequent thermal process to further enhance surface passivation. Fixed charge in the dielectric layers can also form electric fields at the surface, which can effectively reduce the concentration of minority carriers near the silicon surface, thus reducing undesirable recombination. This is referred to as field-effect passivation. Chemical and field-effect passivation realised from hydrogenated dielectric layers have been the key in improving the V_{OC} of industrial silicon solar cells and are critical for the fabrication of world record SHJ solar cells at 26.7% using a-Si:H [9]. Similarly, hydrogen passivation of the surfaces is critical for emerging solar cell technologies with passivated contacts such as tunnelling oxide passivating contact (TOPCon).

The adoption of hydrogenated dielectric layers in silicon solar cell fabrication has also been key for bulk defect passivation. During the high-temperature firing step for metallisation, hydrogen released from dielectric layers can also diffuse from the dielectric layer to the bulk. There are a variety of defects in the bulk that can be passivated by hydrogen. These include structural defects such as grain boundaries and dislocation clusters that are particularly relevant for multi-crystalline and cast-mono silicon. As such, the replacement of hydrogen-lean low-pressure chemical vapour deposited (LPCVD) SiN_x with hydrogen-rich PEVCD was responsible for significant efficiency enhancements for such solar cells and deemed essential for the material. Although in some sense the passivation of structural defects is relatively easy given that such defects are often incorporated into the silicon wafer during crystallisation, significant challenges in the passivation of these defects are reported in the literature such as the impact of grain boundary angle on passivation and impact of impurity decoration. Hydrogen can also be used to passivate defects caused by impurities or process-induced defects such as those introduced by laser processing. However, it should be noted that defects may not just fall under one category, such as oxygen precipitates, which are caused by impurities, evolve during solar cell processing, and are therefore process-induced, and can lead to structural defects. For process-induced defects, depending on when the defects are introduced, a natural passivation of the defects may occur, at least in part, for oxygen precipitates that form prior to the firing step during high-temperature emitter diffusion and oxidation. Similarly, for laser-induced defects introduced prior to firing, such as for the industrial implementation of the laser-doped selective emitter (LDSE) in PERC, a natural passivation of defects can occur. In other instances, a deliberate process to passivate the defects may be required, such as belt furnace annealing.

However, hydrogen passivation is reversible. As such, a hydrogen passivation process effective for one defect may not be effective, or even detrimental, to the passivation of another. In addition, subsequent thermal processes can potentially undo hydrogen passivation achieved in a prior process.

For silicon solar cells, a holistic approach should be given to defect engineering, and hydrogen passivation should not be the only method considered for the fabrication of high-efficiency silicon solar cells. Another key method that is quite complementary to hydrogenation is gettering, which can remove mobile metallic impurities, as well as potential beneficial effects from the thermal processing. The removal of impurities with gettering can improve bulk lifetimes and even enhance the effectiveness of hydrogen passivation. The standard $POCl_3$ emitter diffusion process for industrial PERC solar can substantially reduce interstitial iron concentrations in the material. Together with a standard PECVD SiN_x:H layer and firing process used for PERC solar cells, we can improve the bulk lifetime of p-type Cz silicon material from <20 μs over 500 μs.

Another key attribute of hydrogen passivation is the ability to store hydrogen in the silicon wafer for manipulation during subsequent processing steps. This can enable multi-stage hydrogenation steps to overcome strict thermal processing constraints imposed by dielectric layers such as a-Si:H. Hence, hydrogen can be incorporated during high-temperature processing prior to the deposition of thermally sensitive dielectric layers, followed by an additional hydrogen passivation process at low temperatures to enhance passivation. This method has been used to fabricate hydrogenated heterojunction silicon solar cells using p-type multi-crystalline silicon wafers with V_{OC} values over 700 mV [10].

Over the past ten years, one of the most important industrial p-type solar cells is the boron-oxygen complex that causes light-induced degradation (LID) or, more precisely, carrier-induced degradation in p-type Cz silicon. This can manifest in severe losses in cell performance when the solar cells are operating in the field. Carrier-induced defects are perhaps the most challenging defects to passivate, given that at the end of the solar cell fabrication process (high-temperature firing), the defects are not available for passivation. As such an additional process may be required. Illuminated annealing, as first demonstrated by Herguth *et al.* [11], has enabled the passivation of boron-oxygen defects, both through manipulating hydrogen charge states for passivation and the critical formation of the defects such that they are available for passivation. This method has been utilised to develop techniques involving heating and carrier injection (via illumination of current injection) to passivate boron-oxygen defects in industrial PERC solar cells.

A combination of gettering and advanced hydrogenation has been used to recently achieve a stabilised V_{OC} of over 735 mV for boron-doped Cz silicon wafers, higher than that for the record 26.1% p-type solar cell at 727 mV using float zone material [12]. This is despite an initial wafer lifetime of less than 20 μs, suggesting that current p-type wafers are suitable for high-efficiency solar cell architectures with passivated contacts, and that the efficiencies of such cells could reach over 25% in the near future.

It is well known that hydrogen can be used to passivate defects to improve the performance. However, the presence of hydrogen can also cause unwanted recombination. One challenge associated with understanding hydrogen interactions in the silicon lattice is the fact that it can bond with almost anything to form complexes. One particular defect that can negatively impact industrial silicon solar cells is the defect responsible for light- and elevated temperature-induced degradation (LeTID). This defect is formed in the bulk of the silicon and can cause significant power losses for silicon solar cells, up to $16\%_{rel}$. Although LeTID does go through a degradation and recovery cycle like boron-oxygen-related degradation, the timescales are orders of magnitude slower, meaning that over the 25-year lifetime of a module, the performance may not be recovered. Although the exact mechanism causing LeTID is not currently known, there is significant mounting evidence to suggest that hydrogen is involved in its formation. The extent of LeTID has been shown to be modulated by changing the hydrogen content of the dielectric layer or altering the thermal profile of the firing step. In particular, higher firing temperatures, which are known to release more hydrogen from dielectric layers, result in larger degradation extents. Thankfully, commercially relevant processes have been developed to solve LeTID. For example, Q-Cell's Q.ANTUM technology has reduced the extent of LeTID to $<1\%_{rel}$.

However, attempts to solve LeTID can cause increases in the series resistance of the device, primarily at the front n-type contact, likely due to an accumulation of hydrogen at the surface, neutralisation of dopant atoms and the formation of hydrogen-induced platelets. These effects counteract the belief that hydrogen is purely beneficial for cell performance. As the movement of hydrogen is heavily dependent on the thermal history of the cell, it is increasingly important to understand the balance of hydrogen in the solar cell to counterbalance beneficial passivation and hydrogen-related recombination mechanisms.

10.1.2 Laser doping

The push to higher efficiency industrial silicon solar cells have seen the rise of laser-doping processes. In the world record of PERL cell, the selective emitter approach was achieved using a series of processes involving thermal diffusions and complicated masking steps. This approach is not applicable in an industrial setting. Currently, laser doping is the most promising approach for selective emitter formation. There are multiple reasons that make laser-doping advantageous. Laser doping leads to very localised heating of the silicon so there is negligible undesired thermal load introduced during the doping process. Additionally, the energy from the laser causes localised melting of the silicon, which means the laser-processed region can be heated to well over 1 200 °C, or even melts the silicon. Particularly when laser doping is performed in the molten state, the diffusion coefficient of dopant materials is orders of magnitude higher than in the solid state, such that processes that could take several hours during a standard thermal diffusion process can occur in a matter of microseconds. The dimensions of the laser area are on the scale of microns, which are comparable with the required metallisation widths. For dopants such as boron and phosphorus, the increasing solid solubility with increasing temperature also enables higher active dopant concentrations through the laser-doping process compared to conventional thermal diffusions at lower temperatures.

The laser-doping process requires the optimisation of many aspects. The depth of the laser-doped junction heavily depends on the depth of the molten zone in the silicon. First, the type of laser determines the dopant profile. The laser can either be pulsed, such as for Q-switched lasers, where the laser energy is rapidly pulsed- or continuous-wave (CW). Pulsed lasers lead to much shallower melting zones, which limits the depth of the diffusion profile. In particular, for short nano-second pulse-length lasers, the silicon typically solidifies in between successive pulses, meaning that with each pulse, the same volume is re-molten, leading to a more uniform doping profile. However longer microsecond-pulse-length lasers and CW lasers can lead to a much deeper distribution of dopants in the laser-doped regions, which can lead to more desirable doping profiles under the contacts and better shielding of the metal–Si interface against recombination, creating doped regions with a junction depth of up to 10 μm. This is also the case for mode-locked quasi-continuous-wave (Q-CW) lasers, even with picosecond pulse-lengths, due to the high repetition frequency in the range of 80–120 MHz which does not enable the entire silicon volume to resolidify between successive pulses. The wavelength and exposure time of the laser also heavily impact the doping profile, and generally a 532-nm wavelength is preferred.

The current industrial approach for laser doping in PERC solar cells involves the use of the laser directly after the thermal diffusion process onto the unpassivated silicon, using the phosphosilicate glass from the diffusion as the dopant source. This laser-doping process is generally performed in the solid state to minimise impact to the textured pyramids. The dielectric layers for surface passivation are then subsequently deposited, followed by an aligned screen-printing step for the metallisation. This sequence both reduces the formation of laser-induced defects and allows a natural passivation of the laser-induced defects during the firing step and has seen the fabrication of PERC cells with efficiencies approaching 24% [13]. An alternative approach, which can eliminate the requirement for the aligned metallisation, is performing the laser-doping process after dielectric deposition. Two examples of this approach are the Fraunhofer laser-chemical processing technology [14] and the UNSW LDSE approach. In this case, the laser doping is performed after the dielectric layer deposition and can simultaneously open the dielectric layer for contact formation and perform the melting/dopant diffusion. This is subsequently followed by a self-aligned metallisation process via plating nickel and copper. However, this laser-doping method is highly susceptible to the formation of laser-induced defects. To reduce defect formation due to repeated thermal cycling with the melting and recrystallisation of silicon for successive laser pulses, CW or Q-CW lasers can be used. However, even with this approach, solar cells produced in this manner including PERC and n-PERT can benefit heavily from the addition of a hydrogenation process to passivate defects, which can enhance the efficiency by up to $1.5\%_{abs}$.

In industry, the preferred metallisation approach is currently screen-printing, due to its simplicity and also to avoid potential adhesion or waste management issues faced with plated contacts. The widths of screen-printed contacts have also narrowed significantly, reducing the potential financial benefits of replacing silver screen-printed contacts with copper-plated contacts, although the cost of silver is still significant for PERC with a usage of approximately 90–100 mg of silver per cell and substantially higher for TOPCon and SHJ solar cells. However, as the production capacity of industrial solar cells increases, the use of silver may run into issues with scarcity, which would provide further justification for a transition to the use of copper-plated contacts, along with cost benefits.

Laser processing can also be used for other applications in silicon solar cells, including laser doping for the formation of other high-efficiency cell designs, such as semiconductor finger cells and interdigitated back contact cells. It can also be used to perform edge junction isolation. An elegant process for p-type laser doping is using the passivating AlO_x layers as a dopant source, which can therefore be used to form localised back-surface-field regions in PERC solar cells, or a selective emitter for n-PERT or TOPCon. A challenge with laser doping from AlO_x layers is solid solubility. Above ~1 200 °C, the solid solubility reduces, resulting in a similar solid solubility limit in the molten phase as that at 800 °C.

PERC solar cells represent the current dominant force in the silicon solar cell industry. Due to the significantly lower $\$/W_p$ compared with competing technologies, it is expected that this will remain the case for at least another ten years. This book has detailed the two areas of key importance to silicon solar cells, hydrogen passivation and laser doping, which have been vital for achieving efficiencies approaching 24%.

Looking beyond PERC, there are several technologies on the horizon that may capture a large portion of the market in the future. In the next section, we describe how hydrogenation techniques and laser doping may be adapted for emerging industrial silicon solar cell technologies.

10.2 Future outlook

This book has primarily focused on issues regarding PERC solar cells, as they are the current dominant technology. Further optimisation of PERC will remain relevant for the next decade [7]. In this section, we explore issues that will be at the forefront of future improvements in PERC solar cells. Looking further into the future, we also assess future trends in terms of two groups of silicon solar cells that are currently under investigation and are expected to make up a significant market share in the future [7]. These include n-type passivated contact solar cells and silicon tandem solar cells.

10.2.1 Future issues for industrial PERC solar cells

10.2.1.1 Next-generation laser-doped selective emitters for PERC

Recombination at the front surface is still a key issue for PERC and TOPCon solar cells. For PERC, this is also the case for cells using an LDSE structure due to the large coverage of laser doping (~10%) for the current industrial implementation with screen-printed contacts. Reducing the width of the laser-doped region as alignment tolerances improve will reduce the contribution of the laser-doped region to the total dark saturation current density of the device. In addition, if screen-printing can be performed with intermittent fire-through regions, reduced laser-doped coverage area could be achieved. Figure 10.1(left) shows an example of the impact of changing the laser-doped coverage area on device performance, achieved with an improvement in alignment tolerance from ±30 μm to ±10 μm, corresponding in a reduction in laser-doped coverage area for the fingers from 7.7% down to 4.5%. Further reductions in coverage would be achieved with the development of fine-line screen-printing such as 20-μm-wide lines, which could then reduce laser-doping coverage area to 3%.

Alternatively, a reduction in the metal–Si interface area could be achieved using an approach with laser doping performed through the dielectric layer, followed by an aligned screen-print process using non-fire-through pastes. This could allow a complete re-optimisation of laser-doping profile, without worrying about the formation of a ‘dead layer’. This may allow further reductions in the dark saturation current density and contact resistance. The ongoing development of screen-printed pastes targeting lower contact resistances will also enable the use of smaller contact areas to be used and achieve higher efficiencies (see Figure 10.1(right)).

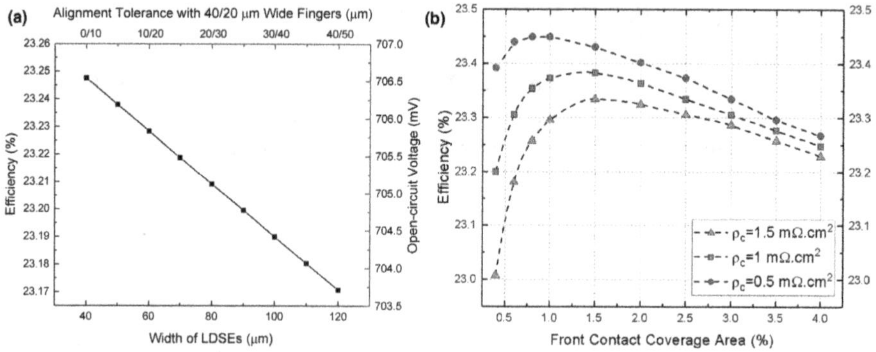

Figure 10.1 *Sentaurus simulations highlighting the gain in efficiency that can be achieved through (left) a reduction in laser-doped line width for screen-printed solar cells with 40 μm or 20 μm wide fingers. (right) Impact of contact resistance and coverage area on device efficiency.*

10.2.1.2 Surface-related degradation

Chapter 6 details the occurrence of LeTID in PERC solar cells. This degradation, which occurs at elevated temperatures, is related to the formation of a defect in the bulk of the silicon. Although the precise origin of the defect is not currently known, several strategies for the mitigation of LeTID have been developed, therefore, this is no longer a primary concern for manufacturers. Instead, the degradation of the surface passivation quality under similar elevated-temperature testing regimes represents the next challenge for the industry. This is particularly important for cell architectures such as PERC, which rely on the superior surface passivation. Sperber *et al.* were the first to study this phenomenon of deteriorated surface passivation under elevated temperature regimes in 2016 [15]. In their study, they investigated a variety of dielectric stacks on float-zone (FZ) silicon wafers. The observed surface-related degradation (SRD) was shown to occur both under illumination and in the dark, and the rates of degradation were dependent on the temperature. C–V measurements showed that the deterioration in surface passivation was primarily related to a reduction in the chemical passivation at the interface rather than a reduction in the fixed charge in the dielectric layer.

The timescale for the observed SRD is different to that of LeTID. Interestingly, the onset of the SRD appears to correlate with a recovery of the LeTID [16, 17]. This is illustrated in Figure 10.2, which shows the effective lifetime and J_0 for n-type Cz samples that have undergone annealing at 160 °C [17].

This correlation suggests that there could be an interdependent relationship between the two phenomena. Although the exact mechanics of LeTID are not fully understood, one hypothesis indicates that the recovery of LeTID is related to a diffusion of hydrogen away from the bulk, thus reducing the density of LeTID defects in the bulk. If this is the case, it could indicate that the SRD is caused by hydrogen-induced extended defects at the silicon surface, caused by the out diffusion of

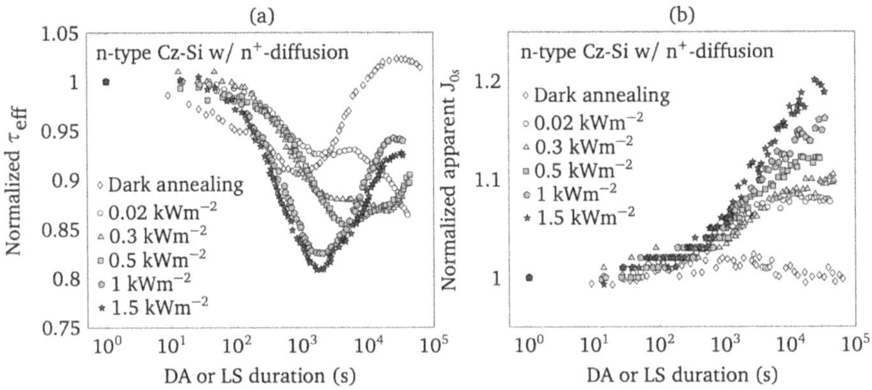

Figure 10.2 *Normalised effective lifetime (a) and apparent J_0 for n-type Cz silicon subjected to annealing at 160 °C under a variety of illumination intensities. Figure reproduced from [17].*

hydrogen from the bulk. The diffusion of hydrogen from the bulk depends on internal electric fields, such as the p-n junction, and the fractional distribution of hydrogen in different charge states, which depends on temperature and carrier injection. This analysis paints a picture of a delicate balance that must be achieved in industrial silicon solar cells. The control of hydrogen will be critical to mitigate both B-O, LeTID and surface-related defects; this will become more critical as the surface passivation of industrial silicon solar cells improves in the future, inadvertently leaving the solar cell performance more susceptible to performance degradation.

10.2.1.3 A shift to gallium-doped silicon wafers

The solar cell industry has predominantly used p-type wafers doped with boron. As discussed in Chapter 5, these wafers are susceptible to LID due to B-O defects. Various stabilisation approaches have been developed to mitigate this degradation, including illuminated annealing and current injection. However, it is advantageous to investigate alternative doping atoms to avoid B-O defects. There are various group III elements that could theoretically be used. Of these, doping with aluminium suffers from severe recombination caused by aluminium-oxygen complexes [18] and doping with indium leads to incomplete ionisation due to the undesirable acceptor level of indium in silicon [19]. Of all the group III elements, gallium is the most appropriate to replace boron. Glunz *et al.* showed that gallium-doped wafers do not suffer from the same LID often associated with boron-doped wafers [20]. The commercial progress of gallium-doped wafers has been limited over the past 20 years, due to a number of core patents held by Shin Etsu Chemical [21]. However, as these patents expired in 2020, the commercial interest in the use of gallium-doped wafers is now substantial. Due to the lack of LID related to B-O defects, the assumption is that switching to gallium-doped wafers will remove all light instabilities; however, recent work by Grant *et al.* suggests that this is not the case [22]. They compared

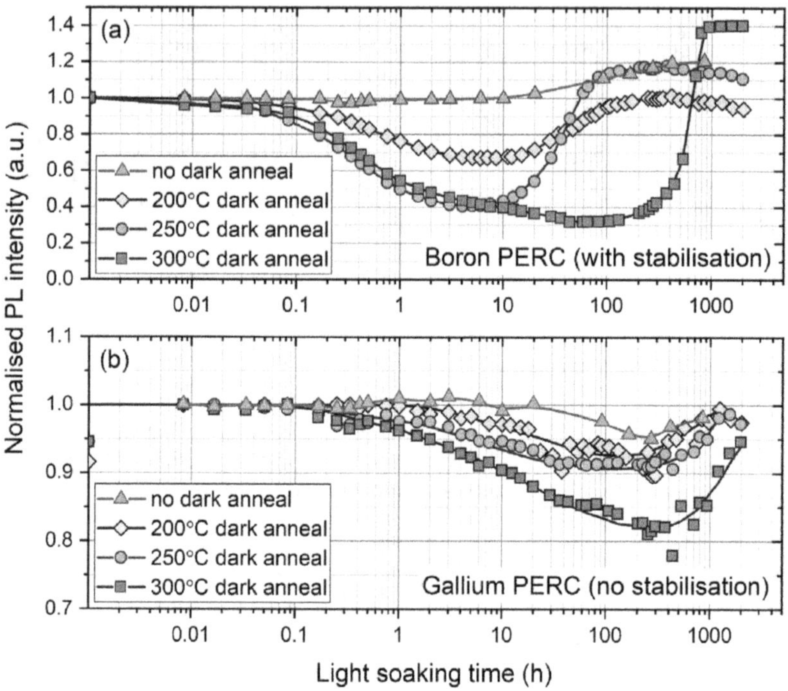

Figure 10.3 *Normalised PL intensity during light soaking at 75 °C, 1-sun for (a) boron-doped and (b) gallium-doped PERC solar cells. A current-injection stabilisation process was applied to the boron-doped cell but not to the gallium-doped cell. Reproduced from [22] under Creative Commons CC-BY. © 2019 The Authors. Published by Elsevier B.V. https://doi.org/10.1016/j.solmat.2019.110299.*

the stability of PERC solar cells that were fabricated with either boron-doped or gallium-doped p-type wafers. To do so, they track degradation by analysing the photoluminescence (PL) counts during standard LeTID testing conditions of 75 °C at 1-sun illumination for 1 000 hours. The results are displayed in Figure 10.3. The boron-doped PERC cell was subjected to an industry-relevant current-injection stabilisation process, while the gallium-doped PERC cell was untreated. For the case where the samples were not exposed to a dark anneal before light soaking, the treated boron-doped PERC cell did not exhibit any LeTID; instead, the lifetime actually increased slightly during light soaking. In contrast, the gallium-doped samples display a slight degradation and recovery. When the samples were subjected to a dark anneal prior to the process, significant LeTID was observed in both the boron-doped and gallium-doped PERC cells, although the degradation extent in that study was less for gallium-doped wafers. Following the experiment, the metallisation was stripped off and super-acid passivation was applied to form lifetime structures. Lifetime analysis of the gallium-doped group showed that the degradation occurs

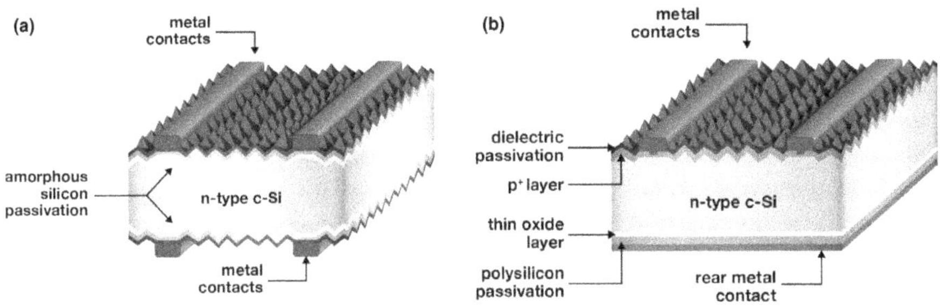

Figure 10.4 *Schematic diagram displaying the device architecture for (left) SHJ and (right) TOPCon solar cells*

in the bulk, indicating that gallium-doped wafers do indeed suffer from LeTID. In addition to this, interstitial iron can cause instabilities due to the formation of Fe–Ga pairs [23], similar to the Fe–B pairs that form in boron-doped silicon. As such, gallium-doped wafers likely require the application of the same defect-engineering processes such as gettering and hydrogenation as their boron-doped counterparts. This observation indicates that more studies are required to fully assess potential instabilities in gallium-doped PERC solar cells, in particular, how potential LeTID will affect the performance of modules in the field and how hydrogenation will play a role in the development of stabilisation treatments for gallium-doped cells. As the industry shifts beyond PERC to improved passivation schemes, bulk defects such as LeTID will become increasingly problematic, due to the increased sensitivity to bulk-related recombination.

10.2.2 N-type passivated contact solar cells

Most commercial silicon solar cells over the past 20 years have been fabricated using a p-type silicon substrate. One of the main justifications for using p-type substrates is that they are cheaper compared to their phosphorous doped n-type counterparts. However, it is expected that over the next ten years, n-type wafers will have a much larger share of the market, with up to 40% by 2030 [7]. This change will be driven by the adoption of new high-efficiency cell architectures. Of these, the two most important will be the TOPCon and SHJ solar cell technologies. Figure 10.4 shows a schematic overview of these device architectures. SHJ solar cells currently hold the world record for the highest efficiency single junction solar with 26.7% [24] and have achieved voltages as high as 750 mV [25]. Hanergy also recently demonstrated efficiencies of >25% by fabricating SHJ cells in an industrial environment, indicating that the excellent results achieved in the laboratory are transferrable to commercial production [26]. These results are facilitated by the excellent surface passivation provided by the hydrogenated amorphous silicon thin films used. The hydrogen from these films provides excellent chemical passivation for dangling bonds at the silicon surface. One unique feature of this approach is that all processing occurs at low temperature (<200 °C) [27]. This is necessary as the surface passivation from the a-Si layers is unstable at elevated temperatures. Due to the low-temperature processing, the process flow does not allow for defect engineering of the silicon, as such, high lifetime n-type Cz wafers are used. One barrier for SHJ market entry is the high cost of the CAPEX, which is primarily related to the PECVD tools used to deposit the passivation layers. Another key barrier for the SHJ technology is the high silver usage, approximately double that for PERC and indium usage for the transparent conductive oxide layer of indium-tin-oxide (ITO). This will become increasingly challenging as the industry continues to expand rapidly and consumes a significant percentage of the world's supply.

The structure of TOPCon solar cells is closer to that of PERC cells; however, the dielectric passivation on the rear is replaced by the tunnelling oxide-doped polysilicon contact. This can lead to very low SRV, and thus TOPCon cells can achieve higher V_{OC}. Additionally, the diffused emitter layer is formed using boron rather than phosphorous.

These cell architectures are only now beginning to have an impact on the solar cell market. The processes discussed in this book will play a role in the development of these cells. In particular, we focus on hydrogen treatments to modify the surface in SHJ cells and the potential use of laser doping for selective emitter formation for industrial TOPCon solar cells.

10.2.2.1 Selective emitter for TOPCon solar cells

In the TOPCon cell architecture, the passivated contact is applied only to the rear surface. The reason for not including the polysilicon layer on the front surface is to avoid parasitic absorption in this layer. As such, the front surface consists of a boron emitter and a dielectric stack that passivates the surface and acts as an anti-reflection layer. This stark difference in the front and rear surface means there is a significant discrepancy in the J_0 associated with each surface. For this case, the efficiency of the cell is limited by losses on the front surface, which is particularly true when high-quality n-type wafers with high bulk lifetimes are used. To significantly drive performance enhancements, this front surface recombination can be addressed by the application of the selective boron emitter. This optimisation, which is currently driving performance improvements in industrial PERC cells, should be applied to the boron emitter in TOPCon cells. By modelling using Sentaurus, Chen *et al.* suggest that an efficiency gain of $1\%_{abs}$ is achievable by switching from a homogeneous boron emitter with sheet resistance of 90 Ω/ϒ, to a selective emitter with sheet resistance of 20 Ω/ϒ and 150 Ω/ϒ for the contact and non-contact regions, respectively [28]. This gain is primarily related to increased V_{OC} due to the lower J_0 contribution from the front surface.

In 2017, Tao *et al.* demonstrated a laboratory TOPCon cell with a selective boron emitter using the chemical etch back approach [29]. By carefully controlling the etch conditions, they achieved a sheet resistance of 180 Ω/ϒ in the non-contacted regions and 65 Ω/ϒ in the contacted area, which reduced the J_0 on the front surface. However, this process requires a lengthy and energy intensive, high-temperature boron-diffusion to achieve such low sheet resistances in the contact region. The schematic device structure and microscopy images of the passivating contact on the rear

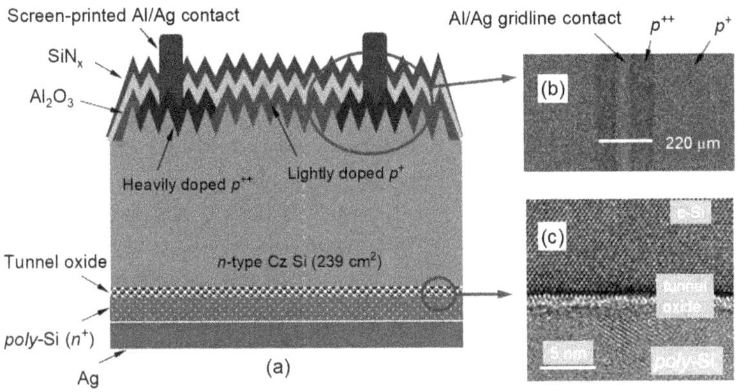

Figure 10.5 *(a) Schematic diagram displaying the selective emitter TOPCon device. The inset shows (b) optical microscope image of the front selective emitter metallisation and (c) TEM image of the rear passivating contact. Reprinted from [29] with the permission of AIP Publishing. © 2017 Author(s).*

and selective emitter on the front surface are shown in Figure 10.5. In the study by Tao *et al.*, the implementation of the selective emitter resulted in a $0.2\%_{abs}$ efficiency gain compared to devices with a homogeneous boron emitter. The improvements were related to both a 6 mV increase in V_{OC}, due to reduced front surface recombination, and a 0.3 mA/cm² increase in J_{SC}, due to improved blue response compared to the homogeneous emitter. This was counter balanced by a slightly reduced fill factor (FF), due to increased series resistance (R_s). This report represents a positive first step; however, more work is required for the development of an industrially relevant selective emitter approach in TOPCon solar cells. The application of laser doping to form a selective emitter in TOPCon cells represents an interesting pathway to high efficiency, which will pose many scientific challenges in coming years, particularly using the AlO_x layer as dopant source. However, with the limited solid solubility of aluminium in silicon, it is unclear on whether an additional dopant source is required, or whether care needs to be taken to minimise the spreading of the existing boron dopants within the emitter to retain a sufficiently high peak active dopant concentration, or subsequent annealing at more favourable temperatures can be used to activate additional aluminium dopants.

10.2.2.2 Instability in n-type passivated contact solar cells

One significant advantage of using n-type wafers is that it eliminates boron-oxygen LID, as the dopant is phosphorous not boron. For this reason, there was a general consensus that these high-efficiency n-type passivated contact cells are immune to the same kind of LID observed in p-type cells/modules. However, n-type wafers are not immune to all types of degradation. Chen *et al.* showed that n-type silicon wafers are also susceptible to bulk LeTID degradation [16]. Specific to passivated contact designs, recent evidence has presented some cases where degradation was present. Jordan *et al.* studied field data for n-type SHJ cells. Their results indicate that n-type SHJ modules did indeed exhibit significant degradation in the field, with degradation rates of up to 1%/year [30, 31]. These results were reported for systems that were deployed many years ago, so the degradation is not a true representation of state-of-the-art industrial cells. On a cell level, Madumelu *et al.* observed significant degradation of up to $5\%_{rel}$ on n-type SHJ cells that were fabricated using copper plating rather than screen-printed contacts [32]. Karas *et al.* also observed degradation in copped-plated SHJ cells; however, cells screen-printed with silver did not exhibit the same degradation [33]. It should be noted that these accounts of degradation require elevated temperature or damp heat exposure, which are not always representative of the environment experienced by modules in the field. There are also many reports of stable SHJ cells under field conditions. Further extensive study is required to unravel the mystery of degradation in these systems, which will be important to help the progression of these high-efficiency technologies.

Several studies have also recently shown degradation in the passivation schemes used in n-type TOPCon solar cells. Winter *et al.* first reported a firing-related degradation and recovery in n-doped polysilicon/tunnelling oxide passivation schemes, deposited on both n-type and p-type silicon wafers [34]. They show that this occurs both in the light and in the dark. This observation was recently verified by Kang *et al.* [35] at ANU. Figure 10.6 displays the effective lifetime and J_0 of fired symmetrical polysilicon passivated samples, light soaked at 1-sun at 140 °C. This indicates a clear degradation in lifetime, which directly correlates with an increase in J_0. To verify whether the degradation was related to a deterioration of the passivation, samples passivated with a superacid passivation scheme were also tested. As shown in Figure 10.6 (stars), these samples were very stable during the light soaking, indicating that the observed changes were not likely caused by a bulk degradation. Grazing incidence X-ray diffraction analysis indicated that no microstructural changes were induced in the polysilicon layer during degradation. The cause of this degradation is still not understood. It is important to understand the role of the movement of hydrogen and whether suitable mitigation strategies can be developed so that this surface degradation does not impact the performance of industrial n-type TOPCon solar cells. However, the degradation and partial recovery have a striking resemblance to degradation mechanisms such as B-O LID and LeTID common to p-type devices. As such, illuminated hydrogenation processes may prove vital for the elimination of these instabilities in TOPCon solar cells.

10.2.2.3 Post-cell treatments for SHJ solar cells

It is currently standard in industry to incorporate some kind of post-cell processing to improve the stability of either Al-BSF or PERC solar cells. The most commonly used approach is a current-injection tool, whereby cells are stacked together and held at a specific current. While there is a lot of knowledge about these treatments for p-type industrial cells, much less work has been directed at n-type passivated

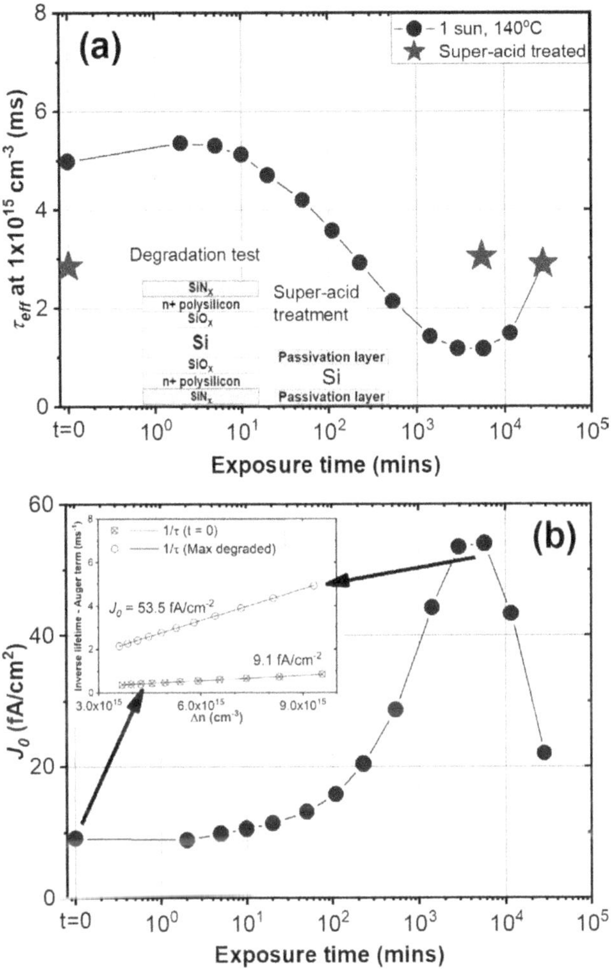

Figure 10.6 *Evolution of effective lifetime (a) and J_0 (b) for symmetrical TOPCon lifetime test structures with symmetrical tunnelling oxide/n-doped polysilicon passivation, capped with SiN_x. The light soaking was performed at 140 °C under 1-sun illumination. The structure of the samples is shown in the inset of (a). The changes in lifetime for a sample passivated with a superacid passivation scheme are also shown. Reprinted from [35], Copyright (2012) with permission from Elsevier.*

contact cells. This is more challenging for SHJ cells, due to the tight temperature constrictions.

Kobayashi *et al.* showed that both light-soaking and current-injection treatments can improve the performance of n-type SHJ solar cells [36]. They showed that

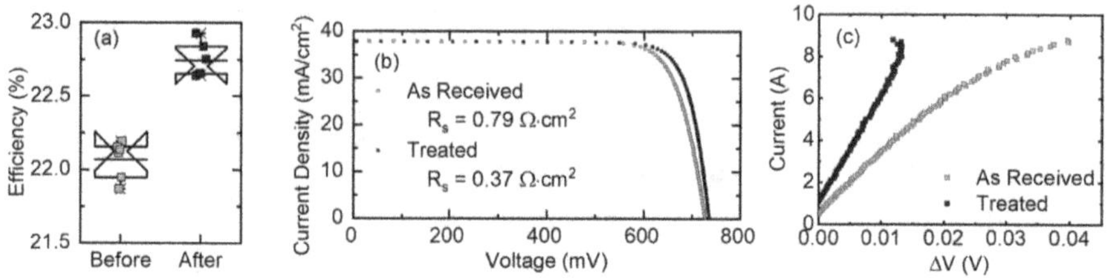

Figure 10.7 *(a) Influence of post-cell treatment on the efficiency of n-type solar cells (b) current–voltage characteristic curves before and after the treatment. The inset shows the reduction in series resistance (c) injection-dependent R_s. Reprinted from [37] with the permission of AIP Publishing. © 2019 Author(s).*

these treatments could lead to a gain in efficiency of 0.3%$_{abs}$, due to improvements in both V_{OC} and FF. Analysis of symmetrical lifetime test structures indicated that the improvements resulted from a reduction in recombination active interface states. These results are surprising, as a-Si:H layers are well known for LID, referred to as the Staebler–Wronski effect. This indicates that further study is required to develop appropriate post-cell treatments to apply to industrial n-type SHJ cells, while keeping in mind the strict temperature constraints.

UNSW has developed an illuminated annealing post-cell treatment approach that has shown efficiency improvements as large as 0.7%$_{abs}$ in industrially fabricated n-type SHJ solar cells [37]. The changes in efficiency are displayed in Figure 10.7a. The efficiency improvements were related to an increase in both the V_{OC} and FF. A comparison of the I–V curves before and after the treatment for one n-type SHJ cell is shown in Figure 10.7b. The inset shows the significant reduction in R_s after the treatment. Additionally, Figure 10.7c indicates that the post-cell process addresses the injection level dependence of R_s.

As the production of industrial SHJ increases over time, the search for an industrially relevant post-cell process to enhance the efficiency will intensify. The post-cell processing approaches that have become common for p-type technologies are more rigorously being explored for high-efficiency n-type technologies such as SHJ; however, significant understanding and optimisation are still required. Additionally, n-type passivated SHJ solar cells are forerunners as the chosen silicon bottom cell for silicon tandem cells. This is discussed in Section 10.2.3.

10.2.3 Silicon-based tandem solar cells

All discussions thus far have described single-junction industrial silicon solar cells. As the predominant technology transitions from Al-BSF to PERC to passivated contact solar cells, the upper limit on the practically achievable efficiency increases. However, the upper limit of single-junction architectures will ultimately reach a threshold beyond which no further efficiency enhancements will be achievable. When this occurs, the industry will need to transition to tandem solar cells, whereby

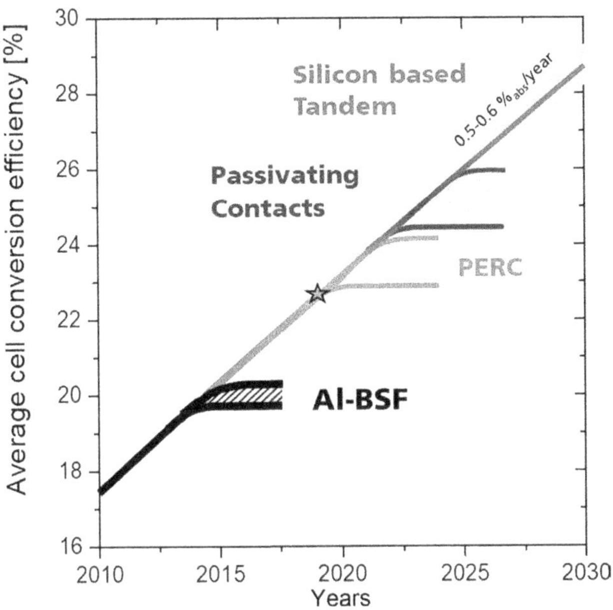

Figure 10.8 *Technology road map for various industrial silicon solar cell technologies. Reprinted from [41], with the permission of AIP Publishing. © 2020 Author(s).*

two subcells are stacked together, to further increase the efficiency. Using two cells with different absorber materials, each with a unique band gap, has a higher efficiency potential due primarily to the more efficient utilisation of the solar spectrum [38]. This approach has a theoretical limiting efficiency of ~46% [39], as opposed to practical efficiency of 29% for single-junction cells [40]. In the near future, tandem solar cells will be composed of a silicon solar cell as the bottom cell, with some variety of thin-film solar cell on top to absorb the short-wavelength photons. The two current front-runner materials for the top cell include perovskite thin films and –V materials, such as GaAs. The exact time frame in which efficiency improvements are exhausted using the single-junction approach is not clear, as technological advances in industry progress so quickly. Hermle *et al.* describe the time frame schematically, as shown in Figure 10.8 [41]. This aggressive estimation indicates that this will occur in the year 2025, after that, silicon-based tandem solar cells will begin to take over. They predict that the steady efficiency improvements of 0.5–0.6$\%_{abs}$/year will continue with silicon-based tandem solar cells. Oxford PV has already begun commercial production of perovskite/silicon tandem solar cells and has already achieved efficiencies of 28% [42].

The design requirements for a silicon solar cell as the bottom cell for a tandem cell are vastly different to that of a conventional single-junction cell. Some of the main differences include:

- The top cell absorbs short wavelength photons, it is as if the silicon cell is operated using an optical filter. As such, it is crucial that the silicon cell efficiently absorbs long wavelength light and its blue response is unimportant.
- The top cell and associated charge transfer layers can place tight temperature constraints on the processing.
- There is no metallisation on the front surface; however, the passivation layers on the front must also be able to conduct charge.
- Front surface reflection not only impacts photocurrent generation in the silicon cell but also impacts the internal reflection and therefore controls light management throughout the whole device.

These stark differences mean that a paradigm shift in processing is required. It is not currently clear which technology will be most appropriate for fabrication of the silicon bottom cell in a silicon-based tandem. Over the past five years, most of the world record efficiencies for perovskite/silicon solar cells have been achieved using a n-type SHJ solar cell design for the silicon bottom cell. This approach has several advantages: (i) superior passivation which leads to high V_{OC}; (ii) the main drawback of this approach, parasitic absorption in the a-Si:H films, is not a problem for tandem architectures, as short wavelength photons are absorbed in the top cell and (iii) the a-Si passivation layers are conductive, unlike dielectric layers used in PERC cells.

Although n-type SHJ solar cells with a-Si passivation yield the highest efficiency single-junction-device values, there are advantages to using a more conventional diffused homojunction approach. The highest efficiency perovskite solar cells are formed using TiO_2 as the electron transport layer. This requires sintering at 400 °C, which means TiO_2 is not compatible with the SHJ solar cells, which degrade at temperatures >200 °C. These temperature constraints are not a problem for homojunction cells; however, the design of PERC and TOPCon cells is also not directly applicable, as the dielectric layers used for front surface passivation are electrical insulators and thus do not allow for charge transport to the intermediate layers. Wu *et al.* [43] developed an innovative approach to employ a diffused homojunction cell in a perovskite/silicon tandem structure. They use an n-type wafer with a boron emitter to form the homojunction cell. Both the front and rear surfaces are passivated using dielectric layers. To allow for charge to move from the boron emitter to the ITO recombination layer, locally deposited Cr/Pd/Ag metal stacks were formed by thermal evaporation. These stacks act as intermediate contacts between the emitter in the silicon cell and the rest of the device. In doing so, they achieved an efficiency of 22.5% for a monolithic silicon/perovskite solar cell. This is a promising approach; however, the V_{OC} in the silicon cell was limited to only ~620 mV. This was related to the surface passivation and the thickness of the wafer; however, the main limiting factor was the use of the large area emitter. Optimisation of the emitter design would allow for significantly improved device performance. Emitter design via laser doping may be a key to allow for the use of conventional homojunction solar cells to be utilised in perovskite/silicon tandem solar cells.

Further areas of study for tandem solar cells will include hydrogen passivation and ensuring that effective hydrogen passivation of bulk and surface defects of the silicon bottom cell can be achieved within the thermal constraints imposed by the top cell, if performed on the finished device. Advanced hydrogenation process incorporating illumination at low temperatures, as already demonstrated for SHJ solar cells, may be a promising candidate. Alternatively, the tandem devices may require a multi-stage approach whereby a high-temperature hydrogenation process is performed prior to the deposition of the top cell, in a similar method as was used in Chapter 4 for hydrogenated heterojunction solar cells fabricated on low-quality p-type wafers.

References

[1] Green M.A. 'How did solar cells get so cheap?' *Joule*. 2019;**3**(3):631–3.
[2] Offer G.J. 'Automated vehicles and electrification of transport'. *Energy & Environmental Science*. 2015;**8**(1):26–30.
[3] Hosseini S.E., Wahid M.A. 'Hydrogen from solar energy, a clean energy carrier from a sustainable source of energy'. *International Journal of Energy Research*. 2020;**44**(6):4110–31.
[4] Haegel N.M., Margolis R., Buonassisi T., *et al*. 'Terawatt-scale photovoltaics: trajectories and challenges'. *Science*. 2017;**356**(6334):141–3.
[5] Haegel N.M., Atwater H., Barnes T., *et al*. 'Terawatt-scale photovoltaics: transform global energy'. *Science*. 2019;**364**(6443):836–8.
[6] Fertig F., Lantzsch R., Mohr A., *et al*. 'Mass production of P-type Cz silicon solar cells approaching average stable conversion efficiencies of 22 %'. *Energy Procedia*. 2017;**124**(13):338–45.
[7] VDMA. International technology roadmap for photovoltaic (ITRPV). 11th edition. Frankfurt, Germany: VDMA; 2020.
[8] Hallam B.J., Hamer P.G., Wenham S.R., *et al*. 'Advanced bulk defect Passivation for silicon solar cells'. *IEEE Journal of Photovoltaics*. 2014;**4**(1):88–95.
[9] Yoshikawa K., Yoshida W., Irie T., *et al*. 'Exceeding conversion efficiency of 26% by heterojunction interdigitated back contact solar cell with thin film Si technology'. *Solar Energy Materials and Solar Cells*. 2017;**173**(17):37–42.
[10] Chen D., Kim M., Shi J., Vicari Stefani B., Yu Z., Liu S. 'Defect engineering of P-type silicon heterojunction solar cells fabricated using commercial-grade low-lifetime silicon wafers'. *Progress in Photovoltaics: Research and Applications*. 2018;**2019**:1–15.
[11] Herguth A., Schubert G., Käs M., Hahn G., Kaes M., Hahn G. 'A new approach to prevent the negative impact of the metastable defect in boron doped cz silicon solar cells'. 4th IEEE World Conference on Photovoltaic Energy Conversion. IEEE; 2006. pp. 940–3.
[12] Haase F., Hollemann C., Schäfer S., *et al*. 'Laser contact openings for local poly-Si-metal contacts enabling 26.1%-efficient POLO-IBC solar cells'. *Solar Energy Materials and Solar Cells*. 2018;**186**(6):184–93.

[13] Chen R., Tong H., Zhu H., Ding C., Li H., Chen D. '23.83% efficient mono-PERC incorporating advanced hydrogenation'. *Progress in Photovoltaics: Research and Applications*. 2019;**2020**:1–9.

[14] Kray D., Fell A., Hopman S., *et al.* 'LCP) – A versatile tool for microstructuring applications'. *Applied Physics A: Materials Science & Processing*. 2008;**93**(1):99–103.

[15] Sperber D., Herguth A., Hahn G. 'Instability of dielectric surface Passivation quality at elevated temperature and illumination'. *Energy Procedia*. 2016;**92**(1):211–17.

[16] Chen D., Hamer P.G., Kim M., *et al.* 'Hydrogen induced degradation: a possible mechanism for light-and elevated temperature-induced degradation in N-type silicon'. *Solar Energy Materials and Solar Cells*. 2018;**185**:174–82.

[17] Chen D., Hamer P., Kim M., *et al.* 'Hydrogen-induced degradation: explaining the mechanism behind light- and elevated temperature-induced degradation in N- and P-type silicon'. *Solar Energy Materials and Solar Cells*. 2020;**207**:110353.

[18] Rosenits P., Roth T., Glunz S.W., Beljakowa S. 'Determining the defect parameters of the deep aluminum-related defect center in silicon'. *Applied Physics Letters*. 2007;**91**(12):122109.

[19] Murphy J.D., Pointon A.I., Grant N.E., *et al.* 'Minority carrier lifetime in indium doped silicon for photovoltaics'. *Progress in Photovoltaics: Research and Applications*. 2019;**27**(10):844–55.

[20] Glunz S.W., Rein S., Knobloch J., Wettling W., Abe T. 'Comparison of boron- and gallium-doped P-type Czochralski silicon for photovoltaic application'. *Progress in Photovoltaics: Research and Applications*. 1999;**7**(6):463–9.

[21] Abe Y., Hirasawa T., Tokunaga K., Igarashi T., Yamaguchi M. 'Silicon Single Crystal and Wafer Doped with Gallium and Method for Producing Them'. United States Patent US 6,815,605; 2000.

[22] Grant N.E., Scowcroft J.R., Pointon A.I., Al-Amin M., Altermatt P.P., Murphy J.D. 'Lifetime instabilities in gallium doped monocrystalline PERC silicon solar cells'. *Solar Energy Materials and Solar Cells*. 2020;**206**:110299.

[23] Nærland T.U., Bernardini S., Haug H., *et al.* 'On the recombination centers of iron-gallium pairs in Ga-doped silicon'. *Journal of Applied Physics*. 2017;**122**(8):85703.

[24] Yoshikawa K., Kawasaki H., Yoshida W., *et al.* 'Silicon heterojunction solar cell with interdigitated back contacts for a photoconversion efficiency over 26%'. *Nature energy*. 2017;**2**(5):17032

[25] Masuko K., Shigematsu M., Hashiguchi T., *et al.* 'Achievement of more than 25% conversion efficiency with crystalline silicon heterojunction solar cell'. *IEEE Journal of Photovoltaics*. 2014;**4**(6):1433–5.

[26] Ru X., Qu M., Wang J., *et al.* '25.11% efficiency silicon heterojunction solar cell with low deposition rate intrinsic amorphous silicon buffer layers'. *Solar Energy Materials and Solar Cells*. 2020;**215**:110643.

[27] De Wolf S., Descoeudres A., Holman Z.C., Ballif C. 'High-Efficiency silicon heterojunction solar cells: a review'. *Green*. 2012;**2**(1):7–24.

[28] Chen C.-W., Hermle M., Benick J., *et al.* 'Modeling the potential of screen printed front junction cz silicon solar cell with tunnel oxide passivated back contact'. *Progress in Photovoltaics: Research and Applications*. 2017;**25**(1):49–57.

[29] Tao Y., Madani K., Cho E., Rounsaville B., Upadhyaya V., Rohatgi A. 'High-efficiency selective boro*n* emitter formed by wet chemical etch-back for *n*-type scree*n*-printed Si solar cells'. *Applied Physics Letters*. 2017;**110**(2):021101.

[30] Jordan D.C., Kurtz S.R., VanSant K., Newmiller J. 'Compendium of photovoltaic degradation rates'. *Progress in Photovoltaics: Research and Applications*. 2016;**24**(7):978–89.

[31] Jordan D.C., Deline C., Johnston S., *et al.* 'Silicon Heterojunction system field performance'. *IEEE Journal of Photovoltaics*. 2018;**8**(1):177–82.

[32] Madumelu C., Wright B., Soeriyadi A., *et al.* 'Investigation of light-induced degradation in N-type silicon heterojunction solar cells during illuminated annealing at elevated temperatures'. *Solar Energy Materials and Solar Cells*. 2020;**218**(1):110752.

[33] Karas J., Sinha A., Buddha V.S.P., *et al.* 'Damp heat induced degradation of silicon Heterojunction solar cells with Cu-Plated contacts'. *IEEE Journal of Photovoltaics*. 2020;**10**(1):153–8.

[34] Winter M., Bordihn S., Peibst R., Brendel R., Schmidt J. 'Degradation and regeneration of n+-doped poly-si surface passivation on p-type and n-type Cz-Si under illumination and dark annealing'. *IEEE Journal of Photovoltaics*. 2020;**10**(2):423–30.

[35] Kang D., Sio H.C., Yan D., *et al.* 'Long-term stability study of the passivation quality of polysilicon-based passivation layers for silicon solar cells'. *Solar Energy Materials and Solar Cells*. 2020;**215**(6):110691.

[36] Kobayashi E., De Wolf S., Levrat J., *et al.* 'Light-induced performance increase of silicon heterojunction solar cells'. *Applied Physics Letters*. 2016;**109**(15):153503.

[37] Wright M., Kim M., Dexiang P., Xin X., Wenbin Z., Wright B. 'Multifunctional process to improve surface passivation and carrier transport in industrial n-type silicon heterojunction solar cells by 0.7% absolute'. *15th International Conference on Concentrator Photovoltaic Systems*. 2019;**2149**:110006.

[38] Shen H., Walter D., Wu Y., *et al.* 'Monolithic perovskite/Si tandem solar cells: pathways to over 30% efficiency'. *Advanced Energy Materials*. 2020;**10**(13):1902840–18.

[39] Bremner S.P., Levy M.Y., Honsberg C.B. 'Analysis of tandem solar cell efficiencies under AM1.5G spectrum using a rapid flux calculation method'. *Progress in Photovoltaics: Research and Applications*. 2008;**16**(3):225–33.

[40] Richter A., Hermle M., Glunz S.W. 'Reassessment of the limiting efficiency for crystalline silicon solar cells'. *IEEE Journal of Photovoltaics*. 2013;**3**(4):1184–91.

[41] Hermle M., Feldmann F., Bivour M., Goldschmidt J.C., Glunz S.W. 'Passivating contacts and tandem concepts: approaches for the highest silicon-based solar cell efficiencies'. *Applied Physics Reviews*. 2020;**7**(2):021305.

[42] Green M.A., Dunlop E.D., Hohl-Ebinger J., Yoshita M., Kopidakis N., Hao X. 'Solar cell efficiency tables (version 56)'. *Progress in Photovoltaics: Research and Applications*. 2020;**28**(7):629–38.
[43] Wu Y., Yan D., Peng J., *et al*. 'Monolithic perovskite/silicon-homojunction tandem solar cell with over 22% efficiency'. *Energy & Environmental Science*. 2017;**10**(11):2472–9.

Index

aerosol jet printing 348–9
aluminium anneal process 86–7
alneal aluminium back surface field (Al-BSF) solar cells 11–12, 147–8, 438
  efficiency 179, 180
  gettering and hydrogenation 147–52
  industrial 438
  manufacturing process flow 438
  production sequences
    dielectric antireflection layer deposition 57
    emitter diffusion 56–7
    firing process 57–8
    metallisation 57
    phosphosilicate glass removal 57
    process flow 55, 56
    saw damage removal 55–6
    starting material 55
  surface recombination velocity (SRV) 438
aluminium dopant 194
aluminium oxide thin films 99–101
amorphous silicon thin films
  bulk defects 96
  capping layers 98–9
  hydrogenated amorphous silicon 96
  industrial SHJ solar cells 98
  minority carrier lifetime 99
  post-deposition annealing temperature 97
  Staebler–Wronski effect 96
anti-reflection coating (ARC) 57
atomic layer deposition (ALD)
  aluminium oxide thin films 100–1
  silicon oxide thin films 90–1
atom probe tomography (APT) 43
Auger recombination 341–2

back surface field (BSF) 83
balance of systems (BOS) costs 12
beam optics 329, 330
belt furnace anneal (BFA) 143, 282
boron dopant passivation 136
boron-doped Czochralski-grown silicon 46
boron-oxygen defect system
  passivation stability 223–4
  rapid laser-based processes 222–3
boron-oxygen light-induced degradation
  boron-doping concentration reduction 194–5
  casted multi-crystalline and quasi-mono-crystalline silicon 193
  commercial B-O LID mitigation solutions
    co-firing process 225
    illumination 225–6
    illumination intensity setting 226–7
    production process 225
    stabilised efficiency 228
  dark-annealed lifetime recovery 188–9
  defect formation 220–2
  defect theories 189–90
  generalised model

iron-boron pairs (Fe-B) 191
LeTID 192
three-state model 190
germanium and carbon co-doping 195
hydrogen charge states
donor and acceptor energy levels 218
excess carrier concentration 219–20
illumination intensity 219
negative-U impurities 217–18
regeneration rate 218
Sah-Shockley statistics 218
hydrogen role
film and dielectric stack composition 213–15
firing 210–13
modulating kinetics 215–16
unintentionally introduced hydrogen 216–17
V_{oc} and efficiency loss 209–10
interstitial oxygen concentration reduction 192
module lamination 228
permanent deactivation process 180
precursor annihilation and reformation 200–9
p-type Cz silicon degradation
boron and oxygen concentration 182–4
boron-doping concentration 184–5
carrier injection 188
deep level transient spectroscopy 183–4
doping concentration and illumination intensity 187–8
normalised defect density (NDD) 182–3
quasi-steady-state photoconductance measurements 182
two-stage degradation 185–6
p-type Cz silicon solar cells 179, 180
p-type silicon heterojunction solar cells
efficiency 229–30
fabrication 232
high-temperature firing process 232
high-temperature hydrogenation process 231
I–V characteristics 230–1
light-induced degradation 230
in situ open-circuit voltage 231–2
three-key-processing regimes 230
self-recovery 228–9
thermal processes
four-state model 196–7
high-temperature tube-furnace processes 197
intermediate temperatures 198
rapid thermal processing 198–9
wafer alternatives 195–6
bulk defects
gettering and hydrogenation (*See* gettering and hydrogenation)
highly defected multi-crystalline silicon material 115
impurity-related defects 130–7
light-and carrier-induced defects 143–4
process-induced defects 137–43
p-type mono-crystalline silicon solar cells 115
structural defects 115–30
additional thermal hydrogenation treatments 128–30
cast-mono-crystalline silicon material 126–8
cast silicon wafers 116–17
dehydrogenation 125–6
grain boundaries and dislocation clusters 117–25

carbon-hydrogen (CH) complexes 250–2
carbon-oxygen-hydrogen (COH) complexes
DLTS/MCTS peak formation 254–5
ground-state structures 253
thermal stability studies 252–3
carrier-induced degradation (CID) 250, 257–8
cast-mono-crystalline silicon material 126–8
cast silicon wafers 116–17
charge-assisted passivation 83–4
chemical anodisation 91
chemical edge junction isolation process 149
chemical passivation 440
commercial silicon solar cells
performance-limiting defects
boron-doped p-type Cz wafer 15
crystallographic defects 15
float-zone (FZ) p-type wafers 16
hydrogen-containing dielectrics 14–15
light-induced degradation 16
PERC technologies 15–16
selective emitter design 17–18
worldwide market share 14
contact resistivity 343–4
continuous wave (CW) lasers 322
conventional hydrogenation processes (CHPs) 120
copper contamination 267–8
crystalline silicon-based solar panels 5, 10

deep-junction laser doping 371–2
deep-level transient spectroscopy (DLTS) 43, 134, 183–4, 251
carbon-hydrogen (CH) complexes 251
carbon-oxygen-hydrogen (COH) complexes 254–5
laser-induced defect generation 358, 364
oxygen-hydrogen (OH) complexes 252
dielectric hydrogenation processes 102
dielectric layer deposition 30–1
diode-pumped solid-state lasers 327
direct hydrogen plasma 29
dopant atoms neutralisation
binding energy 36
initial high-temperature firing process 35
ionised boron and phosphorus dopants 34
minimum energy configuration 35, 36
shielded hydrogen plasma 35
dopant diffusion
conventional doping methods
gas-phase diffusion 307, 309
ion implantation 307
phosphorus concentration profile 308–9
phosphorus diffusion 307
SIMS measurements 308
thermal diffusion 307
emitter 306
laser doping 310–12
P-n junction formation 306
dopants passivation 135–7

Elastic Recoil Detection Analysis (ERDA) 44–5
electric vehicles 6
electron back-scattered diffraction (EBSD) technique 120
electron beam-induced current (EBIC) measurement technique 119
etch-back technique 350
external quantum efficiency (EQE) 340

float zone (FZ) silicon 192, 251
forming gas annealing (FGA) 26, 86, 379

Fourier Transform Infrared Spectroscopy (FTIR) 92, 93
fractional defect density (FDD) 198

gallium dopant 194
gallium-doped silicon wafers 447–9
gas immersion laser doping (GILD) 312
gettering and hydrogenation
  Al-BSF and PERC solar cells
    chemical edge junction isolation process 149
    dielectric layer depositions 150
    high-temperature firing process 149
    hydrogen release 151
    phosphorus diffusion 148–9
    production sequence 147–8
  defect engineering processes 144
  heterojunction silicon solar cells 157–64
  heterojunction solar cells 144–5
  high-efficiency n-type cell technologies 155–6
  lifetime
    background lifetime improvements 145–6
    grain boundaries recombination 145
    iV_{oc} mapping 146–7
    solar-grade silicon 145
    spatially resolved LBIC measurements 146
  solar cell architectures and processes
    considerations 154–5
    gettering layers removal 154
    hydrogen-defect systems 152
    selective emitter structures 153
    thermal oxidation processes 153–4
  TOPCon solar cells 144–5
grain boundaries and dislocation clusters 117
  characterisation
    EBIC measurement technique 119
    light beam-induced current (LBIC) 119
    photoluminescence (PL) imaging approach 118–19
    silicon wafer size 117
    SiN_X:H deposition process 120
  dielectric layer composition and thermal processing
    atomic layer deposition (ALD) 122
    firing temperature 124
    hydrogen desorption kinetics 122–3
    recombination activity 124
    Si–H and N–H bond densities 123–4
  passivation challenges
    deep energy level structural defects 122
    dislocation densities 121
    electron back-scattered diffraction technique 120
    recombination activity 120

Hanwha Q-Cells 12
heterojunction silicon (SHJ) solar cells 41, 92
  advantage 157
  boron-oxygen light-induced degradation 229–32
  gettering and hydrogenation
    AHP incorporating minority carrier injection 160
    current–voltage (I–V) results 160, 161
    defect-engineered wafers 158
    defect engineering processes 159
    n-type SHJ solar cells 164
    pre-fabrication gettering 159–60
    pre-fabrication hydrogenation 158
    thermal processing 160, 162

UMG silicon 163
light-and elevated temperature-induced degradation (LeTID) 280–1
manufacturing cost 158
performance 157
post-cell treatments 452–4
surface recombination 157
high performance multi-crystalline (HPM) material 128
high-temperature tube-furnace processes 197
hydrogenated dielectric films 269–70
hydrogen behaviour, negative effects
charge and dopants neutralisation 285–7
hydrogen-induced contact resistance
belt furnace anneal (BFA) 282, 283
multi-crystalline solar cells 283–4
PERC cells 284–5
process reversibility 283
p-type mc-Si PERC solar cells 282
transmission line measurements (TLMs) 283, 284
platelet formation 287–9
hydrogen complexes
carbon-hydrogen (CH) complexes 250–2
carbon-oxygen-hydrogen (COH) complexes 252–5
oxygen-hydrogen (OH) complexes 252
transition metal-hydrogen complexes 255
vacancy-hydrogen complexes 255–7
hydrogen defect modelling 54–5
hydrogen detection, silicon
atom probe tomography 43
complex detection 43
DLTS and vibrational spectroscopy techniques 43
dopant impurities deactivation 43, 45, 46
hydrogen concentrations 44–5
hydrogen-metal interactions 44
NRRA and ERDA 44–5
Raman spectroscopy measurements 44, 45
vacuum-based techniques 43
hydrogen diffusivity
charge state 41–2
silicon heterojunction (SHJ) solar cells 41
temperature range 40
hydrogen interactions
complications 39
high temperatures 39
hydrogen-carbon complexes 38–9
interstitial oxygen 39
vacancies 38
hydrogen migration 40–3
hydrogen passivation
bulk defect passivation 441
carrier-induced defects 441
charge state 440
chemical passivation 440
dielectric layers 440
gettering 441
high-efficiency silicon solar cells
emitter saturation current density 26, 27
forming gas annealing 26
polysilicon-based passivating contacts 27
silicon nitride dielectrics 27
high-temperature firing 441
hydrogen introduction
dielectric layer deposition 30–1
forming gas annealing (FGA) 28
hydrogen containing plasma 29–30
hydrogen molecule splitting 2
unintentional incorporation 32

hydrogen storage 441
illuminated annealing 442
interstitial hydrogen behaviour (*see* interstitial hydrogen behaviour)
laser-induced defects 441
multi-stage hydrogenation steps 441–2
process-induced defects 441
silicon surfaces (*see* silicon surface passivation)
surface passivation 440
uncertainties 42–3

impurity-related defects
DLTS measurements 134
dopants passivation 135–7
metallic impurities 130–1
n-type Cz wafers 133
oxygen-related defects 131
recombination active defects 130
upgraded metallurgical grade (UMG) silicon 131–3
indium dopant 194
industrial silicon solar cells
climate change and fossil fuels
average global temperature 1–2
ecological and health impact 2
global CO_2 emissions 2–3
greenhouse gases 2
radiative forcing 3
performance 1
production sequences
Al-BSF solar cells 55–8
PERC solar cells 58–60
injection-dependent lifetime spectroscopy (IDLS) analysis 261
inkjet printing 348–9
inline Selective Emitter ConcepT (inSECT) approach 350
interdigitated back-contact (IBC) solar cells
advantages 412–13
bifaciality 414
commercial success 413
emitter diffusion 416
interconnection process 413
interdigitated buried back-contact (IBBC) solar cell 414
laser-doped 416–17
laser-doped compensated contact 420–1
light phosphorus diffusion 414
manufacturing challenges 414–16
phosphorus diffusion 414
polarities of contact 412
processing sequence 415
reduced cell series resistance 413
SunPower technology 413
superior front surface passivation 413
thermal diffusion 416
interdigitated buried back-contact (IBBC) solar cell 414
internal quantum efficiency (IQE) 340
interstitial hydrogen behaviour
donor and acceptor levels 32–4
energy levels 32, 33
heavily doped n-type silicon 33
negative charge state 34
positive charge state 33
strain energy 34
transitions 34
dopant atoms neutralisation 34–6
fractional charge state concentrations
band-gap narrowing 47
excess carrier concentration 50–2
H^0 concentrations 48–9
high illumination intensities 52–3
materials lifetime 52–3
Sah-Shockley statistics 46
and solar cell structure impact 53
thermal equilibrium 46, 47
hydrogen detection 43–6
hydrogen dimer formation 36–8

hydrogen interactions 38–9
hydrogen migration 40–3
laser ablation 143, 347–8
laser-based advanced hydrogenation process (AHP) 383
laser chemical processing (LCP) 313, 354
laser contact opening (LCO) process 142
laser-doped selective emitter (LDSE) PERC solar cells
aligned screen-printing process
co-firing step 351
industrial LDSE PERC solar cell 351, 352
peak efficiencies 353
solar cell structure 351, 352
locally heavily doped regions 351
self-aligned plated contacts
advantage 354–5
dielectric layers 354
laser chemical processing (LCP) 354
laser doping 354
plating 355
laser-doped selective emitter (LDSE) structure 337
laser doping 17–18
beam optics 329, 330
gas immersion laser doping (GILD) 312
homogeneous emitter 311
industrial approach 444
interdigitated back-contact solar cells 412–18
ion implantation 312
ion-implanted semiconductors 310
laser annealing 310–11
laser chemical processing (LCP) 313
laser-doped junction depth 443
laser-doped regions
diffusion profile 322–4
ECV/SIMS 317–19
SEM/EBIC 319
sheet resistance 317
laser-enhanced gettering (LEG) 423–7
laser irradiation 312
laser processing 444
laser speed 324–7
laser types 310–12
LIMPID process 314, 315
local back surface field formation, PERC
aluminium/silicon (Al/Si) alloying process 396, 398–403
laser-doped back surface field (BSF) region 396
laser-doped rear contacts 397
nomenclature 310
opposite polarity of dopants
contacts forming process 418
edge isolation 421–3
laser-doped IBC cell 420–1
p-n-p transistor structure 420
p-type dopants mix 418
SEM/EBIC images 419
tunnel junction 419
origin, silicon solar cell manufacturing 305–6
principle 310
pulsed lasers 443
screen-printing 444
selective emitter formation
LDSE PERC solar cells 351–5
locally heavily doped regions creation 351
semiconductor finger solar cells 407–12
SIMS profiles 311
wavelength choice 327–9
laser-enhanced gettering (LEG)
Cz-Si and mc-Si wafers 425, 426
integrated circuit 423
interstitial iron concentrations 425
laser damage depth 424
laser-irradiation 424–5
LDSE diffusion process 425–6

qualitative and quantitative assessment 426–8
solar cell fabrication 424
laser-induced defect generation
detrimental impact 356
DLTS-based characterisation 364
electron beam microscopy-based characterisation 358
hydrogen passivation
efficiency gain 380, 382
fast-firing proces 374
forming gas annealing (FGA) 379
injection-dependent inverse minority carrier lifetime curves 374–5, 377
laser-doped Al-BSF cell 380
lifetime parameters 376, 378
and light-induced degradation mitigation 382–5
low-temperature belt furnace-annealing processes 375, 377, 380
low-temperature metallisation scheme 382
non-radiative recombination pathways 373
processing speeds 379
laser- and thermal-annealing 372–3
laser-doped region 357
laser doping
advantage 367
deep-junction 371–2
dielectric layers 369–71
and grooving 368
high-precision screen-print alignment 365
localised passivated contact structures 369
prior to dielectric deposition 365–6
two-step laser-doping/LCO contact formation 367
laser-doping coverage reduction 365–6
laser pulse overlap 356
metal/Si interface area reduction 365–6
QSSPC and photoluminescence-based characterisation 358–62
spectral photoluminescence-based characterisation 362–4
voids formation 357
laser-induced melting of pre-deposited impurity doping (LIMPID) process 314, 315
laser thermal annealing post-implantation 312
lateral emitter resistance 344–5
levelised cost of electricity (LCOE) 12
light-and carrier-induced defects 143–4
light-and elevated temperature-induced degradation (LeTID) 192, 250
apparent defect energy levels 263
capture cross section ratios 263
carrier-induced degradation (CID) 257–8
cells and modules 258–9
dielectrics 264
firing process 259–61
FZ silicon wafers 261–2
hydrogen
concentration 271
deep-level transient spectroscopy (DLTS) 271–2
direct correlation 270–1
higher temperature firing 268–9
hydrogenated dielectric films 269–70
microwave-induced remote hydrogen plasma 270–1
plasma-enhanced chemical vapour deposition 271
injection-dependent lifetime spectroscopy analysis 261
k-value 264–5
lifetime spectroscopy 264
metallic impurities
copper precipitates 267–8
diffusion coefficient 267

elemental micro-analysis techniques 267
gettering procedures 268
high-temperature firing 266
proposed model 266
rapid cooling 266
mitigation
contact firing profile 273
dielectric layer 273
phosphorus gettering process 273
re-firing approach 273–4
wafer thickness 272–3
models
B–H pair concentration 279
four-state model 277, 278
NDD of 278
three-state defect model 275–7
N-type silicon wafers 261
p-type silicon heterojunction solar cells 280–1
temperature-and injection-dependent lifetime spectroscopy 264, 265
light beam-induced current (LBIC) 119
light-induced degradation (LID) 16
boron-oxygen (*see* boron-oxygen light-induced degradation)
laser-induced defect 382–285
local vibrational modes (LVM) 256
LONGi Solar 1, 17, 192
low-pressure chemical vapour deposition (LPCVD) 42, 180
low-temperature metallisation approaches 143

Marangoni effect 402, 403
metal catalysed chemical etching (MCCE) 56
metallic impurities 130–1
metallisation firing 198–9
microwave-induced remote hydrogen plasma 270
mode-locked quasi-continuous wave (Q-CW) lasers 322

neodymium-doped laser 327, 328
normalised defect density (NDD) 182–3, 385
n-type passivated contact solar cells
instability 451–2
SHJ solar cells 449
TOPCon solar cells 450–1
Nuclear Resonant Reaction Analysis (NRRA) 44–5

ordinary differential equations (ODEs) 204
oxygen-hydrogen (OH) complexes 252, 253
oxygen-related defects 131
conventional firing process 139
Cz silicon wafer growth 137
firing proces 138–9
gettering process 140
high-temperature process 137–8
industrial n-PERT solar cells 141
NH_3 plasma process 140
n-type Cz wafers 140
recombination activity reduction 138
tabula rasa process 138

passivated emitter and rear cell (PERC) 1
aluminium/silicon (Al/Si) alloying process
aluminium oxide dopant 403–7
boron laser-doped LBSFs 401, 402
doping concentration 396
firing step 396
laser-doped rear contacts 397
laser doping 400
Marangoni effect 402, 403

metallisation methods 401
scanning electron microscope (SEM) images 398, 399
silicon concentration 399
solid solubilities 398
voids 398
cells and modules 258–9
efficiency 179, 180
gallium-doped silicon wafers 447–9
gettering and hydrogenation 147–52
laser-doped selective emitters 445–6 (*see also* laser-doped selective emitter PERC solar cells)
optimal hydrogenation 92
production sequences 58–60
solar cell technology 12, 13
surface-related degradation 446–7
passivated emitter, rear totally diffused (PERT) cell 438
performance-limiting defects 116
perovskite solar cells 11
phosphorous emitter technologies 19
phosphorus diffusion 307
phosphorus diffusion gettering (PDG) 268
phosphosilicate glass (PSG) removal 57
photoconductance decay (PCD) 80, 82
photolithography 346–7
photoluminescence (PL) imaging approach 118–19
plasma-enhanced chemical vapour deposition (PECVD) 180
aluminium oxide thin films 100
laser doping and grooving 368
LeTID formation 271
silicon nitride thin film 92
chemical hydrogenation and FEP 94
hydrogen effect 94
silicon oxide film 88–90
SiN_x:H films deposition 57
platelet formation, hydrogen 287–9
polysilicon-based passivating contacts 27
process-induced defects
laser-induced defects 142–3
oxygen-related defects 137–42
p-type silicon heterojunction solar cells
boron-oxygen light-induced degradation
efficiency 229–30
fabrication 232
high-temperature firing process 232
high-temperature hydrogenation process 231
I–V characteristics 230–1
light-induced degradation 230
in situ open-circuit voltage 231–2
three-key-processing regimes 230
light-and elevated temperature-induced degradation (LeTID) 280–1
pulsed lasers 443

Q-switched lasers 310–12
quasi-continuous-wave (Q-CW) lasers 443
quasi-mono technology 193
quasi-steady-state photoconductance (QSS-PC) 134
charge-carrier recombination 361–2
injection-dependent inverse lifetime curves 359–61
laser-doped regions 317–19
laser doping/laser contact opening (LCO) 361
lifetime deterioration 361
measurement 182
recombination rate 361

Raman spectroscopy, hydrogen-induced platelets 288
reactive ion etching (RIE) 56
recombination active defects 25
re-firing approach 273–4
remote plasma hydrogenation 30

Sah-Shockley statistics 218
screen-printed silicon solar cells 9–13
screen-printed solar cells (SPSCs) 407–8
secondary ion mass spectrometry (SIMS) 288
  dopant diffusion 308
  laser-doped regions 317–19
  phosphorus-and boron-implanted silicon 311
selective emitters 1
  Auger recombination 341–2
  benefits 339, 344
  contact resistivity 343–4
  design 17–18
  emitter recombination 340
  internal quantum efficiency 340, 341
  lateral emitter resistance 344–5
  LDSE PERC solar cells
    aligned screen-printed contacts 351–3
    self-aligned plated contacts 354–5
  metal/Si interface recombination 343–4
  PERL solar cell architecture 338
  phosphosilicate glass (PSG) 337
  thermal diffusion steps 345
  total emitter recombination current density 343
  UNSW PERL cell
    dielectric patterning 346–9
    doped silicon inks 349
    etch-back technique 350
    ion implantation 349–50
    oxide-masking layers 345
    processing sequence 346
semiconductor finger solar cells (SFSC)
  advanced laser-doped 411–12
  fabrication processes 408
  future potential 412
  laser-doping
    advantages 409–10
    disadvantages 410
    metal–Si interface 410
    SIMS profile 411
  lateral conductive paths 407
  lightly doped emitter region 407
  screen-printed solar cells (SPSCs) 407–8
  semiconductor finger (SF) design 408
sheet resistance 317
Shockley-Read-Hall (SRH) recombination 76–7
silicon-based tandem solar cells 454–7
silicon heterojunction (SHJ) solar cells. *See* heterojunction silicon solar cells
silicon nitride thin films
  chemical hydrogenation and FEP 94
  chemical passivation 95
  double-layer stacks 95
  effusion experiments 92, 93
  FTIR measurements 92–4
  interfacial hydrogenation 94–5
  origin 91
  SiN_x–Si interface 91
  surface recombination velocity (SRV) 95, 96
silicon surface passivation
  aluminium oxide thin films 100–1
  amorphous silicon thin films 96–9
  characterisation
    bulk and surface lifetime 80–1
    measured lifetime 81, 82

photoconductance decay (PCD) 80, 82
surface recombination velocity (SRV) 81
surface saturation current density 81
dielectric coatings 82–4
hydrogenated silicon nitride 84–5
recombination and passivation
dangling bonds 75–6
one-dimensional lattice, Shockley's description 76
Shockley-Read-Hall (SRH) recombination 76–7
solid-air interface 76
surface recombination velocity (SRV) 76–8
silicon nitride thin films 91–6
silicon oxide thin films
atomic layer deposition (ALD) 90–1
chemical anodisation 91
PECVD 88–90
thermal silicon dioxide 86–8
thermal oxidation 84
tunnelling oxide passivating contact layers 101–2
solar photovoltaics
cost reductions
ambitious 'Million Roof' program 8
average PV module prices 7–8
Chinese manufacturers 8–9
Flat Plate Solar Array Project 8
soft costs 9
crystalline silicon-based solar panels 5
disruptive technology 6
electricity generation 5–6
electricity installations 4–5
electric vehicles 6
energy transition model 5
screen-printed silicon solar cells 9–13
Staebler–Wronski effect 96, 454
structural defects
additional thermal hydrogenation treatments 128–30
cast-mono-crystalline silicon material 126–8
cast silicon wafers 116–17
dehydrogenation 125–6
grain boundaries and dislocation clusters 117–25
Sunpower's rear interdigitated back-contact (IBC) solar cells 1–2
SunPower technology 413
surface recombination velocity (SRV) 76–80, 340, 438
capture cross sections 78
effective 81
excess minority carrier concentration 77–9
positive and negative dielectric charge concentration 80
surface-related degradation (SRD) 446–7

tabula rasa process 138
temperature and injection-dependent lifetime spectroscopy (TIDLS) 264
temperature programmed desorption (TPD) 122
thermal diffusion 307
thermal silicon dioxide
alnealed SiO_2 87
aluminium anneal process 86–7
electrical techniques 86
forming gas anneal (FGA) 86
hydrogen-containing plasma 88
interface defect states density 86, 87
refractive index 88
Si–SiO_2 interface 86, 88, 89
surface orientation 87
thin-film solar cells
Al-BSF cells 11–12

compound semiconductor 10
market share 10–11
PERC solar cells 12, 13
perovskite solar cells 11
thin films materials 11
three-state LeTID defect model
B-O defect system 275–7
dark annealing 277
three-state model, B-O defect system
degradation rate 202
high-level behaviour 200–1
kinetic modelling
ordinary differential equations (ODEs) 204
processing time 208–9
reaction rates 207
temperature-dependent reaction rates 207, 208
minority carrier lifetime 201
normalised defect density (NDD) 202–3
reaction rates 205–6
recovery 200
temperature 203–4
transition metal-hydrogen complexes 255
transmission electron microscopy (TEM) 358
transmission line measurements (TLMs) 283, 284
TRINA Solar cells 17
tunnelling oxide passivating contact layers 101–2
tunnel oxide passivated contact (TOPCon) solar cell 1–2, 101–2, 144–5, 156, 450–1
two-stage degradation 185–6

UNSW passivated emitter, rear locally diffused (PERL) cell
dielectric patterning
inkjet or aerosol jet printing 348–9
laser ablation 347–8
photolithography 346–7
doped silicon inks 349
etch-back technique 350
ion implantation 349–50
oxide-masking layers 345
processing sequence 346
upgraded metallurgical grade (UMG) silicon 131, 132

vacancy-hydrogen complexes 255–7

www.ingramcontent.com/pod-product-compliance
Lightning Source LLC
LaVergne TN
LVHW011550200926
848484LV00001B/299

* 9 7 8 1 7 8 5 6 1 6 2 3 5 *